THE IMPORTANT PLANT AREAS OF MOZAMBIQUE

THE IMPORTANT PLANT AREAS OF MOZAMBIQUE

Compiled and edited by:

Iain Darbyshire, Sophie Richards, Jo Osborne, Hermenegildo Matimele, Clayton Langa, Castigo Datizua, Alice Massingue, Saba Rokni, Jenny Williams, Tereza Alves & Camila de Sousa

A publication of the
Tropical Important Plants Areas Programme

Kew Publishing
Royal Botanic Gardens, Kew

First published in 2023 by Royal Botanic Gardens, Kew, Richmond, Surrey, TW9 3AB, UK
www.kew.org

ISBN 978 1 84246 788 6
eISBN 978 1 84246 789 3

Distributed on behalf of the Royal Botanic Gardens, Kew in North America by the University of Chicago Press, 1427 East 60th Street, Chicago, IL 606037, USA.

British Library Cataloguing in Publication Data
A catalogue record for this book is available from the British Library.

Design: Nicola Thompson, Culver Design
Production Manager: Georgie Hills
Proofreading: Sharon Whitehead

Printed in Great Britain by Halstan & Co Ltd

For information or to purchase all Kew titles please visit shop.kew.org/kewbooksonline or email publishing@kew.org

Kew's mission is to understand and protect plant and fungi, for the wellbeing of people and the future of all life on Earth.

Kew receives approximately one third of its funding from Government through the Department for Environment, Food and Rural Affairs (Defra). All other funding needed to support Kew's vital work comes from members, foundations, donors and commercial activities, including book sales.

Citations:
Darbyshire, I., Richards, S., Osborne, J., Matimele, H., Langa, C., Datizua, C., Massingue, A., Rokni, S., Williams, J., Alves, T. & Sousa, C. de (2023). *The Important Plant Areas of Mozambique.* Royal Botanic Gardens, Kew.

Author(s) for the individual IPA site assessments are listed at the beginning of each account; these can be cited as follows (e.g.):
Richards, S., Darbyshire, I., Matimele, H. & Datizua, C. (2023). Bilene-Calanga Important Plant Area. Pp. 342–347 in: Darbyshire, I., Richards, S., Osborne, J., Matimele, H., Langa, C., Datizua, C., Massingue, A., Rokni, S., Williams, J., Alves, T. & Sousa, C. de, *The Important Plant Areas of Mozambique.* Royal Botanic Gardens, Kew.

Author affiliations:
Iain Darbyshire, Sophie Richards, Jo Osborne, Saba Rokni, Jenny Williams – Royal Botanic Gardens, Kew (Kew)

Hermenegildo Matimele – Instituto de Investigação Agrária de Moçambique (Agricultural Research Institute of Mozambique – IIAM) & Durrell Institute of Conservation and Ecology, University of Kent

Clayton Langa, Castigo Datizua, Tereza Alves, Camila de Sousa – Instituto de Investigação Agrária de Moçambique (Agricultural Research Institute of Mozambique – IIAM)

Alice Massingue – Department of Biological Sciences, Faculty of Science, Eduardo Mondlane University / Universidade Eduardo Mondlane (UEM)

CONTENTS

FOREWORD

Plants have profoundly changed the history of life on this planet. They perform a predominant role in the cycles of water and carbon dioxide, recycling oxygen, and influence the chemistry of the atmosphere and how landscapes reflect or absorb the light from the sun.

Plants have created the perfect conditions for many species to thrive, including us, humans. It is from them that people obtain food, medicine to treat diseases and improve health, wood for the construction of shelters, fuel, as well as places for spiritual healing.

Science and research have been the driving forces for much of the knowledge and understanding of the world of plants. From the expeditions to collect, identify and catalogue the vast diversity of plants, to the intense studies on their properties and uses to improve human well-being. Until today, scientists and researchers are still making astonishing discoveries on plants. How they live, compete, and cooperate, how they transmit messages to each other on threats and opportunities in their environment, mimic pheromones of predatory insects, produce energy from sunbeams, draw prodigious quantities of water from the ground and send it to the atmosphere. Plants have transformed life on this planet, and with it, it gave humans the opportunity to exist and prosper.

Today, the world's human population has now reached 8 billion people. The pressures over our planet's resources and on its very diverse habitats will be greater than ever. In Mozambique, the population has more than tripled since independence in 1975, reaching over 30 million people. This growth entails greater demand for land, for food, for energy, and a variety of other important needs. The challenges are aggravated, in a country that has still to deal with high levels of malnutrition and poverty. The expansion of areas for agriculture has been one of the main drivers of the current conversion and loss of about 267,000 hectares of forest every year. The projections from our REDD+ National Strategy forecast an increase to about 500,000 hectares per year from 2030.

It is a delicate balance to find the path towards a healthy and sustainable development, for which researchers constitute one of the most important players. The production of information and knowledge is key to help guide the tasks of designing interventions that can satisfy the needs of the population, whilst not putting at risk the fundamental resources that safeguard human life. Researchers are helping to identify and understand the critical areas of national and global relevance, due to their diversity of species and habitats, and the ecosystem services they provide.

This publication is a remarkable accomplishment of such an effort. Compiling data from almost all the Provinces of Mozambique, it identifies 57 areas of plant importance describing in detail each one of them.

The information includes reference to some of the rare and threatened species they contain, the conservation issues facing each area as well as its key ecosystem services. It registers and broadens our knowledge on the diversity of plant species in the country and our understanding on their current levels of threats and global relevance. It provides a strong basis for decision making, to help guide adequate regulation and promotion of land uses in the country as well as effective prioritising of conservation efforts.

The fruit of decades of scientific collaboration between the Agricultural Research Institute of Mozambique, the Royal Botanic Gardens, Kew and Eduardo Mondlane University, this publication is a reference and inspiration for current and future research in the country. It is a celebration of the  diversity of plant species Mozambique has, and a recognition of what is possible to achieve when humans come together to produce knowledge to improve human well being.

Celso Ismael Correia
Minister of Agriculture and Rural Development

FOREWORD

International reports on the conservation status of the natural world draw the attention of governments to the accelerated loss of biodiversity at a global level. These losses, exacerbated by climate change and natural disasters, put at risk the survival of species and natural habitats, in some cases even before they are scientifically identified and before the risks are perceived by governments and local communities that could intervene to preserve them.

In the 6th CBD Report and the Strategy and Action Plan for the Conservation of Biological Diversity in Mozambique 2015 – 2035, the country recognises that it has an incomplete knowledge of its plant diversity, which limits the ability to identify actions to be taken to achieve the targets defined in the international conventions to which Mozambique is a signatory.

In this context, conservation efforts have been oriented towards conservation areas, which are indicative of biological wealth and cover 26% of the national territory, although there is still not enough knowledge about their effectiveness in the conservation of the diversity of flora and habitats in Mozambique.

The 57 Important Plant Areas identified, demarcated and described in this work (of which only 18 are within conservation areas) cover about 3% of Mozambique's land surface and incorporate about 82% of the threatened plant species in Mozambique. Therefore, they must be protected and conserved, for which they must be integrated into land management plans.

The knowledge and dissemination of the national flora and its state of conservation presented in this work will make it possible to optimise its sustainable use and prioritise research and conservation efforts. It will allow BIOFUND to present Mozambique as the guardian of its valuable natural heritage.

This publication contributes significantly to BIOFUND's strategy of contributing to the protection and monitoring of known biodiversity, as well as supporting the continuity of research in places identified as being of botanical interest but where there is still insufficient botanical information to validate them as IPAs, such as the Maputo National Park, Serra Choa and Levasflor.

Professor Narciso Matos
Presidente do Conselho de Administração /
Chairman of the Board of Directors
BIOFUND

EXECUTIVE SUMMARY

This publication presents the results of the five-year project "Tropical Important Plant Areas: Mozambique", a collaboration between the Agricultural Research Institute of Mozambique (IIAM), the Royal Botanic Gardens, Kew (Kew) and Eduardo Mondlane University (UEM). Building on several decades of scientific collaboration, this work has combined botanical research and field surveys, collation of existing data, building of in-country capacity and engaging with conservationists and land use planners to document and promote the conservation of Important Plant Areas (IPAs) in Mozambique.

IPAs are an internationally recognised standard for identifying sites of national and global importance for preserving plant diversity and supporting the highest priority species and habitats, as well as the important ecosystem services that they provide. IPAs in Mozambique were identified based on the presence of globally threatened plant species (IPA criterion A), exceptional richness in species of high conservation importance (criterion B(ii)) and nationally threatened and range-restricted habitats (criterion C(iii)). The IPA assessment is supported by a revised IUCN Red List of globally threatened species in Mozambique. This has focused on the ca. 670 endemic (unique) and near-endemic taxa (species, subspecies and varieties), with 462 having been assessed for their Red List status to date, of which 55% are threatened with extinction.

The resultant network of IPAs comprises 57 sites totaling an area of 22,950 km², or fewer than 3% of the total land area of Mozambique. All 57 sites qualify under criterion A, with 83% of the total threatened taxa being represented in one or more IPAs and with 49 of the sites containing threatened taxa that are not represented elsewhere in the IPA network. 12 sites also qualify under criterion B(ii), having exceptional richness in taxa of high conservation importance, whilst 26 IPAs qualify under criterion C(iii), with 12 different threatened or range-restricted habitats represented. Of particular note are the eight IPAs that meet all three criteria: Chimanimani Lowlands, Chimanimani Mountains, Lower Rovuma Escarpment, Mount Gorongosa, Mount Namuli, Quiterajo, Ribáuè-M'paluwe and Tsetserra.

All the provinces of Mozambique are represented in the IPA network except for Tete and Maputo City. However, threatened and endemic species are particularly concentrated in the four cross-border Centres of Plant Endemism (CoEs): Rovuma in the northeast and Maputaland (including Inhambane and Lebombo Mountains Sub-Centres) in the southeast, both CoEs consisting largely of coastal plains with some hilly terrain inland; and the two montane CoEs of Chimanimani-Nyanga in the west along the Zimbabwe border, and Mulanje-Namuli-Ribáuè in northern Mozambique and southern Malawi. With high levels of both threatened and endemic species, IPAs are largely concentrated within these CoEs.

Of the 57 IPAs identified, only 18 fall fully within Mozambique's Protected Area (PA) network at present, whilst a further 10 partially overlap with PAs. Hence, just over half of the IPAs have no formal protection at present. None of the sites is entirely free from threats, and some of the sites (including several that are protected under law) are facing severe threats from a range of factors, including unsustainable slash-and-burn agriculture and over exploitation of natural resources.

There is therefore much work to be done to effectively protect and manage the IPAs and the rare and threatened species, habitats and resources they contain. IPAs are an important tool in enabling effective prioritisation of these conservation efforts. This work will help fulfil Mozambique's important commitments to conserving the natural world under the Ministério da Terra e Ambiente's *National Strategy and Action Plan of Biological Diversity of Mozambique 2015–2035* and as a signatory to the Convention on Biological Diversity. IIAM, Kew and UEM are committed to continuing to work together, and in co-ordination with policy makers, conservation practitioners, land use planners and local communities, to help protect and sustainably manage the IPAs and the plant life they support for the wellbeing of future generations.

ACKNOWLEDGEMENTS

This work could not have been undertaken without the generous support of a wide range of contributors and collaborators to whom we are most grateful. First and foremost, we are deeply indebted to the Oppenheimer Generations Foundation and Stephen and Margaret Lansdown for their generous financial support that has enabled the TIPAs Mozambique project to happen, and for their continued encouragement.

We thank Olga Faftine, Director of Instituto de Investigação Agrária de Moçambique (IIAM), for supporting this project and related initiatives on the conservation of Mozambique's flora. We thank the Ministério da Terra e Ambiente (MTA) and the Administração Nacional das Áreas de Conservação (ANAC) for their stewardship of Mozambique's natural resources and for their positive engagement in this project. We would particularly like to thank Cidália Mahumane (ANAC) for facilitating the permits needed to undertake fieldwork in protected areas.

We thank all the Mozambican botanists, biologists and conservation practitioners who have contributed to this work through expert input. These include Salomão Bandeira and Celia Macamo (Eduardo Mondlane University), Marcelino Inácio Caravela (Lúrio University, Pemba), Pita Sito (Pegagological University, Maputo) and João Massunde (Micaia Foundation). The staff at the LMA herbarium, IIAM are thanked, in particular Aurélio Banze, Ines Chelene and Josias Zandemela who have contributed significantly in terms of fieldwork and/or herbarium-based research. We particularly thank Papin Mucaleque (IIAM Centro Zonal Nordeste) who, alongside undertaking fieldwork, completed a number of IUCN Red List assessments and participated in both Red List and IPA workshops that supported the identification of IPAs in northeast Mozambique. Also, from the regional IIAM offices, we would like to thank Aristides Mamba (IIAM Centro Zonal Nordeste), Tome Rachide (IIAM Centro Zonal Noroeste) and Valdemar Fijamo (IIAM Centro Zonal Centro) for their contributions to field research. We also thank all the field support staff and local community leaders and members who welcomed and guided our field visits.

We are most grateful to the Micaia Foundation (particularly Milagre Nuvunga and Andrew Kingman), Legado (Majka Burhardt), Nitidae (Jean-Baptiste Roelens), Parque Nacional da Gorongosa (Marc Stalmans) and Santuário Bravio de Vilanculos (Taryn Gilroy) who have provided technical input and support in the identification of specific IPAs. Biofund, the Foundation for the Conservation of Biodiversity in Mozambique, are thanked for their important role in promoting biodiversity and its conservation nationally. We also thank Biofund for giving us the opportunity to promote our work at their exhibitions, with particular thanks to Alexandra Jorge for supporting this project. We thank the Wildlife Conservation Society (WCS) Mozambique, and in particular Hugo Costa and Eleutério Duarte, for sharing knowledge and draft publications on the Key Biodiversity Areas initiative in Mozambique.

We are highly grateful to the international experts who have contributed botanical knowledge at all stages of this work, from co-authoring site accounts, to reviewing the assessments and participating in the TIPAs workshops. In particular, Jonathan Timberlake (independent botanist, formerly Kew), John and Sandie Burrows (formerly Buffelskloof Nature Reserve), Ton Rulkens (independent botanist), Bart Wursten (associate, Meise Botanic Garden), Warren McCleland (SLR Consulting, formerly ECOREX Consulting Ecologists) and Mervyn Lötter (Mpumalanga Tourism and Parks Agency) have been hugely generous in sharing their expert knowledge of the Mozambican flora and sites of interest. We also thank Petra Ballings, Obety Baptiste, Julian Bayliss, Frances Chase, Phil Clarke, Meg Coates Palgrave, Colin Congdon, Timothy Harris, Mark Hyde, Linda Loffler, Quentin Luke, David Roberts, Ernst Schmidt and Douglas Stone for their helpful contributions and willingness to share information.

At Kew, we thank the following botanists for their contributions to the naming of plant specimens collected during the project expeditions: Henk Beentje, Renata Borosova, Andrew Budden, Xander van der Burgt, Martin Cheek, Phil Cribb, Nina Davies, Aaron Davis, David Goyder, Aurélie Grall (now University of Basel), Nicholas Hind, Isabel Larridon, Gwil Lewis, Mike Lock, Alan Paton, Roger Polhill,

Brian Schrire, Andre Schuiteman, Kaj Vollesen and Martin Xanthos. A spatial dataset for Mozambique was collated in ArcGIS by Julia Thorley, Kew project intern, to whom we are most grateful. We also thank Tim Wilkinson for additional GIS support.

We thank the members of the IUCN-SSC Southern African Plant Specialist Group (SAPSG) for their valuable contributions to the plant Red Listing and species conservation efforts in Mozambique. In particular, we thank Domitilla Raimondo of the South African National Biodiversity Institute (SANBI), former chair of SAPSG, who has been a great supporter of the conservation initiatives in Mozambique, and a great advocate for building in-country capacity in conservation planning. Lize von Staden and Hlengiwe Mtshali of SANBI are also thanked as the former and current Red List Authority Co-ordinators for SAPSG; respectively; they have played key roles in reviewing Mozambican assessments and in training Mozambican scientists in Red Listing. Quentin Luke, Roy Gereau and Kirsty Shaw of the IUCN-SSC East African Plant Red List Authority are also thanked for sharing updates on Red List data relevant to Mozambique, and we thank Emily Beech of Botanic Gardens Conservation International for helping to co-ordinate Red List efforts for Mozambique's trees.

Botanical data capture at Kew was led by the authors of this publication together with Jeneen Hadj-Hammou (1-year sandwich student from the University of Leeds; currently researching for her Ph.D. at Lancaster University), Toral Shah (project intern, currently researching for her Ph.D. at Imperial College, London and Kew), and Sarekha (Sonia) Dhanda (project intern, currently "Scientific Officer – CITES" at Kew). We are most grateful to the curators of the following herbaria for permitting us access to their collections: BM, BNRH, EA, K, NH, LISC, LISU, LMA, LMU, P, PRE and SRGH (herbarium codes follow Thiers [continuously updated]). In particular, we thank Barbara Turpin at BNRH for kindly sharing data whenever requested, and Maria Cristina Duarte and Maria Romeiras for hosting research visits to LISC.

The GBIF Biodiversity Information for Development (BID) fund supported the project BID-AF-2017-0047-NAC (2017–2019): "Mobilizing primary biodiversity data for Mozambican species of conservation concern", which enabled the compilation of data on endemic and near-endemic plants species held at the Maputo herbaria. We also acknowledge the important contribution to capturing and collating data on Mozambique's biodiversity made by the SECOSUD II "Conservation and equitable use of biological diversity in the SADC region" project, Biodiversity Network of Mozambique (BioNoMo), a partnership between UEM, the Italian Agency for Development Cooperation, Sapienza University and MTA. In particular, we thank Luca Malatesta and Delcio Odorico for their valuable contributions to the TIPAs workshops.

Previous projects involving IIAM and Kew that have provided significant field data for this current work include: the Darwin Initiative Award 15/036 "Monitoring and Managing Biodiversity Loss in South-East Africa's Montane Ecosystems" completed in 2009 and Award 2380: "Balancing Conservation and Livelihoods in the Chimanimani Forest Belt, Mozambique" completed in 2017; the Critical Ecosystems Partnership Fund Grant 63512 "In from the cold: providing the knowledge base for comprehensive biodiversity conservation in the Chimanimani Mountains, Mozambique; botanical survey component" completed in 2016; and the Pro-Natura International-led project on "Coastal Forests of Mozambique", supported by the Prince Albert II of Monaco Foundation, the Stavros Niarchos Foundation and the Muséum National d'Histoire Naturelle in Paris, completed in 2011. The Bentham-Moxon Trust are also thanked for kindly providing pilot funding for the TIPAs: Mozambique project.

We would also like to thank Plantlife International for their support of the TIPAs programme, in particular Karen Inwood (International Strategy Lead), Elizabeth Radford (Eden Rivers Trust, formerly Plantlife), Seona Anderson (formerly Plantlife), and Ben McCarthy (National Trust, formerly Plantlife) for generously sharing their advice and expertise on IPA identification and conservation efforts globally.

For their stewardship, logistical support and encouragement throughout the project, we thank Bridget Fury, Ashleigh Fynn-Munda, Kim Porteous and Ashleigh Williamson of Oppenheimer Generations, Fionnuala Carvill of Pula Ltd, and Meredith Pierce Hunter, Jonathan Kuhles, Rosemary Sawyer, Joanna Ellams and Marta Lejkowski of Kew Foundation (past and present).

PHOTO CREDITS

We are highly grateful to all who provided photographs of the sites and species featured in this book. All images are credited with the photographers' initials, as per the following list of contributors:

AM	Alice Massingue
AMR	Andrew McRobb / RBG Kew
AR	Anne Robertson
BW	Bart Wursten
CS	Camila de Sousa
CD	Castigo Datizua
CL	Clayton Langa
CC	Colin Congdon
DN	Denise Nicolau/BIOFUND
FC	Frances Chase
HM	Hermenegildo Matimele
ID	Iain Darbyshire
JM	Jacinto Mafalacusser
JO	Jo Osborne
JEB	John Burrows
JT	Jonathan Timberlake
JB	Julian Bayliss
JP	Jose Paula
LL	Linda Loffler
MS	Marc Stalmans
MIC	Marcelino Inácio Caravela
MH	Mark Hyde
MC	Martin Cheek
OB	Obety Baptiste
PM	Papin Mucaleque
PC	Phil Clarke
PP	Phil Platts
QL	Quentin Luke
SB	Salomão Bandeira
SV	Santuario Bravio de Vilanculos
TH	Tim Harris
TB	Tomás Buruwate
TR	Ton Rulkens
TS	Toral Shah
TP	Tracey Parker
WM	Warren McCleland

LIST OF ACRONYMS AND ABBREVIATIONS

Acronyms and general abbreviations

ANAC	Administração Nacional das Áreas de Conservação (National Administration of Conservation Areas)
AOO	Area of Occupancy
AZE	Alliance for Zero Extinction
BGCI	Botanic Gardens Conservation International
BID	Biodiversity Information for Development
BioNoMo	Biodiversity Network of Mozambique
CBD	Convention on Biological Diversity
CEPF	Critical Ecosystems Partnership Fund
CoE	Centre of Endemism
EAPRLA	IUCN SSC Eastern African Plant Red List Authority
EOO	Extent of Occurrence
FR	Forest Reserve
F.T.E.A.	Flora of Tropical East Africa
F.Z.	Flora Zambesiaca
GBIF	Global Biodiversity Information Facility
GIS	Geographic Information System
GSPC	Global Strategy for Plant Conservation
HRE	Highly Restricted Endemic
IIAM	Instituto de Investigação Agrária de Moçambique (Agricultural Research Institute of Mozambique)
IBA	Important Bird Area
IPA	Important Plant Area
IUCN	International Union for the Conservation of Nature
IUCN SSC	International Union for the Conservation of Nature Species Survival Commission
K	Herbarium of the Royal Botanic Gardens, Kew
Kew	Royal Botanic Gardens, Kew
KBA	Key Biodiversity Area
LMA	National Herbarium of Mozambique, Instituto de Investigação Agrária de Moçambique
LMU	Herbarium of the Universidade Eduardo Mondlane (Eduardo Mondlane University)
MAE	Ministério da Administração Estatal (Ministry of State Administration)
MCP	Minimum Convex Polygon
MICOA	Ministério para a Coordenação da Acção Ambiental (Ministry for Coordination of Environmental Action)

MITADER	Ministério da Terra, Ambiente e Desenvolvimento Rural / Ministry of Land, Environment and Rural Development (now replaced by MTA, see below)
MOZTIPA	Mozambique Important Plant Area site code
ms.	Unpublished manuscript
MTA	Ministério da Terra e Ambiente (Ministry of Land and Environment); formerly MITADER
PA	Protected Area
Pers. comm.	Personal communication
Pers. obs.	Personal observation
Pop'n	Population
POWO	Plants of the World Online
RRE	Range Restricted Endemic
SABONET	Southern African Botanical Diversity Network
SANBI	South African National Biodiversity Institute
SAPSG	IUCN SSC Southern African Plant Specialist Group
SBV	Santuario Bravio de Vilanculos Lda.
SIS	IUCN Species Information Service
TIPAs	Tropical Important Plant Areas Programme
UEM	Universidade Eduardo Mondlane (Eduardo Mondlane University)
UNEP	United Nations Environment Programme
UNESCO	United Nations Educational, Scientific and Cultural Organisation
WCS	Wildlife Conservation Society
WWF	World Wildlife Fund

IUCN Red List Categories

CR	Critically Endangered
DD	Data Deficient
EN	Endangered
LC	Least Concern
NE	Not Evaluated
NT	Near Threatened
VU	Vulnerable

Botanical abbreviations

ined.	Ineditus; taxonomic name is not yet published
sp.	Species (singular)
sp. nov.	Species novum; new species
spp.	Species (plural)
subsp.	Subspecies
var.	Variety

INTRODUCTION

Plants and plant-based habitats are vital to the survival and prosperity of life on Earth. As well as underpinning wider biodiversity, they provide essential resources, foods and medicines that support human health and livelihoods, and important ecosystem services such as the regulation of climate, water quality and soil fertility. On a global scale, the sequestration of carbon within plant-based habitats will make a central contribution towards limiting global temperature rise to below 2°C following the Paris Climate Agreement. Moreover, rich and intact natural habitats provide inspiration and wellbeing to mankind globally, while many plant species and the sites that support them have important cultural and spiritual significance that should be passed on, intact, for future generations. Valuing and protecting our plant resources and using them in a responsible and sustainable way is, therefore, an important part of our stewardship of the natural world.

Mozambique's rich plant diversity

Mozambique occupies a land area of approximately 800,000 km² in southern tropical and sub-tropical Africa between the latitudes of 10° 28' S and 26° 52' S (-10.47° to -26.87°) and the longitudes of 30° 13' E and 40° 50' E (30.22° to 40.84°). It is bordered to the north by Tanzania, to the northwest by Malawi, to the west by Zambia and Zimbabwe, to the southwest and south by South Africa and Eswatini and to the east by the Indian Ocean, with over 2,700 km of coastline.

Our knowledge of the flora of Mozambique is still incomplete but current estimates suggest a total vascular plant diversity of 6,284 native or naturalised species (Hyde et al. 2021). However, this number continues to grow as targeted botanical surveys of previously unbotanised and botanically interesting areas are conducted, adding new records and new species to science (Darbyshire et al. 2019a, 2020a). This marks Mozambique as one of the most exciting countries for botanical research in tropical Africa, yet to many even amongst the botanical community, these rich plant resources are poorly known.

This varied plant life stems, in part, from Mozambique's diverse geography, geology and climates, which have resulted in a wide range of habitats and biogeographical affinities (Darbyshire et al. 2019a). The country supports four biomes and thirteen terrestrial ecoregions (Burgess et al. 2004a; https://ecoregions2017.appspot.com), and a recent ecosystem assessment records the presence of over 150 ecosystems nationally (Lötter et al., in prep.).

Whilst the majority of Mozambique is characterised by miombo and mopane woodland of the Zambezian Regional Centre of Plant Endemism (White 1983a), which is widely distributed across southern tropical Africa, it also features four much more localised cross-border Centres of Endemism (CoEs) which support high numbers of range-restricted species and habitats (Darbyshire et al. 2019a; see map in chapter "The Important Plant Areas of Mozambique: an overview"). In summary, these are:

(1) the Rovuma CoE of northeast Mozambique and southeast Tanzania, which extends along the coast and eastern lowlands through Cabo Delgado, Nampula and Zambézia provinces as far south as Quelimane (Burrows & Timberlake 2011);

(2) the Maputaland CoE, shared with South Africa and Eswatini, which extends along the coastal lowlands of southern Mozambique. Maputaland sensu stricto is taken to extend from KwaZulu-Natal province in South Africa to the Limpopo River in Mozambique (van Wyk 1996; van Wyk & Smith 2001), whilst Maputaland sensu lato, which also includes the proposed Inhambane sub-Centre, extends north to the Save River (Darbyshire et al. 2019a; A. Massingue, unpubl. data). The Lebombo Mountains, which straddle the border of the three countries, can also be considered a further sub-Centre of Maputaland (van Wyk and Smith 2001; Loffler and Loffler 2005);

(3) the Chimanimani-Nyanga (or Manica Highlands) CoE, a montane region that runs along the border with Zimbabwe in Manica province but also extends eastwards to the isolated massif of Mount Gorongosa in Sofala province;

(4) the Mulanje-Namuli- Ribáuè CoE, a series of inselbergs and massifs running from southern Malawi to Zambézia and Nampula provinces of northern Mozambique (Bayliss et al. 2014) of which

the most significant peaks are Mount Mulanje and the Zomba Plateau in Malawi, and Mounts Namuli, Mabu, Inago and the Ribáuè-M'paluwe in Mozambique.

A recent annotated checklist of the endemic vascular plants of Mozambique (Darbyshire *et al.* 2019a), the first comprehensive review of botanical endemism in the country, revealed 271 strict-endemic taxa (235 species) and 387 near-endemic taxa (337 species), together constituting nearly 10% of the total flora. That study also noted the presence of five strict-endemic genera (*Baptorhachis*, *Emicocarpus*, *Gyrodoma*, *Icuria* and *Micklethwaitia*) and two near-endemic genera (*Triceratella* and *Oligophyton*). Analysis of Mozambican endemic and range-restricted taxa revealed that 69% could be assigned to one of four cross-border CoEs noted above, highlighting their high botanical importance. As one of the only, and in many cases the only, nation to host these unique plant taxa, Mozambique is particularly responsible for the stewardship of these plants and the habitats that support them, and it is therefore imperative that conservation schemes and protected area networks take these into account both in terms of site delimitation and management, to ensure their survival for future generations.

Protection of and threats to plant diversity in Mozambique

Mozambique has 56 formally declared Protected Areas (PAs) including Ramsar Sites, with terrestrial PAs together covering 233,249 km^2 or 29.48% of the total land area of Mozambique (UNEP-WCMC 2021). These PAs are managed by the National Administration of Conservation Areas (ANAC) under the Ministry of Land and Environment (MTA). They include eight National Parks (Banhine, Bazaruto Archipelago, Chimanimani, Gorongosa, Quirimbas, Limpopo, Mágoè and Zinave), six National Reserves, three Community Conservation Areas, the São Sebastião Total Protection Area, the Primeiras and Segundas Islands Environmental Protection Area and a series of Hunting Reserves/ Wildlife Utilisation Areas and Forest Reserves, plus two Ramsar Sites. Several of these sites are vast flagship PAs for which Mozambique is internationally famous. The Niassa Reserve alone is 42,000 km^2 in area, or nearly 20% of the total terrestrial Pas (UNEP-WCMC 2021). The Mozambique PA network continues to grow, with the Quirimbas National Park being established as

recently as 2002 and the Chimanimani PA being elevated from National Reserve to National Park status in 2020 (Cabo 2020; UNEP-WCMC 2021). In late 2021, ministerial approval was given to merge Maputo Special Reserve and Ponta d'Ouro Partial Marine Reserve to form a new protected area, Maputo National Park, which will bring the total number of national parks to nine.

However, no assessment has been made to date as to how effectively the current PA network conserves Mozambique's plant diversity and, in particular, its unique and threatened flora and habitats. Many of the PAs were established primarily to protect the fauna and/or wilderness landscapes they contain. Plants have rarely featured highly in the establishment and management decisions. Many of Mozambique's PAs suffer from limited funding and resourcing and some of the key sites for plant diversity within this PA network are not effectively managed at present. Most notable is the network of 13 Forest Reserves which are not currently managed for their biodiversity with most facing significant threats from habitat loss and degradation (Müller *et al.* 2005).

Outside of PAs, many of Mozambique's habitats and the species they support face significant and accelerating threats from human activities. Unsustainable agricultural practices, with short-term slash-and-burn subsistence farming most prevalent, are resulting in significant destruction of natural vegetation and lowering of soil fertility. Further to this, in 2020, Mozambique recorded a population growth rate of 2.9%, the joint 11th highest rate globally (World Bank 2021). With a largely rural population, pressures on habitats are expected to increase in the coming years with many Mozambican people relying heavily on natural resources for their livelihoods. Other threats, such as excessive wood harvesting for fuel, construction and commercial sale, industrial and urban development and the spread of invasive species are all contributing to environmental degradation. Data from *Global Forest Watch* (World Resources Institute 2021) indicate that Mozambique lost 3.52 Mha of tree cover between 2001 and 2020, equivalent to a 12% decrease nationally. Data for tree cover gain are only available for the period 2001–2012, but the estimate of 145 kha gain over that period indicates that tree cover loss is greatly outstripping gains. This will, in turn, contribute to the longer-term impacts of human-induced climate change, with the severe flooding and resultant humanitarian crisis in central Mozambique triggered by Cyclone Idai in March 2019 serving as a stark indication of this future threat. Whilst intact ecosystems cannot protect against such events in the future, they can mitigate some of the worst impacts of such events by, for example, slowing the flow of and absorbing flood waters and preventing excessive soil erosion.

These wide-ranging human activities are having a significant impact upon Mozambique's unique plant diversity. Darbyshire *et al.* (2019a) noted that, based on taxa assessed to that point, over half of the endemic and near-endemic flora was threatened with extinction; the figures are updated in the chapter "The threatened plants of Mozambique" within this work. There is therefore an urgent need to effectively prioritise conservation efforts and change land use practices in order to manage Mozambique's natural resources more sustainably.

Motivation for the current study: conservation of Mozambique's flora

The government of Mozambique is committed to conserving and sustainably managing the biodiversity over which it has stewardship and is a signatory to the Convention on Biological Diversity (CBD). Driven by these commitments, Mozambique's National Strategy and Action Plan of Biological Diversity of Mozambique 2015–2035 (MITADER 2015) sets out a series of detailed national targets for documenting and conserving this rich biodiversity. Target 6 of this strategy aims to "by 2025, have at least 30% of habitats of endemic and/or threatened flora and fauna species with strategies and action plans for their conservation in place" with a series of related priority actions, including:

- **Action 6.1** – establish and implement coordinated programs for the systematic assessment of the conservation status of endemic and endangered species;

- **Action 6.2** – identify and describe the **Areas of Plant Importance**;

- **Action 6.3** – disseminate the Red data Book on national flora and fauna.

To address these targets and actions in order to effectively protect Mozambique's plant

diversity for future generations, the Instituto de Investigação Agrária de Moçambique (the Agricultural Research Institute of Mozambique – IIAM) and the Royal Botanic Gardens, Kew (Kew), with support from Eduardo Mondlane University (UEM), launched the **Tropical Important Plant Areas: Mozambique** project (https://www.kew.org/science/projects/tropical-important-plant-areas-tipas-mozambique). This project set out to combine existing data and expertise with targeted field survey data to identify and document Important Plant Areas in Mozambique, and to promote the conservation and sustainable management of these critical sites. This work builds on a long-term collaboration between Kew and Mozambican botanical institutions, stemming from the "Flora Zambesiaca" programme (1960 – present) and, more recently, the series of botanical surveys in sites of high biodiversity interest across Mozambique that have been conducted by IIAM, Kew and collaborators over the past 15 years. This work also draws on the national plant Red Listing programme and working group, established in 2011 through the IUCN-SSC Southern African Plant Specialist Group.

The Important Plant Areas concept and Tropical Important Plant Areas Programme

Important Plant Areas (IPAs) are defined as the most important places in the world for wild plant and fungal diversity that can be protected and managed as specific sites (Plantlife International 2004). The concept was developed by Plantlife International in the early 2000s to provide a systematic approach to identifying sites of high botanical value (Anderson 2002; Plantlife International 2004). The focus on plant diversity is of particular importance as there are known to be varying levels of congruence between the distributions of faunal and botanical diversity at a national level (Radford & Ode 2009; Byfield *et al.* 2010; Willis 2017). Reliance on faunal taxa alone to identify areas of national conservation importance, therefore, risks overlooking key sites for plant conservation while, even where plant and animal diversity co-occur, sites identified for their faunal taxa rarely consider plant diversity within their management plans (Darbyshire *et al.* 2017). The identification of IPAs provides an opportunity to address the under-representation of plants in conservation prioritisation and to effectively focus plant conservation efforts where they are most urgently needed. Whilst strong progress was made in the identification of IPAs in Europe and the Mediterranean region over the decade and a half since the launch of the IPA programme, very limited progress had been made in the tropics. Progress was hindered by challenges with IPA identification in tropical regions, due to the combination of high plant diversity and highly limited data availability. However, given the severe threats to the rich biodiversity in many tropical countries, there is a clear and urgent need to accelerate IPA identification and protection. With this in mind, the Tropical Important Plant Areas (TIPAs) programme was launched in 2015 (https://www.kew.org/science/our-science/projects/tropical-important-plant-areas). At the same time, Kew and Plantlife International carried out a global consultation on proposed revisions to the IPA criteria which built on the collective knowledge and experiences of the past decade and a half of IPA identification and also factored in some of the issues and challenges facing plant diversity in tropical countries. The post-consultation consensus on the revised IPA criteria was published by Darbyshire *et al.* (2017), with a user guide prepared by Plantlife (2018). Mozambique was amongst the seven countries / regions selected for the first phase of the TIPAs programme. So far, two national IPA assessments have been completed under this programme: the British Virgin Islands (Dani Sanchez *et al.* 2019) and Guinea (Couch *et al.* 2019). The network of IPAs in Mozambique presented here is, therefore, only the second IPA assessment to be completed in tropical Africa.

The impact of Important Plant Area networks

Although IPAs are not legally designated conservation areas, they can be used to enable maximum impact in environmental planning at the national, regional and international levels, prompting and reinforcing the protection and management of the identified areas (Dani Sanchez *et al.* 2019). They can provide an assessment of the importance of existing protected areas for plant conservation, highlight gaps in the national protected areas network and form the focus for community-led and citizen science-led conservation initiatives. IPAs can be an important tool in the mitigation hierarchy in industry and development, in particular the "avoidance" and

"offsetting" stages. Through providing evidence of the biodiversity value of a given site, IPA networks can be an important tool for environmental and social impact assessments at the planning stage of major development projects.

IPAs are closely aligned to targets set by the CBD and its Global Strategy for Plant Conservation (GSPC). These are currently being revised for the post-2020 Biodiversity Framework, but it is clear that IPAs will provide a significant contribution to the revised goals and targets. In particular, target 3 of the Framework is drafted as:

Target 3. Ensure that at least 30 per cent globally of land areas and of sea areas, especially areas of particular importance for biodiversity and its contributions to people, are conserved through effectively and equitably managed, ecologically representative and well-connected systems of protected areas and other effective area-based conservation measures, and integrated into the wider landscapes and seascapes.

IPAs also contribute to Goal 15 of the United Nations Sustainable Development Goals (SDGs), part of the 2030 Agenda to stimulate action for people, planet and prosperity:

Goal 15: Protect, restore and promote sustainable use of terrestrial ecosystems, sustainably manage forests, combat desertification, and halt and reverse land degradation and halt biodiversity loss.

Aims of the Important Plant Area project in Mozambique

A number of key aims were realised through this work:

- To bring together in readily accessible formats all available data on the plant diversity of Mozambique, with a particular focus on range-restricted plant species and habitats including the endemic (unique) biodiversity for which Mozambique has a particularly critical role in ensuring its long-term survival

- To establish which plant species in Mozambique are at most risk of extinction through contributing to a national list of globally threatened species using the internationally recognised criteria of the IUCN Red List

- To identify and map a network of priority sites – IPAs – which together can enable the preservation of plant diversity and priority plant species and habitats of Mozambique

- To provide readily accessible maps and data on the IPA network to support informed and positive land use decision-making, including increased sustainability and the avoidance of commercial development in areas rich in biodiversity

- To highlight significant gaps in the existing Protected Area network in Mozambique

- To help Mozambique fulfil its international commitments on biodiversity under the CBD and UN Sustainable Development Goals, and to contribute significantly to Mozambique's National Strategy and Action Plan of Biological Diversity

- To encourage national pride in the rich natural resources of Mozambique, and to support their conservation for future generations of Mozambican nationals

- To draw greater attention to the rich biodiversity of Mozambique to a wide range of audiences, including national and international scientists, students, policy makers, funders of conservation action, and tourists and visitors to Mozambique

- To raise awareness of the critical ecosystem services that healthy natural ecosystems can provide to local communities in Mozambique, including key plant resources for materials, food and medicines, ecosystem services such as the protection of fresh water sources, fertile soils and carbon sequestration

- To build capacity in Mozambique in the identification of plant conservation priorities, and to develop a range of other skills that enable Mozambican scientists to lead on botanical research moving forward, including field-based surveys, collections-based data collation and analysis and data dissemination.

Throughout this process, we have been conscious of the need for a realistic approach to targeting site-based plant conservation efforts, one which can effectively support species and habitats of conservation concern without excessive demands on land area, given the need to balance conservation efforts with the requirements and wellbeing of the growing human population.

IDENTIFYING IMPORTANT PLANT AREAS IN MOZAMBIQUE: METHODS AND RESOURCES

The identification of the network of IPAs in Mozambique presented here is the result of a five-year programme of herbarium, field and desk-based research, consultation with a wide range of stakeholders and experts, building of in-country capacity in the skills needed for IPA assessment, and documentation and review of the findings. The methods employed and the resources used to carry out this assessment are presented below, and it is advised that users read this section before consulting the summary text on threatened species and the IPA network, and the individual site assessments that follow.

Compilation of taxon data

Data compilation for this project was focussed primarily on Mozambique's endemic, near endemic and globally threatened plant taxa (species, subspecies and varieties). Threatened taxa are those assessed as globally Vulnerable, Endangered or Critically Endangered on the IUCN Red List of Threatened Species (IUCN 2012). Endemics refer to taxa that occur only within the political boundaries of Mozambique, while near-endemic taxa are as defined by Darbyshire et al. (2019a), meeting at least one of the following criteria:

(a) the majority of the taxon's range lies within Mozambique, and they are scarce and/ or highly range-restricted beyond; and/or

(b) the global range of the taxon is less than 10,000 km²; and/or

(c) the taxon is known globally from five or fewer localities.

Much of the data used to produce the checklist of endemic and near-endemic taxa (Darbyshire et al. 2019a) was subsequently used to delineate the IPAs of Mozambique. Extensive reviews of floristic and taxonomic literature for Mozambique and neighbouring countries contributed to the formation of this checklist. One of the key sources was the "Flora Zambesiaca" series (F.Z.; 1960 – present). Comprising 15 volumes and 49 parts published thus far, the Flora is over 90% complete, while we have also had access to the completed and partially completed volumes for Commelinaceae, Asteraceae (Compositae) in part, and Hyacinthaceae. Endemic and near-endemic Asteraceae may, however, be under-represented in this checklist as this family has not yet been completed for F.Z. Additional specimen citations and habitat information for Mozambique were derived from species accounts in the discontinued "Flora de Moçambique" series. Other key literature sources on Mozambique's threatened and range-restricted taxa include the seminal "Trees and Shrubs Mozambique" (Burrows et al. 2018) and a range of reports on recent botanical surveys and checklists of key localities (see Darbyshire et al. 2019a and the table on fieldwork below for a full list). The online "Flora of Mozambique" (Hyde et al. 2021), which is based on F.Z. but with regular updates and a wealth of additional useful information on the Mozambique flora, sites and botanical exploration, was widely consulted. Other key online resources were the "African Plants Database" (2021), "Plants of the World Online" (POWO 2021) and the "IUCN Red List of Threatened Species" (IUCN 2021).

Alongside these literature sources, herbarium collections were extensively referenced and collated, most notably those housed at BM, BNRH, EA, K, LISC, LMA, LMU, NH, P, PRE and SRGH (herbarium codes follow Thiers [continuously updated]).

Databasing of collections and sight records of endemic, near-endemic and threatened taxa, alongside fieldwork collections made during the course of the Mozambique TIPAs project, was undertaken in BRAHMS (Version 7.9.15), with over 13,000 records compiled so far. The 2017–2019 GBIF Biodiversity Information for Development (BID) project: *Mobilizing primary biodiversity data for Mozambican species of conservation concern* supported the collation of data from Mozambican herbaria. Where possible, each record was georeferenced, allowing for the mapping of taxon ranges. In turn, this distribution data allowed for a number of subsequent analyses, including: assessment of taxa against the endemism criteria set out above, the identification of potential

Centres of Endemism, the identification of sites that are species-rich with rare and threatened taxa in Mozambique and the completion of IUCN Red List Assessments. Data for endemic and near endemic taxa of Mozambique that have been assessed for the IUCN Red List are published on the Global Biodiversity Information Facility (GBIF.org) by Instituto de Investigação Agrária de Moçambique (IIAM) (Matimele 2021). Our intention is to release the full BRAHMS database as a freely available BRAHMS Online web resource in due course, although data for sensitive species (e.g. some *Encephalartos* spp.) will be withheld.

Accepted names of species and infraspecific taxa generally follow the African Plants Database (2021) and/or Plants of the World Online (POWO 2021), although in a small number of cases we follow alternative sources where they are considered to be most appropriate given our knowledge of the taxa involved. Endemic and near-endemic taxa follow the nomenclature of Darbyshire *et al.* (2019a) with a small number of additional taxa that have been described or uncovered since that publication. For brevity, authors of plant names are not included in the IPA reports. However, in Appendix 1, a complete list of the threatened and range-restricted taxa (IPA criteria A(i) and B(ii)) of Mozambique includes authors.

Targeted field surveys

In addition to collating existing data, this work has been strongly informed by recent botanical fieldwork at a wide range of locations across Mozambique. Sites were specifically targeted, either as areas that were candidate IPAs for which we needed contemporary data on critical species, habitats, management practices and threats, or as sites of potential botanical interest for which we had little or no existing data on which to support an IPA assessment. These are documented in the table opposite, which includes fieldwork conducted under the TIPAs programme together with other important surveys conducted recently (past 15 years) by one or more of the lead institutions in the IPA assessment team (Kew, IIAM and UEM) that have contributed significantly to this work.

This work has also benefitted from the extensive, recent field surveys of many other botanists who have kindly shared their knowledge of sites and species. Most prominent among these is the work of John Burrows, Sandie Burrows, Mervyn Lötter and Ernst Schmidt who travelled widely across Mozambique to study the woody flora in preparation for the "Trees and Shrubs Mozambique" (Burrows *et al.* 2018). Other important field contributions are from Ton Rulkens and Obety Baptiste who have visited many sites whilst studying the succulent flora of Mozambique; Bart Wursten, Petra Ballings, Mark Hyde and Meg Coates Palgrave who have botanised extensively in central Mozambique, notably in the Cheringoma-Gorongosa region and in the Chimanimani Mountains and foothills; and Quentin Luke who conducted fieldwork in Cabo Delgado province. The extensive field knowledge of in-country botanists such as Salomão Bandeira at UEM has also been of great value to this work.

Castigo Datizua and Jo Osborne pressing plant specimens in the field (ID)

Botanical field camp near Panda, Inhambane province (ID)

Province and location(s) visited	Date	Project	Report / resultant publication (if available)
Zambézia: Mount Chiperone	Nov./Dec. 2006	Darwin Initiative Award 15/036: SE Africa Montane Ecosystems	Timberlake *et al.* (2007); Harris *et al.* (2011)
Zambézia: Mount Namuli	May & Nov. 2007	Darwin Initiative Award 15/036: SE Africa Montane Ecosystems	Timberlake *et al.* (2009); Harris *et al.* (2011); Timberlake (2021a)
Zambézia: Mount Mabu	Oct. 2008	Darwin Initiative Award 15/036: SE Africa Montane Ecosystems	Harris *et al.* (2011); Timberlake *et al.* (2012); Bayliss *et al.* (2014)
Zambézia: Mount Inago	May 2009	Darwin Initiative Award 15/036: SE Africa Montane Ecosystems	Bayliss *et al.* (2010)
Cabo Delgado: Lower Rovuma Escarpment, Quiterajo, Lupangua Peninsula	Nov. – Dec. 2008 & Nov. 2009	ProNatura Coastal Forests of Mozambique	Timberlake *et al.* (2010, 2011); Darbyshire *et al.* (2020a)
Gaza/Inhambane/Maputo: coastal habitats of southern Mozambique	Nov. 2013 to Dec. 2015	A. Massingue Ph.D. thesis, Nelson Mandela University	Massingue (2019)
Manica: Chimanimani Mountains	April 2014, Sept. 2014, April/May 2016	CEPF Eastern Afromontane hotspot: Grant 63512, Chimanimani Mountains	Timberlake *et al.* (2016a); Wursten *et al.* (2017)
Gaza/Maputo: coastal habitats of southern Mozambique	Jan. & Oct. 2015	H. Matimele M.Sc. thesis, University of Cape Town	Matimele (2016)
Manica: Chimanimani Lowlands	June/July 2015 & Nov. 2015	Darwin Initiative Award 2380: Conservation and Livelihoods in the Chimanimani Forest Belt	Timberlake *et al.* (2016b); Rokni *et al.* (2019)
Nampula: Ribáuè-M'paluwe and Mount Inago Zambézia: Mount Chiperone	April 2017	CEPF and the National Geographic Society: Hidden under the clouds: Species discovery in the unexplored montane forests of Mozambique to support new Key Biodiversity Areas (H. Matimele participated)	N/A
Nampula: Ribáuè-M'paluwe and Matibane Forest Reserve	Oct. 2017	TIPAs Programme	N/A
Nampula/Zambézia	Nov. 2017	A. Massingue & N. Ngqiyaza; survey of sites for *Icuria dunensis* for Kenmare Moma Mining Ltd	N/A
Maputo: Lebombo Mountains	March 2018	TIPAs Programme	Osborne *et al.* (2018a)
Zambézia: Mount Lico and Pico Muli	May 2018	Scientific expedition to Mt Lico and adjacent mountains (organised by J. Bayliss)	Osborne *et al.* (2018b)
Manica: Serra Choa, Serra Garuzo and Tsetserra	June 2018	TIPAs Programme	Osborne & Matimele (2018)
Inhambane: Panda, Mabote and Lagoa Poelela	Jan. – Feb. 2019	TIPAs Programme	Osborne *et al.* (2019a)
Niassa: Txitonga Mountains and Njesi Plateau	May 2019	TIPAs Programme	Osborne *et al.* (2019b)
Maputo: Maputaland including Lebombo Mountains	Dec. 2019 & Jan. 2020	H. Matimele, Ph.D. studies, University of Kent	N/A
Nampula: Goa and Sena Islands	Sept. 2020	TIPAs Programme	Mucaleque (2020a)
Inhambane: São Sebastião Peninsula	May – June 2021; further surveys to take place in 2022	TIPAs Programme	Massingue *et al.* (2021)

Summary of botanical fieldwork conducted by one or more of the partner institutions and authors that has contributed significantly to the IPA assessment.

IUCN Red List Assessments

Over the course of this Mozambique TIPAs project, 273 global extinction risk assessments were undertaken using the categories and criteria of the IUCN Red List (IUCN 2012). Assessments again focused primarily on endemic and near-endemic taxa (see chapter "The threatened plants of Mozambique").

Many of these assessments were undertaken at workshops in Mozambique with collaborators from the IUCN-SSC (Species Survival Commission) Southern African Plant Specialist Group (SAPSG). National and international experts were brought together to assess batches of species from specific geographic regions or specific groups of plants, with the participants together reaching a consensus view on the appropriate extinction risk status for each taxon. Such workshops also served as a training opportunity in the application of the Red List criteria. Previous to the commencement of this project, further workshops were held on Rovuma and central Mozambican endemics (2014) and Maputaland endemics (2016), both hosted by the SAPSG, and on Chimanimani-Nyanga endemics hosted at Kew, which also greatly informed the identification of IPAs in Mozambique. These workshops are listed in the table on the next page. In addition, once fully trained and experienced in the Red Listing process, members of the IPA assessment team from Kew and IIAM assessed selected species individually or collectively outside of these workshops in order to make further progress on the assessment of Mozambique's priority plant taxa.

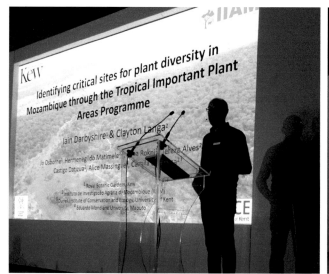

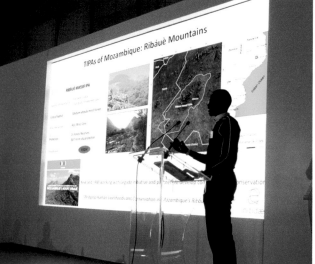

Iain Darbyshire and Clayton Langa present the IPAs of Mozambique project at the Oppenheimer Research Conference, Midrand, South Africa, October 2019 (CL/ID)

IUCN Red List workshop, Maputo, February 2018 (ID)

Location	Date	Number of days	Focus	Project and contributors
Buffelskloof Nature Reserve	July 2014	5	Rovuma Centre of Endemism endemics; central Mozambique endemics	SAPSG – launch of red listing programme in Mozambique IIAM, UEM, SANBI, Buffelskloof N.R., Kew, Q. Luke, M. Hyde, M. Coates-Palgrave
Kew	July 2015; March & June 2016	8	Chimanimani-Nyanga Centre of Endemism endemics	CEPF Chimanimani Project Kew, IIAM, National Herbarium of Zimbabwe (SRGH)
Buffelskloof Nature Reserve	May 2016	2	Maputaland Centre of Endemism endemics	H. Matimele, M.Sc. Research IIAM, UEM, SANBI, Kew, Buffelskloof N.R.
Maputo	Feb. 2017	3	Red List training workshop [SANBI] Mt Namuli endemics	TIPAs Project IIAM, UEM, Kew, SANBI, Pedagogical University Maputo, Legado
Maputo	Feb.-March 2018	5	Mt Namuli, Ribáuè Mts, Chimanimani Mts, Maputaland endemics	TIPAs Project IIAM, UEM, Kew, SANBI, Pedagogical University Maputo, T. Rulkens, J. Timberlake
Maputo	Jan. 2019	4	Endemic trees Maputaland / Inhambane endemics Cabo Delgado endemics	TIPAs Project IIAM, UEM, Kew, SANBI, Pedagogical University Maputo, Buffelskloof N.R.

Summary of IUCN Red Listing assessment and training workshops relevant to the Mozambique IPA work.

The majority of these assessments were based on the georeferenced specimen data compiled in the BRAHMS database but were informed by expert input and observation provided through the workshops. Tools such as GeoCAT (Bachman *et al.* 2011) and rCAT (Moat & Bachman 2020) were used to calculate area of occupancy (AOO) and extent of occurrence (EOO). Together with estimates for the number of locations for each species, these statistics were used to assess species primarily under criteria B and D of the IUCN Red List (IUCN 2012). The results of these assessments were entered and submitted via the IUCN Species Information Service (SIS). Assessments were reviewed by colleagues on the SAPSG and/or, where relevant, the East African Plant Red List Authority (EAPRLA) and were shared with taxonomic specialist groups on the IUCN-SSC where appropriate.

Several other projects have generated data on the extinction risk of Mozambique's flora that have been useful in the current work. In particular, (1) the Global Tree Assessment (BGCI 2021) resulted in a significant number of Red List assessments for Mozambique's tree species, although most of the endemic species were assessed through the TIPAs programme; and (2) the EAPRLA has contributed many assessments of species that occur in East Africa and extend into Mozambique, of particular importance being the assessment of species within the Coastal Forests of Eastern Africa hotspot.

The generation of these Red List assessments, and the use of assessments for Mozambican species produced outside this project, was of great importance in the identification of IPAs under criterion A. Alongside published assessments, taxa with assessments in press (i.e., those that have passed the review stage) and those in need of updating (assessed using a previous iteration of the Red List criteria and/or over 10 years old) were considered in IPA site assessments. In a few cases, we found that the species distribution cited in the published Red List assessments of non-endemic species, and in particular the Mozambique portion of the range, does not fully represent the currently known range. The species in question may need to be reassessed, particularly those previously found to be threatened under criterion B or D2 of the Red List criteria. In such cases, we have used the existing Red List category in IPA assessments if there are only one or few location(s) known in addition to those cited in the assessment. Where additional localities would clearly downgrade the Red List category of a species to Least Concern or Near Threatened, and therefore A(i) of the IPA criteria would not be applicable, species have not been considered when applying the IPA criteria.

IPA criteria

Important Plant Areas (IPAs) provide a systematic approach to identifying sites of high botanical value (Plantlife International 2004). To be identified as an IPA, a site must meet at least one of three criteria based on A) threatened species, B) botanical richness and C) threatened habitats. For each of these criteria, there are sub-criteria and associated thresholds that a site must meet or exceed in order to trigger IPA status. Following the 2015 launch of the Tropical Important Plant Areas (TIPAs) programme by Kew in collaboration with its in-country partners and Plantlife International, the IPA criteria were revised through a global consultation process (Darbyshire *et al.* 2017) to make IPAs more readily applicable globally and to address the difficulties of identifying sites of high botanical importance when data are highly limited. The criteria and thresholds are summarised in the table below; we also followed the guidance of Plantlife (2018) in the application of the criteria and delimitation of sites.

Alice Massingue presenting at the TIPAs: Mozambique project inception workshop, Maputo, 2017 (ID)

CRITERIA AND SUB-CRITERIA	THRESHOLD
A: THREATENED SPECIES	
A(i). Site contains one or more **globally threatened** species	Site known, thought or inferred to contain ≥1% of the global population AND/OR ≥5% of the national population OR the 5 "best sites" for that species nationally, whichever is most appropriate
A(ii). Site contains one or more **regionally threatened** species	Site known, thought or inferred to contain ≥5% of the national population, OR the 5 "best sites" for that species nationally, whichever is most appropriate
A(iii). Site contains one or more **highly restricted endemic** species that are potentially threatened	Site known, thought or inferred to contain ≥1% of the global population AND/OR ≥5% of the national population, OR the 5 "best sites" for that species nationally, whichever is most appropriate
A(iv). Site contains one or more **range-restricted endemic** species that are potentially threatened	Site known, thought or inferred to contain ≥1% of the global population AND/OR ≥5% of the national population, OR the 5 "best sites" for that species nationally, whichever is most appropriate

B: BOTANICAL RICHNESS	
B(i). Site contains a **high number of species** within **defined habitat or vegetation types**	For each habitat or vegetation type: Up to 10% of the national resource can be selected within the whole national IPA network OR the 5 "best sites" nationally, whichever is the most appropriate
B(ii). Site contains an **exceptional number of species of high conservation importance**	Site known to contain ≥3% of the selected national list of species of conservation importance OR the 15 richest sites nationally, whichever is most appropriate
B(iii). Site contains an **exceptional number of socially, economically or culturally valuable species**	Site known to contain ≥3% of the selected national list of socially, economically or culturally valuable species OR the 15 richest sites nationally, whichever is most appropriate

C(i). Site contains **globally threatened or restricted** habitat / vegetation type	Site known, thought or inferred to contain ≥**5%** of the national resource (area) of the threatened habitat type OR site is among the best quality examples required to collectively prioritise **20-60%** of the national resource OR the **5 "best sites"** for that habitat nationally, whichever is the most appropriate
C(ii). Site contains **regionally threatened or restricted** habitat / vegetation type	Site known, thought or inferred to contain ≥**5%** of the national resource (area) of the threatened habitat type, OR site is among the best quality examples required to collectively prioritise **20-60%** of the national resource OR the **5 "best sites"** for that habitat nationally, whichever is the most appropriate
C(iii). Site contains **nationally threatened or restricted** habitat / vegetation type, AND/OR habitats that have **severely declined in extent** nationally	Site known, thought or inferred to contain ≥**10%** of the national resource (area) of the threatened habitat type OR site is among the best quality examples required to collectively prioritise up to **20%** of the national resource OR the **5 "best sites"** for that habitat nationally, whichever is most appropriate

Summary of the IPA criteria (from Darbyshire *et al.* 2017).

Botanical survey team at Ribáuè-M'paluwe, Oct. 2017 (ID)

The definitions below follow Darbyshire *et al.* (2017):

- **Population** – the term "population" refers to the total number of individuals of a species within a discrete geographical unit

- **Range** – the range of a species is the known or inferred limits of its distribution. Taxon ranges were usually calculated using a minimum convex polygon (MCP) approach, aligned with extent of occurrence (EOO) as defined by IUCN. The few exceptions to this are where a taxon has a significant disjunction in its range which results in a highly inflated EOO based on MCP – in such cases, the disjunct population is treated separately in the range calculation. The most notable example of this are Chimanimani-Nyanga highland endemic species that also extend to Mount Gorongosa ca. 120 km to the east of the main highland range.

- **Highly restricted endemic species** – species with a total range of <100 km²

- **Range-restricted endemic species** – species with a total range of <5,000 km² but >100 km²

- **Restricted range species** – species with a total range of <10,000 km².

Application of IPA criteria and associated data

Botanical data for Mozambique are generally sparse and, despite continued efforts to address this and to fill the gaps in our knowledge, a significant proportion of specimens date from pre-independence (pre-1975), with a number of sites not visited by botanists since. The decision was made by the IPA assessment team, therefore, to allow the use of historical data where it was concluded that a taxon was still likely to be extant at the locality. In such cases, the continued presence and intactness of suitable habitat for the trigger species was used to infer the likelihood of continued presence.

Limited data availability also led to the application of only a subset of the IPA criteria, namely:

Criterion A(i) – this was the most frequently applied sub-criterion, as extensive data on threatened taxa had been generated through IUCN Red List assessments undertaken during this and other projects (see chapter "The threatened plants of Mozambique"). Threatened taxa at the rank of species, subspecies and variety are included in the assessments, but no sites were identified based on threatened infraspecific taxa alone.

Criteria A(iii) and A(iv) – these sub-criteria were applied for a small number of taxa that are known to be highly range-restricted or range-restricted endemics but have not been assessed for the Red List or were found to be Data Deficient.

Criterion B(ii) – a list of taxa of high conservation importance was compiled, primarily using the data from Darbyshire *et al.* (2019a). This list comprises taxa with a range of less than 10,000 km² together with all strict Mozambique endemic taxa, which currently total 507 taxa. Hence, to qualify under the threshold of ≥3% *of the selected national*

Discussion groups at IPA workshop, March 2018 (CL)

Papin Mucaleque and Clayton Langa (IIAM) working on specimen identification and data collation at Kew (ID)

list of species of conservation importance, a site needed to contain 16 or more trigger taxa. This sub-criterion is most applicable to sites where reasonably complete botanical inventory work has been undertaken, and with further botanical surveys in future there are likely to be more sites in Mozambique that will be found to meet the B(ii) threshold. However, by including sites within the top 15 for total B(ii) qualifying species, we have allowed a more inclusive list of sites to be considered under this sub-criterion. As nine sites were tied for 13th position in this ranking, only the top 12 were selected for qualification under B(ii) (see Summary IPAs Table in chapter "The Important Plant Areas of Mozambique: an overview").

With regard to criterion B(iii) – richness in species of high social, economic or cultural value – we have begun to gather data relevant to this sub-criterion and have a preliminary list of socio-economically important species in Mozambique in preparation. Data-collection and documentation for this sub-criterion has not, however, been exhaustive. As such, no site has been assessed under this sub-criterion in the current IPA network, although some of the IPAs identified here probably do qualify. Further research should be undertaken, and assessments updated accordingly. For the current assessment, species and habitats which provide important ecosystem services are noted under that section of the IPA reports.

Criterion C(iii) – national priority habitats (potentially threatened and/or range-restricted) were reviewed at an IPA workshop in Maputo in March 2018 (see Identifying IPAs), attended by a range of participants from the biodiversity research and education, conservation and forestry sectors. The following habitats were agreed as priorities for consideration under criterion C(iii):

- MOZ-01. Montane Moist Forest (primarily > 1600 m elevation)
- MOZ-02. Medium Altitude Moist Forest (primarily 900 – 1400 m elevation)
- MOZ-03. Low Altitude Moist Forest (primarily < 600 m elevation)
- MOZ-04. Inhamitanga (or Cheringoma) Sand Forest
- MOZ-05. Maputaland Coastal Dry Forest
- MOZ-06. Dwarf Forest on Coral Rag
- MOZ-07. Licuáti Thicket
- MOZ-08. Montane Scrubland
- MOZ-09. Montane Grassland
- MOZ-10. Seasonally Inundated Grassland
- MOZ-11. Granite Inselbergs
- MOZ-12. Rovuma Coastal Dry Forest

Participants at the IPA training workshop, March 2018 (CL)

Subsequent to this workshop, three further priority habitats were agreed by the IPA assessment team, two of which are distinct subdivisions of the Rovuma Coastal Dry Forest that are dominated by genera endemic to Mozambique and so are particularly noteworthy for conservation, the other being the only sizable area of limestone forest habitat in Mozambique:

- MOZ-12b. Rovuma *Micklethwaitia* Coastal Dry Forest

- MOZ-12c. Rovuma *Icuria* Coastal Dry Forest

- MOZ-13. Cheringoma Limestone Forest

These habitats, their structure and characteristic species, are described in more detail under the individual IPA site assessments and so are not summarised here.

Subsequent attempts to apply criterion C(iii) within IPA assessments proved challenging for some of the above habitats. In particular: (1) MOZ-05 Maputaland Coastal Dry Forest was found to be to too broad in scale, encompassing a range of dune forest and thicket and sand forest habitats that are collectively extensive but are worthy of subdivision; (2) MOZ-11 Granite Inselbergs, although botanically rich, is a frequently encountered and widespread habitat in Mozambique with many small- to medium-sized sites and so it is difficult to apply the thresholds for this habitat. In view of these challenges, Criterion C(iii) was not used for habitats MOZ-05 and MOZ-11 in the current network of sites.

Criterion C(iii) was applied using both calculated or estimated extent (area) of the priority habitat in question and its intactness within a given IPA. To assist with the application of criterion C(iii) to the moist forest categories (MOZ-01 – MOZ-03), Mozambique forest cover was extracted from a 13 year (2005 – 2018) time-series of dry season (June-September) satellite images from 30 m Landsat-7 and Landsat-8 imagery archive (courtesy of the U.S. Geological Survey, Woodcock *et al.* 2008) using Google Earth Engine (Gorelick *et al.* 2017). The forest cover output was refined using the Global Forest Change 2000 – 2019 dataset (Hansen *et al.* 2013), then classified and cleaned in ERDAS Imagine (ERDAS 2018). The gridded forest cover was then grouped into forest types using defined altitudinal thresholds from SRTM (NASA Shuttle Radar Topography Mission). The forest area associated with each altitudinal grouping was calculated per polygon for the IPA polygons using ArcGIS Pro (ESRI 2019). Some challenges were found in separating forest from dense miombo woodland at some sites, notably those in northern Mozambique such as the Txitonga Mountains where woodland cover is particularly dense. However, in general, this GIS analysis gave results that were broadly comparable with previously calculated estimates of forest cover at selected sites.

As part of the Key Biodiversity Areas initiative in Mozambique (WCS *et al.* 2021), a revised vegetation map for Mozambique is being prepared together with a national Red List of Ecosystems assessment (Lötter *et al.* in prep.). This work will provide a fine-scale classification of the historic (or potential) vegetation of the country and proposes over 150 ecosystem types. The subdivision of ecosystems is at a considerably finer scale than previous maps, and several of the criterion C(iii) habitats noted above will be subdivided. With this in mind, the IPA Criterion C(iii) assessment presented here must be considered preliminary and should be revisited once the vegetation map and Red List of Ecosystems are finalised; no IPA sites have been identified using criterion C alone at present.

Identifying IPAs

Training of partners in the assessment of IPAs and the identification of potential IPA sites was undertaken at a workshop hosted by IIAM in Maputo in March 2018. The preliminary IPA network was informed by the distribution of endemic, near endemic and threatened species and threatened and range-restricted habitats. Previous work towards identifying possible IPAs in Mozambique was conducted by a partnership of The Southern African Botanical Diversity Network (SABONET), IIAM and IUCN-Mozambique in a 2001 workshop (Izidine & Cándido 2004; Smith 2005). While these preliminary IPAs were identified under the original IPA criteria, they helped inform which sites may be of interest. Furthermore, analyses aiming to identify Mozambique's Centres of Plant Endemism (CoE) highlighted a number of sites with high concentrations of endemic and near endemic species, further informing which sites could be delineated as IPAs.

Delineation of sites was undertaken using Google Earth Pro (Version 7.3) and ArcGIS Pro (Version 2.8.3). At each potential IPA, distribution data for taxa of interest were extracted from BRAHMS and mapped alongside additional data from the IUCN Red List and relevant literature. A draft version of the historical vegetation map of Mozambique developed by Lötter *et al.* (in prep.; see above) was made available to us, allowing us to capture habitats of interest more accurately within IPA boundaries. Consideration was also given to the present-day quality of habitats; using Google Earth imagery, areas of degraded and converted habitat were excluded from IPAs where possible. In addition, IPA boundaries were drawn to follow natural boundaries (such as rivers and coastlines) and anthropogenic boundaries (such as roads and protected area boundaries) wherever possible, so that sites can be readily understood on the ground. In some cases, in particular at sites that have largely been surveyed along access roads only, but for which apparently contiguous habitat is more widespread, the boundaries of the IPAs have been drawn to encompass areas that have not yet been surveyed but probably host similar species. However, as these boundaries are based on an assumption that species of interest occur elsewhere within the seemingly contiguous areas of habitat, delineation is not exact at present and would require further ground-truthing; such cases are highlighted in the site reports.

Reports to accompany each IPA were compiled in the Tropical Important Plant Areas database (https://tipas-data.kew.org), and a version of each report is included within this publication and on the TIPAs Explorer data portal (https://tipas.kew.org).

Review of IPA site documentation

Reports generated for each IPA assessment were reviewed wherever possible by national and/or international specialists who have knowledge of the site and/or the flora present. Two full-day, online workshops were also held in August 2021 involving partners at Kew, IIAM and UEM. These workshops were an opportunity to review and edit IPA boundaries, site names, and the trigger taxa/habitats of each IPA and, as a result, the criteria under which each IPA qualified.

The IPA network identified here represents a summary of our current knowledge of the most important sites nationally for preserving Mozambique's plant diversity and its contribution to global biodiversity. However, this should not be considered to be a final work; rather the IPA process is iterative and it should be reviewed regularly in light of new information. This could include, but is not limited to, updating trigger species and habitat lists, qualifying criteria, site boundaries and identifying new sites. If a site becomes irrevocably degraded or further information invalidates IPA designation (a trigger species is found to be extinct or is reduced to synonymy within another taxon that does not meet IPA criteria etc.), de-listing of a site would be necessary to ensure that conservation efforts are focused on only the highest priority sites nationally.

Data presentation – IPA reports

The IPA reports included within this publication and on the TIPAs Explorer portal include the following text sections:

- **Site description** – a brief overview of the site, including the geographic location (province, district and any nearby towns and villages), main landscape features and other information of general interest about the site.

- **Botanical significance** – detail of important botanical elements within each site, including taxa and habitats which are considered to be of high conservation value and/or of cultural or economic significance.

- **Habitat and geology** – description of the range of habitats found at the site, detailing dominant and representative species in each where possible, as well as the underlying geology and soils, and climate and precipitation averages.

- **Conservation issues** – an overview of conservation measures at the site, including past and future initiatives, and whether the site falls within a protected area or other conservation designation (a Key Biodiversity Area (KBA) or Important Bird Area (IBA), for example). Comparisons to the KBA network are based upon the revised assessment of KBAs in Mozambique, identified using the new KBA criteria (see WCS *et al.* 2021). This section also describes the threats posed to each site, past, present and future. Faunal taxa of conservation interest, particularly those that depend heavily on the habitats of an IPA, may also be listed within this section.

- **Key ecosystem services** – details the value (natural capital) of the site to local people and beyond, primarily focussed on resources and services provided or supported by the vegetation within each IPA. A list of services is provided in the "**Ecosystem service categories**" section; this uses the typology of ecosystem services proposed by *The Economics of Ecosystems and Biodiversity* (TEEB 2010).

- **IPA assessment rationale** – a summary of which IPA criteria each site qualifies under and the botanical elements which trigger each criterion. This section is intended to be a stand-alone summary of the IPA status for each site, and so in some cases this will repeat (in summary) some of the information presented in the *Botanical significance* section.

In addition to the text accounts, each IPA report is accompanied by a number of data tables:

- **Priority species (IPA Criteria A and B)** – lists the species that (potentially) trigger criteria A and B(ii). Species that potentially trigger criterion A are scored (✓ = meets the threshold) against the three thresholds, ≥ *1% of global population*, ≥ *5% of national population* and/or *1 of 5 best sites nationally*. However, criterion A trigger species are listed even if the site does not meet any of these thresholds (in which case, all three columns are blank), as it is useful to be aware of the presence of such species at a given site. As these columns are relevant to criterion A only, they are not generally scored for criterion B(ii) species. Further columns detailing whether the entire population occurs at a site, the socio-economic importance and the abundance for each taxon are included for information only (i.e., these columns do not denote thresholds under the IPA criteria); these columns are scored for all criterion A and B(ii) trigger taxa. B(ii) qualifying species occurring at each site are included in the priority species table even where the cumulative total of these taxa at the site does not trigger Criterion B, as it is again useful to be aware of their presence.

- **Threatened habitats (IPA Criterion C)** – C(iii) trigger habitats are scored against the thresholds, ≥ *10% of national resource* or *1 of 5 best sites nationally*. The C(i) and C(ii) threshold of ≥ *5% of national resource* is included for completeness, in case of future Red Listing of Ecosystems that allows for a global or regional habitat assessment, but this threshold is not ticked if the ≥ *10% of national resource* threshold is met. As with criterion A trigger species, criterion C(iii) trigger habitats are included in this table even if the thresholds are not met (in which case, the three columns will be blank). An estimation of the current area of the C(iii) habitat is recorded where available.

- **Protected areas and other conservation designations** – lists any formally protected areas (either public or private) as well as other conservation designations (such as Key Biodiversity Areas, Ramsar sites and Important Bird Areas) at the site. We also record how these are related spatially to the IPA, i.e., whether the protected area or other conservation designation matches, overlaps, encompasses or is encompassed by the IPA. If the site is not formally protected then "No formal protection" is listed, but the site can still have other conservation designations, such as KBA status.

- **Threats** – lists all threats faced by the flora at each site, using the standardised IUCN threat classification applied in the Red List of Threatened species. Each threat is given a severity rating (low, medium, high, unknown) and a timing (past - not likely to return, past - likely to return, ongoing - stable, ongoing- increasing, ongoing - declining, ongoing - trend unknown, future - planned activity, future - inferred threat).

Within this publication, each IPA report is accompanied by two site maps:

- a general reference map of the IPA and surrounding areas to indicate the local geography of a site; and

- an imagery map which presents satellite data from ESRI as a visual presentation of habitats within an IPA.

The legend for the maps is included below and references for map layers are provided within the **Bibliography**. All maps and analyses were undertaken using the World Geodetic System 1984 (WGS 84) geographic coordinate system.

- IPA Boundaries (reference map)
- IPA Boundaries (imagery map)
- ● Major Locality
- – – – Provinces
- ——— Roads
- Lakes, Rivers, Sea, Wetlands
- Botanical Reserve

Protected Areas

- Buffer Zone
- Protected Area (Overlapping)
- Protected Area

All references cited in the text are provided in the **Bibliography** at the end of the report. **Appendix 1** provides a summary of the distribution within the IPA network of Mozambique's current IUCN Red List of globally threatened taxa and all the taxa that qualify as being of high conservation importance under criterion B(ii).

THE THREATENED PLANTS OF MOZAMBIQUE

To provide the necessary evidence for identifying and assessing potential IPAs under criterion A(i), work was undertaken to expand the number of Mozambican taxa assessed for the IUCN Red List. Assessments for the TIPAs project were focussed on endemic and near-endemic species. In total, 273 taxa were assessed with 264 of these assessments submitted to the IUCN Red List over the course of the Mozambique TIPAs project, while over 100 additional assessments were compiled and published in preparation for this project (see chapter "Identifying Important Plant Areas in Mozambique: methods and resources" for details of Red List Workshops), representing a significant contribution towards the national Red List.

Overall, around 30% of the flora of Mozambique has been assessed for the IUCN Red List (including assessments that have been submitted but not yet published and published assessments which need updating). Of those assessed to date, 23% (335 taxa) have been found to be threatened.

Endemic and near-endemic taxa
69% (462 taxa) of the endemic and near endemic Mozambican taxa have now been assessed for the IUCN Red List (including assessments that have been submitted but not yet published and assessments that require updating as they were assessed under previous iterations of the IUCN Red List criteria and/or are more than 10 years old). Of these assessed, endemic and near-endemic taxa, 55% are threatened with extinction, almost double the proportion of threatened taxa across all assessed Mozambican native species. The higher level of threat experienced by endemic and near-endemic species is likely due to their more restricted ranges. Range size is a strong predictor for extinction risk in plant taxa as the probability of an environmental perturbation or disturbance impacting the entire taxon range is greater for those taxa with smaller ranges (Gaston & Fuller 2009; Leão *et al.* 2014).

Species of Rubiaceae and Fabaceae represent 30% of these threatened endemics and near endemics, reflecting that these are the two most speciose families within Mozambique's endemic flora (Darbyshire *et al.* 2019a). Annonaceae, Melastomataceae, Acanthaceae, Zamiaceae and Gesneriaceae all show percentages of threatened endemic and near-endemic taxa higher than the 55% average for all families.

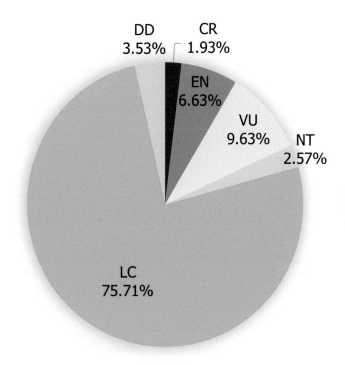

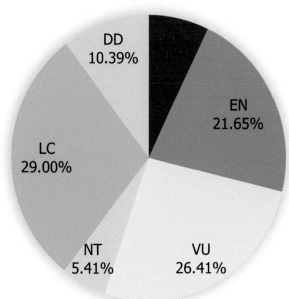

IUCN Red List categories for all assessed, native plant taxa of Mozambique

IUCN Red List categories for all assessed, endemic and near-endemic plant taxa of Mozambique

Family	Total endemic and near-endemic taxa	Total endemic and near-endemic taxa assessed for IUCN Red List	% assessed endemic and near-endemic taxa threatened with extinction
Annonaceae	11	10	100% (10 taxa)
Melastomataceae	15	14	86% (12 taxa)
Acanthaceae	26	21	81% (17 taxa)
Zamiaceae	11	10	80% (10 taxa)
Gesneriaceae	11	9	78% (7 taxa)

Families with the highest levels of over-representation within threatened taxa, excluding those with fewer than nine assessments or with fewer than half of their endemic and near endemic taxa assessed for the IUCN Red List.

While it is possible that, for some taxa, assessment for the IUCN Red List was prioritised if they were thought to be threatened with extinction, given the good coverage of assessments within the five families shown above, we can conclude that this over-representation is due to genuinely higher levels of threat rather than sampling bias.

The higher levels of threat experienced by Annonaceae and Melastomataceae species could be related to the habitats in which many of the taxa within these families reside. Endemic and near-endemic taxa from these families predominantly occur within forest habitats (10 Annonaceae; 11 Melastomataceae), particularly in dry coastal forests (8 Annonaceae; 8 Melastomataceae). The dry coastal forests of Mozambique continue to experience high levels of degradation and habitat loss due to threats such as expanding agriculture and logging for timber and fuelwood (Timberlake *et al.* 2011), which may account for the over-representation of these families within threatened endemics and near-endemics.

Acanthaceae species, however, are less associated with specific habitat types, having diversified in a wide range of ecosystems, but the family is known to contain a high number of range-restricted and scarce species (Manzitto-Tripp *et al.* 2021) which may, in turn, increase the risk of species becoming threatened with extinction. Similarly, many of the endemic and near endemic Zamiaceae of Mozambique have highly limited ranges, in many cases restricted to just one mountain, such as *Encephalartos munchii* (CR) on Mount Zembe and *E. pterogonus* (CR) on Mount Muruwere, or one

mountain range, including *Encephalartos aplanatus* (VU), *E. senticosus* (VU) and *E. umbeluziensis* (EN) in the Lebombo Mountains. The endemics and near endemics within the Gesneriaceae family also include many taxa with restricted ranges including *Streptocarpus brachynema* (CR), found only on Mount Gorongosa, and *S. acicularis* (CR), *S. grandis* subsp. *septentrionalis* (NE), *S. hirticapsa* (VU) and *S. montis-bingae* (DD), which are all restricted to the Chimanimani Mountains.

By contrast, endemics and near-endemics in the Asteraceae family may be less likely to be threatened with extinction than the average. Of the 15 assessed taxa, 73% were found to be Least Concern, compared to the overall average of 29% Least Concern for endemic and near-endemic taxa. However, 12 endemic and near-endemic taxa within this family are yet to be assessed and so further research is needed to confirm that endemic and near-endemic Asteraceae are truly at lower risk of extinction than other families.

Distribution of threatened species

Analyses of the distribution of threatened taxa in Mozambique were undertaken in ArcGIS Pro (ESRI 2019) using occurrence data for threatened species databased in BRAHMS over the course of the Mozambique TIPAs project. The database is largely complete, although distributions for some widespread threatened species may be incomprehensive.

A quarter degree square (QDS) analysis of threatened taxa distributions was undertaken by first dividing Mozambique into a grid of 0.25° x 0.25° cells, then calculating how many threatened

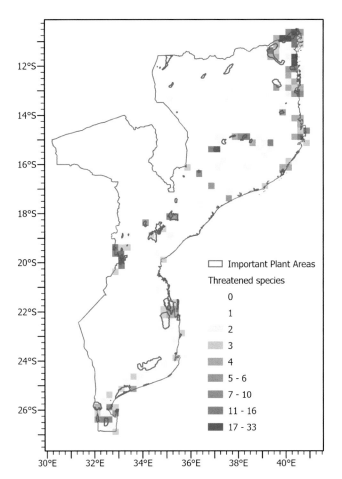

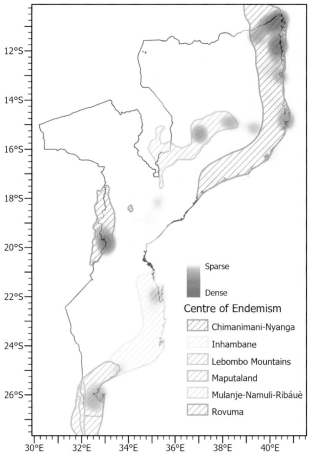

Number of threatened plant species within each cell of a quarter degree square grid of Mozambique, alongside the national network of Important Plant Areas.

Heatmap showing the density of threatened species occurrences in Mozambique alongside the proposed Centres of Endemism adapted from Darbyshire *et al*. (2019a). Density here is a measure of the number of threatened species occurring within a 30 km radius of each cell in a 1 km² cell grid of Mozambique.

taxa occur within each of these cells. By necessity, some of these cells overlap with the ocean or neighbouring countries to ensure that all areas of Mozambique were captured, however only occurrence data from Mozambique were used in these analyses. A heatmap of threatened taxa occurrences was also produced to interpret spatial patterns on a broader scale. As with the quarter degree square analysis, multiple occurrences of the same threatened taxon within a QDS were removed to analyse the number of threatened taxa in an area. Density was then calculated as a measure of the number of threatened taxa occurring within a 30 km radius of each cell in a 1 km² cell grid of Mozambique. Therefore, unlike the QDS analysis, each cell within the heatmap is influenced by the number of threatened taxa in the surrounding area.

Both maps demonstrate a correlation between the areas of highest threatened taxon density and the proposed Centres of Plant Endemism (CoEs) in Mozambique. This association is because Mozambican endemic species are more likely to be threatened, as demonstrated earlier in this chapter, and so with high levels of endemic species concentrated within the CoEs, we can also expect a higher density of threatened species.

The coastal dry forests and thickets of the Rovuma CoE have some of the highest densities of threatened species of all the CoEs. These habitats are particularly threatened by rapid transformation to subsistence agriculture, logging (often as a source of fuelwood) and uncontrolled fires, while exploration for oil and gas may well accelerate the clearance of coastal forests within this CoE (Timberlake *et al.* 2011). Coupled with these threats is the exceptionally high number of endemic and near-endemic taxa within this CoE, with more Mozambican endemics confined to this CoE (over 50 taxa) than to any other CoE in Mozambique (Darbyshire *et al.* 2019a). The combination of high levels of threat to coastal dry forests and thickets, with many species confined to these habitats, therefore, accounts for the high threatened species density in the Rovuma CoE.

The highest density of threatened taxa nationally, based on the QDS analysis, occurs within the Rovuma CoE, south-west of Quiterajo town covering much of Namacubi Forest, with 33 taxa occurring within this QDS cell. This area is followed closely by the eastern Rovuma escarpment and valley area, with the second highest value QDS cell (27 taxa) and several other high-density cells neighbouring.

While both Quiterajo and the lower Rovuma escarpment fall within the proposed Rovuma CoE, the third highest value QDS cell (24 taxa) occurs across the Chimanimani mountains and lowlands, a key area in the Chimanimani-Nyanga (Manica Highlands) CoE, with neighbouring Tsetserra also scoring highly. Similar to the Rovuma CoE, there is a high number of species confined to this area, with over 150 endemic and near-endemic species restricted to the Manica Highlands CoE. While the highland areas remain largely intact save for the localised impact of artisanal gold mining and some increases in fire frequency, the species-rich lowlands of Chimanimani are under threat from habitat loss, largely through agriculture and frequent burns (Timberlake *et al.* 2016a, 2016b; Wursten *et al.* 2017).

The Mulanje-Namuli-Ribáuè CoE shows areas of high threatened taxon density that are highly localised, which is to be expected as species richness in these areas is associated with archipelago-like inselbergs and massifs. This is demonstrated within the QDS map, showing isolated cells of high threatened density, while the heatmap fails to detect such areas in parts of this CoE as it takes into account the areas surrounding each mountain where there are few or no threatened species present. Mount Namuli ranks within the top ten QDS cells nationally for threatened taxa (20 taxa), while Mount Mabu (7 taxa) and the Ribáuè and M'paluwe mountains (12 taxa) also show moderately high levels of threatened taxon density.

Only moderate densities of threatened taxa occur within the Inhambane, Lebombos and Maputaland (in the narrow sense) CoEs. However, the coastal forests and thickets that cover large areas of the Inhambane and Maputaland CoEs, like those of the Rovuma CoE, are under great pressure from habitat loss, particularly for subsistence agriculture (Key Biodiversity Areas Partnership 2020). The lower concentrations of threatened species density in these CoEs possibly reflects the lower richness of endemic and near-endemic taxa (Darbyshire *et al.* 2019a), which are more prone to fall within a threatened category, rather than lower levels of threat within these CoEs.

Threatened species within the IPA network

Threat category	Coverage within IPA network
VU	82.86%
EN	83.06%
CR	77.14%

The IPA network (see chapter "The Important Plant Areas of Mozambique: an overview") incorporates the vast majority of Mozambique's threatened plant taxa, with 82% occurring within at least one IPA.

Most of the taxa that do not occur inside any IPAs are from the more speciose families in the Mozambican flora (Fabaceae – 17%, Euphorbiaceae – 12%, Acanthaceae – 9%, Rubiaceae – 7%).

Many of the threatened taxa that fall outside the IPA network are not known to co-occur with other features of conservation importance such as other threatened taxa, threatened habitats or areas rich in endemic taxa. However, further research may strengthen the case for identifying these areas as IPAs (see "Additional sites of botanical interest" at the end of chapter "Important Plant Area assessments"). Alternatively, these taxa may occur in areas that are already heavily degraded, where the viability of successful conservation or restoration is low.

Aloe mossurilensis, for instance, is a Critically Endangered species known only from a single confirmed locality on the coastal cliffs in the Mossuril area of Nampula Province (Darbyshire *et al.* 2019b). Surrounding tourist areas leave the small area of *A. mossurilensis* habitat isolated and unsuitable for recognition within an IPA. Elsewhere, *Emicocarpus fissifolius*, the only species known from the endemic genus *Emicocarpus*, has been assessed as Critically Endangered (Possibly Extinct) as it is only known from historical records within Maputo City, the localities of which have all been heavily transformed in the decades following the last collection in 1966 (Matimele *et al.* 2016). Recent searches at sites with potentially suitable habitat have not been successful and so it is not possible to incorporate this species into the IPA network unless viable extant populations are found in the future.

For species such as *Aloe mossurilensis*, and any remaining populations of *Emicocarpus fissifolius* within Maputo City, other conservation actions, such as *ex situ* conservation, maybe be more suitable for preventing extinction. However, future research should focus on the identification of sites where threatened species that are not currently within the IPA network reside. Sites with viable conservation opportunities could then be incorporated into a future iteration of the IPA network, with the ultimate aim of including at least one population of each threatened species within the network wherever possible.

Raphia australis palm (globally VU) growing in a reed bed within Chidenguele IPA (HM)

THE IMPORTANT PLANT AREAS OF MOZAMBIQUE: AN OVERVIEW

Distribution, size and complexity of the IPA network

The network of Important Plant Areas (IPAs) of Mozambique comprises **57 sites**, distributed widely across the country, with all of the provinces represented, except Maputo City and Tete. Most of the sites are entirely terrestrial, but small areas of shallow marine environments with mangrove or seagrass communities are included in six coastal IPAs. The 57 sites together cover an area of **22,990 km²**. This represents **fewer than 3%** of the total land area of Mozambique but encompasses important populations of **82%** of the threatened plant taxa of Mozambique (see chapter "The threatened plants of Mozambique"), together with the intact habitats that support these species and the associated ecosystem services they provide. Hence, the conservation and sustainable management of this relatively small land area would have enormous benefits for the preservation of the rare and endemic flora and important habitats over which Mozambique has custodianship. The summary results of the IPA network are tabulated and mapped at the end of this section.

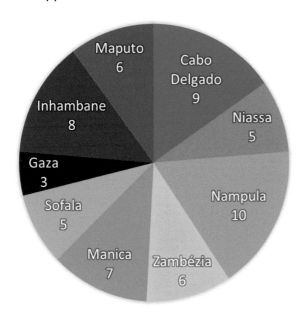

The distribution of IPAs per province of Mozambique (IPAs that cross provincial boundaries are counted twice).

The IPAs vary considerably in size and complexity. The majority of the sites are relatively small in area and are largely defined by discrete dominant habitats (e.g. coastal dry forest, woodland and thicket in the Quiterajo IPA) and/or geographic features (e.g. many of the montane sites such as the Mount Zembe and Ribáuè-M'paluwe IPAs). The majority of IPAs (33 sites) are less than 200 km² in area, and over two-thirds (39 sites) are less than 300 km², hence they are of a scale that is practical for management as a single unit. At the lowest end of the scale, five sites are less than 10 km², including Bobole (0.2 km²) which is designated on the basis of supporting a small but important population of the palm *Raphia australis* along the Bobole River north of Maputo; the Goa and Sena Islands IPA (0.7 km²) which comprises a small area of island coral rag thicket supporting one endemic and two highly range-restricted and threatened species; and the Lúrio Waterfalls at Chiure (6.7 km²) which supports the entire known global population of *Aloe argentifolia*. At the highest end of the scale, three sites cover an area of over 2,000 km²: Panda-Manjacaze (2,599 km²), Mueda Plateau and Escarpments (2,200 km²) and Mapinhane (2,070 km²). These landscape-scale sites contain a mosaic of distinct but inter-connected habitats that together support a rich plant diversity. However, IPA trigger species and habitats are typically only known from small areas within these mosaics, and more extensive fieldwork in each could help to define critical localities within these IPAs that might be most suitable for intensive conservation strategies and measures, whilst the IPA site as a whole might be more suited to "Protected Landscape" or "Protected Area with Sustainable Use of Natural Resources" status (Categories V and VI of the IUCN protected area management classification; Dudley 2013).

Qualifying criteria

Criterion A: Threatened Species

All of the 57 sites within the IPA network qualify under criterion A and all but one of these contain at least one globally threatened species as assessed for the IUCN Red List, and hence qualify under criterion A(i). The exception is the Txitonga Mountains which currently qualifies only under criterion A(iii) due to the presence of *Hartliella txitongensis*.

A highly restricted endemic that has only recently been described (Osborne *et al*. 2022), this species has not yet been assessed on the IUCN Red List but is highly likely to be threatened based on current knowledge. These mountains are amongst the least-explored areas of Mozambique botanically and more criterion A species are likely to be found following further surveys, as is the case for many of the IPAs documented here. On average, each site contains populations of over seven globally threatened A(i) taxa, but this varies widely between sites. Fourteen IPAs contain populations of 10 or more threatened taxa, with the richest sites for threatened species being the Lower Rovuma Escarpment (54), Quiterajo (38), Chimanimani Mountains (29) and Mount Namuli (22). At the other end of the scale, seven sites contain only one globally threatened taxon (species).

Criterion B: Botanical Richness

Botanical richness in Mozambique, as measured using criterion B(ii), is highly variable across the IPA network. Eight IPAs meet the threshold of containing >3% (≥16 taxa) of the total list of conservation priority trigger taxa (i.e., national endemic taxa and/or those species with a global range of less than 10,000 km²). A further four sites contain >10 of these taxa (>2%) and trigger criterion B(ii) under the threshold of the "15 richest sites". Of the sites that do not currently meet this criterion, eight are currently known to support 10 trigger taxa and some of these sites will probably qualify under criterion B(ii) following more exhaustive botanical surveys. By far the richest site recorded in the current study is the Chimanimani Mountains, with 96 B(ii) trigger taxa, or nearly 20% of the total B(ii) taxon list. Three other montane sites, Mount Namuli (40 taxa), Tsetserra (36 taxa) and Ribáuè-M'paluwe (22 taxa) are also exceptionally rich, whilst the two coastal Cabo Delgado sites with the highest concentrations of threatened species, the Lower Rovuma Escarpment and Quiterajo, are also amongst the richest sites with 22 trigger taxa each. The strong positive correlation between high botanical richness and concentrations of threatened species is unsurprising given that restricted range is typically a strong predictor of extinction risk (Gaston & Fuller 2009; Leão *et al*. 2014).

Criterion C: Threatened Habitats

Under the preliminary assessment of nationally threatened and range-restricted habitats of Mozambique, 26 IPAs qualify under criterion C(iii) as supporting nationally important examples of one or more of 12 different habitat types. No IPAs are identified on the basis of Criterion C alone (see Methods for discussion on this point). Given that we do not yet have accurate figures for the current areal extent of these critical habitats, in most cases they are identified on the basis of being amongst the "five best sites" for a given habitat. However, for some habitats, such as the Rovuma *Icuria* Coastal Dry Forest, we have sufficient information available on the total extent and the area within each site to calculate whether the site meets the C(iii) threshold: "site known, thought or inferred to contain ≥10% of the national resource (area) of the threatened habitat type". In other cases, individual sites are known to be well in excess of that 10% threshold and so are assessed as such, for example the extensive Medium Altitude Moist Forest at Mount Mabu IPA and Lowland Moist Forest in the Chimanimani Lowlands IPA, both of which are overwhelmingly the largest examples of these habitats in Mozambique. Mount Gorongosa is the only site to contain nationally important extents of three C(iii) habitats: Montane Moist Forest, Medium Altitude Moist Forest and Montane Grassland. Six further sites meet the threshold for two C(iii) habitats.

As noted in the Methods section, the status of habitats within the IPA network will need to be revisited once the revised vegetation map is completed (Lötter *et al*. in prep.) and a national ecosystem threat assessment is conducted, using the categories and criteria of the IUCN Red List of Ecosystems. It is considered likely that (a) all of the habitats highlighted in the current assessment will be assessed as threatened; and (b) many IPA sites will contain additional threatened habitats beyond those highlighted in the current work. In particular, the range of dry coastal forests, thickets and woodlands of southern Mozambique are likely to be highlighted as threatened and the IPAs identified in this region will contain nationally significant extents of these habitat types. A good example is the Pande Sand Thicket within the Mapinhane and Temane IPAs, which is a range-restricted habitat type that is highly threatened by agricultural activity (Lötter *et al*. in prep.) and the large majority of this habitat is located within these two IPAs.

Sites meeting all three criteria

Of particular note are the eight IPAs that meet all three criteria: Chimanimani Lowlands, Chimanimani Mountains, Lower Rovuma Escarpment, Mount Gorongosa, Mount Namuli, Quiterajo, Ribáuè-M'paluwe and Tsetserra. These sites contain important populations of multiple threatened species (10 or more in all cases), exceptional botanical richness (15 or more B(ii) trigger species in all cases, and 20 or more in all but one case) and between one and three threatened habitats. Whilst all of the 57 IPAs identified in the Mozambique network are of high significance for maintaining both national and global plant diversity, and we have refrained from attempting any ranking of their importance, these eight IPAs are clearly of critical importance to protecting Mozambique's unique biodiversity. As such, they should ideally be given high priority in future conservation strategies, particularly at sites where interventions are urgently needed, such as on Mount Namuli and at Quiterajo, both of which are currently unprotected and facing significant threats.

Complementarity and representativity within the IPA network

49 IPAs (85%) support populations of globally threatened taxa that are not found elsewhere within the Mozambique IPA network, and, as noted above, over 80% of the total threatened taxa of Mozambique are represented. The network of sites selected, therefore, provides effective representation for Mozambique's threatened flora whilst avoiding excessive repetition of similar biodiversity units. The sites with the highest number of "unique" threatened taxa (unique here referring to species found at only one site within the IPA network) are similar to those with the highest number of threatened species: Lower Rovuma Escarpment (25), Chimanimani Mountains (21), Mount Namuli (16) and Quiterajo (12).

Similarly for botanical richness, 49 IPAs (85%) support populations of B(ii) trigger taxa that are not found elsewhere within the IPA network and 21 sites have 3 or more "unique" B(ii) taxa. This indicates that, whilst complementarity was not explicitly built into the criteria applied for measuring botanical richness, the network nevertheless has a good representation of the full range of B(ii) taxa – 73% of the B(ii) trigger taxa (370 taxa) are

represented in total - with sites complementing one another in terms of their species assemblages. This reflects a conscious effort by the IPA assessment team to identify sites that support species and habitat complements that represent a broad range of Mozambique's ecosystems and biogeographical units. The richest sites for "unique" B(ii) trigger taxa within the IPA network are again similar to those with the highest total number of B(ii) species: Chimanimani Mountains (71), Mount Namuli (30), Tsetserra (18) and Lower Rovuma Escarpment (14).

Comparison to the provisional 2005 SABONET IPA network

The preliminary review of potential IPAs in Mozambique conducted under the Southern African Botanical Diversity Network (SABONET) programme identified 28 sites across the country (Smith 2005). This exercise was based on a workshop of national and regional experts on the Mozambican flora, and attempts were made to identify which criteria each of the identified sites was likely to meet, informed in part by the first attempt at a national Red List of threatened species for Mozambique (Izidine & Bandeira 2002). The current IPA assessment includes (in full or in part) only 16 of the 28 IPAs proposed in 2005. Those that are omitted have not been found to meet the criteria in the current study, although several are noted as sites of interest at the end of the site accounts and may qualify as IPAs in the future, such as Serra Choa and Zitundo. For several of these sites, there is currently insufficient botanical data to support an IPA assessment – a good example is the extensive Tchuma Tchato community conservation area in Tete province (Filimão *et al.* 1999) which harbours a rich fauna and intact habitats, and qualifies as a Key Biodiversity Area (WCS *et al.* 2021), but for which there are currently very limited botanical data.

Biogeographical links

Strong links are evident between the distribution of IPAs in Mozambique and the local Centres and Sub-Centres of Plant Endemism (CoEs) that have been proposed within the country (van Wyk 1996; van Wyk & Smith 2001; Loffler & Loffler 2005; Burrows & Timberlake 2011; Bayliss *et al.* 2014 & in prep.; Darbyshire *et al.* 2019a, 2020a). These CoEs are subject to ongoing research to more accurately define their limits and endemic floras, aided by our growing understanding of plant and habitat distributions in Mozambique. Based on current evidence, 47 of the sites (>80%) are considered to fall within a local CoE, as per the table below.

Within the Rovuma CoE, a region of coastal lowlands supporting a range of dry forest, woodland and thicket habitats, the majority of sites fall within the northern portion of this phytogeographical region in Cabo Delgado and northern Nampula Provinces south to the Matibane Forest IPA. This area could be defined as the "core" area of the Rovuma CoE, extending northwards into southeast Tanzania, and is particularly rich in range-restricted species, with high species turnover in some plant groups. A small number of Rovuma IPAs, associated primarily with the distribution of the endemic forest-forming tree *Icuria dunensis*, extend into southern Nampula and Zambézia Provinces.

South of the Zambezi River, and in particular south of the Save River, the coastal lowland habitats support a markedly different endemic flora which can broadly be defined as the Maputaland CoE. This phytogeographical region has previously been considered to have a much smaller range, essentially south of the Limpopo River and extending into northern KwaZulu Natal Province. However, there are clear phytogeographical links between the core Maputaland region and lowland areas further north in Gaza and Inhambane Provinces, suggesting that the delimitation of Maputaland should be extended significantly. There are, however, some notable differences in the endemic elements between the northern and southern limits of this broadly defined Maputaland which lend support to the recognition of an Inhambane Sub-Centre, but the boundary of this region still requires better definition. For example, recent fieldwork within the Panda-Manjacaze IPA, currently included within the Inhambane Sub-Centre, has revealed several species (*Cola dorrii*, *Psydrax fragrantissima*, *Xylopia torrei*) that were previously considered to be endemic to Maputaland s.s.

Inland in the extreme south of Mozambique, the low hilly region of the Lebombo Mountains running along the border with Eswatini and South Africa, represented by two IPAs in Mozambique (Goba and Namaacha), has a small but notable set of endemics, and also contains habitats that differ from those of the rest of Maputaland, supporting its recognition as a discrete phytogeographical unit.

The majority of Mozambique's montane IPAs fall clearly within either the proposed Mulanje-Namuli-Ribáuè CoE or the Chimanimani-Nyanga CoE. Both these regions have been treated previously within a broader Eastern Afromontane phytochorion but, whilst many habitats and even some of the locally dominant species overlap between these two phytogeographical regions, there is very little overlap in terms of their endemic and range-restricted floras, hence recognition as two discrete CoEs is most appropriate. The five IPAs in Niassa Province (sites 11 – 15 on the map) currently lie outside any of the local CoEs and, given the incomplete botanical inventory within these sites, their phytogeographical affinities are still not fully understood. However, all of these five sites are montane or submontane and could be included within the broad Eastern Afromontane region.

The only area of Mozambique in which there is a notable concentration of IPAs outside of the proposed local CoEs is on the Cheringoma Plateau and adjacent Urema depression (rift valley) in Sofala Province, where four IPAs (sites 31 – 35 excluding Mount Gorongosa) are located. This area has a small but notable endemic flora, some of which also extends just north of the Zambezi River, and it could represent a further, as yet unrecognised local CoE for Mozambique (J.E. Burrows, pers. comm. 2019).

(Sub) Centre of Plant Endemism	Number of IPAs	Sites on Map
Rovuma	17	1–10, 17–21, 30
Maputaland *sensu lato*	15	43–57
Maputaland *sensu stricto*	3	53, 54, 57
Inhambane	10	43–52
Lebombo Mountains	2	55, 56
Mulanje-Namuli-Ribáuè	8	22–29
Chimanimani-Nyanga	7	33, 36–42

IPAs in relation to local Centres and Sub-Centres of Plant Endemism in Mozambique.

Conservation and threat status of IPAs

IPAs and formal protection
Of the 57 IPAs identified, only 18 are currently fully protected under national Mozambican law, whilst a further 10 IPAs are partially protected.

Hence, just over half of the IPAs have no formal protection at present. Furthermore, even for many of the sites that are protected on paper, there is little or no effective management in place and threats are often high within these protected areas. Of particular note is the Forest Reserve network, in which five IPAs coincide. Most Forest Reserves (FRs) in Mozambique are not currently managed for their biodiversity, and some - such as the Maronga FR in the Chimanimani Lowlands and the Ribáuè and M'paluwe FRs within the Ribáuè-M'paluwe IPA - are not carefully managed at present, with no controls over agricultural encroachment or wood harvesting. The forests of Ribáuè-M'paluwe IPA, for example, are predicted to be exhausted within the next 35 years unless interventions are made urgently (Montfort 2020). Hence, there is a major challenge ahead to protect and effectively manage the IPAs. Without such interventions, several of these sites may lose their botanical value imminently.

Threats faced by IPAs

Although the threat status of the sites varies enormously, no single site is considered to be entirely unthreatened, with all sites either currently experiencing or being potentially threatened by some level of habitat destruction or degradation. That said, several of the sites face only few or low-level threats. For example, the Catapú IPA is managed effectively as a sustainable forestry and ecotourism concession and is currently unthreatened except for a moderate risk from increased fire frequency in parts of the site. Elsewhere, the forests of Mount Mabu, whilst unprotected by official means, are largely intact and are protected in part by the spiritual importance of the site for local communities together with the rugged and often inaccessible terrain, and so are facing only low-level threats at present. Conversely, several sites are severely threatened and are on the brink of losing their botanical importance because of habitat destruction, for example the Mueda

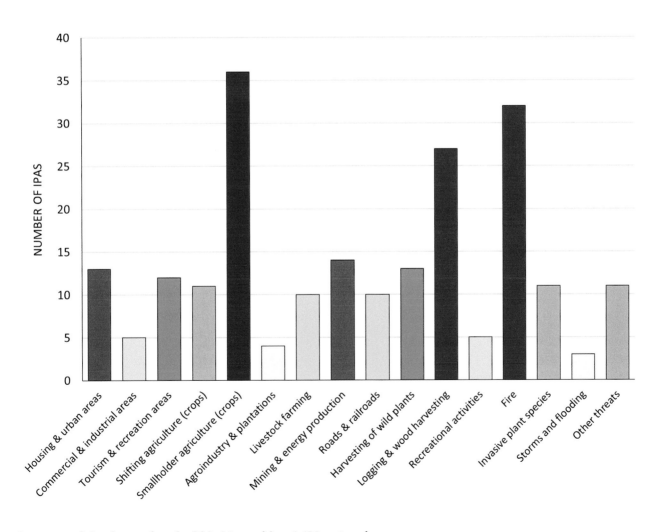

Summary of the threats faced within Mozambique's IPA network.

Mature forest trees being burnt to clear land for agriculture within the Ribáuè-M'paluwe IPA (ID)

Plateau and Escarpments where high population levels and associated settlement, slash-and-burn agriculture and wood harvesting have resulted in extensive losses of the woodland and dry forest habitats that support rare and threatened species.

Habitat destruction and degradation across the IPA network is the result of a wide variety of threats, but several are prevalent. Of particular concern are unsustainable farming methods, most notably the use of slash-and-burn techniques for short-term crop growth, this being the single biggest threat to IPAs. Although fire plays a natural role in many of the seasonally dry habitats of Mozambique, human-induced changes in fire regime, and in particular increased fire frequency and intensity above natural levels, are a major factor in habitat degradation. This is often associated with the agricultural techniques noted above, but fire is also used more widely, for example for habitat clearance for settlement and to aid the hunting of wild animals. Harvesting of wood for charcoal production, fuelwood, local construction and commercial timber exploitation is also a major threat to woody species and habitats in many IPAs as it is currently conducted at unsustainable levels, with some of Mozambique's most widespread timber trees experiencing serious declines, such as *Afzelia quanzensis*, *Bobgunnia madagascariensis* and *Pterocarpus angolensis* (Burrows *et al.* 2018; Hills 2019). Other widespread threats to IPAs include the expansion of settlements and urban

areas, unsustainable tourism practices, particularly at coastal sites, and unsustainable harvesting of wild plant resources other than timber. Invasive plant species impact a small but significant number of sites at present (11 sites), with the rapid spread of *Vernonanthura polyanthes* in montane sites of Manica and Nampula being particularly problematic (Timberlake *et al.* 2016a). The rising threat of climate change has not yet been assessed in relation to Mozambique's IPA network, but the risk of increased extreme weather events such as tropical storms and flooding has been noted for three IPAs to date, for example the low-lying coral islands of Goa and Sena in Nampula Province and the Quirimbas Archipelago of Cabo Delgado. Extreme dry-season temperatures and prolonged drought events, particularly in inland areas of the country, could potentially impact IPAs in the future by resulting in local population losses and/or changing vegetation dynamics.

The type of intervention required to overcome these threats and best benefit the IPAs will vary considerably between sites, depending on their current protection status, human population size, land uses and the perceived benefits of supporting nature conservation. Considerable research has been carried out at several IPAs as to the optimal management practices and land use changes to protect biodiversity whilst benefitting local communities, for example at Mount Mabu where a management plan has been drafted and where a consortium is now working on the gazettement of a community conservation area (CEPF 2021; Biofund 2021). However, at the majority of IPAs, this research and planning is yet to be done, and much further research is needed to identify how best to protect the IPA network moving forward.

IPAs and Key Biodiversity Areas

Key Biodiversity Areas (KBAs) are defined as sites that contribute significantly to the global persistence of biodiversity (IUCN 2016). As with IPAs, this is a criterion-based system, measured using multiple organism groups (including both fauna and flora) in terrestrial, freshwater, marine and underground systems. The recent KBA assessment for Mozambique (WCS *et al.* 2021) identified 29 KBAs across Mozambique that collectively occupy 139,947 km² including both marine and terrestrial areas. The KBAs were triggered by 180 species of which the majority (57%) are plants, with 18 of the 29 KBAs having been triggered in part or in full by plant species. This is considerably higher than the average representation of plant taxa in KBA analyses globally due, in part, to botanical data collation and Red Listing efforts of the TIPAs programme, alongside other botanical initiatives in Mozambique, with data from these initiatives being made available via GBIF (Matimele 2021). A comparison of the IPA and KBA networks reveals that 23 IPAs are fully included within the KBA network and a further 11 are partially included (together comprising nearly 60% of the IPAs). In several cases, multiple IPAs have been identified within a single KBA. For example the vast Gorongosa and Marromeu Complex KBA fully or largely encompasses five IPAs, Catapú, Cheringoma Limestone Forests, Inhamitanga Forest, Mount Gorongosa and Urema Valley and Sangarassa Forest), whilst the Quiterajo KBA includes the much smaller Quiterajo IPA plus the Muàgámula River IPA and the northern portion of the Quirimbas Archipelago IPA. Such differences in scale are driven partially by the differing range of organisms assessed, with critical sites for plants often being focussed on a smaller area as plants are sedentary and highly habitat-specific whereas many faunal taxa are more mobile, migratory or reliant upon a wider range of habitats.

Given that all IPA sites are triggered, at least in part, by the presence of important populations of threatened species (and the vast majority of these are inferred to represent >1% of the global population of these trigger species), and given that the large majority of IPAs identified also contain one or more plant species with a restricted range of <10,000 km², we are confident that the vast majority of IPAs will meet the required thresholds for KBA status. It is important that all such sites are included within the KBA network in the next iteration of that work, to strengthen the case for protecting these sites and their unique biodiversity.

Ecosystem Services

Although the natural capital of the IPA network has not been surveyed extensively to date, the initial assessment carried out in this study reveals that the sites provide a wide range of ecosystem services, from provisioning of foods and resources, to a range of regulatory services and cultural services, as well as providing ecological support services for wider biodiversity. These ecosystem services are highly important to local communities living in and around the IPAs, while many of the services provided also have a wider importance, including national income generation, through income from (eco)tourism for example, or regional regulatory services such as sequestration of carbon as a means of ameliorating climate change and its negative impacts. Other particularly frequent ecosystem services provided by IPAs include the provision and maintenance of fresh water supplies, the prevention of soil erosion and maintenance of fertility, and the provision of critical habitat for fauna, including species of conservation and socio-economic interest.

Future work on IPAs

At the end of the site accounts, we also report on 12 localities that could potentially qualify as IPAs following further study. As noted in the introduction, it is highly likely that further sites of high botanical importance in Mozambique will be uncovered in the future following more exhaustive botanical surveys, and so the network documented here should be considered a first iteration. Moving forward, it will be important to review and assess potential additions to the IPA network as new data become available, e.g. through botanical fieldwork in under-explored parts of the country.

It is important that the IPA network and its sites are regularly monitored to track changes, including progress towards conserving species and protecting and restoring habitats alongside monitoring of any degradation of sites. Changes in IPA condition should then be considered in future rounds of IPA assessment in Mozambique (see "Review of IPA site documentation" in chapter "Identifying Important Plant Areas in Mozambique: methods and resources"). Stakeholder engagement will also be important in understanding how local people interact with sites, particularly the social and economic value of habitats and species. This will allow a future, in-depth assessment of taxa of socio-economic importance under B(iii) of the IPA criteria and further documentation of ecosystem services provided by sites. Engagement with local communities will also be central to developing any future conservation management plans within the IPAs, to ensure that the livelihoods of local people are considered and that communities feel involved and invested in site management, thereby enhancing the long-term sustainability of conservation efforts.

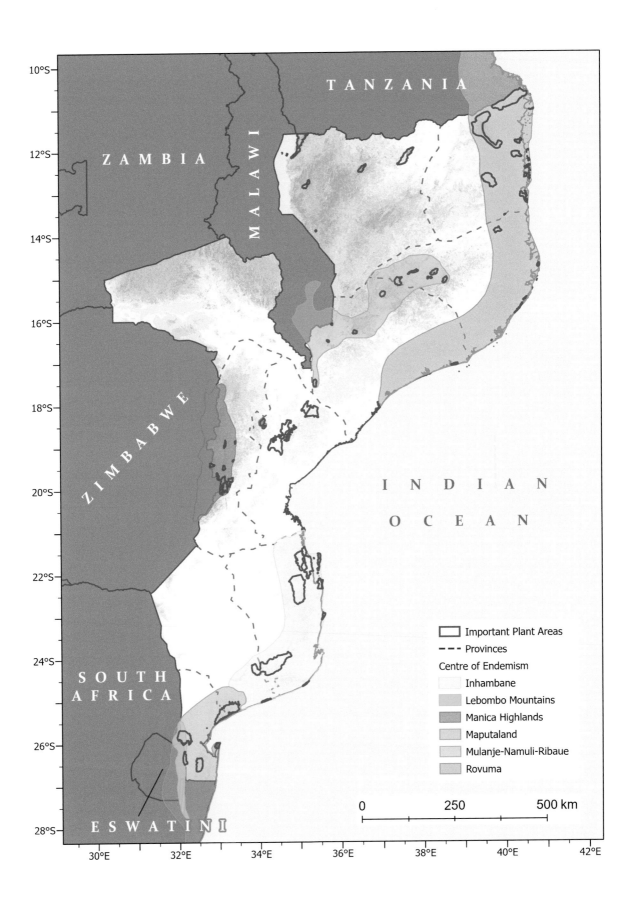

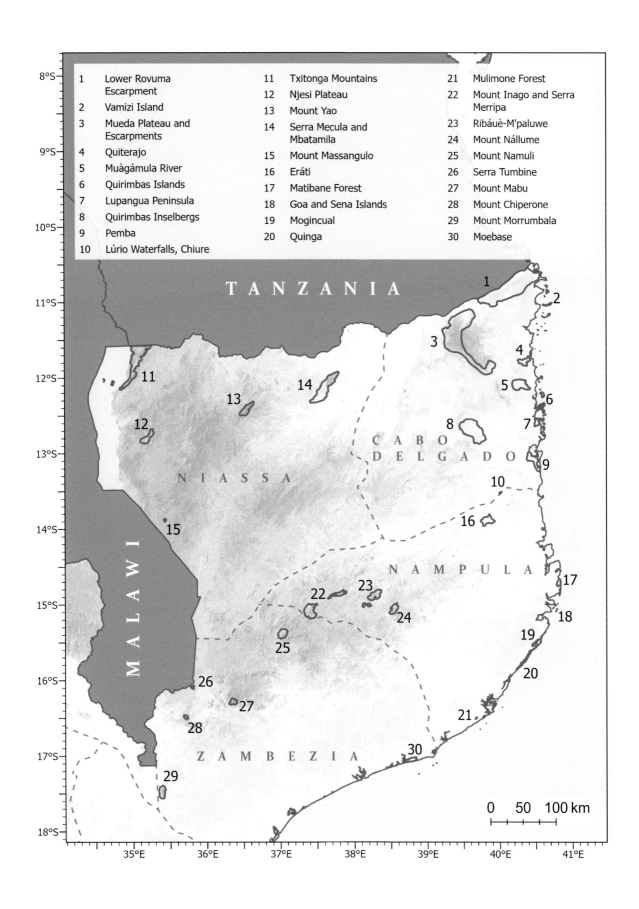

1	Lower Rovuma Escarpment	11	Txitonga Mountains	21	Mulimone Forest
2	Vamizi Island	12	Njesi Plateau	22	Mount Inago and Serra Merripa
3	Mueda Plateau and Escarpments	13	Mount Yao	23	Ribáuè-M'paluwe
4	Quiterajo	14	Serra Mecula and Mbatamila	24	Mount Nállume
5	Muàgámula River	15	Mount Massangulo	25	Mount Namuli
6	Quirimbas Islands	16	Eráti	26	Serra Tumbine
7	Lupangua Peninsula	17	Matibane Forest	27	Mount Mabu
8	Quirimbas Inselbergs	18	Goa and Sena Islands	28	Mount Chiperone
9	Pemba	19	Mogincual	29	Mount Morrumbala
10	Lúrio Waterfalls, Chiure	20	Quinga	30	Moebase

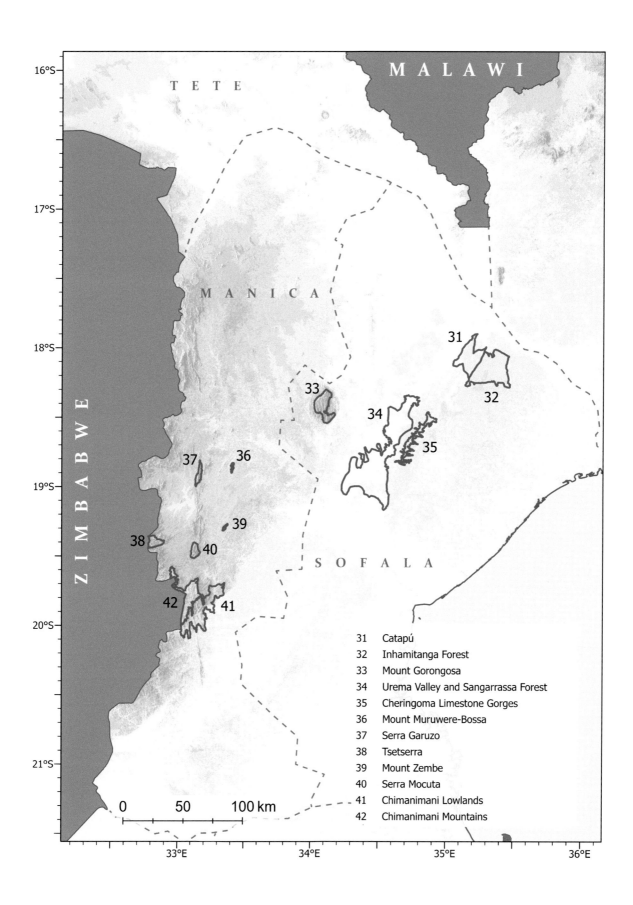

31 Catapú
32 Inhamitanga Forest
33 Mount Gorongosa
34 Urema Valley and Sangarrassa Forest
35 Cheringoma Limestone Gorges
36 Mount Muruwere-Bossa
37 Serra Garuzo
38 Tsetserra
39 Mount Zembe
40 Serra Mocuta
41 Chimanimani Lowlands
42 Chimanimani Mountains

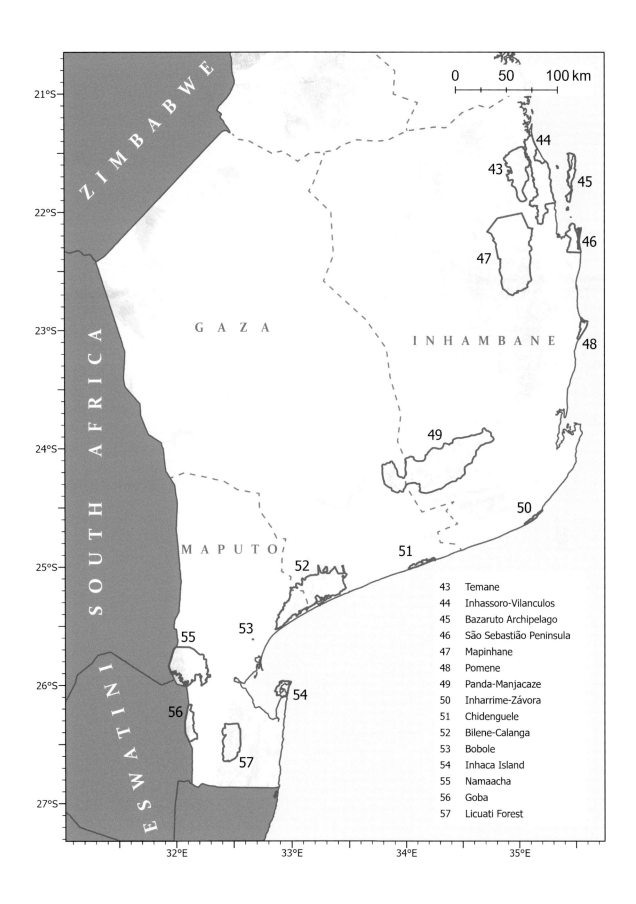

43   Temane
44   Inhassoro-Vilanculos
45   Bazaruto Archipelago
46   São Sebastião Peninsula
47   Mapinhane
48   Pomene
49   Panda-Manjacaze
50   Inharrime-Závora
51   Chidenguele
52   Bilene-Calanga
53   Bobole
54   Inhaca Island
55   Namaacha
56   Goba
57   Licuati Forest

Site no. on map	Site name	Province	Area (km²)	Protection status	Criteria met	Total Criterion A(i) taxa	Total Criterion A(iii) taxa	Total Criterion A(iv) taxa	Criterion A taxa unique to site*	Total Criterion B(ii) taxa	Criterion B(ii) taxa unique site*	Criterion C(iii) habitats
1	Lower Rovuma Escarpment	CD	1,999	Un	A(i), B(ii), C(iii)	54			25	**22**	14	12a
2	Vamizi Island	CD	16	P [pr]	A(i), C(iii)	3			2	1	0	06
3	Mueda Plateau and Escarpments	CD	2,200	Un	A(i)	19			9	10	3	
4	Quiterajo	CD	129	Un	A(i), A(iv), B(ii), C(iii)	38		1	12	**22**	8	12a, 12b
5	Muàgámula River	CD	291	P	A(i)	11			4	10	3	
6	Quirimbas Archipelago	CD	108	P	A(i)	3			1	4	0	
7	Lupangua Peninsula	CD	57	P	A(i), C(iii)	3			0	1	0	12b
8	Quirimbas Inselbergs	CD	812	P	A(i)	6			3	3	1	
9	Pemba	CD	231	Un	A(i)	11			2	8	1	
10	Lúrio Waterfalls, Chiure	CD/Na	6.7	Un	A(i)	1			1	1	1	
11	Txitonga Mountains	Ni	741	P	A(iii)		1		1	1	1	
12	Njesi Plateau	Ni	165	P	A(i), C(iii)	1			1	1	1	01, 09
13	Mount Yao	Ni	183	P	A(i)	1			1	1	1	
14	Serra Mecula and Mbatamila	Ni	626	P	A(i), C(iii)	1			1	2	1	02
15	Mount Massangulo	Ni	11	Un	A(i)	2			1	3	2	
16	Eráti	Na	174	Un	A(i), A(iii), A(iv)	4	2	1	5	5	5	
17	Matibane Forest	Na	45	P [FR]	A(i), C(iii)	14			4	10	4	12b, 12c
18	Goa and Sena Islands	Na	0.7	Un	A(i)	3			3	3	3	
19	Mogincual	Na	21	Un	A(i), C(iii)	2			0	2	0	12c
20	Quinga	Na	63	Un	A(i), C(iii)	3			1	3	1	12c
21	Mulimone Forest	Na	3.2	P-P	A(i), C(iii)	3			1	3	1	12c
22	Mount Inago and Serra Merripa	Na	373	Un	A(i)	6			1	10	3	
23	Ribáuè-M'paluwe	Na	221	P-P [FR]	A(i), A(iii), B(ii), C(iii)	15	1		7	**22**	7	02
24	Mount Nállume	Na	120	Un	A(i)	3			1	5	1	

Site no. on map	Site name	Province	Area (km²)	Protection status	Criteria met	Total Criterion A(i) taxa	Total Criterion A(iii) taxa	Total Criterion A(iv) taxa	Criterion A taxa unique to site*	Total Criterion B(ii) taxa	Criterion B(ii) taxa unique site*	Criterion C(iii) habitats
25	Mount Namuli	Z	146	Un	A(i), A(iii), B(ii), C(iii)	22	4		16	**40**	30	01, 09
26	Serra Tumbine	Z	14	Un	A(i)	3			1	1	1	
27	Mount Mabu	Z	75	Un	A(i), C(iii)	8			2	10	5	02
28	Mount Chiperone	Z	24	Un	A(i), C(iii)	3			0	0	0	01
29	Mount Morrumbala	Z	135	Un	A(i), C(iii)	2			1	4	1	03
30	Moebase	Z	71	P	A(i), C(iii)	3			1	3	1	12c
31	Catapú	S	352	P [sf]	A(i), C(iii)	9			2	10	1	04
32	Inhamitanga Forest	S	622	P	A(i), C(iii)	11			2	10	1	04
33	Mount Gorongosa	S	216	P	A(i), B(ii), C(iii)	10			5	**16**	8	01, 02, 09
34	Urema Valley and Sangarassa Forest	S	1594	P-P	A(i)	3			1	**13**	6	10
35	Cheringoma Limestone Gorges	S	182	P-P	A(i), C(iii)	1			1	6	3	13
36	Mount Muruwere-Bossa	Mc	10	Un	A(i)	1			1	2	1	
37	Serra Garuzo	Mc	51	Un	A(i)	2			1	1	0	
38	Tsetserra	Mc	77	P	A(i), A(iii), A(iv), B(ii), C(iii)	14	1	3	11	**36**	18	01, 09
39	Mount Zembe	Mc	7.6	Un	A(i)	3			2	3	2	
40	Serra Mocuta	Mc	62	Un	A(i), C(iii)	2			0	4	1	02
41	Chimanimani Lowlands	Mc	514	P`	A(i), A(iii), A(iv), B(ii), C(iii)	14	1	3	12	**20**	11	02, 03
42	Chimanimani Mountains	Mc	319	P	A(i), A(iv), B(ii), C(iii)	29		6	21	**96**	71	09
43	Temane	In	678	Un	A(i)	5			1	6	1	
44	Inhassoro-Vilanculos	In	953	Un	A(i), A(iv)	7		1	1	**12**	2	
45	Bazaruto Archipelago	In	190	P	A(i)	4			2	8	2	
46	São Sebastião Peninsula	In	227	P	A(i)	4			0	9	0	
47	Mapinhane	In	2,070	Un	A(i)	4			0	8	1	

Site no. on map	Site name	Province	Area (km²)	Protection status	Criteria met	Total Criterion A(i) taxa	Total Criterion A(iii) taxa	Total Criterion A(iv) taxa	Criterion A taxa unique to site*	Total Criterion B(ii) taxa	Criterion B(ii) taxa unique site*	Criterion C(iii) habitats
48	Pomene	In	74	P-P	A(i), A(iv)	3		1	1	**11**	1	
49	Panda-Manjacaze	In/G	2,599	Un	A(i), A(iv)	5		1	3	8	3	
50	Inharrime-Závora	In	32	Un	A(i)	3			1	5	1	
51	Chidenguele	G	60	Un	A(i)	3			0	6	1	
52	Bilene-Calanga	G/Mp	1,366	Un	A(i)	4			1	7	2	
53	Bobole	Mp	0.2	P-P	A(i)	1			0	0	0	
54	Inhaca Island	Mp	132	P-P	A(i)	8			2	**12**	4	
55	Namaacha	Mp	854	Un	A(i), A(iv)	4		2	3	6	3	
56	Goba	Mp	217	Un	A(i), A(iv), B(ii)	7		4	7	10	7	
57	Licuáti Forest	Mp	470	P-P [FR]	A(i), C(iii)	10			5	9	6	07

Summary table of the Important Plant Areas of Mozambique. Provinces of Mozambique are abbreviated as follows CD = Cabo Delgado; G = Gaza; In = Inhambane; Mc = Manica; Mp = Maputo; Na = Nampula; Ni = Niassa; S = Sofala; T = Tete; Z = Zambezia. C(iii) habitat codes are explained in the Methods section. Protection status is abbreviated as follows: P = site is fully protected within a formal protected area that is included within Mozambique's national Protected Area network; P-P = site is partially protected within a formal protected area that is included within Mozambique's national Protected Area network; P [FR] or P-P [FR] = site is wholly or partially within a Forest Reserve, but is not managed for its biodiversity; P [pr] = site is protected with a private protected area or reserve; P [sf] = site receives some protection within a sustainable forestry area; Un = site is unprotected at present. Numbers highlighted in bold in the column "Total Criterion B(ii) taxa" indicate that the sites qualifies as an IPA under that criterion.

* for the columns on criterion A and B(ii) species "unique to this site", this reflects uniqueness within the IPA network; it does not imply that the species are only found at that site globally.

IMPORTANT PLANT AREA ASSESSMENTS

CABO
DELGADO
PROVINCE

LOWER ROVUMA ESCARPMENT

Assessor: Iain Darbyshire

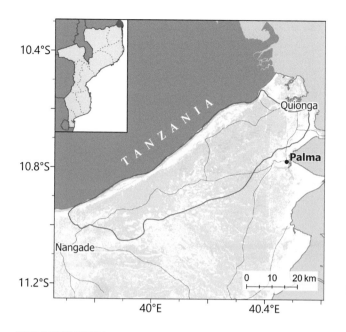

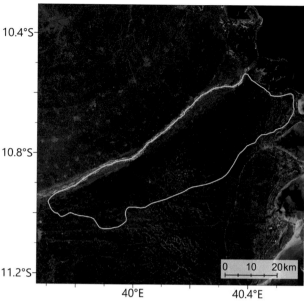

INTERNATIONAL SITE NAME		Lower Rovuma Escarpment	
LOCAL SITE NAME (IF DIFFERENT)		Escarpas do Baixo Rovuma	
SITE CODE	MOZTIPA023	PROVINCE	Cabo Delgado

LATITUDE	-10.79189	LONGITUDE	40.12267
ALTITUDE MINIMUM (m a.s.l.)	0	ALTITUDE MAXIMUM (m a.s.l.)	200
AREA (km²)	1999	IPA CRITERIA MET	A(i), B(ii), C(iii)

Site description

The Lower Rovuma Escarpment IPA is situated in the Palma and Nangade Districts of northeast Cabo Delgado Province. This landscape-scale site extends for ca. 90 km WSW to ENE between the towns of Nangade inland and Quionga and Palma on the coast, parallel to the Rovuma River valley which here forms the international border with Tanzania. The IPA covers the Mozambique side of the Rovuma floodplain and escarpment and the adjacent low undulating plateau. It contains a range of habitats, some of which are scarce elsewhere in Mozambique or in East Africa more generally, and includes the largest contiguous extent of dry coastal forest in East Africa (Clarke 2011). It is therefore a key site within the Coastal Forest of Eastern Africa biodiversity hotspot (CEPF 2020). This large area could potentially be subdivided into

discrete management units but maintaining the integrity of the landscape as a whole is critical to its conservation (Timberlake *et al.* 2010).

Botanical significance

Of the rich and varied mosaic of habitats represented within this IPA, the extensive intact areas of coastal dry forest and thicket are of primary botanical significance (Timberlake *et al.* 2011). The most extensive areas of forests are within the proposed Rovuma Centre of Plant Endemism (Burrows & Timberlake 2011). Three main forest blocks are recorded within the IPA: (1) the Nhica do Rovuma – Macanga River Block, an extensive area of up to 300 km², containing two core forest areas; (2) the Pundanhar Block which contains ca. 120 km² of forest; and (3) the Nangade Block, a much smaller and more disturbed area with only ca. 5 km² of undisturbed forest remaining (Clarke 2011). Together, these forest blocks comprise two of the four "high priority" sites for the conservation of coastal dry forest in northeast Mozambique proposed by Timberlake *et al.* (2010). The extensive intact forest and woodland vegetation within the IPA is in marked contrast to the northern side of the Rovuma Escarpment in southeast Tanzania where much of the natural vegetation has been heavily denuded or replaced by farmland and settlement.

These forests are characterised by high species turnover and a high number of highly range-restricted and threatened species (Timberlake *et al.* 2011; Darbyshire *et al.* 2019a, 2020a). Over 50 globally threatened plant species are present, including several for which this site contains the majority of the global population, notably *Casearia rovumensis* (EN), *Crossopetalum mossambicense* (EN), *Garcinia acutifolia* (VU), *Pyrostria* sp. nov. "*makovui*" (EN), *Vangueria domatiosa* (EN), *Vitex francesiana* (EN) and *Xylopia lukei* (EN). It is also the only known Mozambican site for *Coffea schliebenii* (VU) which is otherwise scarce in the Lindi region of Tanzania, and for *Combretum lindense* (not yet evaluated on the Red List), *Didymosalpinx callianthus*

(EN) and *Diospyros magogoana* (currently CR but without the Mozambique population included in the assessment), which are otherwise known only from a single location each in Tanzania.

The seasonal pan landscape that dominates parts of the IPA also supports rare and threatened species, including the recently described endemics *Convolvulus goyderi* (EN) and *Ochna dolicharthros* (VU) (Crawford & Darbyshire 2015; Darbyshire *et al.* 2020a).

Botanical exploration of this vast area is incomplete and has focussed on only small sections to date. The likelihood of further discoveries of new species to science is high, particularly among the under-explored herbaceous flora (Darbyshire *et al.* 2020a), whilst undescribed species that are already known to occur at this site include *Combretum* sp. A and *Deinbollia* sp. A of Burrows *et al.* (2018). Several scarce species that have so far only been recorded from the environs of Palma may also be found within this IPA following a more complete survey; these include *Ammannia pedroi* (VU) and *Striga diversifolia* (DD) in more open habitats and *Pavetta lindina* (EN) in forested areas.

Overlooking the Rovuma valley near Pundanhar, with *Borassus* palm savanna (JT)

Hexalobus mossambicensis (QL)

Berlinia orientalis (JEB)

Habitat and geology

The geology and landscape of this region is dominated by a gentle isocline of Quaternary or Neogene sedimentary deposits which runs northwest from the coast towards the Mueda Plateau in Mozambique and the Makonde Plateau in Tanzania, with the Rovuma River cutting a sharp channel of ca. 10 km wide into the deposits. On the Mozambique side, the slopes and top of the steep Rovuma escarpment are freely drained and support a dense woody vegetation on red-brown sandy/clay loam soils. Some outcrops of iron-rich sandstones, likely of the Mikindani Formation of mid-Neogene origin (ca. 10–15 mya), are recorded in the east of this region and these may be more widespread than currently documented along the escarpment given that sandstone outcrops were observed in association with some of the dry forest patches across this region (Timberlake *et al.* 2010). Further south in Cabo Delgado, for example at Quiterajo [MOZTIPA021], there is a close association between Mikindani sandstone and dry forest patches. This rock gives rise to a coarsely sandy well-drained red soil. In the southern and central portion of the IPA and continuing southwards, the gentle undulations in the sedimentary deposits give rise to a series of large shallow seasonal pans which support a much more open grassland/savanna landscape underlain by more clay-rich soils (Timberlake *et al.* 2010; Clarke 2011).

A detailed description of the main vegetation types of this region is provided by Timberlake *et al.* (2010) and Clarke (2011), with summaries of the woody vegetation provided by Burrows *et al.* (2018) under their "Rovuma Basin Coastal Thicket-Forest" and "Rovuma Coastal Woodland" vegetation types; what follows is a brief summary.

The dry forest patches often occur as small lenses within a mosaic of miombo woodland. In areas of intact forest, the canopy is typically 8–20 m tall with emergent trees up to 40 m in some areas. The composition of these forests varies considerably across the IPA. In the western portion of the site in Nangade District, forest dominated by *Scorodophloeus fischeri* and *Guibourtia schliebenii* occurs, sometimes with *Hymenaea verrucosa*, but this forest type is not recorded further east in Palma District, where a more mixed tree assemblage occurs. Here, *Manilkara sansibarensis*, *M. discolor*, *Terminalia* (formerly *Pteleopsis*) *myrtifolia* and *Ochna mossambicensis* can be common. Canopy emergents include *Afzelia quanzensis*, *Berlinia orientalis*, *Dialium holstii*, *Hymenaea verrucosa*, *Milicia excelsa* and *Terminalia myrtifolia*. A range of Rubiaceae and *Diospyros* spp. are important in the understorey (Timberlake *et al.* 2010; Clarke 2011). Whilst some extensive and intact patches occur, much of the forest appears to be secondary in nature, and is believed to have regenerated over the past 50–60 years, as evidenced by the numerous multi-trunked larger trees, indicating widespread coppicing, and by the frequent occurrence of charcoal in forest soil profiles (Clarke 2011). Frequent throughout the region are large termitaria up to 20 m in diameter that support patches of dense woody vegetation that can include dry forest species. *Hirtella zanzibarica* is particularly characteristic of termitaria woodland, with *Hymenaea verrucosa* and *Berlinia orientalis* also frequent (Timberlake *et al.* 2010).

In areas with a high water table, both in the lowlands towards the coast and in the ecotone between the seasonal pans and the well-drained wooded areas, the range-restricted tree *Berlinia orientalis* (VU) can dominate, often in association with *Brachystegia spiciformis*. This assemblage is somewhat intermediate between a miombo woodland and a dry forest (Clarke 2011). Typical miombo woodland is frequent on well-drained soils throughout the region, with dominant species including *B. spiciformis*, *Parinari curatellifolia* and *Uapaca nitida*, and other common components including *Bobgunnia madagascariensis*, *Julbernardia globiflora*, *Pterocarpus angolensis*, and *Sclerocarya caffra* (Timberlake *et al.* 2010; Clarke 2011).

Significant areas of the vegetation have experienced varying degrees of disturbance and subsequent fallow periods, which have given rise to extensive seral scrub forest and thicket assemblages, containing a mixture of miombo and pioneering dry forest species (Clarke 2011).

Extensive edaphic grasslands and lightly wooded grasslands occur both on the Rovuma floodplain, where a palm savanna with large trees of *Borassus aethiopum* is found, and in the pan landscape where the smaller palms *Hyphaene compressa* and *Phoenix reclinata* occur together with scattered miombo tree species.

The climate of this region is highly seasonal, with a prolonged dry season from May to November and a short hot and wet season mainly between December and April. Annual rainfall, at 900–1100 mm per year, is amongst the lowest along the East African coastline (Clarke 2011).

Conservation issues

At present, none of this extensive site is protected for nature conservation. A portion of the Nangade (western) forest block is designated as a Hunting Concession, with a camp and some staff in place at least in the late 2000s, which afforded some protection for the forest in order to preserve hunting stocks (Timberlake *et al.* 2010). The eastern-most parts of the site were included in the proposed Palma National Reserve, which was intended mainly to protect the rich marine and coastal resources of this area but with the inclusion of some of the terrestrial habitats. However, this reserve has not come to fruition, nor to date has the proposed Rovuma River mouth Trans-Frontier Conservation Area shared with Tanzania. The entirety of the IPA is included within the vast Palma Key Biodiversity Area.

Timberlake *et al.* (2010) estimate approximately 65% of the forest cover has been lost in the area of Nangade-Palma-Mocímboa da Praia. However, the most severe losses have been seen outside of the IPA boundary, around and between Palma and Mocímboa where the landscape is now severely degraded with only small pockets of high-biodiversity-value habitat remaining. It is for this reason that this area is excluded from the IPA, although there are still some important forest patches there that would benefit from conservation efforts. Whilst woody vegetation is extensive within the IPA, much of the forest appears to be secondary in nature (see Habitat and Geology above). The first wave of deforestation is likely to have occurred in the Portuguese colonial period when there was extensive timber exploitation. This region witnessed heavy military action during the war for Independence and the post-Independence Civil War (1960s – 1991) which led to significant depopulation, and it is during this time that the extensive woody vegetation appears to have reestablished. Since the 1990s, parts of this region have experienced rapid repopulation, driven in part by improved transport routes and in part by exploration for oil and gas across the region (Timberlake *et al.* 2010, 2011; Darbyshire *et al.* 2020a). However, the population remains low relative to other parts of the coastal lowlands, with particularly low numbers of people in the areas with seasonally inundated soils, and Palma District has one of the lowest population densities in East Africa (Clarke 2011). The recent violent insurgency in coastal Cabo Delgado since 2017 has temporarily halted much of the migration into the region, but in the longer term there is likely to be a continuing trend of population growth and increased pressure on resources once stability returns.

The most significant threat to this IPA is the ongoing and widespread clearance of forest and woodland for shifting subsistence agriculture, aided by burning. This is particularly evident along transport routes and in the western portion of the IPA where large areas are being actively cleared. Uncontrolled fires primarily impact the miombo woodland as these have a much higher fuel-load due to the abundance of grasses. However, they can also penetrate the seral scrub forests and thickets and in the longer term can result in a gradual erosion of the dry forest margins (Clarke 2011).

Some charcoal production and firewood extraction occur in this region but these are not considered to be a severe threat. The main concern is that exhaustion of wood supplies closer to Palma and around the city of Mtwara in Tanzania may result in increase exploitation of the woodlands and forests of the Rovuma Escarpment in the future. Commercial and illegal logging have not been considered a major threat until now (Clarke 2011). There are some logging concessions in parts of the IPA, particularly in the west, but these are intended to be sustainable. Most illegal logging to date has targeted widespread woodland species such as *Afzelia quanzensis*, *Millettia stuhlmannii* and *Pterocarpus angolensis* (Timberlake *et al.* 2010). However, there is a concern that the growing lawlessness in northeast Cabo Delgado associated with the insurgency may result in increased illegal logging in the forests.

Oil and gas exploration in 2007–2008 resulted in an extensive network of cut lines being made across the landscape to allow for vehicle access. These were each 3–5 m wide and avoided the felling of larger tree species, with the smaller species being coppiced to promote regrowth. The lines were closed in late 2008 and are showing good signs of regeneration. Subsequent industrial activity in the region has focused offshore, with two large liquefied natural gas (LNG) extraction operations underway. Onshore infrastructure is centred on the Afungi Peninsula to the southeast of Palma and so does not directly impact the Lower Rovuma Escarpment area, but it is likely to result in accelerated migration into the region once the regional security concerns are overcome; this is likely to be a significant future threat.

Given the significant threats to the future of these critical habitats, there is an urgent need to protect the Lower Rovuma Escarpment landscape and to ensure that any exploitation of its resources is sustainable in the long term.

Key ecosystem services

This area provides a wide range of important ecosystem services of both local and regional importance. The large, contiguous extent of woodland and forest is a significant carbon sink due to the vast woody biomass (Clarke 2011). This area is also reported to be a major water source for the towns of Palma and Mocímboa da Praia (Timberlake *et al.* 2010). The open savanna landscape around the extensive pan systems in the south of the IPA support a rich wildlife, including populations of elephant, roan and sable antelope, African wild dog, and lion, and these species find dry-season refuge in the more densely wooded and forested areas along the Rovuma escarpment, with well-established migration routes between the Rovuma floodplain and the pan landscape (Timberlake *et al.* 2010, Clarke 2011). This dense woody vegetation also protects the underlying sedimentary deposits and soils from excessive erosion, and so protects the ecosystem of the lower Rovuma River and its mouth. If and when the security concerns in this region of Mozambique are overcome, the Rovuma Escarpment should also have high ecotourism potential, given its striking landscape heterogeneity, relatively intact habitats and rich wildlife (Clarke 2011).

Ecosystem service categories

- Provisioning – Raw materials
- Provisioning – Fresh water
- Regulating services – Local climate and air quality
- Regulating services – Carbon sequestration and storage
- Regulating services – Erosion prevention and maintenance of soil fertility
- Habitat or supporting services – Habitats for species
- Habitat or supporting services – Maintenance of genetic diversity
- Cultural services – Tourism

IPA assessment rationale

The Lower Rovuma Escarpment is one of the most important sites for plant diversity and local endemism in Mozambique and qualifies as an IPA under all three criteria. Under criterion A(i), it holds important populations of 54 globally threatened plant species, of which 19 are assessed as Endangered and one (*Diospyros magogoana*) is currently assessed as Critically Endangered. Other globally threatened species are likely to be added to this list when a full Red List for the region is finalised. The site holds at least 22 qualifying species under criterion B(ii) and hence significantly exceeds the 3% threshold for this criterion. It also holds nationally important areas of Rovuma coastal dry forest, a threatened habitat, and this IPA holds the largest extent of continuous woody coastal vegetation in the whole of Cabo Delgado Province, hence qualifying under criterion C(iii).

Didymosalpinx callianthus (JEB)

Priority species (IPA Criteria A and B)

FAMILY	TAXON	IPA CRITERION A	IPA CRITERION B	≥ 1% OF GLOBAL POP'N	≥ 5% OF NATIONAL POP'N	IS 1 OF 5 BEST SITES NATIONALLY	ENTIRE GLOBAL POP'N	SPECIES OF SOCIO-ECONOMIC IMPORTANCE	ABUNDANCE AT SITE
Annonaceae	*Hexalobus mossambicensis*	A(i)	B(ii)	✓	✓	✓			occasional
Annonaceae	*Monanthotaxis suffruticosa*	A(i)		✓	✓	✓			unknown
Annonaceae	*Monanthotaxis trichantha*	A(i)		✓	✓	✓			unknown
Annonaceae	*Xylopia lukei*	A(i)	B(ii)	✓	✓	✓			occasional
Apocynaceae	*Landolphia watsoniana*	A(i)			✓	✓			unknown
Araceae	*Gonatopus petiolulatus*	A(i)		✓	✓	✓			scarce
Celastraceae	*Crossopetalum mossambicense*	A(i)	B(ii)	✓	✓	✓			occasional
Celastraceae	*Salacia orientalis*	A(i)		✓	✓	✓			unknown
Clusiaceae	*Garcinia acutifolia*	A(i)		✓	✓	✓			unknown
Combretaceae	*Combretum lindense*	A(i)	B(ii)	✓	✓	✓			unknown
Combretaceae	*Combretum stocksii*		B(ii)						unknown
Connaraceae	*Vismianthus punctatus*	A(i)	B(ii)	✓	✓	✓			unknown
Convolvulaceae	*Convolvulus goyderi*	A(i)	B(ii)	✓	✓	✓	✓		scarce
Cucurbitaceae	*Peponium leucanthum*	A(i)			✓	✓			scarce

Priority species (IPA Criteria A and B) cont.

FAMILY	TAXON	IPA CRITERION A	IPA CRITERION B	≥ 1% OF GLOBAL POP'N	≥ 5% OF NATIONAL POP'N	IS 1 OF 5 BEST SITES NATIONALLY	ENTIRE GLOBAL POP'N	SPECIES OF SOCIO-ECONOMIC IMPORTANCE	ABUNDANCE AT SITE
Ebenaceae	*Diospyros magogoana*	A(i)		✓	✓	✓			unknown
Ebenaceae	*Diospyros shimbaensis*	A(i)			✓	✓			unknown
Euphorbiaceae	*Mildbraedia carpinifolia*	A(i)			✓	✓			unknown
Fabaceae	*Acacia latistipulata*	A(i)		✓	✓	✓			occasional
Fabaceae	*Baphia macrocalyx*	A(i)		✓	✓	✓			frequent
Fabaceae	*Berlinia orientalis*	A(i)		✓	✓	✓			common
Fabaceae	*Guibourtia schliebenii*	A(i)			✓	✓			occasional
Fabaceae	*Millettia impressa* subsp. *goetzeana*	A(i)		✓	✓	✓			unknown
Fabaceae	*Millettia makondensis*	A(i)		✓	✓	✓			frequent
Fabaceae	*Ormocarpum sennoides* subsp. *zanzibaricum*	A(i)			✓	✓			unknown
Fabaceae	*Platysepalum inopinatum*	A(i)			✓	✓			scarce
Fabaceae	*Tephrosia reptans* var. *microfoliata*		B(ii)						unknown
Fabaceae	*Xylia africana*	A(i)			✓	✓			unknown
Hypericaceae	*Vismia pauciflora*	A(i)		✓	✓	✓			scarce
Lamiaceae	*Clerodendrum lutambense*	A(i)		✓	✓	✓			scarce
Lamiaceae	*Premna hans-joachimii*	A(i)	B(ii)	✓	✓	✓			occasional
Lamiaceae	*Premna tanganyikensis*	A(i)		✓	✓	✓			scarce
Lamiaceae	*Vitex carvalhi*	A(i)		✓	✓	✓			scarce
Lamiaceae	*Vitex francesiana*	A(i)	B(ii)	✓	✓	✓			occasional
Loganiaceae	*Strychnos xylophylla*	A(i)			✓	✓			scarce
Loranthaceae	*Erianthemum lindense*	A(i)		✓	✓	✓			unknown
Malvaceae	*Grewia limae*	A(i)	B(ii)	✓	✓	✓			occasional
Malvaceae	*Sterculia schliebenii*	A(i)			✓	✓			scarce
Melastomataceae	*Memecylon torrei*	A(i)	B(ii)	✓	✓	✓			unknown
Ochnaceae	*Ochna dolicharthros*	A(i)	B(ii)	✓	✓	✓	✓		occasional
Rubiaceae	*Chassalia colorata*	A(i)	B(ii)	✓	✓	✓			occasional
Rubiaceae	*Coffea schliebenii*	A(i)	B(ii)	✓	✓	✓			occasional
Rubiaceae	*Didymosalpinx callianthus*	A(i)	B(ii)	✓	✓	✓			occasional
Rubiaceae	*Leptactina papyrophloea*	A(i)			✓	✓			occasional
Rubiaceae	*Oxyanthus biflorus*	A(i)	B(ii)	✓	✓	✓			scarce
Rubiaceae	*Oxyanthus strigosus*	A(i)	B(ii)	✓	✓	✓			occasional
Rubiaceae	*Pavetta macrosepala* var. *macrosepala*	A(i)			✓	✓			unknown
Rubiaceae	*Psydrax micans*	A(i)			✓	✓			unknown

Ochna dolicharthros (FC)

Chassalia colorata (ID)

Priority species (IPA Criteria A and B) cont.

FAMILY	TAXON	IPA CRITERION A	IPA CRITERION B	≥ 1% OF GLOBAL POP'N	≥ 5% OF NATIONAL POP'N	IS 1 OF 5 BEST SITES NATIONALLY	ENTIRE GLOBAL POP'N	SPECIES OF SOCIO-ECONOMIC IMPORTANCE	ABUNDANCE AT SITE
Rubiaceae	*Pyrostria* sp. D of F.T.E.A. "*makovui*" ined.	A(i)	B(ii)	✓	✓	✓			occasional
Rubiaceae	*Rothmannia macrosiphon*	A(i)			✓	✓			scarce
Rubiaceae	*Tricalysia schliebenii*	A(i)		✓	✓	✓			occasional
Rubiaceae	*Tricalysia semidecidua*	A(i)		✓	✓	✓			occasional
Rubiaceae	*Vangueria domatiosa*	A(i)	B(ii)	✓	✓	✓			occasional
Rutaceae	*Vepris allenii*	A(i)	B(ii)	✓	✓	✓			occasional
Rutaceae	*Zanthoxylum lindense*	A(i)		✓	✓	✓			unknown
Salicaceae	*Casearia rovumensis*	A(i)	B(ii)	✓	✓	✓			occasional
Sapotaceae	*Vitellariopsis kirkii*	A(i)			✓	✓			unknown
		A(i): 54 ✓	B(ii): 22 ✓						

Threatened habitats (IPA Criterion C)

HABITAT TYPE	IPA CRITERION C	≥ 5% OF NATIONAL RESOURCE	≥ 10% OF NATIONAL RESOURCE	IS 1 OF 5 BEST SITES NATIONALLY	ESTIMATED AREA AT SITE (IF KNOWN)
Rovuma Coastal Dry Forest [MOZ-12a]	C(iii)		✓	✓	

Protected areas and other conservation designations

CONSERVATION AREA TYPE	CONSERVATION AREA NAME	RELATIONSHIP OF IPA TO CONSERVATION AREA
No formal protection	N/A	
Key Biodiversity Area	Palma	Protected/conservation area encompasses IPA

Seasonal wetlands and surrounding woodland and forest at Nhica do Rovuma (JEB)

Dry forest rich in local endemics near Pundanhar (JEB)

Threats

THREAT	SEVERITY	TIMING
Small-holder farming	medium	ongoing – increasing
Oil & gas drilling	low	past, not likely to return
Gathering terrestrial plants	low	ongoing – increasing
Increase in fire frequency/intensity	low	ongoing – increasing

VAMIZI ISLAND

Assessors: Iain Darbyshire, John Burrows

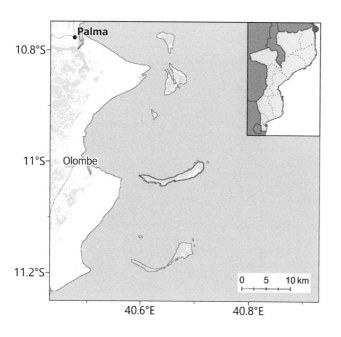

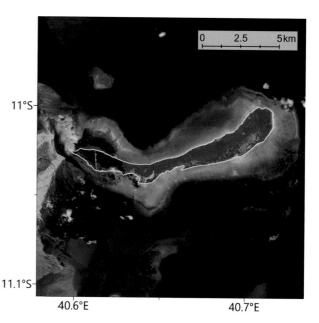

INTERNATIONAL SITE NAME		Vamizi Island	
LOCAL SITE NAME (IF DIFFERENT)		Ilha Vamizi	
SITE CODE	MOZTIPA017	PROVINCE	Cabo Delgado

LATITUDE	-11.02840	LONGITUDE	40.66830
ALTITUDE MINIMUM (m a.s.l.)	0	ALTITUDE MAXIMUM (m a.s.l.)	17
AREA (km²)	15.5	IPA CRITERIA MET	A(i), C(iii)

Site description

Vamizi Island lies in the north of the Quirimbas Archipelago, off the coast of Palma District in Cabo Delgado Province, northeast Mozambique. This narrow, low-lying island lies ca. 4 km offshore from the mainland and is 13.3 km long and less than 2 km wide at its widest point, totaling 15.5 km² in size. It is formed from ancient raised coral rock deposits (coral rag) and coral-derived sands. The island is surrounded by coral reefs and other marine areas that are of international conservation importance for their marine life.

Botanical significance

Vamizi is the only known site in Mozambique to support dwarf coastal forest on coral rag (Burrows & Burrows 2012; Burrows *et al.* 2018), a habitat that is highly localised, scattered and often threatened elsewhere in Eastern Africa. This habitat supports a low-diversity but unique assemblage of plant species that are adapted to this rather harsh, water-scarce environment. The forests include important populations of the globally threatened tree *Olea woodiana* subsp. *disjuncta* (EN) and the understorey herb *Barleria whytei* (EN), Vamizi being the only known site for these species in Mozambique (Darbyshire *et al.* 2015; Burrows *et al.* 2018). Three potential endemic, undescribed species are recorded from the island, which may prove to be additional threatened species: *Cordia* sp. ?nov. aff. *ovalis* (J.E. Burrows, pers. obs.), *Pleurostylia* sp. aff. *opposita* (R.H. Archer, pers. comm.) and *Psydrax* sp. A (Burrows *et al.* 2018). It also holds an important population of the recently described species *Acacia quiterajoensis* (LC), which is a co-dominant in the coral rag forest and for which the Vamizi plants differ from the mainland populations – which do not occur on coral rag – in fruit and leaf size (Burrows *et al.* 2018) such that they may represent a distinct taxon. The island is also the only known site in Mozambique for a number of other species, including the coastal East African climbing shrub *Capparis schefflera*, and some Pacific drift-seed species such as *Cycas thouarsii* (LC), *Morinda citrifolia* and *Hernandia nymphaefolia* (Burrows & Burrows 2012).

Vamizi Island, eastern side from the air (JEB)

The flora of Vamizi has so far been incompletely surveyed (Silveira & Paiva 2009; Burrows & Burrows 2012), and a full inventory is highly desirable in view of the unusual coral rag forest habitat and the high potential for further new discoveries.

Habitat and geology

The island is formed from exposed coral rocks (coral rag), overlain in some areas by thin sandy soils. Coral-sand beaches fringe the whole island apart from the north-western coast which is composed of an undercut coral rock shelf. The site has a highly seasonal climate with the rainy season in December to April and dry season from May to November; temperatures remain high all year, however, with monthly maxima averaging 27–30°C at Palma on the Cabo Delgado mainland.

Burrows & Burrows (2012) provide a summary of the vegetation with dominant species, and a preliminary species inventory. The dominant vegetation in the interior of the island is dwarf coastal forest on coral rag, which is a low-diversity deciduous or semi-evergreen assemblage with a canopy of 10–12 m and occasional emergents to 15 m, and with an understorey shrub and small tree layer. Dominant species include *Diospyros consolatae*, *Pleurostylia* sp. aff. *opposita*, *Sideroxylon inerme* and *Terminalia boivinii*, with *Suregada zanzibarensis* dominant in the understorey. The coral sand beaches and dunes support a low littoral scrub which includes a nationally important population of *Xylocarpus moluccensis*, with other frequent species including *Bourreria petiolaris*, *Grewia glandulosa*, *Sophora tomentosa* and *Suriana maritima* amongst others. The southern portion of the island supports a low intermediate scrub on shallow soils between the littoral vegetation and the forest, with e.g. *Commiphora* spp., *Pemphis acidula* and *Sideroxylon inerme*. Three inlets on the southern side of the island support mangroves dominated by *Brugueira gymnorrhiza* and *Rhizophora mucronata*, although these are not well developed as there are no rivers on the island to provide nutrient-rich silts during floods. The western portion of the islands is much more impacted by humans, with areas of cultivation and patches of open, disturbed woodland.

Conservation issues

Although not formally a protected area, Vamizi Island is managed as a conservation area. The Friends of Vamizi project and charitable Trust were established in 2002 and 2012, respectively, to protect Vamizi's terrestrial and marine ecosystems through combining biodiversity conservation with community development and tourism (Friends of Vamizi 2020). The Vamizi Marine Conservation Research Center conducts research and community outreach to protect this area, and the eastern two-thirds of the island are run as a tourism and conservation concession. The natural vegetation of this concession area is remarkably well preserved, impacted only by a few roads and limited tourism infrastructure. The main concerns within the tourist areas are (1) that the siting of future infrastructure be carefully planned to minimise impact and to prevent damage to the sensitive littoral vegetation; (2) that the sewerage is managed appropriately as the coral rag and its water table are highly susceptible to pollution; and (3) that exotic plant species are not allowed to spread. *Casuarina* trees are well established along the beaches but are not a threat to the biodiversity so long as seedlings are not allowed to establish in the natural habitats (Burrows & Burrows 2012).

The western third has been settled for several centuries, and much of the vegetation here has been largely degraded or transformed. Today, there are around 2,000 inhabitants on the island, but the population has fluctuated historically. The main occupation is fishing, which is putting some pressure on the rich marine life of the offshore reefs. A private airstrip has been constructed in the western portion.

A notable threat, assuming that gas drilling and extraction continues offshore in Cabo Delgado, is pollution from the offshore rigs and shipping that will pass nearby north of Vamizi. Although it may not directly impact the vegetation, this poses a threat to the integrity of the Vamizi ecosystem (T. Hempson, pers. comm.) including the Vamizi Key Biodiversity Area, which as currently defined is entirely marine and does not overlap with the IPA.

Key ecosystem services

The white coral sand beaches, offshore marine life and coral reefs, together with the island's tranquility and natural beauty, draw tourists to Vamizi Island, but the Friends of Vamizi and the Vamizi Marine Conservation Research Center are working to

Barleria whytei, photographed in Kenya (AR)

Dwarf forest on coral rag on Vamizi (JEB)

ensure responsible tourism and environmental education at this site (Friends of Vamizi 2020). The recent civil unrest in northeast Cabo Delgado associated with the violent insurgency by Islamist militants has temporarily halted tourism in this region but it is likely to recover in the future. The island provides important habitat for a range of species and these habitats also provide important regulatory services, particularly in the prevention of coastal erosion.

Ecosystem service categories

- Regulating services – Erosion prevention and maintenance of soil fertility
- Habitat or supporting services – Habitats for species
- Cultural services – Recreation and mental and physical health
- Cultural services – Tourism
- Cultural services – Education

IPA assessment rationale

Vamizi Island qualifies as an IPA under criteria A(i) and C(iii). At present, it qualifies under A(i) on the basis of three globally threatened taxa, two of which are Endangered and for which Vamizi is the only known site in Mozambique: *Barleria whytei*, a coral rag forest specialist that is known from only five localities globally, and *Olea woodiana* subsp. *disjuncta*, a coastal dry forest tree otherwise known from Kenya and Tanzania. However, as the flora of the site becomes more thoroughly researched, it is likely that other criterion A species will be found, such as *Psydrax* sp. A which, if confirmed as a distinct species, is likely to qualify as threatened under IUCN criterion D2. The site qualifies under criterion C(iii) due to the presence of extensive areas of intact dwarf coastal forest on coral rag, a nationally scarce and range-restricted habitat for which Vamizi is the only known example in Mozambique.

Priority species (IPA Criteria A and B)

FAMILY	TAXON	IPA CRITERION A	IPA CRITERION B	≥ 1% OF GLOBAL POP'N	≥ 5% OF NATIONAL POP'N	IS 1 OF 5 BEST SITES NATIONALLY	ENTIRE GLOBAL POP'N	SPECIES OF SOCIO-ECONOMIC IMPORTANCE	ABUNDANCE AT SITE
Acanthaceae	*Barleria whytei*	A(i)		✓	✓	✓			occasional
Fabaceae	*Acacia quiterajoensis*		B(ii)						common
Oleaceae	*Olea woodiana* subsp. *disjuncta*	A(i)			✓	✓			scarce
Rutaceae	*Zanthoxylum lindense*	A(i)			✓	✓			unknown
		A(i): 3 ✓	B(ii): 1						

Threatened habitats (IPA Criterion C)

HABITAT TYPE	IPA CRITERION C	≥ 5% OF NATIONAL RESOURCE	≥ 10% OF NATIONAL RESOURCE	IS 1 OF 5 BEST SITES NATIONALLY	ESTIMATED AREA AT SITE (IF KNOWN)
Dwarf Forest on Coral Rag [MOZ-06]	C(iii)		✓	✓	

Protected areas and other conservation designations

CONSERVATION AREA TYPE	CONSERVATION AREA NAME	RELATIONSHIP OF IPA TO CONSERVATION AREA
Private nature reserve	Vamizi Island concession	IPA encompasses protected/conservation area

Threats

THREAT	SEVERITY	TIMING
Tourism & recreation areas	low	Ongoing – stable
Oil & gas drilling	medium	Future – inferred threat
Climate change & severe weather – storms & flooding	unknown	Future – inferred threat

MUEDA PLATEAU AND ESCARPMENTS

Assessor: Iain Darbyshire

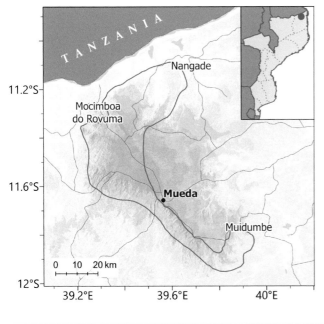

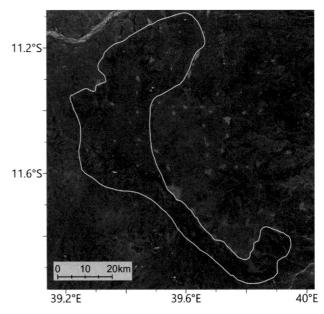

INTERNATIONAL SITE NAME		Mueda Plateau and Escarpments	
LOCAL SITE NAME (IF DIFFERENT)		Planalto e Escarpas de Mueda	
SITE CODE	MOZTIPA025	PROVINCE	Cabo Delgado

Tarenna sp. 53 of Degreef (= *Cladoceras rovumense*) (QL)

Monodora carolinae (QL)

LATITUDE	-11.51000	LONGITUDE	39.40000
ALTITUDE MINIMUM (m a.s.l.)	97	ALTITUDE MAXIMUM (m a.s.l.)	1,027
AREA (km²)	2200	IPA CRITERIA MET	A(i)

Site description

The Mueda Plateau (sometimes referred to as the Makonde or Maconde Plateau) is located in northern Cabo Delgado Province close to the border with Tanzania. The IPA is located mainly within Mueda and Muidumbe Districts, but also extends into Nangade District in the northeast section. This low plateau rises rather gradually on its eastern side and reaches just over 1,000 m elevation at its highest point around Chomba in the west. The northern, southern and particularly the western scarps of the plateau are typically steep and with complex gulley systems. It is bounded to the north by the broad valley of the Rovuma River and to the south by the Messalo River valley. The IPA covers an area of 2,251 km², extending for ca. 100 km from north to south, and is slightly under 30 km wide at its widest point.

The Mueda Plateau has been settled for at least two centuries and was the cradle of Mozambique's independence movement in the 1960s. A large number of settlements are well established across the site, including the towns of Mueda in the southwest and Mocimboa do Rovuma in the northwest of the escarpment. The plateau is traversed by route 509 which connects Mueda to the coastal town of Mocimboa da Praia, before continuing inland to Montepuez.

This site was previously covered by extensive thickets and dry forests of high botanical importance. The natural vegetation is, however, under severe threat due to the high population and associated habitat transformation, with the key habitats now largely restricted to steeper slopes and inaccessible areas along the plateau escarpments. For this reason, the IPA primarily covers the escarpments, as well as the somewhat less densely populated northern section of the plateau; the main towns and the heavily populated central and southern parts of the plateau are excluded. For the Mueda Plateau to retain its biodiversity value, there is an urgent need for protection and management of the remaining natural habitats, and potentially for a programme of habitat restoration in the less densely populated areas.

Botanical significance

The botanical significance of the Mueda Plateau is associated primarily with the extensive thickets, woodlands and dry forests that are believed to have originally covered much of this site. These habitats contain a number of rare and range-restricted species of the proposed Rovuma Centre of Plant Endemism (Burrows & Timberlake 2011; Darbyshire *et al.* 2019a). It is a critical site for several species, notably *Hugonia grandiflora* (EN), for which the majority of known localities are on the Mueda Plateau (Wabuyele *et* al. 2020), as well as *Monodora carolinae* (EN) and *Paropsia grewioides* var. *orientalis* (EN), for which this IPA is the only known Mozambican site. *Tarenna* sp. 53 of Degreef (2006), which is currently under description as *Cladoceras rovumense* (I. Darbyshire *et al.*, in press) is also present in the remnant forest patches. Although the herbaceous flora is not well documented, the discovery in 2009 of a population of *Celosia patentiloba* (CR) on the plateau (A. Banze #106) is of particular note as this is only the second confirmed record of this Critically Endangered species, the type being from Newala on the adjacent Makonde Plateau in Tanzania where it is highly threatened by extensive losses of suitable habitat (Howard *et al.* 2020). Potential records of this species from the Rondo Plateau in Tanzania in fact refer to a closely related but distinct and apparently undescribed species (I. Darbyshire, pers. obs.).

The southeastern escarpment near Muidumbe holds an important outlier population of the rock-dwelling *Aloe ribauensis* (EN), otherwise known only from the Ribáuè Mountains in Nampula Province (McCoy *et al.* 2014; Osborne *et al.* 2019c).

Also included within the IPA boundary is an area of lowland woodland-dry forest mosaic along the northeastern side of the plateau and south of the town of Nangade, which supports the only known population globally of the Critically Endangered species *Uvaria rovumae* (Deroin & Lötter 2013). Although heavily disturbed, there are still reasonably extensive patches of this lowland mosaic away from the Nangade to Namau road.

In total, 19 globally threatened plant taxa are known to occur on the Mueda Plateau and adjacent footslopes, although in some cases their continued existence at this site requires confirmation given the scale of habitat transformation. The loss of most of the natural woody vegetation is likely to have had a profound impact on many of these species. This site is also of interest as the only known locality in Mozambique for a number of taxa including *Ancylobothrys tayloris* (LC), *Cassia angolensis*, *C. burttii*, *Vernonia* (*Jeffreycia*) *zanzibarensis* (LC) and *Whitfieldia orientalis*. It is one of only two known Mozambican localities for the rare *Streblus usambarensis*, and a noted locality for the scarce and overexploited timber tree *Pterocarpus megalocarpus* (J. Burrows, pers. comm. 2021). It is also worth noting that some highly localised species have been recorded from the lowlands to the west of the Plateau and towards Negomano; this locality includes the only Mozambican site for two species, *Blepharispermum brachycarpum* (EN) and *Crotalaria misella* (DD), as well as populations of *Paranecepsia alchorneifolia* (VU) and *Stylochaeton*

Mueda Plateau on the road from Mocimboa do Rovuma to Mueda (QL)

euryphyllus (VU). It is possible that these species will be recorded in the lower elevation areas of the Mueda Plateau IPA in the future.

Habitat and geology

Mueda is one of a series of low plateaux in the Mozambique-Tanzania coastal border region, including the Makonde Plateau which is separated from Mueda only by the Rovuma River valley. Further north, inland from Lindi in Tanzania, lies the Rondo Plateau which is renowned for its botanical importance (Clarke 2001). The dominant underlying geology of the Mueda Plateau is thought to comprise iron-rich sandstone and conglomerates of the Mikindani Formation of mid-Neogene origin (ca. 10–15 mya), giving rise to red soils that are well-drained, sand-rich and poorly structured. Along the escarpments there are outcrops of the Cretaceous Maconde Formation conglomerates and sandstones (I.N.G. 1987, reproduced in Timberlake *et al.* 2010; Hancox *et al.* 2002). The older formations are overlain by Quaternary or Neogene sedimentary deposits which form a gentle isocline running northwest from the coast, and reaching its highest points on the Mueda Plateau in Mozambique and the Makonde Plateau in Tanzania, with the Rovuma River cutting a sharp channel of ca. 10 km wide into the deposits (Clarke 2011). Elsewhere within the coastal Cabo Delgado region, outcrops of the Mikindani sandstones are associated with dry forest patches of high botanical importance (Timberlake *et al.* 2010).

Given the long history of human impact on this area (see Conservation issues), the original vegetation of the plateau is difficult to ascertain with certainty. The vegetation map of Wild & Barbosa (1968) indicates that much of the Mueda Plateau, particularly on the eastern side, was dominated by a formation of Dry Deciduous Lowland Forest on sandstone and conglomerates (their type 6), a vegetation type that was largely confined in this region to Mueda. Dominant species in this community included *Adansonia digitata*, *Balanites maughamii*, *Bombax rhodognaphalon*, *Cordyla africana*, *Dialium holtzii*, *Milicia excelsa*, *Millettia stuhlmannii* and *Sterculia* spp. (Wild & Barbosa 1968). Much of the woody vegetation of the plateau has been removed and replaced with farmland and areas of fallow, and this forest vegetation type is now reduced to small remnants (Lötter *et al.*, in prep.). Even more severely impacted appears to be moist semi-deciduous forest on the highest parts of the plateau, which is today evident only by the presence of scattered moist forest indicator species, such as *Casearia gladiiformis*, *Dracaena mannii*, *Erythrophleum suaveolens*, *Harungana madagascariensis* and *Rinorea ferruginea* (Lötter *et al.*, in prep.). A patch of swamp forest is recorded on the western edge of the plateau, this being the source of water for the town of Mueda. This patch is dominated by *Albizia adianthifolia*, *Synsepalum brevipes*, *Syzygium owariensis* and *Voacanga thouarsii* together with the climbing swamp fern, *Stenochlaena tenuifolia* (Lötter *et al.*, in prep.).The remaining woodland and thicket is mostly secondary in nature, with denser and more intact areas largely confined to steeper slopes and gulleys along the escarpments. The areas of the escarpment near Mocimboa da Rovuma and towards Ngapa appear more intact than elsewhere. Miombo woodland is widespread along the escarpments, typically dominated by *Julbernardia globiflora* with *Brachystegia* spp., *Diplorhynchus condylocarpon*, *Oxytenanthera abyssinica*, *Pericopsis angolensis*, *Pterocarpus angolensis*, *Sterculia quinqueloba* and *Terminalia stenostachya* (Lötter *et al.*, in prep.).

The climate of the Mueda Plateau is highly seasonal, with a prolonged dry season from May to November and a short hot and wet season mainly between December and April. Annual rainfall at Mueda town is approximately 1,100 mm per year (Timberlake *et al.* 2010), which is comparable to that of the Rondo Plateau in Tanzania.

Conservation issues

The Mueda Plateau and the surrounding lowlands are not currently under any formal protection and this is one of the most severely threatened and degraded IPAs in Mozambique. The plateau has a long history of settlement, starting at least as early as the beginning of the nineteenth century. This was driven in part by the establishment of slaving routes along the Rovuma, which drove local populations onto the adjacent plateaus which were much less accessible and densely wooded (Israel 2006). Indeed, the plateau settlers were named "Makonde" after the densely vegetated uplands. The Makonde (colloquially named the Mavia, or "the nervous") were mentioned in Livingstone's journals and were visited in 1882 by Henry O'Neill, British Consul to Mozambique (Timberlake *et al.* 2010). Timberlake *et al.* (2010) note that much of the Dry Deciduous Forest of the eastern plateau is likely to have been destroyed through agricultural expansion and logging in the pre-independence period. During the 1960s, the Mueda Plateau became the focus point of the independence struggles, with FRELIMO establishing their main base there, supported by the Makonde and their strong ties to independent and socialist Tanzania immediately to the north (Israel 2006). The area saw much military action, with associated environmental impacts. Following independence, larger and more formal settlements were established on the plateau and it is from this time that clearance of the remaining dense woodlands and thickets is believed to have accelerated.

The reasonably fertile soils and reliable rainfall mean that the plateau is attractive for agriculture, with a variety of grains grown both for subsistence and some export, including millet, vegetables and particularly maize, as well as cashew cultivation. As a result, the vast majority of the original wooded vegetation has been cleared in all but the steeper and less inaccessible areas. Dense vegetation cover (woodland, thicket and forest) is estimated to have declined from an estimated 2,332 km² historically to only 89 km² at present, a decline of over 96% (Timberlake *et al.* 2011), with losses particularly severe on the eastern slopes of the plateau. The steeper escarpments and some areas of the northern portion of the plateau have escaped the worst of the clearance, and a portion of the northwest plateau, escarpment and footslopes (within the current IPA boundary) was proposed as a potential conservation area by Timberlake *et al.* (2010). However, even the northern parts of the plateau have experienced recent heavy logging following increased settlement there (J. Burrows, pers. comm. 2021).

Studies on the adjacent Makonde Plateau of Tanzania have revealed that the sandy soils have a weakly developed structure and are highly prone to gully erosion in areas where the vegetation has been denuded (Achten *et al.* 2008; Kabanza *et al.* 2013). This situation is likely to be equally applicable to the Mueda Plateau which has similar soils (Achten *et al.* 2008). A further threat is from increased frequency of uncontrolled wildfires due to deliberate burning; such fires are noted to be impacting the populations of *Aloe ribauensis* in the vicinity of Muidumbe on the southern edge of the plateau (Osborne *et al.* 2019c).

The most urgent conservation priorities on the Mueda Plateau are to raise community awareness and support for sustainable management of the existing remnants of dry forest and thicket vegetation and, potentially, to develop a restoration scheme for these habitats in areas that are not so densely inhabited. Some optimism for such an approach can be taken from the Rondo Nature Forest Reserve in southeast Tanzania, where considerable regeneration of forest has occurred since the cessation of logging in the 1980s. Without such conservation schemes, the Mueda Plateau may soon lose its remaining biodiversity value. *Ex situ* conservation measures are also required for some of the most range-restricted species that occur on the plateau and adjacent lowlands, such as *Celosia patentiloba*, *Hugonia grandiflora* and *Uvaria rovumae*, given the high extinction risk they face in the wild.

Key ecosystem services
The ecosystem services provided by this IPA are yet to be fully documented. The natural vegetation appears to provide important provisioning services to local communities, including harvesting of wood for charcoal production and construction, although these practices have clearly been unsustainable in the past. The more intact areas of the IPA are also likely to provide important regulatory services, particularly in terms of preventing erosion of the rather vulnerable, poorly structured soils. The area also has some cultural importance given the long history of the Makonde settlement and the role that this site and its people played in the independence movement.

Ecosystem service categories

- Provisioning – Raw materials
- Regulating services – Moderation of extreme events
- Regulating services – Erosion prevention and maintenance of soil fertility
- Habitat or supporting services – Habitats for species
- Habitat or supporting services – Maintenance of genetic diversity
- Cultural services – Cultural heritage

IPA assessment rationale

The Mueda Plateau and Escarpments qualify as an Important Plant Area under criterion A(i) as they hold populations of 19 globally threatened plant taxa, of which eight are assessed as Vulnerable, nine as Endangered and two as Critically Endangered. This IPA includes the entire known global population of *Uvaria rovumae*, and is the only Mozambican IPA known to contain populations of *Celosia patentiloba*, *Lannea welwitschii* var. *ciliolata*, *Momordica henriquesii*, *Monodora carolinae* and *Paropsia grewioides* var. *orientalis*. Mueda does not yet qualify under criterion B as only ten (ca. 2%) of the B(ii) qualifying species have so far been recorded from this locality, although this figure is likely to increase with further exploration of the remnant patches of natural vegetation on the plateau and escarpment. Given the extent of transformation and fragmentation of the natural habitats on the plateau, and the very limited extent of Rovuma forest and thicket still present, this site does not qualify as an IPA under criterion C.

Vismianthus punctatus (JEB)

Priority species (IPA Criteria A and B)

FAMILY	TAXON	IPA CRITERION A	IPA CRITERION B	≥ 1% OF GLOBAL POP'N	≥ 5% OF NATIONAL POP'N	IS 1 OF 5 BEST SITES NATIONALLY	ENTIRE GLOBAL POP'N	SPECIES OF SOCIO-ECONOMIC IMPORTANCE	ABUNDANCE AT SITE
Amaranthaceae	*Celosia patentiloba*	A(i)	B(ii)	✓	✓	✓			unknown
Anacardiaceae	*Lannea welwitschii* var. *ciliolata*	A(i)			✓	✓			unknown
Annonaceae	*Monodora carolinae*	A(i)	B(ii)	✓	✓	✓			unknown
Annonaceae	*Uvaria rovumae*	A(i)	B(ii)	✓	✓	✓	✓		scarce
Asphodelaceae	*Aloe ribauensis*	A(i)	B(ii)	✓	✓	✓			scarce
Capparaceae	*Maerua andradae*		B(ii)						unknown
Celastraceae	*Salacia orientalis*	A(i)		✓	✓	✓			unknown
Combretaceae	*Combretum stocksii*		B(ii)	✓	✓	✓			unknown
Connaraceae	*Vismianthus punctatus*	A(i)	B(ii)	✓	✓	✓			unknown
Cucurbitaceae	*Momordica henriquesii*	A(i)		✓	✓	✓			unknown
Fabaceae	*Acacia latistipulata*	A(i)		✓	✓	✓			unknown
Fabaceae	*Acacia quiterajoensis*		B(ii)						unknown

Priority species (IPA Criteria A and B)

FAMILY	TAXON	IPA CRITERION A	IPA CRITERION B	≥ 1% OF GLOBAL POP'N	≥ 5% OF NATIONAL POP'N	IS 1 OF 5 BEST SITES NATIONALLY	ENTIRE GLOBAL POP'N	SPECIES OF SOCIO-ECONOMIC IMPORTANCE	ABUNDANCE AT SITE
Fabaceae	*Baphia macrocalyx*	A(i)		✓	✓	✓			unknown
Linaceae	*Hugonia grandiflora*	A(i)		✓	✓	✓			scarce
Loranthaceae	*Erianthemum lindense*	A(i)		✓	✓	✓			unknown
Malvaceae	*Sterculia schliebenii*	A(i)				✓			unknown
Passifloraceae	*Paropsia grewioides* var. *orientalis*	A(i)		✓	✓	✓			unknown
Rubiaceae	*Cuviera schliebenii*	A(i)		✓	✓	✓			unknown
Rubiaceae	*Cuviera tomentosa*	A(i)		✓	✓	✓			unknown
Rubiaceae	*Oxyanthus biflorus*	A(i)	B(ii)	✓	✓	✓			scarce
Rubiaceae	*Rothmannia macrosiphon*	A(i)			✓	✓			unknown
Rubiaceae	*Tarenna* sp. 53 of Degreef (= *Cladoceras rovumense*)		B(ii)						unknown
Rubiaceae	*Tricalysia semidecidua*	A(i)		✓	✓	✓			occasional
		A(i): 19 ✓	B(ii): 10						

Threatened habitats (IPA Criterion C)

HABITAT TYPE	IPA CRITERION C	≥ 5% OF NATIONAL RESOURCE	≥ 10% OF NATIONAL RESOURCE	IS 1 OF 5 BEST SITES NATIONALLY	ESTIMATED AREA AT SITE (IF KNOWN)
Rovuma Coastal Dry Forest [MOZ-12a]	C(iii)				

Protected areas and other conservation designations

CONSERVATION AREA TYPE	CONSERVATION AREA NAME	RELATIONSHIP OF IPA TO CONSERVATION AREA
No formal protection	N/A	

Threats

THREAT	SEVERITY	TIMING
Housing & urban areas	medium	Ongoing – trend unknown
Small-holder farming	high	Ongoing – trend unknown
Logging & wood harvesting	high	Ongoing – trend unknown
Increase in fire frequency/intensity	high	Ongoing – trend unknown
Soil erosion, sedimentation	unknown	Ongoing – trend unknown

QUITERAJO

Assessor: Iain Darbyshire

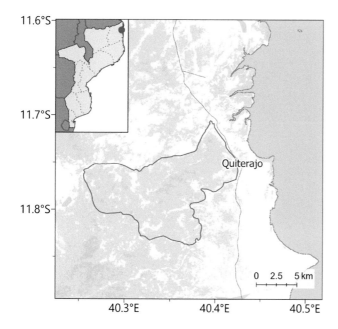

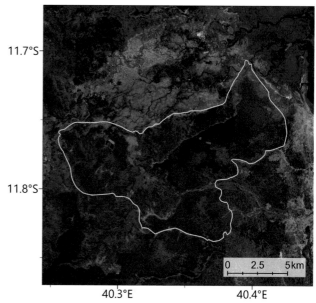

INTERNATIONAL SITE NAME		Quiterajo	
LOCAL SITE NAME (IF DIFFERENT)		–	
SITE CODE	MOZTIPA021	PROVINCE	Cabo Delgado

LATITUDE	-11.76450	LONGITUDE	40.39660
ALTITUDE MINIMUM (m a.s.l.)	5	ALTITUDE MAXIMUM (m a.s.l.)	173
AREA (km²)	129	IPA CRITERIA MET	A(i), A(iv), B(ii), C(iii)

Site description

The Quiterajo IPA covers an area of 129 km² inland from the coastal village of Quiterajo in Macomia District of Cabo Delgado Province. It lies to the west of highway 247, ca. 45 km south of the port of Mocimboa da Praia. The site primarily covers dry forest and dense woodland that occupy a low plateau (mainly 90–150 m a.s.l.) to the south of the floodplain of the Messalo River. The main, eastern, block of the IPA contains the ca. 31 km² Namacubi Forest, sometimes referred to as "The Banana" Forest because of its shape. Also included are the forests and woodlands around and west of Lake Macungue which are separated from Namacubi by floodplain grasslands, open palm savanna and seasonal wetlands, and the Namparamnera Forest to the south of Namacubi. These forests are home to a rich flora including many globally rare and

Xylopia tenuipetala (JT)

Guibourtia schliebenii tree, Quiterajo (JEB)

threatened species, and this site must be considered an urgent priority for conservation action. Indeed, the Namacubi Forest is of such high botanical importance that the major publication on the *Trees and Shrubs [of] Mozambique* (Burrows *et al.* 2018) was dedicated to this unique site in the hope that it would promote international recognition and its formal conservation.

This IPA could be expanded in the future to include the heavily wooded and forested areas of the Sakaje Plateau to the southwest of the current site. To our knowledge, this area has not yet been botanised, but the vegetation is largely intact and looks similar in composition to some of the important patches of woody vegetation at Quiterajo. This would add an extra ca. 200 km² to the IPA, or the Sakaje Plateau could be recognised as a separate IPA.

Botanical significance

Quiterajo was listed as one of four "high priority" sites for the conservation of coastal dry forest in northeast Mozambique (Timberlake *et al.* 2010). This site contains globally important examples of intact dry forest of the proposed Rovuma Centre of Plant Endemism (CoE), a threatened habitat type known for its high rates of local endemism and high species turnover between patches (Timberlake *et al.* 2010, 2011; Burrows and Timberlake 2011; Darbyshire *et al.* 2019a). The ca. 31 km² Namacubi Forest is dominated by *Guibourtia schliebenii*, a globally Vulnerable species for which this is believed to be one of the most important sites. It contains a significant number of species not known elsewhere in Mozambique, many of which are rare and/or threatened Rovuma CoE endemics, such as *Drypetes sclerophylla* (EN), *Omphalea mansfeldiana* (EN) and *Xylopia tenuipetala* (EN). The lattermost of these is a Mozambican endemic for which this is the most important site globally. The aroid *Stylochaeton tortispathum* (VU) is currently considered to be endemic to Namacubi. The diversity of woody Melastomataceae is particularly impressive: Namacubi contains two endemic species, *Warneckea albiflora* (CR) and *Warneckea cordiformis* (CR), as well as being the only Mozambican site for *Memecylon rovumense* (EN), otherwise known from three sites in southeast Tanzania, and the prime locality for the Mozambican endemic species *Memecylon torrei* (EN). The adjacent Namparamnera Forest is the only known locality for *Memecylon aenigmaticum* (CR). The sacred forest west of Lake Macungue is dominated by *Micklethwaitia carvalhoi*, a globally Vulnerable Mozambican endemic species and genus, with ca. 5,000 individuals present in an area of approximately 1 km². Whilst the surrounding floodplain grasslands and seasonal wetlands are of lesser botanical importance, this is the only Mozambican site for the rare labiate herb *Orthosiphon scedastophyllus* (CR), otherwise known from Tendaguru in Tanzania.

Several undescribed taxa are known from Quiterajo, some of which are potentially further endemic species. These include a new *Asparagus* sp. currently under description by S.M. and J.E. Burrows (to be named *Asparagus inopinatus*); a possible new species of succulent *Euphorbia* allied to *E. ambroseae*; a species of *Vepris* also known from one site in Zambézia Province; *Deinbollia* sp. A of Burrows *et al.* (2018); and several members of the coffee (Rubiaceae) family: a *Coffea* sp.; two species of *Pyrostria* currently under description; *Tarenna* sp. 53 of Degreef (2006) (= *Cladoceras rovumense*; I. Darbyshire *et al.*, in press), also known from the Rondo Plateau in Tanzania and Mueda Plateau [MOZTIPA025]; and *Rytigynia* sp. M of Burrows *et al.* (2018).

Habitat and geology

The low plateau above the Messalo floodplain, capped by dry forest, is composed of iron-rich sandstones of the Mikindani Formation of mid-Neogene origin (ca. 10–15 mya). This rock gives rise to a coarsely sandy well-drained red soil. A 50 x 50 m plot surveyed in the Namacubi Forest (Timberlake *et al.* 2010) revealed that 50–60% of the canopy is dominated by *Guibourtia schliebenii*. Other common species in the canopy and subcanopy include *Manilkara discolor*, *Rinorea angustifolia*, *Terminalia myrtifolia*, *Xylopia tenuipetala* and a range of woody Melastomataceae, notably *Memecylon torrei*, *Warneckea cordiformis* and *W. sansibarica*. *Lannea antiscorbutica* and *Vitex carvalhi* are important emergent trees. Timberlake *et al.* (2010) estimate a richness of ca. 50–60 woody species per ha. The geology underlying the *Micklethwaitia*-dominated dry forest west of Lake Macungue is not known but it may differ from that of Namacubi given that it has a very different species assemblage.

The forests have a strong deciduous element and significant numbers of sclerophyllous species. This is in response to the regional climate, which has a prolonged dry season from May to November/December, with a single rainy season from December to April; annual rainfall is approximately 1,000 mm/yr.

Miombo woodland is frequent, particularly on the lower slopes away from the Mikindani sandstone. It is dominated by widespread species including *Brachystegia spiciformis*, *Julbernardia globiflora* and the heavily exploited *Afzelia quanzensis*, as well as the more range-restricted species *Berlinia orientalis* (Timberlake *et al.* 2010). The surrounding floodplains and gentle depressions are underlain by more recent Quaternary deposits and alluvial soils. These areas support open floodplain grassland and savanna, with dominant grasses including *Panicum coloratum*, *Pennisetum polystachion* in disturbed areas, and *Hyparrhenia* spp., and trees including

Vitex mossambicensis (JEB)

Warneckea cordiformis (JEB)

Acacia seyal, A. sieberiana, Faidherbia albida, Kigelia africana and the palms *Hyphaene compressa, Phoenix reclinata* and occasional *Borassus aethiopum*, together with seasonal wetlands (Timberlake *et al.* 2014). These latter areas are of lesser importance for plants but provide critical habitat for other wildlife including elephants.

Conservation issues

There is no formal conservation or biodiversity management in place at Quiterajo. The eastern portion of the site, including Namacubi Forest, was previously included within the ca. 300 km² Messalo Wilderness Area of the Maluane Conservancy (or Cabo Delgado Biodiversity and Tourism Project), a privately run tourism concession. Much of the management focus of this concession was on controlling illegal poaching, and conserving the elephant population on the Messalo floodplain, but there were also efforts to prevent illegal logging in the forests. However, activity within this concession appears to have diminished since 2012, with the Maluane Conservancy focusing more on Vamizi Island to the north (see MOZTIPA017).

The greatest threat posed to this site is from the steady immigration into northeast Cabo Delgado since the end of the post-independence civil war from the 1990s onwards. This has resulted in the expansion of settlement and subsistence agriculture, increased logging of woody species for construction and charcoal, and the increased frequency of wildfires set intentionally for habitat clearance and hunting (Timberlake *et al.* 2010). Illegal commercial logging for export is also an ongoing problem. Timberlake *et al.* (2014) estimate a ca. 10% reduction in forest cover at Namacubi between 1999 and 2013, and encroachment into the southern portion of the forest in particular is clearly evident on satellite imagery (Google Earth 2021). A significant threat arose in the mid-2010s from the proposed construction of a new road from Mocimboa da Praia to Pemba associated with oil and gas industrial activity which would have run through the Namacubi Forest. Thankfully, this project did not proceed, and the threat appears to have abated. Current petroleum industry activity is focused on offshore liquefied natural gas (LNG) extraction further north on the Cabo Delgado coast and the impact south of Mocimboa da Praia is low at present. A violent insurgency in this region since 2017 has disrupted much of this development and has resulted in significant population displacement away from many of the local villages. However, repopulation is likely to follow any abatement of these security concerns in the future. And, should the new access road again be contemplated, the resulting influx of ribbon development and associated environmental degradation would severely threaten the existence of Namacubi Forest (J.E. Burrows, pers. comm.).

In view of its irreplaceability, formal protection of this globally important site and active management to prevent further encroachment or illegal logging should be considered a national conservation priority.

Key ecosystem services

The site is primarily of importance for its habitat and biodiversity supporting services. It also provides provisioning services for local communities, including the sourcing of timber. The site has local cultural and spiritual significance, notably the sacred forest near Lake Macungue. Owing to its proximity to the coast road between Pemba and Palma, this site has potential as an ecotourism destination for specialist wildlife tours. However, the current extreme security concerns (see above), together with accessibility issues resulting from the destruction of bridges across some of the major rivers in the area, will prevent any such ecotourism development in the short to medium term.

Ecosystem service categories

- Provisioning – Raw materials
- Regulating services – Carbon sequestration and storage
- Regulating services – Erosion prevention and maintenance of soil fertility
- Habitat or supporting services – Habitats for species
- Habitat or supporting services – Maintenance of genetic diversity
- Cultural services – Recreation and mental and physical health
- Cultural services – Tourism
- Cultural services – Spiritual experience and sense of place
- Cultural services – Cultural heritage

IPA assessment rationale

Quiterajo meets all three of the criteria to qualify as an IPA. Under criterion A(i), it holds nationally and, in most cases, internationally important populations of over 30 globally threatened plant species, 11 of which are assessed as Endangered and three which are Critically Endangered: *Memecylon aenigmaticum*, *Warneckea albiflora* and *W. cordiformis* which are all endemic to this site. Other globally threatened species are likely to be added to this list when a full Red List for the region is finalised and when the potentially new species are delimited. The site contains at least 21 qualifying species under criterion B(ii) and hence exceeds the 3% threshold for this criterion. It also holds nationally important areas of Rovuma coastal dry forest, a nationally (and almost certainly globally) threatened habitat, and Quiterajo is considered to be one of the five best sites nationally for this habitat, hence it qualifies under criterion C(iii).

Warneckea albiflora (JEB)

Priority species (IPA Criteria A and B)

FAMILY	TAXON	IPA CRITERION A	IPA CRITERION B	≥ 1% OF GLOBAL POP'N	≥ 5% OF NATIONAL POP'N	IS 1 OF 5 BEST SITES NATIONALLY	ENTIRE GLOBAL POP'N	SPECIES OF SOCIO-ECONOMIC IMPORTANCE	ABUNDANCE AT SITE
Annonaceae	*Hexalobus mossambicensis*	A(i)	B(ii)	✓	✓	✓			unknown
Annonaceae	*Monanthotaxis trichantha*	A(i)			✓	✓			unknown
Annonaceae	*Xylopia tenuipetala*	A(i)	B(ii)	✓	✓	✓			unknown
Araceae	*Stylochaeton euryphyllus*	A(i)			✓	✓	✓		unknown
Araceae	*Stylochaeton tortispathus*	A(i)	B(ii)	✓	✓	✓	✓		scarce
Combretaceae	*Combretum stocksii*		B(ii)						unknown
Connaraceae	*Vismianthus punctatus*	A(i)	B(ii)	✓	✓	✓			occasional
Erythroxylaceae	*Nectaropetalum carvalhoi*	A(i)	B(ii)	✓	✓	✓			unknown
Euphorbiaceae	*Croton kilwae*	A(i)			✓	✓	✓		unknown
Euphorbiaceae	*Mildbraedia carpinifolia*	A(i)			✓	✓			common
Euphorbiaceae	*Omphalea mansfeldiana*	A(i)			✓	✓	✓		scarce
Fabaceae	*Acacia latispina*	A(i)	B(ii)	✓	✓	✓			occasional
Fabaceae	*Acacia latistipulata*	A(i)			✓	✓	✓		frequent
Fabaceae	*Acacia quiterajoensis*		B(ii)						frequent

Priority species (IPA Criteria A and B)

FAMILY	TAXON	IPA CRITERION A	IPA CRITERION B	≥ 1% OF GLOBAL POP'N	≥ 5% OF NATIONAL POP'N	IS 1 OF 5 BEST SITES NATIONALLY	ENTIRE GLOBAL POP'N	SPECIES OF SOCIO-ECONOMIC IMPORTANCE	ABUNDANCE AT SITE
Fabaceae	*Berlinia orientalis*	A(i)		✓	✓	✓			frequent
Fabaceae	*Guibourtia schliebenii*	A(i)		✓	✓	✓			abundant
Fabaceae	*Micklethwaitia carvalhoi*	A(i)	B(ii)	✓	✓	✓		✓	frequent
Fabaceae	*Millettia impressa* subsp. *goetzeana*	A(i)			✓	✓			unknown
Hypericaceae	*Vismia pauciflora*	A(i)		✓	✓	✓			unknown
Lamiaceae	*Orthosiphon scedastophyllus*	A(iv)	B(ii)	✓	✓	✓			unknown
Lamiaceae	*Premna schliebenii*	A(i)			✓	✓			unknown
Lamiaceae	*Vitex carvalhi*	A(i)		✓	✓	✓			unknown
Lamiaceae	*Vitex mossambicensis*	A(i)		✓	✓	✓			frequent
Loganiaceae	*Strychnos xylophylla*	A(i)			✓	✓			scarce
Malvaceae	*Grewia limae*	A(i)	B(ii)	✓	✓	✓			scarce
Malvaceae	*Sterculia schliebenii*	A(i)			✓	✓			unknown
Malvaceae	*Thespesia mossambicensis*		B(ii)						unknown
Melastomataceae	*Memecylon aenigmaticum*	A(i)	B(ii)	✓	✓	✓	✓		occasional
Melastomataceae	*Memecylon rovumense*	A(i)	B(ii)	✓	✓	✓			unknown
Melastomataceae	*Memecylon torrei*	A(i)	B(ii)	✓	✓	✓			common
Melastomataceae	*Warneckea albiflora*	A(i)	B(ii)	✓	✓	✓	✓		unknown
Melastomataceae	*Warneckea cordiformis*	A(i)	B(ii)	✓	✓	✓	✓		frequent
Putranjivaceae	*Drypetes sclerophylla*	A(i)		✓	✓	✓			occasional
Rubiaceae	*Chassalia colorata*	A(i)	B(ii)	✓	✓	✓			unknown
Rubiaceae	*Leptactina papyrophloea*	A(i)		✓	✓	✓			unknown
Rubiaceae	*Oxyanthus strigosus*	A(i)	B(ii)	✓	✓	✓			scarce
Rubiaceae	*Pavetta lindina*	A(i)	B(ii)	✓	✓	✓			scarce
Rubiaceae	*Psydrax micans*	A(i)				✓			unknown
Rubiaceae	*Tarenna* sp. 53 of Degreef (= *Cladoceras rovumense*)		B(ii)						unknown
Rubiaceae	*Tricalysia schliebenii*	A(i)		✓	✓	✓			unknown
Rubiaceae	*Tricalysia semidecidua*	A(i)		✓	✓	✓			occasional
Rutaceae	*Vepris sansibarensis*	A(i)			✓	✓			unknown
Rutaceae	*Zanthoxylum lindense*	A(i)		✓	✓	✓			scarce
		A(i): 38 ✓ A(iv): 1 ✓	B(ii): 21 ✓						

Threatened habitats (IPA Criterion C)

HABITAT TYPE	IPA CRITERION C	≥ 5% OF NATIONAL RESOURCE	≥ 10% OF NATIONAL RESOURCE	IS 1 OF 5 BEST SITES NATIONALLY	ESTIMATED AREA AT SITE (IF KNOWN)
Rovuma Coastal Dry Forest [MOZ-12a]	C(iii)		✓	✓	35
Rovuma *Micklethwaitia* Coastal Dry Forest [MOZ-12b]	C(iii)			✓	1

Protected areas and other conservation designations

CONSERVATION AREA TYPE	CONSERVATION AREA NAME	RELATIONSHIP OF IPA TO CONSERVATION AREA
No formal protection	N/A	
Key Biodiversity Area	Quiterajo	protected/conservation area encompasses IPA

Threats

THREAT	SEVERITY	TIMING
Small-holder farming	medium	Ongoing – trend unknown
Roads & railroads	unknown	Past, not likely to return
Gathering terrestrial plants	medium	Ongoing – increasing
Increase in fire frequency/intensity	medium	Ongoing – trend unknown

The Messalo River floodplain at Quiterajo (JEB)

Inside Namacubi Forest at Quiterajo (CS)

MUÀGÁMULA RIVER

Assessor: Iain Darbyshire

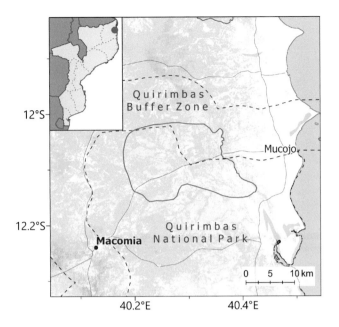

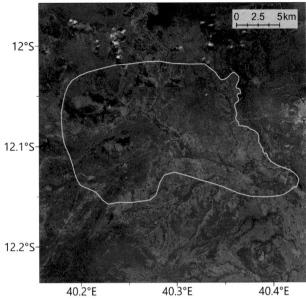

INTERNATIONAL SITE NAME		Muàgámula River	
LOCAL SITE NAME (IF DIFFERENT)		Rio Muàgámula	
SITE CODE	MOZTIPA027	PROVINCE	Cabo Delgado

LATITUDE	-12.09200	LONGITUDE	40.30420
ALTITUDE MINIMUM (m a.s.l.)	15	ALTITUDE MAXIMUM (m a.s.l.)	190
AREA (km²)	291	IPA CRITERIA MET	A(i)

View over the Muàgámula valley (JT)

Oxyanthus strigosus (JEB)

Heinsia mozambicensis (JEB)

Site description

This IPA encompasses the broad valley of the Muàgámula River and its tributaries and adjacent lower slopes, approximately 10–35 km inland from the Indian Ocean coastline in Macomia District of Cabo Delgado Province. This area is referred to as the Mucojo flats by Timberlake *et al.* (2010). It is bisected by the main road from Mucojo to Macomia, which passes through a mosaic of woodland and grassland habitats of high botanical interest. The site lies within the Quirimbas National Park and buffer zone. The exact boundary of this IPA, particularly to the north and south, is incompletely delimited at present as botanical exploration of this site has so far been concentrated in the vicinity of the Mucojo-Macomia road.

Botanical significance

Timberlake *et al.* (2014) note that the varied geological substrates and landforms across the broad valley of the Muàgámula River result in a rich mosaic of habitats that are largely intact and undisturbed, resulting in a high conservation significance. It is of primary importance for its areas of *Acacia* woodland on calcareous clay-rich soils, together with seasonally inundated grasslands rich in suffrutices. These are unusual habitats in northeast Mozambique which hold several rare and threatened plant species, for which the Muàgámula River IPA is a critical site. Of particular note are globally important populations of *Acacia latispina* (VU), *Duosperma dichotomum* (VU), *Grewia filipes* (EN), *Heinsia* (formerly *Pseudomussaenda*) *mozambicensis* (EN), *Tarenna pembensis* (EN) and *Terminalia* (formerly *Pteleopsis*) *barbosae* (VU). This site also contains small areas of raised sandstone outcrops with more dense woody vegetation including small patches of dry forest on sand that support localised and threatened species, including *Premna schliebenii* (VU) and *Oxyanthus strigosus* (EN). All these species are endemic to the proposed Rovuma Centre of Plant Endemism (Burrows & Timberlake 2011; Darbyshire *et al.* 2019a).

To date, only very limited botanical exploration has taken place within this IPA and most or all of this has been concentrated along the Mucojo-Macomia road, in part due to access difficulties across much of the rest of the site. A more thorough and extensive survey is required to fully document the plant diversity of this interesting area; it is likely that this will result in the discovery of more interesting and rare species within this IPA. For example, a potentially new species of *Hygrophila* from edaphic grassland at this sitewas noted by Timberlake *et al.* (2014).

Habitat and geology

This IPA contains a rich mosaic of habitats which are summarised by Timberlake *et al.* (2014), from which

Acacia latispina (JEB)

Thorns of *Acacia latispina* (JEB)

the following information is derived; however, it should be noted that much of the IPA has not yet been surveyed. Outcrops of marl and limestone from the lower Cenozoic (Tertiary) Period occur along the lower slopes of the valley. In combination with Quaternary clays, these give rise to calcareous-rich clay soils that support an *Acacia*-dominated woodland. Important species include *Acacia gerrardii*, *A. polyacantha* and *A. robusta* subsp. *usambarensis*, together with *Dalbergia melanoxylon* and *Spirostachys africana*. This habitat, rare in northeast Mozambique, also supports a number of rare species (see above) including *Acacia latispina*.

Ridges of sandstone are encountered across the landscape and these hold a mixture of miombo woodland, dominated by *Julbernardia globiflora* together with *Afzelia quanzensis*, *Berlinia orientalis* and *Diplorhynchus condylocarpon*, and on deeper sands, a woodland dominated by *Hymenaea verrucosa*, *Millettia stuhlmannii* and *Terminalia* (formerly *Pteleopsis*) *myrtifolia* (Timberlake *et al.* 2014). Some small patches of dry forest occur within these woodlands but these are not well documented at this site.

The floodplain holds Quaternary black or greyish clays supporting an extensive wooded grassland and open grassland, which is inundated during the wet season but is frequently burnt in the dry season. The dominant tree is *Acacia seyal*, with shrubs of *Combretum* spp. and the striking Mozambican endemic shrub *Thespesia mossambicensis* being common. The dominant grasses are *Panicum coloratum* or, on patches of heavier soil, *Setaria incrassata*. River and stream channels, usually dry in the prolonged dry season, are lined by dense thickets.

The climate is characterised by a prolonged dry season from May to November/December, with a single rainy season December to April; annual rainfall is approximately 1,000–1,150mm per year. The rivers and streams are mainly seasonal. Dry season fires are frequent across the floodplain.

Conservation issues

Much of this IPA lies within the wilderness zone of the Quirimbas National Park and UNESCO Biosphere Reserve, although the northeast portion north of the Mucojo-Macomia road is within the Park's buffer zone. This park was established in 2002, initially with support from WWF Mozambique and French and Danish development agencies. However, active management and conservation within the park is limited due to insufficient resources, and the Muàgámula River IPA is not considered to be well protected at present. This has been exacerbated by the recent violent insurgency in Cabo Delgado Province which has resulted in large displacements of populations from north of Pemba and major security concerns across the region. There are now serious problems with wildlife poaching and illegal timber extraction in coastal Cabo Delgado. Ecotourism, which could greatly benefit the Quirimbas wilderness zone, is not viable in the current political situation. Wood harvesting for charcoal and timber has been noted to be degrading some of the woodland habitats along the Mucojo-Macomia road, particularly targeting timber species such as *Millettia stuhlmannii*. However, extensive areas of habitat at this site remain largely intact and undisturbed.

A significant threat arose in the mid-2010s from the proposed construction of a new road from Mocimboa da Praia to Pemba associated with oil and gas industrial activity which would have run through the Muàgámula floodplain. Thankfully, this project did not proceed, and the threat appears to have abated. Current petroleum industry activity is focused on offshore liquefied natural gas (LNG) extraction further north on the Cabo Delgado coast.

The Muàgámula River IPA falls within the vast Quiterajo Key Biodiversity Area, which was designated based upon the range of threatened and range-restricted plant species in this region.

Key ecosystem services

The ecosystem services provided by this IPA are not well documented. However, it is of importance for its habitat and biodiversity supporting services, and is known to be of importance for supporting populations of large mammals which could attract ecotourism once the political situation and infrastructure improve.

Ecosystem service categories

- Habitat or supporting services – Habitats for species
- Habitat or supporting services – Maintenance of genetic diversity
- Cultural services – Tourism

IPA assessment rationale

The Muàgámula River valley qualifies as an IPA under criterion A(i) as it contains important populations of 11 globally threatened species, four of which are assessed as Endangered and seven as Vulnerable. Of particular note, this is considered to be the most secure locality globally for *Terminalia barbosae* (VU) and it is one of only two known localities for both *Heinsia mozambicensis* (EN) and *Duosperma dichotomum* (VU). At present, these three species are only represented at this site within Mozambique's IPA network, as is *Grewia filipes* (EN). The site does not yet qualify under criterion B as it contains only ten of the B(ii) qualifying species (ca. 2%) but with further exploration it is possible that this site may meet the 3% threshold in the future. The small areas of Rovuma Coastal Dry Forest are not considered sufficient to trigger Criterion C(iii) but are nevertheless of high interest at this site.

The Muàgámula valley before the onset of the rainy season (JT)

Priority species (IPA Criteria A and B)

FAMILY	TAXON	IPA CRITERION A	IPA CRITERION B	≥ 1% OF GLOBAL POP'N	≥ 5% OF NATIONAL POP'N	IS 1 OF 5 BEST SITES NATIONALLY	ENTIRE GLOBAL POP'N	SPECIES OF SOCIO-ECONOMIC IMPORTANCE	ABUNDANCE AT SITE
Acanthaceae	*Duosperma dichotomum*	A(i)	B(ii)	✓	✓	✓			unknown
Capparaceae	*Maerua andradae*		B(ii)						common
Combretaceae	*Combretum caudatisepalum*	A(i)	B(ii)	✓	✓	✓			unknown
Combretaceae	*Terminalia barbosae*	A(i)	B(ii)	✓	✓	✓			unknown
Fabaceae	*Acacia latispina*	A(i)	B(ii)	✓	✓	✓			occasional
Fabaceae	*Acacia latistipulata*	A(i)		✓	✓	✓			scarce
Fabaceae	*Acacia quiterajoensis*		B(ii)						occasional
Fabaceae	*Millettia makondensis*	A(i)		✓	✓	✓			unknown
Lamiaceae	*Premna schliebenii*	A(i)			✓	✓			unknown
Malvaceae	*Grewia filipes*	A(i)		✓	✓	✓			unknown
Malvaceae	*Thespesia mossambicensis*		B(ii)						common
Rubiaceae	*Heinsia mozambicensis*	A(i)	B(ii)	✓	✓	✓			scarce
Rubiaceae	*Oxyanthus strigosus*	A(i)	B(ii)	✓	✓	✓			unknown
Rubiaceae	*Tarenna pembensis*	A(i)	B(ii)	✓	✓	✓			unknown
		A(i): 11 ✓	B(ii): 10						

Threatened habitats (IPA Criterion C)

HABITAT TYPE	IPA CRITERION C	≥ 5% OF NATIONAL RESOURCE	≥ 10% OF NATIONAL RESOURCE	IS 1 OF 5 BEST SITES NATIONALLY	ESTIMATED AREA AT SITE (IF KNOWN)
Rovuma Coastal Dry Forest [MOZ-12a]	C(iii)				

Protected areas and other conservation designations

CONSERVATION AREA TYPE	CONSERVATION AREA NAME	RELATIONSHIP OF IPA TO CONSERVATION AREA
National Park	Quirimbas National Park	protected/conservation area overlaps with IPA
UNESCO Biosphere Reserve	Quirimbas Biosphere Reserve	protected/conservation area overlaps with IPA
Key Biodiversity Area	Quiterajo	protected/conservation area encompasses IPA

Threats

THREAT	SEVERITY	TIMING
Gathering terrestrial plants	medium	Ongoing – trend unknown
Roads & railroads	high	Past, not likely to return

QUIRIMBAS ARCHIPELAGO

Assessor: Iain Darbyshire

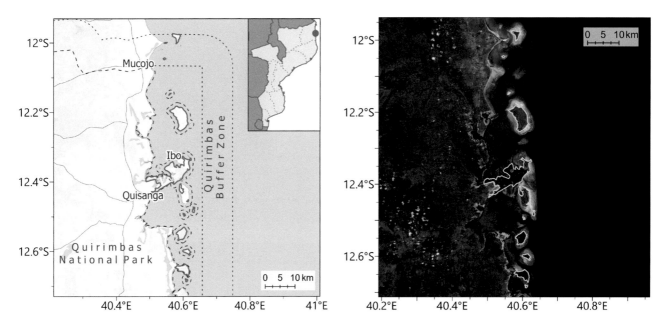

INTERNATIONAL SITE NAME		Quirimbas Archipelago	
LOCAL SITE NAME (IF DIFFERENT)		Arquipélago das Quirimbas	
SITE CODE	MOZTIPA028	PROVINCE	Cabo Delgado

LATITUDE	-12.35050	LONGITUDE	40.61000
ALTITUDE MINIMUM (m a.s.l.)	0	ALTITUDE MAXIMUM (m a.s.l.)	20
AREA (km²)	108	IPA CRITERIA MET	A(i)

The western side of Mefunvo Island, looking towards the Lupangua Peninsula (JT)

Site description

The Quirimbas Archipelago IPA is situated in the Ibo, Macomia and Quissanga Districts of Cabo Delgado Province along the Indian Ocean coastline, between -11.97° and -12.69° latitude. It comprises the southern islands of the Quirimba Archipelago, with a total of 12 islands and islets of coralline rock included or partially included within the IPA, namely from north to south: Makaloe, Mogudala, Rolas, Matemo, Ibo, Quirimba, Sencar, Quilalea, Mefunvo, Quisiva, Situ and Quipaco. All of these islands are located within the Quirimbas National Park (QNP) and UNESCO Biosphere Reserve, with the exception of Makaloe Island which falls within the QNP buffer zone. Some of the larger islands have a long history of occupation, including Matemo (the largest of the islands at ca. 25 km²), Ibo and Quirimba. The latter two islands are extensively transformed, and so only the more intact habitats on those islands are included within the IPA boundary. The large mangrove forests that stretch west of Ibo towards the continental coastline near Quissanga are also included within the IPA boundary, as is the Pangane Peninsula, a coral rag outcrop attached to the mainland to the west of Makaloe Island. Together, these islands and peninsula contain some of the best examples of coral rag thicket and mangroves in Mozambique and support a number of scarce and threatened species. Several of the islands are yet to be botanised, but are included within the IPA as they support most of the same habitats as the islands that have been explored botanically.

Botanical significance

The southern Quirimbas Islands are notable for the presence of extensive intact thickets on coral rag; these are particularly well developed on the sparsely populated and uninhabited islands. Whilst coral rag thicket is relatively widespread along the coast of northern Mozambique, these islands hold some of the best examples of this habitat type nationally. These thickets contain a number of noteworthy species. Of particular importance, this IPA is likely to be the most important site globally for *Nectaropetalum carvalhoi* (VU), a shrub or small tree that is noted to be easily seen on some of these islands (Burrows *et al.* 2018), with records from the Pangane Peninsula and from Makaloe Island in the north of the IPA (E. Schmidt, pers. comm. 2020). Ibo Island is the type locality for *Pavetta mocambicensis* (EN), recorded there by Manuel Rodrigues de Carvalho in the late 19th Century but not collected from the islands since that time. These islands are also the only known site in Mozambique for *Barleria rhynchocarpa* (VU), an attractive yellow- or orange-flowered herb or subshrub that favours coastal thickets, grasslands and foreshores – the type locality is from Quirimbas Island whilst a more recent collection, from 1948, was made from Ibo Island (*Pedro & Pedrogão* #5046). The continued presence of the latter two species within this IPA requires confirmation but they are likely to still be present given the extensive suitable habitat still intact here. Other noteworthy species include the scarce Mozambique endemic *Ochna angustata* (NT). Botanical surveying of these islands has been incomplete to date and the likelihood of recording further species of conservation concern in the future is high. A full botanical inventory should be considered a high priority for this IPA.

Barleria rhynchocarpa, photographed in Tanzania (ID)

Whilst mangroves are generally low in plant diversity and most of the species present within this habitat are widespread, the mangroves in the vicinity of Quissanga are noteworthy for the presence of the parasitic mangrove shrub *Viscum littorum* (NT). This species is a highly localised endemic of northern Mozambique, otherwise known only from the vicinity of Pemba [MOZTIPA024]. It is considered to be Near Threatened due to the ongoing loss of mangrove habitat within its small range, although it is likely to be under-recorded given the limited botanical survey to date within these extensive and often inaccessible mangrove communities (Alves *et al.* 2014a).

Habitat and geology

The low-lying islands and peninsulas of the Quirimbas Archipelago IPA are formed from outcrops of coral rag of Pleistocene age (Carvalho & Bandeira 2003), which support thin, sand-rich soils with frequent areas of exposed, sharp rock. Some of the islands have small coralline sea cliffs of up to 8 m high. Coral rag thicket dominates the undisturbed vegetation of the islands and the Pangane Peninsula. This is a short, dense thicket of 2–7 m tall with a rather low diversity of woody species. Dominant species in this habitat include *Cassipourea mossambicensis*, *Coptosperma littorale*, *Diospyros consolatae*, *Erythroxylum platyclados*, *Euclea* spp., *Mimusops obtusifolia* and *Olax dissitiflora* amongst others (Burrows *et al.* 2018). Borghesio & Gagliardi (2015) also note *Commiphora* spp. and *Salvadora persica* as frequent. Herbaceous species are rather scarce. The upper beach margins often have a distinctive thicket assemblage of species with Indo-Pacific distributions, such as *Colubrina asiatica*, *Pemphis acidula* and *Suriana maritima* (Burrows *et al.* 2018).

In disturbed areas, a more open thicket and grassland occurs. Whilst the coral rag soils are generally unsuitable for agriculture, some farming occurs in areas with better-developed sandy soils, and there has also been some planting of exotic trees such as coconut palms.

Extensive mangrove communities are recorded along the coastline mainly on the western side of the islands, with the vast mangrove forests west of Ibo towards Quissinga – the "Ibo stand" (ca. 17 km²) – included within the IPA boundary. Smaller mangrove communities occur around the other islands. Eight species of mangrove are noted to occur here, the dominant species being *Rhizophora mucronata* (Barnes 2001).

The eastern shores of the islands are typically fringed by coral reefs whereas the more sheltered western shores are generally fringed by shallow waters with sandy seabeds and extensive seagrass communities. A total of 10 seagrass species have been recorded from Montepuez Bay to the west of Quirimba Island, with the dominant species including *Thalassia hemprichii* in the intertidal areas and *Enhalus acorioides* and *Thalassodendron ciliatum* in the subtidal zones (Bandeira & Gell 2003); no threatened seagrass species have been noted to date. The extensive intertidal areas also support diverse macroalgal communities; recent surveys of the seaweed flora revealed 27 new records for Mozambique out of a total of 101 taxa recorded around the islands, mainly occurring in coral reef habitats but also amongst the seagrass beds (Carvalho & Bandeira 2003). These marine communities are not included within the IPA boundary at present but may be added in future once a full threat assessment of these habitats and their plant species has been carried out.

The climate is warm throughout the year; temperatures peak in December with an average high of 30.4°C, and are at their lowest in July when the average high is 26.7°C at Ibo. Average annual rainfall at Ibo is 1,047 mm whilst at Quissanga it is 1,320 mm; the rainy season peaks in December to March, with a prolonged dry season from May to November. However, humidity remains high throughout the year at over 70% (climate-data.org).

Conservation issues

The Quirimbas National Park (QNP), an extensive area of ca. 9,013 km² of both marine and terrestrial environments, was established in 2002 following a consultation process with local communities who recognised the need to preserve the natural resources on which they depend (Harari 2005). A multi-stakeholder approach was taken from the outset, including national and provincial government, NGOs, private investors and local communities. The aim of the QNP is to balance biodiversity conservation with improved local livelihoods through securing useful natural resources and developing income-generating opportunities from the park for local communities, particularly through ecotourism. Supported by the French Development Agency AFD (2002–2017), a management plan was developed by WWF who managed the park until 2010, together with the establishment of infrastructure, training of park staff, and funding for community-based projects. The site was also designated as a UNESCO Biosphere Reserve in 2018.

Much of the conservation focus to date has been on the rich marine environments within the QNP, in particular the protection of fisheries against over-harvesting (Harari 2005). The coastal waters of the park, and associated extensive coral reefs, seagrass communities and mangroves, provide feeding and/or nesting grounds for sea turtles, dugongs, cetaceans and a high diversity of fish, many of which are of conservation importance (Harari 2005). The islands and their coasts are also of international importance for migratory Palearctic birds, supporting their recognition as a wetland of international importance based on the Ramsar criteria (Borghesio & Gagliardi 2015).

Despite a range of projects having been implemented within the QNP, little is known about their effectiveness in contributing to biodiversity conservation in the park, in part due to lack of baseline data and monitoring, and many projects are not considered to have been successful in achieving their desired outcomes or have been stopped prematurely (Chevallier 2018). It has also been noted that the lack of inventories for many groups of terrestrial organisms hinders the development of a comprehensive management plan for the Park or access to funding (Harari 2005). The recent violent insurgency in Cabo Delgado Province has made on-the-ground management more difficult as well as halting tourist revenue streams for the QNP.

Threats to the terrestrial environments on the islands are not considered to be severe at present. Most of the inhabitants of the islands are reliant on fishing as their main source of subsistence and income. Some of the more accessible and long-inhabited islands have experienced habitat transformation for subsistence agriculture, although this is limited by the thin, low fertility soils and the lack of available freshwater (Chevallier 2018). Agriculture most notably

impacts the islands of Matemo, Ibo and Quirimba; much of the lattermost island is excluded from the IPA as extensive areas are given over to coconut plantations. Although in accessible areas there is some cutting of mangroves for poles, the mangroves within the QNP are largely intact and are actually experiencing net gains across the Quirimbas landscape (Shapiro *et al.* 2020).

A significant future threat to the islands is from climate change, including rising sea levels and increased frequency of extreme weather events. The islands were severely impacted by Cyclone Kenneth in 2019, the most severe tropical cyclone in Mozambique since modern records began.

Key ecosystem services

The islands supported a resident population of ca. 9,000 inhabitants in 1998, but this number is increased significantly by the transient presence of fishermen, attracted by the rich fishing waters (Harari 2005). Fish provides the major source of protein for local residents; fishing is mainly carried out at a subsistence and local market scale because of lack of access to larger commercial markets. The extensive mangrove stands, seagrass communities and coral reefs are all important areas for marine biodiversity and support the rich fisheries. The intact coastal habitats are also an important buffer to coastal erosion and extreme weather events in this region.

There is high potential for sustainable ecotourism as a source of local revenue and to support the QNP and its biodiversity. Prior to the recent insurgency in Cabo Delgado, tourism from South Africa and Zimbabwe was increasing, with the main attractions including the coral sand beaches, diving and snorkeling activities. Resorts and lodges have been developed on several of the islands. Ibo Island is also of historical importance, as a major early Indian Ocean trading port until the early 20th Century. There is much evidence of its significance within the Portuguese empire, including 17th century fortifications in Vila de Ibo, that could attract tourism. However, there are a number of challenges to realising the tourism potential of this site. These include the difficulties of balancing tourism development with nature conservation, and the need to better involve local communities in Park activities and management and in the economic benefits they bring.

Ecosystem service categories

- Provisioning – Food
- Regulating services – Moderation of extreme events
- Regulating services – Erosion prevention and maintenance of soil fertility
- Habitat or supporting services – Habitats for species
- Cultural services – Tourism
- Cultural services – Cultural heritage

IPA assessment rationale

The Quirimbas Archipelago islands qualify as an IPA under criterion A(i) as they contain a globally important population of *Nectaropetalum carvalhoi* (VU), and are likely to contain a globally important population of *Pavetta mocambicensis* (EN). These islands are also the only known site in Mozambique for *Barleria rhynchocarpa* (VU) and so are of national importance for this species.

Priority species (IPA Criteria A and B)

FAMILY	TAXON	IPA CRITERION A	IPA CRITERION B	≥ 1% OF GLOBAL POP'N	≥ 5% OF NATIONAL POP'N	IS 1 OF 5 BEST SITES NATIONALLY	ENTIRE GLOBAL POP'N	SPECIES OF SOCIO-ECONOMIC IMPORTANCE	ABUNDANCE AT SITE
Acanthaceae	*Barleria rhynchocarpa*	A(i)			✓	✓			unknown
Erythroxylaceae	*Nectaropetalum carvalhoi*	A(i)	B(ii)	✓	✓	✓			common
Ochnaceae	*Ochna angustata*		B(ii)	✓					unknown
Rubiaceae	*Pavetta mocambicensis*	A(i)	B(ii)			✓			unknown
Santalaceae	*Viscum littorum*		B(ii)	✓	✓	✓			unknown
		A(i): 3 ✓	B(ii): 4						

Quipaco Island in the Quirimbas Archipelago (JT)

Protected areas and other conservation designations

CONSERVATION AREA TYPE	CONSERVATION AREA NAME	RELATIONSHIP OF IPA TO CONSERVATION AREA
National Park	Quirimbas National Park	protected/conservation area encompasses IPA
UNESCO Biosphere Reserve	Quirimbas Biosphere Reserve	protected/conservation area encompasses IPA

Threats

THREAT	SEVERITY	TIMING
Tourism & recreation areas	medium	Ongoing – trend unknown
Small-holder farming	unknown	Ongoing – trend unknown
Fishing & harvesting aquatic resources	low	Ongoing – trend unknown
Logging & wood harvesting	low	Ongoing – trend unknown

LUPANGUA PENINSULA

Assessors: Iain Darbyshire, Phil Clarke

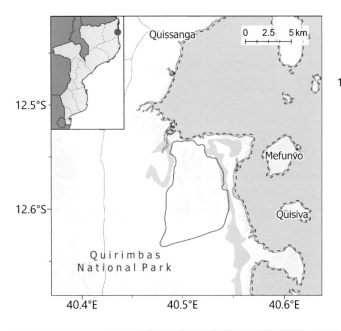
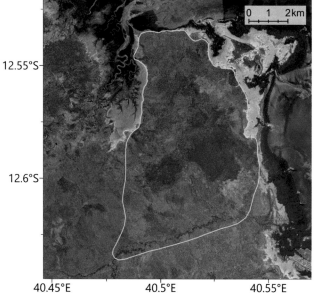

INTERNATIONAL SITE NAME		Lupangua Peninsula	
LOCAL SITE NAME (IF DIFFERENT)		Península de Lupangua	
SITE CODE	MOZTIPA026	PROVINCE	Cabo Delgado

LATITUDE	-12.57770	LONGITUDE	40.51470
ALTITUDE MINIMUM (m a.s.l.)	4	ALTITUDE MAXIMUM (m a.s.l.)	132
AREA (km²)	57	IPA CRITERIA MET	A(i), C(iii)

Aerial view of coastal dry forest on the Lupangua Peninsula (JT)

Site description

The Lupangua Peninsula IPA is located on the coast of Quissanga District in Cabo Delgado Province, northeastern Mozambique. It lies ca. 12 km to the south of the coastal fishing village of Quissanga, close to the village of Mahate. This site, with an area of 57 km², is within the Quirimbas National Park and is adjacent to the islands of Mefunvo and Quisiva in the Quirimbas Archipelago. The peninsula comprises coastal lowlands and a series of low hills, rising up to 132 m elevation on Lupangua Hill, but with the western ridge of the peninsula reaching 80 m elevation. It contains an important example of coastal dry forest of the proposed Rovuma Centre of Plant Endemism (CoE) (Burrows & Timberlake 2011; Darbyshire *et al.* 2019a).

Botanical significance

Lupangua is of botanical importance primarily for its ca. 25 km² of near-pristine coastal dry forest, dominated by the globally threatened tree species *Micklethwaitia carvalhoi* (VU), which is endemic to Mozambique (Clarke 2010). The population of this species at Lupangua is estimated at over 10,000 individuals (Burrows *et al.* 2014a) and is believed to

be the largest and most secure population of this species globally (Clarke 2010). It was on this basis that this site was highlighted as one of the four highest priority sites for conservation in a recent review of coastal dry forests of Cabo Delgado (Timberlake *et al.* 2010). It is one of the few sizable remnant dry forest areas within the proposed Rovuma CoE. This site had apparently not been explored by biologists prior to a reconnaissance survey by a small team in November 2009. In addition to discovering the important *Micklethwaitia* population, this brief survey focusing on the western ridge also found the first known site in Mozambique for the globally Endangered *Hildegardia migeodii* (since also found at Pemba Bay) and a population of *Premna schliebenii* (VU). Other interesting discoveries included a potential new species of *Erythrina*, which requires further investigation, and the second known Mozambican population of *Kabuyea hastifolia*, a monotypic genus endemic to East Africa (Clarke 2010). The botanical inventory of this site is highly incomplete and further species of high conservation concern are likely to be uncovered following more exhaustive surveys. A provisional species list, comprising only 28 taxa, is provided by Clarke (2010).

Habitat and geology

Clarke (2010) provides a preliminary assessment of the main vegetation types of this site. The peninsula is surrounded to the north, west and east by extensive mangrove swamps and adjacent salt flats; these are not included within the IPA. Above the salt flats, the low-lying coastal areas support a coastal woodland dominated by *Terminalia sambesiaca* (which is common in all habitats throughout the site), *Acacia nigrescens* and *A. robusta* subsp. *usambarensis*; some of the woodland and wooded grassland on the peninsula appears to be secondary in nature. Further inland and upslope, there are areas of scrub forest with a broken canopy at ca. 8 m and a more complete lower canopy of ca. 3–4 m, with a mixed species assemblage including *Dobera loranthifolia*, *Manilkara mochisia* and *Monodora junodii*. On the ridges and hill tops there are large stands of *Micklethwaitia*-dominated coastal dry forest, typically with a canopy at ca. 8 m; *Monodora junodii* is co-dominant with *Micklethwaitia* in the understorey layer. Occasional emergent trees including *Adansonia digitata* are noted on parts of the peninsula. This forest type also extends down to sea level in steep gullies that are protected from fire. The soils on the surveyed parts of the peninsula are heavy clays with numerous interspersed calcareous rocks, which are also abundantly scattered on the soil surface. The *Micklethwaitia* forest favours well-drained soils (Clarke 2010).

Average annual rainfall at nearby Quissanga is 1,320 mm per year, with the main rainy season being December to April, and a prolonged dry season between May and November (climate-data.org).

Conservation issues

The whole of the Lupangua peninsula is contained within the Quirimbas National Park and UNESCO Biosphere Reserve, but there is no active conservation management of the site at present. The human population is currently small and the peninsula is not easily accessible as it is not served by any sizable roads. Some small non-permanent fishing settlements are found on the coast, and a large village is located to the southeast of the main forest block (Clarke 2010). Some notable expansion of agricultural activity is visible on recent satellite imagery (Google Earth 2021) which has, post-2003, encroached into the forest in the northeast and, in particular, in the southeast of the site. However, the soils on the hills are thin and rocky and so unlikely to be of agricultural value. A single cut-line running north-south along the peninsula is also clearly visible on satellite imagery and probably dates back to oil exploration in the early 1980s (Clarke 2010).

The *Micklethwaitia* is used here as a source of poles for construction, as its timber is hard and termite resistant. However, this species coppices well and the 2009 survey found evidence of it regenerating from stumps (Clarke 2010; Burrows *et al.* 2018). A more significant threat is from fire, to which this species appears to have very little resistance. Fires are set deliberately by people primarily to control wild animals (Clarke 2010).

The Lupangua Peninsula is included within the Quiterajo Key Biodiversity Area (WCS *et al.* 2021).

Hildegardia migeodii (TR)

Seedlings of *Micklethwaitia carvalhoi* (PC)

Within the *Micklethwaitia carvalhoi* forest (PC)

Key ecosystem services

The site is primarily of importance for its habitat and biodiversity supporting services. It is also a locally important source of materials for local communities, notably timber. Tourism potential at this site is low in view of its inaccessibility, and it would be favourable to keep this as such because any increase in access is likely to lead to accelerated loss of forest.

Ecosystem service categories

- Provisioning – Raw materials
- Habitat or supporting services – Habitats for species
- Habitat or supporting services – Maintenance of genetic diversity

IPA assessment rationale

The Lupangua Peninsula qualifies as an Important Plant Area under criterion A(i), as it contains important populations of one Endangered species, *Hildegardia migeodii*, and two Vulnerable species, *Micklethwaitia carvalhoi* and *Premna schliebenii*. It is believed to be the most important site globally for the *Micklethwaitia*, and so also qualifies under criterion C(iii) in containing over 10% of the national resource of Rovuma *Micklethwaitia*-dominated Coastal Dry Forest. It is considered highly likely that this under-explored locality will contain other globally threatened and/or range-restricted species.

Priority species (IPA Criteria A and B)

FAMILY	TAXON	IPA CRITERION A	IPA CRITERION B	≥ 1% OF GLOBAL POP'N	≥ 5% OF NATIONAL POP'N	IS 1 OF 5 BEST SITES NATIONALLY	ENTIRE GLOBAL POP'N	SPECIES OF SOCIO-ECONOMIC IMPORTANCE	ABUNDANCE AT SITE
Fabaceae	*Micklethwaitia carvalhoi*	A(i)	B(ii)	✓	✓	✓		✓	abundant
Lamiaceae	*Premna schliebenii*	A(i)			✓	✓			unknown
Malvaceae	*Hildegardia migeodii*	A(i)		✓	✓	✓			occasional
		A(i): 3 ✓	B(ii): 1						

Threatened habitats (IPA Criterion C)

HABITAT TYPE	IPA CRITERION C	≥ 5% OF NATIONAL RESOURCE	≥ 10% OF NATIONAL RESOURCE	IS 1 OF 5 BEST SITES NATIONALLY	ESTIMATED AREA AT SITE (IF KNOWN)
Rovuma *Micklethwaitia* Coastal Dry Forest [MOZ-12b]	C(iii)		✓	✓	25

Protected areas and other conservation designations

CONSERVATION AREA TYPE	CONSERVATION AREA NAME	RELATIONSHIP OF IPA TO CONSERVATION AREA
National Park	Quirimbas National Park	protected/conservation area encompasses IPA
UNESCO Biosphere Reserve	Quirimbas Biosphere Reserve	protected/conservation area encompasses IPA
Key Biodiversity Area	Quiterajo	protected/conservation area encompasses IPA

Threats

THREAT	SEVERITY	TIMING
Small-holder farming	low	Ongoing – increasing
Gathering terrestrial plants	low	Ongoing – trend unknown
Increase in fire frequency/intensity	low	Ongoing – trend unknown

QUIRIMBAS INSELBERGS

Assessors: Iain Darbyshire, Marcelino Inácio Caravela

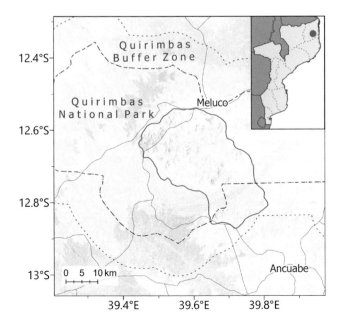

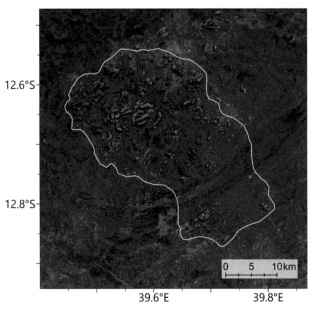

INTERNATIONAL SITE NAME		Quirimbas Inselbergs	
LOCAL SITE NAME (IF DIFFERENT)		Montes-Ilha das Quirimbas	
SITE CODE	MOZTIPA022	PROVINCE	Cabo Delgado

LATITUDE	-12.81450	LONGITUDE	39.69280
ALTITUDE MINIMUM (m a.s.l.)	215	ALTITUDE MAXIMUM (m a.s.l.)	766
AREA (km²)	812	IPA CRITERIA MET	A(i)

Site description

This IPA encompasses an extensive series of impressive gneissic inselbergs in Ancuabe and Meluco Districts of Cabo Delgado Province in northeast Mozambique. This area lies ca. 100 km inland from the Indian Ocean coastline and from the port city of Pemba, and is contained within the Quirimbas National Park which was established in 2002, with the inland portion of this Park having been established specifically to protect these inselbergs. The most well-surveyed of these inselbergs are at Taratibu in the south of the IPA, approximately 25 km NW of the town of Ancuabe. Taratibu and the neighbouring peaks in the southern portion of this IPA are separated from the larger concentration of inselbergs to the north by the Montepuez River which flows ENE through the site. The northern part of the IPA lies adjacent to highway 525 which passes through the small town of Meluco; the town and neighbouring agricultural lands are excluded from the IPA.

Botanical significance

This site is of importance for its extensive areas of xerophytic inselberg flora and for significant intact stands of lowland semi-deciduous forest. Although the botanical inventory of these habitats is far from complete, this IPA is known to contain important populations of rare and threatened species. It is the only location globally for the succulent shrub *Euphorbia unicornis* (EN), which is known only from the inselbergs in the vicinity of

View of the Quirimbas Inselbergs, looking north from Tarabitu (JEB)

Pouteria pseudoracemosa (TP)

Lithophytic flora at Taratibu (MIC)

Meluco. It is one of only two known locations for the Mozambique endemic shrub *Rytigynia torrei* (EN) and the only known location in Mozambique for the impressive forest canopy tree *Pouteria pseudoracemosa* (VU), which is otherwise scarce and scattered in the coastal forests of Tanzania and southeast Kenya. It also holds a population of wild Ibo coffee, *Coffea zanguebariae* (VU). Other notable species include the Mozambican endemic succulent shrub *Euphorbia corniculata* (LC), which is a common constituent of the lithophytic flora, and the striking near endemic species *Euphorbia* (formerly *Monadenium*) *torrei* and *Aloe mawii* (LC).

To date, only a small area of this IPA around the inselbergs of the privately owned Taratibu Concession has been botanised and, with the exception of *Euphorbia unicornis*, most of the species noted above are known only from that area at present, although they are all likely to also occur in the inselbergs north of the Montepuez River. A more complete botanical inventory of this IPA is sorely needed, and the likelihood of further discoveries of rare, threatened and new species is very high at this site. A rapid survey at Taratibu in early 2017, primarily to investigate the population of *Pouteria pseudoracemosa*, led to the discovery of a new shrubby species of *Pavetta*, *Pavetta* sp. J of Burrows *et al.* (2018), a highly distinctive new *Asparagus* species ("*Asparagus procerus*" S.M.Burrows & J.E.Burrows, ms.) and an unusual tree species of Euphorbiaceae that is unknown and not yet placed to genus (J.E. Burrows, pers. obs.). This latter species has also been recorded from an inselberg near Nampula ca. 250 km to the south.

Habitat and geology

The gneissic inselbergs are of Paleoproterozoic to Neoproterozoic age. They vary greatly in size but rise to a maximum of over 700 m a.s.l. The region has a dry to sub-humid climate, with an average annual rainfall of 800–1,200 mm, with a short wet season mainly in December to March and a prolonged intervening dry season. The xerophytic flora of the rock faces and crevices is dominated by *Xerophyta pseudopinifolia* and *X. suaveolens*, together with a range of succulent species including several *Aloe* and *Euphorbia* spp. Other rock-loving plants, including *Myrothamnus flabellifolius* and *Strophanthus hypoleucos,* are also common. Some seasonal seepage areas with peaty soils are observed but these have not yet been surveyed for their plant diversity. The flora on these slopes is mainly herbaceous and shrubby but scattered trees of *Brachystegia* and *Ficus* spp. occur.

The vegetation of the Taratibu Concession, focusing on the southwest portion of the reserve in the vicinity of the inselbergs, has recently been characterised by Joaquim & Caravela (2019) who documented five vegetation types: (1) mixed riverine fringing forest with frequent *Ancylobotrys petersiana*, *Pseudobersama mossambicensis* and *Rawsonia lucida*; (2) a semi-closed dry forest/ thicket of *Oxytenanthera abyssinica* and *Millettia stuhlmannii*; (3) inselberg xerophytic habitat with abundant *Xerophyta* and *Euphorbia*; (4) miombo woodland dominated by *Julbernardia globiflora*, with *Brachystegia spiciformis* and *Diplorhynchus condylocarpon* amongst other miombo species; and (5) closed seasonally moist semi-deciduous forest with large trees of *Pouteria pseudoracemosa* and *Parkia filicoidea*, along with a mixed tree and shrub assemblage including *Englerophytum natalense*, *Rawsonia lucida* and *Rinorea arborea*. Based on a review of satellite imagery, these and similar habitats are believed to occur across the IPA, including some extensive areas of intact forest between the more remote inselbergs north of the Montepuez River. Much of the lowland plain is occupied by miombo woodland of varying density. The semi-deciduous forest and thicket appear to be largely confined to the foot of the inselbergs, being particularly well developed in intervening ravines and sheltered areas. Riverine forest is best developed along the Montepuez RIver and its tributaries and this would be worthy of further botanical exploration.

Conservation issues

The large majority of the IPA lies within the western extension of the Quirimbas National Park (QNP), although the northern-most and southern-most inselbergs lie outside the park boundary within the buffer zone. The QNP was designated in 2002, primarily to protect a region of coastal forest, mangroves and coral reefs including the southern 11 islands of the Quirimbas Archipelago, but a large inland extension was included within the gazetted site, primarily to protect the inselbergs. Active conservation and management within the QNP are very limited at present. Taratibu, whilst within the QNP boundary, is managed as a private ecotourism concession.

Poaching of fauna is a major problem at this site and has led to the decimation of the local elephant population, which may result in significant ecological changes. However, the vegetation on and around the larger inselbergs appears to be largely intact and the human population is low in much of the central and southern portion of the IPA. The inselbergs in the vicinity of Meluco and highway 525 in the north of the site appear from Google Earth (2021) imagery to be more heavily impacted; most woody vegetation appears to have been removed and there is intensive agriculture in the lowlands surrounding the peaks, with clearance aided by frequent burning. Elsewhere, miombo woodland in particular is being cleared for agricultural land and as a source of fuelwood, and

Euphorbia unicornis (TR)

Strophanthus hypoleucos (ID)

some of the lowland forests may be vulnerable to timber extraction given the relative ease of access. Population pressure is likely to increase within this IPA: Ancuabe and Meluco Districts have experienced over 80% and over 50% population increases, respectively, between 1997 and 2017 (Instituto Nacional de Estatistica Moçambique 2021).

The Taratibu portion of this IPA is included within the Key Biodiversity Areas network due to the presence of an endemic frog, the Quirimbas Mongrel Frog (*Nothophryne unilurio*, CR) and the population of *Rytigynia torrei*. This IPA would also qualify as an Alliance for Zero Extinction site given that it contains the entire known global populations of both the frog species and *Euphorbia unicornis*.

Key ecosystem services

Taratibu Reserve and Bush Camp has been run as an ecotourist site with walking wildlife safaris, rock climbing and the stunning scenery as key attractions, and several trekking trails have been established (Paula *et al.* 2015). However, tourist numbers at the Camp have declined sharply due to the rise in elephant poaching, which has all but wiped out the elephant population as well as causing security concerns as the poachers are heavily armed (WWF Mozambique 2016). The intact vegetation protects the thin soils over the rock outcrops from excessive erosion and provides important habitat and supporting services for a range of biodiversity.

Ecosystem service categories

- Regulating services – Erosion prevention and maintenance of soil fertility
- Habitat or supporting services – Habitats for species
- Habitat or supporting services – Maintenance of genetic diversity
- Cultural services – Recreation and mental and physical health
- Cultural services – Tourism

IPA assessment rationale

The Quirimbas Inselbergs qualify as an Important Plant Area under criterion A(i) on the basis of containing important populations of six globally threatened species: *Euphorbia unicornis* (EN), *Rytigynia torrei* (EN), *Englerina triplinervia* (VU), *Pouteria pseudoracemosa* (VU), *Strophanthus hypoleucos* (VU) and *Coffea zanguebariae* (VU). Of these, it contains the entire known global population of *Euphorbia unicornis* and is the only known site in Mozambique for *Pouteria pseudoracemosa*. Further threatened and range-restricted species are expected to be found in these inselbergs and their rich and varied habitats following more complete botanical surveys.

Inselbergs at Taratibu Reserve (JEB)

Moist lowland forest at Taratibu (JEB)

Priority species (IPA Criteria A and B)

FAMILY	TAXON	IPA CRITERION A	IPA CRITERION B	≥ 1% OF GLOBAL POP'N	≥ 5% OF NATIONAL POP'N	IS 1 OF 5 BEST SITES NATIONALLY	ENTIRE GLOBAL POP'N	SPECIES OF SOCIO-ECONOMIC IMPORTANCE	ABUNDANCE AT SITE
Apocynaceae	*Strophanthus hypoleucos*	A(i)		✓	✓	✓			frequent
Euphorbiaceae	*Euphorbia corniculata*		B(ii)						unknown
Euphorbiaceae	*Euphorbia unicornis*	A(i)	B(ii)	✓	✓	✓	✓		scarce
Loranthaceae	*Englerina triplinervia*	A(i)			✓	✓			unknown
Rubiaceae	*Coffea zanguebariae*	A(i)		✓	✓	✓		✓	occasional
Rubiaceae	*Rytigynia torrei*	A(i)	B(ii)	✓	✓	✓			occasional
Sapotaceae	*Pouteria pseudoracemosa*	A(i)		✓	✓	✓			frequent
		A(i): 6 ✓	B(iii): 3						

Protected areas and other conservation designations

CONSERVATION AREA TYPE	CONSERVATION AREA NAME	RELATIONSHIP OF IPA TO CONSERVATION AREA
National Park	Quirimbas National Park	protected/conservation area overlaps with IPA
Private nature reserve	Taratibu Reserve	IPA encompasses protected/conservation area
UNESCO Biosphere Reserve	Quirimbas Biosphere Reserve	protected/conservation area overlaps with IPA
Key Biodiversity Area	Taratibu	IPA encompasses protected/conservation area

Threats

THREAT	SEVERITY	TIMING
Small-holder farming	low	Ongoing – trend unknown
Hunting & collecting terrestrial animals	high	Ongoing – trend unknown
Increase in fire frequency/intensity	unknown	Ongoing – trend unknown

PEMBA

Assessors: Iain Darbyshire, Marcelino Inácio Caravela

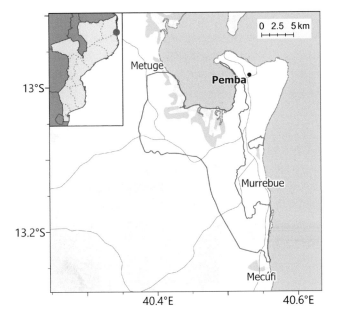

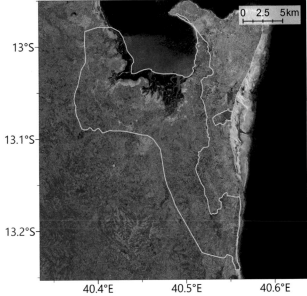

INTERNATIONAL SITE NAME		Pemba	
LOCAL SITE NAME (IF DIFFERENT)		–	
SITE CODE	MOZTIPA024	PROVINCE	Cabo Delgado

LATITUDE	-13.09145	LONGITUDE	40.48880
ALTITUDE MINIMUM (m a.s.l.)	0	ALTITUDE MAXIMUM (m a.s.l.)	148
AREA (km²)	231	IPA CRITERIA MET	A(i)

Site description

The Pemba IPA is located in the Pemba, Metuge and Mecúfi Districts of Cabo Delgado Province. It comprises the coastal lowlands of Pemba Bay from Metuge southwards, parts of the Pemba Peninsula (excluding the urban and residential areas of the port city of Pemba, the capital of Cabo Delgado Province, and associated villages) and extends along the Indian Ocean coastline south towards the town of Mecúfi. The site contains a mosaic of terrestrial coastal habitats, much of which are heavily transformed but with some intact vegetation remaining. It also includes extensive mangroves in Pemba Bay. Although delimited largely as a single unit, with an area of ca. 231 km², only small and isolated pockets of this area are still of high botanical value. The inland boundary of this IPA is only vaguely delimited at present and may require future refinement.

Botanical significance

Although much of this IPA is heavily degraded and with large extents of the original vegetation now lost, it still contains a number of rare and threatened plant species within fragments of the original coastal vegetation of the proposed Rovuma Centre of Plant Endemism (Burrows & Timberlake 2011; Darbyshire *et al.* 2019a). Of primary importance, this site contains the entire known global population of *Eriolaena rulkensii* (EN), the only continental African member of this predominantly Asian genus. This is an attractive yellow-flowered shrub or treelet which occurs in heavy clay over coral-rag in coastal scrub and forest, sometimes at the upper margin of mangrove communities (Dorr & Wurdack 2018; Darbyshire *et al.* 2019c). This species occurs on the west (bay) side of the Pemba Peninsula and in remnant forest patches south of the peninsula towards Mecúfi. This latter area is also of importance for the rare endemic tree *Acacia latispina* (VU) which grows in open woodland on both dark clays and on gravelly and pebbly soils immediately behind coastal dunes. The habitat for the population between Pemba and Mecúfi is severely degraded due to wood-harvesting and overgrazing by cattle, although this species is able to withstand moderate habitat disturbance (Burrows *et al.* 2014b). Only the third known Mozambican site for the globally Endangered tree *Hildegardia migeodii* (EN) was discovered in 2012 on the Pemba peninsula. This IPA also holds some small populations of the Mozambican endemic tree genus *Micklethwaitia carvalhoi* (VU); these have been impacted by firewood cutting, but as this species does coppice well, it can withstand heavy harvesting pressure (Burrows *et al.* 2014a). Pemba Bay holds an important population of the mangrove parasite *Viscum littorum* (NT) which grows here on both *Sonneratia alba* and *Ceriops tagal*; this species is endemic to northern Mozambique. In total, 11 globally threatened species have been recorded from the Pemba IPA, although the continued viability of some of these populations requires confirmation.

The Pemba Peninsula is one of only two known localities historically for the striking herb *Justicia niassensis* (EN), which was recorded near the lighthouse at Maringanha Point. However, the record at this site is from 1960, hence its continued presence on the peninsula requires confirmation given that much development has taken place in the meantime; the area in which this historic collection was made is not included within the IPA boundary but it is hoped that this striking species can be found elsewhere within the IPA in future.

Habitat and geology

The area supports a mosaic of habitats with much farmland and settlement. Away from the coast on the Pemba Peninsula and continuing southwards beyond Murrébue, the land rises rapidly to a low flat ridge up to 150 m a.s.l., comprising iron-rich sandstones of the Mikindani Formation of mid-Neogene origin (ca. 10–15 mya). This rock gives rise to a coarsely sandy well-drained red soil. Elsewhere in Cabo Delgado, these Mikindani sandstones hold important areas of dry forest (Timberlake *et al.* 2010) but this whole area now is highly transformed and with no areas of forest remaining. Wild & Barbosa (1968) indicate on their vegetation map that this may have once supported *Guibourtia schliebenii* thicket (their mapping unit 14) but none of this remains. Elsewhere, the IPA is dominated by more recent Quaternary deposits including littoral dunes and recent alluvial deposits. Areas of heavy clay soils are found both around Pemba Bay and on the coastal lowlands between Murrébue and Mecúfi, and these support an open *Acacia*-dominated woodland. There are also areas of raised coral rag which support a thicket vegetation. The south side of Pemba Bay supports extensive mangrove communities which are included in the IPA.

The coastline here has a highly seasonal humid tropical climate, with the wet season from

Fruits of *Hildegardia migeodii* (TR)

Eriolaena rulkensii (TR)

December to April, usually peaking in March, and with a prolonged dry season from May to November. Annual rainfall is ca. 870 mm at Pemba.

Conservation issues

This IPA is currently unprotected. Threats are considerable and varied, and large areas have already been heavily degraded or transformed. The Pemba Peninsula is impacted by continuing expansion of Pemba city and port. The population of this city is now over 200,000 and has more than doubled in 20 years. Further increases in population, at least in the short term, have resulted from displacement of populations from the north due to the recent violent insurgency. Away from urban areas, the major impact is from agricultural activity with extensive areas given over to growing crops and, particularly in areas of clay soils, high grazing pressure. Woody vegetation is also severely impacted by harvesting for charcoal and construction. The vegetation of the coastal fringe is being impacted by beach tourism; this is particularly prevalent at present on the Pemba Peninsula but it is also a threat to the less built-up coastline south of Murrubue towards Mecúfi (Darbyshire *et al.* 2019c). Regular flights now arrive into Pemba from Maputo and from South Africa, catering to wealthy tourists. Security issues associated with the insurgency to the north are impacting tourism in the short-term but this is likely to be only a temporary hiatus.

There is an urgent need to delimit and protect the remaining areas of natural vegetation and the surviving populations of the conservation-priority species within this IPA. One possible channel of support may be through Lúrio University which has a campus in Pemba with an active interest in biodiversity and conservation.

The Pemba IPA would qualify as an Alliance for Zero Extinction (AZE) site on the basis of *Eriolaena rulkensii*. It is not currently included within Mozambique's Key Biodiversity Areas (KBA) network.

Key ecosystem services

The remaining wild habitats within the Pemba IPA provide a range of important ecosystem services. The extensive mangroves in Pemba Bay are particularly important for the prevention of coastal erosion, provisioning of materials for building and fuelwood, and provision of habitat for productive fisheries. Woodlands and bushlands provide a range of provisioning services, including the supply of building materials and wild fruits. Wild habitats neighbouring agricultural areas also provide important habitat for pollinators of crops and for beekeeping. Finally, the natural habitats such as sand dunes and mangroves contribute to the touristic appeal of this area.

Ecosystem service categories

- Provisioning – Food
- Provisioning – Raw materials
- Regulating services – Moderation of extreme events
- Regulating services – Erosion prevention and maintenance of soil fertility
- Regulating services – Pollination
- Habitat or supporting services – Habitats for species
- Habitat or supporting services – Maintenance of genetic diversity
- Cultural services – Recreation and mental and physical health
- Cultural services – Tourism

IPA assessment rationale

The remnant pockets of natural coastal habitats within the Pemba Bay region and south to Mecúfi qualify as an IPA under criterion A(i). This site contains internationally important populations of 11 globally threatened plant species, five of which are assessed as Endangered and six as Vulnerable. Pemba IPA contains the total known global range of *Eriolaena rulkensii* (EN). As noted above, it may also contain *Justicia niassensis* (EN). The area is designated as an IPA in the hope that the small patches of intact coastal vegetation can be conserved and that degraded areas away from settlements can be restored such that its botanical importance can be protected and enhanced. Given the extent of fragmentation, the remaining areas of Rovuma Coastal Dry Forest are not considered to trigger criterion C(iii) at this site.

Mangroves at Pemba Bay (TR)

Priority species (IPA Criteria A and B)

FAMILY	TAXON	IPA CRITERION A	IPA CRITERION B	≥ 1% OF GLOBAL POP'N	≥ 5% OF NATIONAL POP'N	IS 1 OF 5 BEST SITES NATIONALLY	ENTIRE GLOBAL POP'N	SPECIES OF SOCIO-ECONOMIC IMPORTANCE	ABUNDANCE AT SITE
Acanthaceae	Justicia niassensis	A(i)	B(ii)	?	?	?			unknown
Combretaceae	Combretum caudatisepalum	A(i)	B(ii)	✓	✓	✓			unknown
Fabaceae	Acacia latispina	A(i)	B(ii)	✓	✓	✓			occasional
Fabaceae	Micklethwaitia carvalhoi	A(i)	B(ii)	✓	✓	✓		✓	unknown
Lamiaceae	Vitex carvalhi	A(i)			✓	✓			unknown
Lamiaceae	Vitex mossambicensis	A(i)			✓	✓			unknown
Loranthaceae	Oncella curviramea	A(i)			✓	✓			unknown
Malvaceae	Eriolaena rulkensii	A(i)	B(ii)	✓	✓	✓	✓		occasional
Malvaceae	Hildegardia migeodii	A(i)		✓	✓	✓			unknown
Malvaceae	Thespesia mossambicensis		B(ii)						common
Rubiaceae	Afrocanthium vollesenii	A(i)			✓	✓	✓		unknown
Rubiaceae	Pavetta mocambicensis	A(i)	B(ii)	✓	✓	✓			unknown
Rubiaceae	Tarenna pembensis	A(i)	B(ii)	✓	✓	✓			scarce
Santalaceae	Viscum littorum		B(ii)						unknown
		A(i): 11 ✓	B(ii): 8						

Threatened habitats (IPA Criterion C)

HABITAT TYPE	IPA CRITERION C	≥ 5% OF NATIONAL RESOURCE	≥ 10% OF NATIONAL RESOURCE	IS 1 OF 5 BEST SITES NATIONALLY	ESTIMATED AREA AT SITE (IF KNOWN)
Rovuma Coastal Dry Forest [MOZ-12a]	C(iii)				

Protected areas and other conservation designations

CONSERVATION AREA TYPE	CONSERVATION AREA NAME	RELATIONSHIP OF IPA TO CONSERVATION AREA
No formal protection	N/A	

Threats

THREAT	SEVERITY	TIMING
Housing & urban areas	high	Ongoing – increasing
Commercial & industrial areas	medium	Ongoing – trend unknown
Tourism & recreation areas	medium	Ongoing – increasing
Small-holder farming	high	Ongoing – increasing

Coral rag thicket at Pemba Bay (TR)

Tarenna pembensis (JEB)

Pemba Bay with *Eriolaena rulkensii* in the foreground (TR)

LÚRIO WATERFALLS, CHIÚRE

Assessors: Iain Darbyshire, Jo Osborne

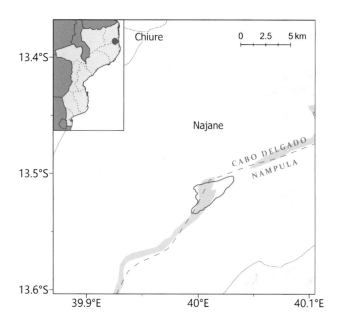

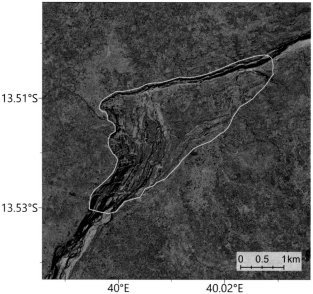

INTERNATIONAL SITE NAME		Lúrio Waterfalls, Chiúre	
LOCAL SITE NAME (IF DIFFERENT)		Quedas do Rio Lúrio	
SITE CODE	MOZTIPA013	PROVINCE	Cabo Delgado

LATITUDE	-13.51310	LONGITUDE	39.99940
ALTITUDE MINIMUM (m a.s.l.)	91	ALTITUDE MAXIMUM (m a.s.l.)	172
AREA (km²)	6.7	IPA CRITERIA MET	A(i)

Site description

The Quedas do Rio Lúrio, or Lúrio Waterfalls, are a dramatic series of waterfalls and rapids at the head of a gorge on the Lúrio River, approximately 70 km upstream from the mouth of the river. The Lúrio is one of the major rivers of northern Mozambique flowing for over 500 km west to east before reaching the Indian Ocean south of Mecufi. It forms the border between Nampula Province to the south and Cabo Delgado and Niassa Provinces to the north, although the lattermost province is many kilometers upstream of this IPA. The falls can be approached from the Cabo Delgado (Chiúre District) side, via an unpaved road from near Najane. The river drops from an elevation of approximately 160 m above the falls to 120 m in the gorge, with a series of large, exposed rock outcrops between the falls that support a well-developed succulent

vegetation, including the only known population globally of *Aloe argentifolia*.

Botanical significance

The Lúrio Waterfalls are of global importance as the only known locality for the striking rosette-forming shrub aloe, *Aloe argentifolia*. A large population of this species, estimated at between 300 and 500 mature individuals, is known from the head of the river gorge (Martínez Richart *et al.* 2019). Despite extensive surveys of aloe diversity in Mozambique in recent years, no other populations of *A. argentifolia* have been located and it appears that it is a very narrow endemic, confined to this unique locality created by the waterfall system (McCoy *et al.* 2017).

This site is otherwise not well known botanically and may well contain other species of interest. The

Aloe argentifolia (OB)

Habitat for *Aloe argentifolia* (OB)

falls are difficult to access during the main rainy season because of impassable roads and the high water level of the Lúrio, hence to our knowledge, it has never been botanised at that time. It would be desirable to conduct a general botanical inventory, focusing on the succulent flora on the exposed rock outcrops and investigating the possibility of a rheophytic flora associated with the waterfalls.

Habitat and geology

The principle habitat of interest is the rocky outcrops between the waterfalls and rapids. During the rainy season these rocks receive substantial mist-spray from the adjacent falls but during the dry season the water levels are low and the rocks rapidly dry out, and hence support a drought-tolerant flora including succulent species. McCoy *et al.* (2017) noted that *Aloe argentifolia* grows in association with the terrestrial orchid *Eulophia petersii*, and with species of *Commiphora*, *Cynanchum*, *Kalanchoe*, *Sansevieria* (= *Dracaena*) and a large caespitose *Xerophtya* species. Botanical collections made here in 1948 by E.C. Andrada recorded the localised *Millettia bussei* (LC; *Andrada* #1280) among the woody flora, with *Adansonia digitata* and *Sterculia* and *Acacia* spp. also noted in the dry woodland. The waterfalls themselves may provide habitat for rheophytic plant species; this requires further investigation.

Average rainfall is estimated at 1,487 mm per year at the nearby town of Namapa, with a marked peak in December to March and a prolonged dry season in May to October (worldweatheronline. com), resulting in the marked seasonal changes in water level on the Lúrio noted above.

Conservation issues

The Lúrio Waterfalls are not currently protected. However, threats appear to be minimal at present. The site is quite isolated and inaccessible for parts of the year, offering some protection for the natural vegetation. Further, the rock outcrops with thin soils are not impacted by agricultural activity or fire. The site may in future become an ecotourist destination due to its natural beauty, but as it is so isolated, it is unlikely to ever receive significant tourist pressure (T. Rulkens pers. obs.).

The future integrity of the site is less certain. There are plans to construct a major hydroelectric plant on the Lúrio River to supply electricity to Cabo Delgado and Nampula Provinces (Macauhub 2014; McCoy *et al.* 2017). Whilst the proposed locality for this plant is understood to be a considerable distance upriver from the waterfalls, the hydroelectric dam could significantly alter downstream flow and this could impact the ecology of the waterfalls site, particularly through reducing amounts of mist-spray onto the rock outcrops from the falls during the wet season. It is unclear as to how reliant the aloe population is on the river as a moisture source.

A more pressing concern at this site is the apparent lack of recruitment of new plants of *Aloe argentifolia*, as all the plants observed in 2013 were mature individuals (McCoy *et al.* 2017). There is no obvious cause for this lack of seedling recruitment, and in view of the natural protection at the site from the waterfalls and rapids, grazing by animals would not seem to be a contributing factor. A key initial conservation action would therefore be to conduct a more thorough population survey at the site and to establish the causes of the recruitment issues.

Key ecosystem services

As noted above, the waterfalls are a natural beauty spot and so have some potential as an ecotourism destination, but at present this is limited due to their isolation from the main tourist destinations in northern Mozambique. The Lúrio River is a major source of fresh water for people living within its catchment.

Ecosystem service categories

- Provisioning – Fresh Water
- Habitat or supporting services – Habitats for species
- Cultural services – Tourism

IPA assessment rationale

The Lúrio Waterfalls qualify as an IPA under criterion A(i) as this site contains the entire known population of *Aloe argentifolia*, which is assessed as globally Vulnerable (VU D1) due to its extremely small population size and distribution (Martínez Richart *et al.* 2019).

Priority species (IPA Criteria A and B)

FAMILY	TAXON	IPA CRITERION A	IPA CRITERION B	≥ 1% OF GLOBAL POP'N	≥ 5% OF NATIONAL POP'N	IS 1 OF 5 BEST SITES NATIONALLY	ENTIRE GLOBAL POP'N	SPECIES OF SOCIO-ECONOMIC IMPORTANCE	ABUNDANCE AT SITE
Asphodelaceae	*Aloe argentifolia*	A(i)	B(ii)	✓	✓	✓	✓		frequent
		A(i): 1 ✓	B(ii): 1						

Protected areas and other conservation designations

CONSERVATION AREA TYPE	CONSERVATION AREA NAME	RELATIONSHIP OF IPA TO CONSERVATION AREA
No formal protection	N/A	

Threats

THREAT	SEVERITY	TIMING
Energy production & mining – Renewable energy	unknown	Future – inferred threat

A section of the Lúrio Waterfalls, Chiúre (OB)

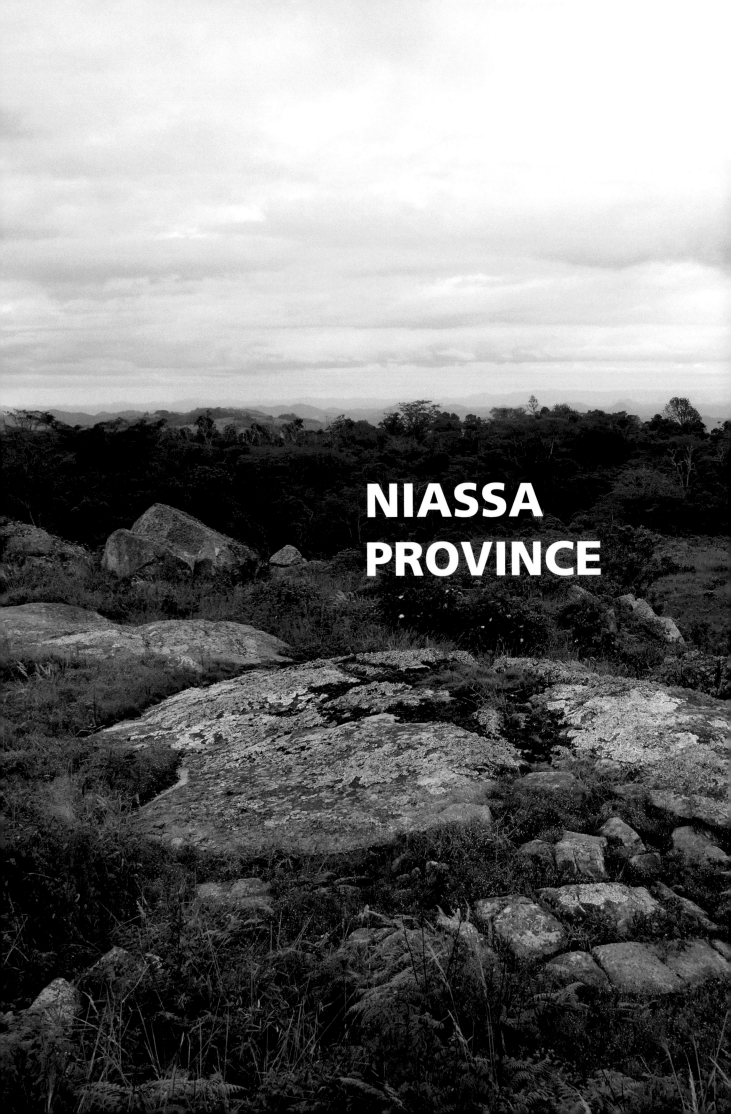

NIASSA
PROVINCE

TXITONGA MOUNTAINS

Assessor(s): Jo Osborne, Sophie Richards, Iain Darbyshire

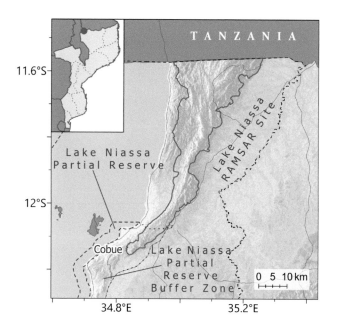

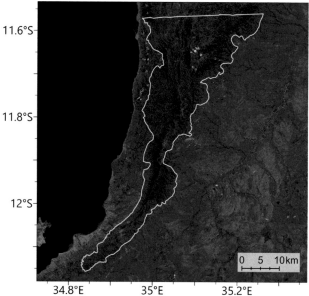

INTERNATIONAL SITE NAME		Txitonga Mountains	
LOCAL SITE NAME (IF DIFFERENT)		Montanhas de Txitonga	
SITE CODE	MOZTIPA020	PROVINCE	Niassa

LATITUDE	-11.78280	LONGITUDE	35.06750
ALTITUDE MINIMUM (m a.s.l.)	500	ALTITUDE MAXIMUM (m a.s.l.)	1,848
AREA (km²)	741	IPA CRITERIA MET	A(iii), C(iii)

Site description

The Txitonga Mountains lie in the northwest of Lago district in Niassa Province. The mountains extend south from the southern end of the Kipengere Range in Tanzania and form part of the eastern escarpment of the East African rift. These mountains are isolated from the other highland areas in Niassa province — the broad Lichinga plateau area and outlying Mount Mecula and Mount Yao. They are therefore likely to support a distinct biodiversity for Mozambique, closely allied to that in southwest Tanzania. Mount Txitonga (or Chitonga) is the highest peak in the range at ca. 1,848 m elevation. The IPA extends south from the Tanzanian border and includes a core zone of montane habitat over 1,200 m elevation situated within a larger area of foothills extending towards the Messinge river plain to the east and the shore of Lake Niassa to the west. The IPA is within the Lake Niassa Reserve Ramsar site, that includes both Lake Niassa and the adjacent terrestrial zone, while the southern tip of this IPA also falls within the buffer zone of Lake Niassa Partial Reserve (Ramsar 2011).

Botanical significance

A significant area of montane habitat can be found in the Txitonga Mountains including montane grassland, a restricted and threatened habitat of Mozambique. There is also montane scrubland present within this IPA, which may also be of national conservation importance; however, this habitat type is difficult to define spatially, and so further research would be required.

The site is not well-studied botanically and several possible plant species new to science were recorded during an expedition in 2019 (Osborne *et al.* 2019b). These include a species of *Streptocarpus* allied to *S. michelmorei* but apparently unmatched, a potentially new species of *Bothriocline*, and a small rosette herb, *Hartliella txitongensis*. The latter genus is, interestingly, an indicator of metal-rich soils and the genus as a whole was previously known only from the Katanga region of D.R. Congo, and northern Zambia. *Hartliella txitongensis* is highly localized, known only from a single location where it was locally common. As *Hartliella txitongensis* occupies a range of less than 100 km² and is yet to be assessed for the IUCN Red List, it triggers A(iii) of the IPA sub-criteria. This species may well be assessed as Critically Endangered in future as habitat is being lost through mining.

A number of other interesting taxa were collected during the 2019 expedition, including several records new to Mozambique – *Barleria holstii* (LC), *Erica woodii*, *Leptoderris brachyptera* (LC), *Polygala gossweileri* and *Protea micans* subsp. *trichophylla* – and two records of species not previous collected in northern Mozambique – *Plectranthus kapatensis* and *Vernonia holstii*. With further investigation, many more notable plant species are likely to be collected within this IPA.

Habitat and geology

The Txitonga Mountains are dominated by metasedimentary rocks, predominantly meta-greywacke, meta-sandstone and schists, the Txitonga geological group also hosts the Niassa Gold Belt the presence of which has motivated goldmining within the IPA (Bingen *et al.* 2007). Soils have not been fully categorised; however, the presence of *Hartliella* potentially indicates metal-rich soils. The site experiences average annual precipitation of around 1,312 mm on the slopes and 1,330 mm on the higher ridges, with the vast majority falling during the wet season between November and April. Temperatures average at 21°C on the escarpments and 17.8°C on the upper ridges (Lötter *et al.*, in prep.).

The vegetation on the foothills and slopes of the mountains is predominantly woodland with narrow strips of moist gallery forest growing along deep stream gullies. At higher altitudes a more open montane savannah, montane scrubland and montane grassland occur. The site is underexplored botanically and the habitat information below was gathered during an expedition into the mountains in 2019 (Osborne *et al.* 2019b).

Woodland on the Txitonga Mountains (JO)

Riverine forest in the Txitonga Mountains (JO)

Hartliella txitongensis (JO)

The woodland that covers the mountain slopes and foothills of the Txitonga Mountains is extensive, supporting a high diversity of trees, shrubs, herbs and grasses. There is also much local variation in the woodland vegetation across the landscape. A large part of the woodland is miombo, dominated by *Brachystegia spiciformis* and *B. boehmii*. *Uapaca kirkiana* is also among the dominant species, in places forming dense, monospecific stands. *Uapaca nitida*, *Faurea rochetiana*, *Parinari curatellifolia*, *Diplorhynchus condylocarpon*, *Monotes engleri* and *Pericopsis angolensis* are frequent, while locally common shrubs include *Droogmansia pteropus* and *Cryptosepalum maraviense*.

Deep stream gullies are frequent and conspicuous on satellite imagery (Google Earth 2021). The stream gullies support humid gallery forest and a very different flora from the wooded hillsides. *Brachystegia tamarindoides* subsp. *microphylla* is dominant on the upper level of the gallery forest, growing with *Burkea africana* and reaching a canopy height to ca. 25 m. At the bottom of the stream gullies, *Breonadia salicina* and stilt-rooted *Uapaca lissopyrena* are common, growing with *Treculia africana*, *Erythrophleum suaveolens* and *Bridelia micrantha*. The understorey includes *Erythroxylum emarginatum* and abundant lianas.

Shaded rocks along the streams support moisture-loving species including ferns and *Streptocarpus* sp.

At higher altitudes the woodland becomes more open in places and the canopy lower, grading into montane savannah grassland and scrubland with rocky outcrops. Woody species are sparse in the savanna and include *Uapaca kirkiana*, *Faurea rochetiana*, *Protea* spp., *Erica* sp., *Psorospermum febrifugum* and *Myrica pilulifera*, mostly to less than 3 m tall. Along the mountain ridge, the low shrubs *Kotschya strigosa* and *Cryptosepalum maraviense* are common. The montane grassland is mostly short, to ca. 50 cm in height and rich in dwarf shrubs, herbs and geophytes.

Conservation issues

The Txitonga Mountains are encompassed by Lake Niassa RAMSAR site, while the southern end of the mountain range also falls within the buffer zone of Lake Niassa Partial Reserve. Although there has been some monitoring of deforestation in the area by WWF (WWF 2011), much of the focus of these conservation areas is on the freshwater rather than montane ecosystems.

Environmental disturbance from gold mining is extensive and ongoing, leaving areas of broken

Montane savanna grassland in the Txitonga Mountains (JO)

rock, deeply eroded fissures, silted rivers and mercury pollution. The mining is legal, licensed through the community mining association. Water is channelled to mining sites over long distances to facilitate the mining process and the water channels are maintained over many years. The risk of mercury pollution to the local people is recognised in Tulo Calanda, to the east of the mountains, where water is piped to the village from a clean site unaffected by the mining. Despite being within the Lake Niassa Reserve, a partial reserve, there appears to be no control over the use of mercury (Osborne *et al.* 2019b).

Across most of Mozambique the expansion of subsistence 'machamba' agriculture is the main cause of habitat loss. In the Txitonga Mountains this is not the case and there is very little cultivation. Here gold mining is more profitable than agriculture and is driving the local economy. The Txitonga Mountains support large areas of valuable natural vegetation. However, gold mining is taking place at multiple sites in the mountains causing habitat loss locally, wider disturbance to the hydrology and environmental pollution (Osborne *et al.* 2019b). The ongoing expansion of gold mining sites poses both a current and future threat to the vegetation.

In addition to the direct impacts of gold mining, the presence of gold miners in the mountains has led to an increased frequency of uncontrolled wildfires that poses a serious threat to the vegetation. Fires are set both intentionally for hunting and unintentionally from cooking fires. Monitoring uncontrolled burning makes up a large proportion of the work of the local government Environmental Officers (Fiscais) and more resources are needed to control this activity (Osborne *et al.* 2019b).

Key ecosystem services

The Txitonga Mountains have a high plant diversity value, in part due to their unique biogeography within Mozambique. The vegetation at this site contributes to carbon sequestration and storage, prevents soil erosion on the slopes and provides habitat for flora and fauna. The mountains also provide a watershed for the local area though this is currently severely threatened by gold mining activity.

Ecosystem service categories

- Provisioning – Fresh Water
- Regulating Services – Carbon sequestration and storage
- Regulating Services – Erosion prevention and maintenance of soil fertility
- Habitat or supporting services – Habitats for species

IPA assessment rationale

The Txitonga Mountains qualify as an Important Plant Area under sub-criterion A(iii) due to the presence of *Hartliella txitongensis*. A highly restricted endemic, the entire global range of *H. txitongensis* falls within this IPA. While there are also areas of montane grassland, a nationally restricted habitat type, within this IPA, this site does not represent one of the five best sites for this habitat type nationally, and so does not qualify under C(iii). This habitat is still of conservation importance at this site, however, particularly with *H. txitongensis* occurring in the ecotone between this habitat and the woodland below.

Miombo woodland in the Txitonga foothills (JO)

Priority species (IPA Criteria A and B)

FAMILY	TAXON	IPA CRITERION A	IPA CRITERION B	≥ 1% OF GLOBAL POP'N	≥ 5% OF NATIONAL POP'N	IS 1 OF 5 BEST SITES NATIONALLY	ENTIRE GLOBAL POP'N	SPECIES OF SOCIO-ECONOMIC IMPORTANCE	ABUNDANCE AT SITE
Linderniaceae	*Hartliella txitongensis*	A(iii)	B(ii)	✓	✓	✓	✓		occasional

Threatened habitats (IPA Criterion C)

HABITAT TYPE	IPA CRITERION C	≥ 5% OF NATIONAL RESOURCE	≥ 10% OF NATIONAL RESOURCE	IS 1 OF 5 BEST SITES NATIONALLY	ESTIMATED AREA AT SITE (IF KNOWN)
Montane Grassland [MOZ-09]	C(iii)				

Protected areas and other conservation designations

CONSERVATION AREA TYPE	CONSERVATION AREA NAME	RELATIONSHIP OF IPA TO CONSERVATION AREA
Partial Reserve	Lake Niassa	protected/conservation area overlaps with IPA
RAMSAR site	Lake Niassa	protected/conservation area encompasses IPA

Threats

THREAT	SEVERITY	TIMING
Mining & quarrying	high	Ongoing – trend unknown
Increase in fire frequency & intensity	high	Ongoing – trend unknown
Seepage from mining	unknown	Ongoing – trend unknown

NJESI PLATEAU

Assessor(s): Jo Osborne, Sophie Richards, Iain Darbyshire

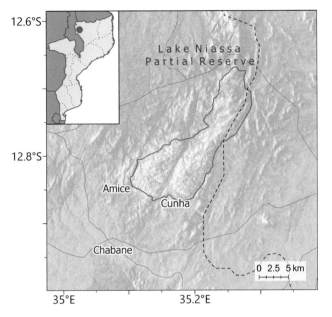

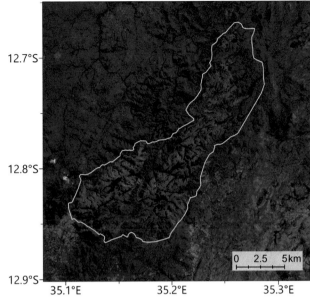

INTERNATIONAL SITE NAME		Njesi Plateau	
LOCAL SITE NAME (IF DIFFERENT)		Planalto de Ndjesi	
SITE CODE	MOZTIPA019	PROVINCE	Niassa

LATITUDE	-12.82840	LONGITUDE	35.18430
ALTITUDE MINIMUM (m a.s.l.)	1,160	ALTITUDE MAXIMUM (m a.s.l.)	1,848
AREA (km²)	165	IPA CRITERIA MET	A(i), C(iii)

Wooded grassland on the slopes of Njesi (JO)

Gallery forests featuring *Albizia gummifera* (JO)

Barleria torrei at the base of Njesi (JO)

Site description

The Njesi Plateau is a montane site in the west of Niassa Province, in Sanga District, ca. 50 km north of Lichinga and 15 km south-east of Maniamba. It lies within the broader Lichinga Plateau, an area of highlands over 1,000 m elevation, of which Njesi is the most significant montane site, covering an area of over 160 km² with much of the Njesi Plateau over 1,700 m. It reaches 1,848 m elevation on Serra Jeci (also known as Jec, Jesi or Gesi). These mountains in the far north of Mozambique are considered biogeographically distinct from the group of mountains in north-central Mozambique that includes Mounts Namuli, Inago and Mabu (Bayliss *et al.* 2014, Jones *et al.* 2020). There is no cultivation or communities living on the Njesi plateau although it does overlook a number of small villages including Amice at the south-western border and Cunha to the south-west.

Botanical significance

Significant areas of montane habitat can be found on the Njesi Plateau including grassland, scrub, rock outcrop, gallery forest and small patches of moist montane forest. Montane grassland and mid-altitude montane moist forest habitats, both restricted and threatened in Mozambique, are present at this site, with this site representing one of the five best sites nationally for both of these habitat types. The montane scrubland at this site is also likely to be of conservation importance but is difficult to define spatially and as such cannot be assessed under the IPA criteria at present.

There has only been limited botanical collection in this IPA, with most of the previous research focusing on faunal taxa. However, a botanical expedition was undertaken in 2019 to address this information gap (Osborne *et al.* 2019b). One particularly important species that occurs in this IPA is the threatened endemic shrub *Barleria torrei,* growing in woodland and on rocky slopes towards the base of the plateau. This species is assessed as Endangered on the IUCN Red List (Osborne & Rokni 2020). In addition to this threatened species, one new record for northern Mozambique was collected at this site. *Vernonia natalensis* was previously thought to be restricted nationally to the central and southern provinces of Mozambique but was observed to occur in the rockfaces of the montane grassland within this IPA. With further investigation, other notable plant species are likely to be recorded from this site.

Habitat and geology

The Njesi Plateau experiences a wet season between November to April. Although annual average precipitation has not been calculated for this site, the figure is approximated to be between 1,300 and 1,700 mm, while the average temperature is around 18°C, reaching an average annual maximum of around 20–26°C between October and November (Lötter *et al.*, in prep.).

Serra Jeci is the highest peak of this IPA, reaching 1,848 m. Geologically this mountain is categorised as a weakly metamorphosed lens (a body of rock that is broad in the middle with tapering edges) of mostly carbonate rocks positioned within the surrounding granulite Unango Complex; it is thought to have a depositional age of around 600 million years (Melezhik *et al.* 2006). The plateau supports a large area of montane savanna grassland with scattered rock outcrops, raised montane forest/thicket patches and gallery forest along stream gullies. Miombo woodland dominates the lower slopes below the plateau, adjoining areas of settlement and cultivation.

The montane savanna grassland on Njesi is predominantly a tall grass savanna, 1.5 to 2 m in height, growing at over 1,700 m elevation. It is rich in tall herb species, particularly in the families Fabaceae, Asteraceae and Lamiaceae, with several dominant grasses including *Hyparrhenia cymbaria* (Osborne *et al.* 2019b). A small herb layer grows in the sheltered habitat beneath the tall herbs and grasses, with species such as *Hypericum peplidifolium* occurring frequently. Woody species are scattered throughout the savanna including tree species *Acacia amythethophylla, Cussonia arborea, Dombeya rotundifolia* and species of *Protea*. The key ecological factors responsible for this tall grass savanna vegetation are thought to be a combination of fertile soil, regular precipitation and frequent fire (Osborne *et al.* 2019b).

Rock outcrops are scattered throughout the savanna providing habitat diversity. Species that are associated with the rock outcrops include the shrubs *Steganotaenia araliacea* and *Tecomaria nyassae*, the herb *Aeollanthus serpiculoides* and many geophytes.

Numerous moist evergreen forest patches occur on the plateau; each patch is limited in size, but together, this habitat covers around 7 km². Trees in the moist forest include *Albizia schimperiana, Bridelia macrantha, Ficus* sp. and *Zanthoxylum* sp. with *Chassalia parvifolia, Cassipourea malosana* and *Tiliacora funifera* in the understorey (Osborne *et al.* 2019b). The shaded forest floor supports herbs including Acanthaceae and orchid species. The canopy within the forest reaches ca. 35 m tall but is irregular, with frequent sheltered forest gaps dominated by dense and impenetrable growth of herbs and lianas. The irregular canopy and frequent forest gaps may be the result of previous fires spreading into the forest from the savanna in the dry season. Forest margin vegetation is frequent and includes abundant *Dracaena steudneri, Maesa rufescens, Senna petersiana* and *Sparrmannia ricinocarpa* alongside diverse tall herbs and abundant lianas (Osborne *et al.* 2019b).

Gallery forest grows along the stream gullies that drain from the plateau. *Albizia gummifera*, to ca. 25 m tall, is the dominant tree species, growing with *Breonadia salicina, Zanthoxylum* sp., *Rauvolfia caffa* and *Schrebera alata*. Where the canopy is more open *Dracaena steudneri, Solenecio mannii* and tree ferns (*Cyathea* sp.) occur and there are dense growths of herbs, bracken fern (*Pteridium aquilinum*) and lianas including stands of tall, shrubby *Acanthopale pubescens* (Osborne *et al.* 2019b). The gallery forest margin is extensive and species diverse, similar to the moist forest margin.

On the slopes towards the base of the plateau, to the south and east of the site, miombo woodland habitat adjoins areas of settlement and cultivation. There may be some habitat loss here due to

Agricultural land around the footslopes (JO)

Rocky outcrop of the plateau, overlooking moist forest patch (JO)

expanding cultivation. The woodland is also likely to be impacted by fires set by hunters, increasing the natural fire frequency in this habitat (Jones *et al.* 2020).

Conservation issues

Although there is no settlement on Njesi Plateau, people do walk up onto the plateau to hunt wildlife including wild pigs, deer, porcupines, giant rats, rabbits and birds. Hunters include both locals and people from further afield (Osborne *et al.* 2019b). Hunting is seasonal, mostly taking place when the plateau is accessible after fires have cleared the tall grass savanna. Fire is a major factor affecting the vegetation on the Njesi Plateau and it is likely that fire frequency is increased as a result of hunters setting fires to flush out wildlife on the plateau. The tall grass savanna vegetation is well adapted to natural wildfires, though there may be species present that have a limited tolerance and that would be affected by increased fire frequency (Osborne *et al.* 2019b). The moist forest habitat is not fire-adapted and increased fire frequency is likely to affect the forest edge habitat as well as potentially reducing the forest extent.

The Njesi Plateau is protected by local Government and patrolled by Environmental Officers (Ficais), reducing the level of hunting on the plateau. However, more resources are needed to prevent hunting and the setting of wildfires. The value of this IPA is emphasized as it lies mostly within the Lake Niassa Partial Reserve, also a Ramsar site, including both Lake Niassa and the adjacent terrestrial zone (Ramsar 2011). This IPA also falls within the larger Njesi Plateau Important Bird Area (IBA), triggered by species such as the Endangered Mozambique Forest-warbler (*Artisornis sousae*) and the Vulnerable Thyolo Alethe (*Chamaetylas choloensis*) (BirdLife International 2019). For the former species, this IBA contains the entire global population and, as the Mozambique Forest-warbler

is an Endangered species, it also meets Alliance of Zero Extinction site criteria. The wider Njesi Plateau area has also been recognised as a Key Biodiversity Area, triggered by the Mozambique Forest-warbler and the Mecula girdled lizard (*Cordylus meculae*). Despite being assessed as Least Concern, the latter species is limited mostly to the Njesi Plateau (Tolley *et al.* 2019a). With the presence of a number of rare and threatened species, it is clear that the habitats of this IPA are of central importance for conserving a number of faunal taxa.

Key ecosystem services

Njesi Plateau has a high plant diversity value providing an island of montane savanna grassland, montane forest and gallery forest habitats for flora and fauna. The plateau provides a watershed for the local area and the vegetation contributes to carbon sequestration and storage. In addition, local people hunt in the area.

Ecosystem service categories

- Provisioning – Food
- Provisioning – Fresh water
- Regulating services – Carbon sequestration and storage
- Habitat or supporting services – Habitats for species

IPA assessment rationale

Njesi Plateau qualifies as an Important Plant Area under criteria A and C. Under criterion A(i) the site supports a population of the globally threatened shrub *Barleria torrei* (EN). The site also qualifies under criterion C(iii), representing one of the five best sites nationally for both montane grassland and montane forest habitats.

Priority species (IPA Criteria A and B)

FAMILY	TAXON	IPA CRITERION A	IPA CRITERION B	≥ 1% OF GLOBAL POP'N	≥ 5% OF NATIONAL POP'N	IS 1 OF 5 BEST SITES NATIONALLY	ENTIRE GLOBAL POP'N	SPECIES OF SOCIO-ECONOMIC IMPORTANCE	ABUNDANCE AT SITE
Acanthaceae	*Barleria torrei*	A(i)	B(ii)	✓	✓	✓			occasional
		A(i): 1 ✓	B(ii): 1						

Threatened habitats (IPA Criterion C)

HABITAT TYPE	IPA CRITERION C	≥ 5% OF NATIONAL RESOURCE	≥ 10% OF NATIONAL RESOURCE	IS 1 OF 5 BEST SITES NATIONALLY	ESTIMATED AREA AT SITE (IF KNOWN)
Montane Grassland [MOZ-09]	C(iii)			✓	
Montane Moist Forest [MOZ-01]	C(iii)			✓	

Protected areas and other conservation designations

CONSERVATION AREA TYPE	CONSERVATION AREA NAME	RELATIONSHIP OF IPA TO CONSERVATION AREA
Important Bird Area	Njesi Plateau	protected/conservation area encompasses IPA
Key Biodiversity Area	Njesi Plateau	protected/conservation area encompasses IPA
Partial Reserve	Lake Niassa Reserve	protected/conservation area encompasses IPA

Threats

THREAT	SEVERITY	TIMING
Small-holder farming	low	Ongoing – trend unknown
Fire	unknown	Ongoing – trend unknown

MOUNT YAO

Assessors: Sophie Richards, Iain Darbyshire

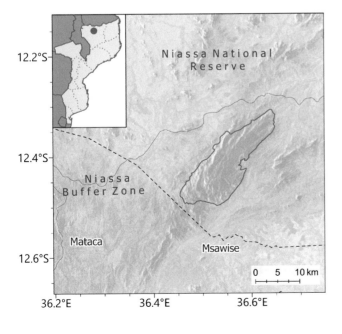

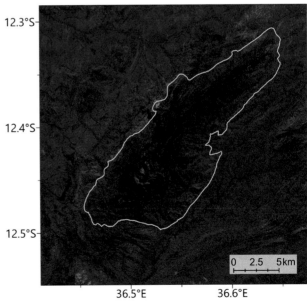

Mount Yao (CC)

View from Mount Yao, overlooking dense miombo woodland (CC)

INTERNATIONAL SITE NAME		Mount Yao	
LOCAL SITE NAME (IF DIFFERENT)		Monte Yao	
SITE CODE	MOZTIPA047	PROVINCE	Niassa

LATITUDE	-12.46006	LONGITUDE	36.50202
ALTITUDE MINIMUM (m a.s.l.)	620	ALTITUDE MAXIMUM (m a.s.l.)	1,313
AREA (km²)	183	IPA CRITERIA MET	A(i)

Site description

Mount Yao (or Jao) IPA is located in Mavago District of Niassa Province. Located ca. 180 km east of Lake Malawi, this inselberg falls within Niassa Special Reserve and peaks at 1,313 m. The area has not been extensively studied; however, available evidence indicates it is botanically interesting as the high altitudes provide a cooler and wetter climate compared to the wooded plains that dominate the surrounding area. The mountain is largely dominated by dense miombo woodlands, with forest and swamp habitat associated with river gullies (Congdon & Bayliss 2013).

The site has been delineated to surround the main montane habitats, covering an area of 183 km². The south-east and eastern boundaries run parallel with the Mataca-Mecula Road, with the south-eastern portion of this IPA falling 2 km from the boundary between Niassa Reserve and the Buffer Zone. The town of Mataca lies 20 km to the south-west while Maswise village is just 9 km south of the IPA boundary. The IPA itself is not populated by people and is largely undisturbed. Alongside the high-quality habitats, there are known to be a number of Mozambican endemic species from various taxa on this mountain, some known from this site alone. Thus far, only a limited number of botanical collections have been made from this site, but it is likely to harbour more plant species of conservation importance than are currently known.

Botanical significance

Mount Yao is primarily of botanical importance due to the presence of *Moraea niassensis*, a Vulnerable species known only from this IPA. This species was collected in 2012 from mid-altitude, dense woodland. There was only one population observed and this was estimated to be below the IUCN Red List D1 threshold of 1,000 individuals (Goldblatt *et al.* 2014), although the true value is probably smaller. Searches for this species were undertaken in similar habitat on the nearest inselberg, Serra Mecula located around 130 km east of this IPA, but to no avail, suggesting that this species is endemic to Mount Yao (Goldblatt *et al.* 2014).

There has only been a handful of botanical collections made at this site, and so it is highly likely that more taxa of conservation interest will be recorded from this site with greater collecting effort. Alongside *Moraea niassensis* there are a small number of faunal taxa that are endemic to this inselberg or found both here and on Serra Mecula. The relative isolation of the mid-altitude habitats on this inselberg may have allowed for the evolution of endemics in several different taxa. There is, therefore, a strong case for a comprehensive inventory of the plant taxa of this IPA.

Although the habitats of this IPA are high-quality and warrant further investigation, they cannot currently be assessed as threatened or restricted. There appears to be no montane forest on this inselberg, while the other habitats, riverine forest and varying densities of miombo, are widespread, despite the unique species associated with this habitat on Mount Yao.

Habitat and geology

Mount Yao is a largely wooded IPA, ranging from lowland miombo to dense montane miombo on the slopes, with some areas of gallery forest around rivers on the mountain (Spottiswoode *et al.* 2016). The inselberg is a granite intrusion of 1,313 m near the border of the Marrupa and Unango geological complexes, both of which are primarily dominated by orthogneisses (Boyd *et al.* 2010). Temperatures for the Mavago district range from 15–22°C in June and July to 21–29°C in October and November, while average annual rainfall is 1,887 mm, with most of this precipitation falling between December and March (World Weather Online 2021). The upper slopes and summit of the mountain also receive moisture through mists, as is evident by the high number of *Usnea* epiphytic lichens.

Despite its interesting botany and pristine habitats, few botanical collections have been made on Mount Yao. One zoological visit was made in 2012 (Congdon & Bayliss 2013), focussing primarily on butterfly taxa, during which the few botanical specimens from this site were collected.

The plain surrounding Mount Yao is categorised by Lötter *et al.* (in prep.) as moist miombo, typical of this part of northern Mozambique, although Congdon & Bayliss (2013) describe this vegetation as more like coastal woodland. There is no species inventory for this area, but the plains and lower slopes are dominated by *Brachystegia*, most likely *B. boehmii* (C. Congdon, pers. comm. 2021).

The montane habitats were documented by Congdon & Bayliss (2013) and the description below is based upon this account and personal communications with C. Congdon (2021).

The slopes of the inselberg are steep around the base with a rocky substrate. Soils here are poor, likely due to natural erosion (C. Congdon, pers. comm. 2021). *Uapaca kirkiana* and *U. sansibarica* dominates miombo here, with a more open canopy, patches of suffritcose *Cryptosepalum* (likely *C. maraviense*) and a short, grassy understorey. Herbaceous understorey species have not yet been documented; however, grass species such as *Hyparrhenia filipendula*, *Themeda triandra*, *Panicum* and *Urochloa* spp. are known from montane miombo in this part of Mozambique (Lötter *et al.*, in prep.). At altitudes of around 1,000 m, this woodland also hosts the only known population of *Moraea niassensis* (VU).

Gallery forests, featuring species such as *Parinari excelsa*, *Bersama abyssinica* and *Anthocleista grandiflora*, occur near rivers and in gullies. The understorey includes shrubs such as *Drypetes gerrardii* while herbs such as *Justicia striolata* and *Afromomum* sp. occur beneath. A red-flowered legume, which appeared similar to a *Desmodium*, was found to dominate the forest floor within these areas. Collection of this legume is recommended in order to identify the species. Due to their strong association with rivers and streams, the boundaries of these riverine forests are well defined, with *Albizia* (probably *A. adianthifolia*) occurring in the ecotone between forest and woodland.

The gallery forests are likely underlain by deep, nutrient-rich, moist soils, as have been reported from similar forests on Serra Mecula (Timberlake *et al.* 2004), with swamp forest occurring in areas of poor drainage. The species composition of these swamps has not yet been recorded, however, one patch was noted to have open pools of water, with little understorey growth, and tree species with aerial and buttress roots. It is possible that *Uapaca lissopyrena*, a swamp tree with stilt roots, occurs in these areas as this species was recorded from swamps on the nearest inselberg, Serra Mecula (Timberlake *et al.* 2004).

Around the summit, vegetation cover is open, with a rocky substrate, and may be categorised as cloud or elfin woodland. The area receives moisture through frequent mists, and thus *Usnea* lichens, known from several moist montane habitats in Mozambique, are common epiphytes in the area. These epiphytes were observed on large, old *Brachystegia spiciformis* trees. The presence of these old trees may suggest that the vegetation in this area has remained undisturbed for some time. The woodland also features species such as *Parinari curatellifolia*, *Uapaca kirkiana*, and *U. sansibarica*, alongside *Bridelia*, *Pericopsis* (likely *P. angolensis*), *Monotes* and *Vitex* species. The understorey features shrubs such as *Maesa lanceolata*, *Annona senegalensis* and *Dombeya* (possibly *D. burgessiae*) with tussocky grasses in crevices in the rock. Hemi-parasitic *Agelanthus* sp. (on *Pericopsis*) and *Viscum shirense* (on *Bridelia*) were observed in this woodland, as were a number of epiphytic orchids. At higher altitudes the woodland thins, the shrub *Protea angolensis* becomes more prevalent, and in rockier areas closer to the peak, *Protea welwitschii* and *Combretum* species were observed.

Conservation issues

Mount Yao IPA falls within Niassa Special Reserve, with the south-western portion, around 100 km² in area, falling inside the reserve buffer zone where a number of small villages are located. In addition, this IPA falls within in the wider Niassa Special Reserve Key Biodiversity Area.

This site and surrounding areas have been categorised as receiving "limited conservation efforts" by the Wildlife Conservation Agency (Luwire Wildlife Conservancy 2019). However, the habitats throughout this IPA are largely pristine and have seen little disturbance – an abandoned Portuguese helicopter base is the only major sign of previous human activity on the inselberg itself (Congdon & Bayliss 2013). This area of Mozambique was largely depopulated due to conflict relating to the independence struggle and later the

Mozambican Civil War (C. Congdon, pers. comm. 2021), so the anthropogenic threats in this area are generally quite low, particularly in comparison to those in other parts of Mozambique.

While anthropogenic disturbance within the IPA is currently minimal, the nearby town of Mataca and the associated agriculture has continued to expand over recent decades, as has Msawise village, on the eastern side, to a lesser degree (Google Earth 2021; World Resources Institute 2021). It is thought that, with continued population expansion in the area, anthropogenic disturbance may increase within the IPA, including cutting of woodland for fuel, clearance of land for agriculture and increased fire frequency (Datizua 2020).

The soils in the reserve are known to be generally of poor fertility and rainfall is low (Timberlake *et al.* 2004). Abandonment of exhausted agricultural land may, therefore, become an issue as it could result in further agricultural expansion, possibly onto the hills in the south of the IPA or on the mountain itself. Soils on the lower slopes of Mount Yao are thin and rocky (C. Congdon, pers. comm. 2021), and are unlikely to be highly productive, however there may be greater or more consistent moisture availability on the mountain, due to frequent mists, which may encourage small-scale cultivation of these areas.

Although there was no evidence of fire within the forests at the time of the 2012 visit (Congdon & Bayliss 2013), fire has been reported elsewhere in the reserve as a method for clearing land for machambas and for subduing bees to allow for honey collection (Timberlake *et al.* 2004; T. Alves, pers. comm. 2021), and so there may be an additional threat of unintentional burning of swathes of land. It is particularly important that, if land is opened up for tourism, the practice of burning vegetation to create and maintain walking trails and vehicle access, as has previously been reported on and around Serra Mecula, is not also employed within this IPA.

Niassa Reserve does not currently receive much tourism, with only 183 visitors in 2013. The reserve, therefore, has limited income for funding conservation projects or supporting alternative livelihoods for local people. As well as providing income, tourism could incentivise the protection

Riverine forest on Mount Yao (CC)

Moraea niassensis (CC)

of wilderness areas such as Mount Yao which contribute to the visitor experience. More recently, there has been government-backed promotion of Niassa, including specific mention of Mount Yao, as a tourist destination (ANAC 2018), which may lead to greater interest or investment into the site.

As well as hosting *Moraea niassensis*, a Vulnerable Iridaceae known only from this IPA, Mount Yao is the only known location for an as yet undescribed butterfly species in the genus *Baliochila* (Congdon & Bayliss 2013). In addition, a species of freshwater crab, *Potamonautes bellarussus*, described in 2014, is thought to be endemic to the Yao and Mecula inselbergs within Niassa Special Reserve (Daniels *et al.* 2014). It is possible that the isolated montane habitats created by the inselbergs in the reserve have allowed for the evolution of highly range-restricted species. As the botany of this site is yet to be inventoried, we may expect that more range-limited, or even site-endemic, species will be documented.

There are a number of sightings of threatened animal taxa around the IPA, including African wild dog (*Lycaon pictus* – EN), African elephant (*Loxodonta africana* – VU) and African lion (*Panthera leo* – VU). However, these species would likely only enter lower altitudes within this IPA and the montane mammalian fauna of this site is yet to be inventoried (van Berkel *et al.* 2019).

Key ecosystem services

Mount Yao contributes to the tourist experience of Niassa Reserve, as reported by the Mozambican government in a document highlighting the investment opportunities for nature-based tourism potential across the country (ANAC 2018). However, at present the site is not easily accessible for tourists (Luwire Wildlife Conservancy 2019).

For local people, the IPA is likely an important source of water, contributing to three different watersheds to the south, north-west and north-east. Yao is also the source of the Chiulezi River, in the lattermost watershed, which serves communities downstream including the villages of Chamba and Matondovela (Luwire Wildlife Conservancy 2019).

Additionally, the lower plains surrounding the inselberg may provide a source of wood for fuel or timber. Elsewhere in the reserve, communities use a wide range of hardwoods but there is no evidence of selective logging of high market value species such as *Millettia stuhlmannii* and *Dalbergia melanoxylon* (Timberlake *et al.* 2004).

Mount Yao shares its name with the Yao ethnic and linguistic group. The waYao (Yao peoples) are thought to have originated from the north of Niassa Province before dispersing through neighbouring areas in the 9th century (Mbalaka 2016). Some sources suggest that the inselberg itself was the nucleus from which the waYao originated. Sacred sites are known from within Niassa Reserve (Wildlife Conservation Society Mozambique 2021), however, any cultural significance of Mount Yao to local communities has not been documented but could be established through interviews. Greater understanding of how local people interact with the ecosystems within the IPA would better inform conservation and the planning of any future tourist activities.

A number of historically significant examples of cave art are known from Niassa Reserve, dating from tens of thousands of years ago, and it is possible that Mount Yao may also be of archaeological and anthropological importance (Wildlife Conservation Society Mozambique 2021).

Ecosystem service categories

- Provisioning – Raw materials
- Provisioning – Fresh water
- Cultural services – Tourism
- Cultural services – Cultural heritage

IPA assessment rationale

Mount Yao qualifies as an IPA under sub-criterion A(i), hosting the only known population of the Vulnerable species *Moraea niassensis*. The vast majority of the botanical diversity of this site has not yet been documented and it is highly likely that further investigation will reveal more plant species of conservation interest.

Priority species (IPA Criteria A and B)

FAMILY	TAXON	IPA CRITERION A	IPA CRITERION B	≥ 1% OF GLOBAL POP'N	≥ 5% OF NATIONAL POP'N	IS 1 OF 5 BEST SITES NATIONALLY	ENTIRE GLOBAL POP'N	SPECIES OF SOCIO-ECONOMIC IMPORTANCE	ABUNDANCE AT SITE
Iridaceae	*Moraea niassensis*	A(i)	B(ii)	✓	✓	✓	✓		occasional
		A(i): 1 ✓	B(i): 1						

Protected areas and other conservation designations

CONSERVATION AREA TYPE	CONSERVATION AREA NAME	RELATIONSHIP OF IPA TO CONSERVATION AREA
National Reserve	Niassa Special Reserve	protected/conservation area encompasses IPA
Key Biodiversity Area	Niassa Special Reserve	protected/conservation area encompasses IPA

Threats

THREAT	SEVERITY	TIMING
Housing & urban areas	low	Future – inferred threat
Small-holder farming	low	Ongoing – increasing
Roads & railroads	low	Past, not likely to return
Logging & wood harvesting	low	Ongoing – trend unknown
Increase in fire frequency/intensity	medium	Future – inferred threat

SERRA MECULA AND MBATAMILA

Assessors: Sophie Richards, Iain Darbyshire

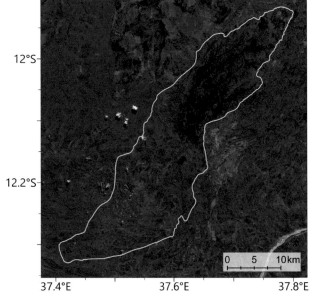

INTERNATIONAL SITE NAME		Serra Mecula and Mbatamila	
LOCAL SITE NAME (IF DIFFERENT)		Serra Mecula e Mbatamila	
SITE CODE	MOZTIPA046	PROVINCE	Niassa

LATITUDE	-12.09954	LONGITUDE	37.62293
ALTITUDE MINIMUM (m a.s.l.)	340	ALTITUDE MAXIMUM (m a.s.l.)	1,442
AREA (km²)	626	IPA CRITERIA MET	A(i)

Mbatamila inselberg, with surrounding miombo woodland and Serra Mecula in the distance (JT)

Site description

Serra Mecula and Mbatamila Important Plant Area falls within Mecula District of Niassa Province. The entire site is encompassed by Niassa Special Reserve, and the inselberg itself, peaking at 1,442 m, is the highest point within this 42,000 km² protected area. To the south, this site covers the smaller inselbergs around Mbatamila, with the southernmost point around 16 km north of the Lugenda River. The Mataca-Mecula road bisects this site, while the 535 road from Marrupa and the Mecula-Naulala road run along the eastern boundary. Along these latter two roads are a number of residential areas, notably the district centre, Mecula town, at the south-eastern foot of the Serra Mecula.

Serra Mecula is unique in that it represents the only area of montane forest within Niassa Reserve (Spottiswoode *et al.* 2016). In addition, the inselbergs around Mbatamila host a number of rare and threatened species, while the dambos within this IPA require further investigation but are also likely to host range-restricted species (Timberlake *et al.* 2004). Many of the habitats of the IPA are largely intact and, despite limited botanical collecting within this site, are thought to be of great botanical interest and warrant further study. Given the contrast between these habitats, the site could be divided into separate montane and lowland IPAs, with scope to also alter the boundaries of each and produce more ecosystem-focused IPAs. However, at this time, data for the site is limited and so they cannot currently be separated and still qualify as IPAs in their own right.

Botanical significance

Serra Mecula has received only limited botanical study to date, and further investigation is required at this site. A full inventory is recommended as observations made on preliminary visits suggest that the flora on this mountain, particularly in the montane forests, is of great botanical interest (J. Burrows, pers. comm. 2021).

Serra Mecula is unique in hosting the only montane forest and montane shrubland across Niassa Reserve, while neighbouring Mount Yao does not reach the necessary altitudes for hosting these habitat types. Compared to other mountains of Mozambique, however, the area of this montane forest is relatively limited, covering only 1.36 km² (Spottiswoode *et al.* 2016). Instead, many of the forested areas of the mountain are riverine and not montane. Timberlake *et al.* (2004) propose that the montane forests of Serra Mecula display greater similarity to those of the Manica Highlands, which run along the Mozambique-Zimbabwe border, than to the closer Mulanje-Namuli-Ribáuè arc of mountains. However, more research is required to support this hypothesis.

In terms of interesting species on the mountain itself, two national endemics have so far been recorded- *Baphia massaiensis* subsp. *gomesii*, and *Rotheca luembensis* subsp. *niassensis*, both limited only to northern Mozambique. The former taxon, *B. massaiensis* subsp. *gomesii*, has also been recorded around Mbatamila. An uncertain record of *Pavetta gurueensis* (*Burrows* #11225), a globally Vulnerable species, from the slopes of

Grassy dambo vegetation north of Mbatamila (TB)

Peak of one of the Mbatamila inselbergs (JT)

Serra Mecula would represent a range extension of a species otherwise limited to northern Zambezia province. However, it is not possible to determine the species with certainty from the collection made and so further investigation should be undertaken to establish which species of *Pavetta* occurs within this IPA.

A small number of interesting species are known from the Mbatamila area, in the southern portion of this IPA, including the globally Vulnerable species, *Justicia attenuifolia*, recorded from the open miombo woodland surrounding the granite outcrops. This species is known from only four locations across northern Mozambique and southern Tanzania, with this location at Mbatamila representing the only population within a protected area (Luke *et al.* 2015a). *Barleria mutabilis*, a species also limited to southern Tanzania and northern Mozambique, occurs on Serra Mecula with the collection on this inselberg representing the first record for Mozambique (Darbyshire 2009), although *B. mutabilis* has since been collected on nearby Mount Yao.

The dambos that occur in grassland between Serra Mecula and Mbatamila are not currently known to be home to rare threatened species, however, it is likely that ground orchids, many of which are limited in their distributions, are present here (Timberlake *et al.* 2004) and so there may be additional range-restricted endemics in this IPA. As little is currently known about the flora of these wetlands, the area of this habitat in this IPA is limited. However, if further investigation finds adjacent areas to be of conservation importance, these should be included within the IPA network in some form.

Habitat and geology

Serra Mecula and Mbatamila IPA hosts a number of habitat types, likely due to the variable topology and range of altitudes - varying from 350 m, in a tributary of the Chiulezi river in the north-east of this site, to 1,442 m, at the peak of Serra Mecula. The site experiences seasonal rainfall, with a wet season running from November to April, and probably receives frequent mists at higher altitudes as have been described for the other mountain of Niassa Reserve, Mount Yao (Congdon & Bayliss 2013). Temperatures are largely stable, between 21.5°C in June and July to 28°C in November.

Although this site requires further research, a survey was undertaken of Niassa Reserve in 2003 by Timberlake *et al.* (2004), including both Serra Mecula and Mbatamila, with further botanical collecting taking place in 2009 on Serra Mecula (J. Burrows) and in 2013 around Mbatamila (T. Parker). The following habitat description is based on the 2003 surveys of Timberlake *et al.* (2004).

Serra Mecula inselberg hosts the most diverse range of habitats in Niassa Reserve. Vegetation on the outer mountain slopes is largely open miombo, with species including *Brachystegia boehmii, B. utilis* and *B. bussei*. On the western-facing slopes, this miombo appears particularly open, possibly because of the steeper incline, which may retain only thin soils (Google Earth 2021). Within gullies, following deeply incised streams, areas of gallery forest are noted, featuring species such as *Khaya anthotheca, Treculia africana* and *Uapaca lissopyrena*. The latter, as a swamp forest species, is associated with areas of poor drainage similar to patches of swamp forest on Mount Yao (Congdon & Bayliss 2013).

At higher altitudes are small patches of mid-montane forest, occurring on a high-altitude plateau around 1,000–1,300 m elevation, and in gullies below the higher peaks. Each patch is around 1–5 ha in area, with a total of 136 ha (1.36 km^2) on Serra Mecula (Spottiswoode *et al.* 2016). Forest patches are dominated by species such as *Peddiea africana* and *Erythroxylum emarginatum* while the understorey commonly includes shrubs such as *Carvalhao macrophylla* and *Rinorea ilicifolia*.

Surrounding the peak, is a sparsely vegetated scrubby plateau. These areas are highly exposed and are populated by low shrubs, succulents and herbs, predominantly grasses. Dominant grasses are *Melinis ambigua* and *Urochloa* while the Asteraceae *Helichrysum kirkii* is also common in these areas. Succulents include *Aloe mawii, Tetradenia riparia* and *Kalanchoe elizae*, while shrubs are occasional and include species such as *Searsia tenuinervis* and *Anthospermum whyteanum*, both of which were observed to be scarce. It was suggested by Timberlake *et al.* (2003) that the species composition of this exposed plateau also has similarities to similar exposed, moist montane habitat in Zimbabwe, although further research would be required to support this hypothesis.

On the southern and eastern slopes of the mountain, land has been cleared for agriculture. Disturbed areas are kept in a sub-climax grassland state of scattered shrubs and trees, predominantly *Strychnos spinosa*, maintained by frequent fires.

South of Serra Mecula, miombo woodland is interspersed with dambos. While the pools and wetlands here are seasonal (Nagy & Watters 2019), the grasslands in this area are a source of moisture year-round. The grass layer of the miombo is well-developed, although the species composition has not been documented. Similarly, the species associated with the dambos in this site are yet to be inventoried. It is thought that these dambos would be suitable habitat for ground orchids. None have thus far been collected, but it is likely that this is because collecting efforts have not coincided with their wet season flowering period.

Miombo species around Mbatamila are of the genera *Brachystegia* and *Julbernardia*, with *B. spiciformis* and *B. boehmii* noted as common. Endemic *Baphia massaiensis* subsp. *gomesii* can be found as a small tree within this miombo and, amongst the grassy understorey, is the endemic woody herb, *Justicia attenuifolia* (VU). Streams within this area are bordered by a narrow strip of evergreen trees including *Syzygium guineense* (likely *S. guineense sensu stricto.*) and *S. cordatum*. The soils are described as sandy.

Associated with the granite inselbergs around Mbatamila are two main habitats: exposed, sparse vegetation on the slopes and denser, sheltered vegetation within gullies or towards the foot of the slopes. The resurrection plant, *Myrothamnus flabellifolius*, and the sedge *Coleochloa setifera* are common on the slopes. There are also a number of succulents on these inselbergs, including *Aloe mawii* and *Euphorbia cooperi*. Woodland on the slopes and ledges has not yet been surveyed, however, aerial observations suggest they are populated by *Brachystegia glaucescens,* and it is also thought that habitat here would be suitable for cycads, although none have been recorded as yet.

The relatively sheltered gullies and foot slopes of the Mbatamila inselbergs have deeper, more nutrient-rich soils and so support denser woodland and forest vegetation. Species in these areas include trees such as *Grewia forbesii*, *Ficus sur* and *Bombax rhodognaphalon* with the small tree *Grewia bicolor* and herbs *Celosia trigyna* and *Ruspolia decurrens* in the understorey. Where foot slopes transition into plane miombo, species composition is similar to that of the planes, with *Brachystegia* and *Julbernardia* miombo. However, as soils are deeper and more clay-rich in these areas, the trees grow to greater heights.

Conservation issues

Serra Mecula and Mbatamila IPA is encompassed by the National Special Reserve. The Mbatamila area falls within the Niassa Wilderness Area which, although not under active conservation management, does receive law enforcement support unlike much of the west of the reserve (Luwire Wildlife Conservancy 2019).

Serra Mecula contributes to tourism within Niassa Reserve, although overall visitor numbers are quite limited, with only 183 in 2013. The income raised through tourism could help to incentivise and support conservation efforts within the reserve and, to this end, ANAC has promoted "nature-based tourism" in the reserve, drawing particular attention to Serra Mecula (ANAC 2018). However, management practices observed during the 2003 survey included the annual burning of secondary grassland and shrubland on the mountain to keep footpaths open, resulting in hectares of land burned and the reduction of forest patch sizes as the edges became degraded (Timberlake *et al.* 2004). Although this practice by rangers may have discontinued in the years since, there probably continues to be a threat of anthropogenic fire, particularly on the eastern slopes where fire has been used to subdue bees and allow the collection of honey (T. Alves, pers. comm. 2021).

Clearing of land on the southern and eastern slopes of Serra Mecula has been recorded since this area was occupied over a 100 years ago by the German Army during the First World War, and the expansion of cleared land has continued in recent years (Timberlake *et al.* 2004; World Resources Institute 2021). The soils within the reserve are known to be poor and rainfall low (Timberlake *et al.* 2004) and so land may be exhausted after few agricultural cycles. Although currently intact, the dambos to the south of Serra Mecula may also be under a high level of threat from agriculture due

Miombo woodland and riverine forest on Serra Mecula (TB)

to the year-round moisture provided in these areas (Timberlake *et al.* 2004). The road running through the dambos makes the site particularly accessible and vulnerable to disturbance. Support for local people is required to develop sustainable agricultural practices, providing food security for communities while also ensuring that key habitats within this site are protected. There is also a threat of further development around Mbatamila, with an airstrip already located to the north-west of the inselbergs. The area should be more fully inventoried to allow careful consideration and planning of any further infrastructure development.

A number of interesting faunal species are also recorded from this site. One species of freshwater fish, *Nothobranchius niassa* (VU), is known only from Niassa Reserve, in the seasonal pools associated with the dambos. Before these pools are lost in the dry season, *N. niassa* lays eggs in the uppermost layer of substrate which then hatch in the subsequent rainy season (Nagy & Watters 2019). Much of this species' range falls within this IPA and therefore the protection of the dambos here is of great importance in preventing the extinction of this species.

Interesting faunal taxa have also been recorded on Serra Mecula itself including two previously undescribed species of butterfly from the genus *Baliochila* (Congdon & Bayliss 2013). This inselberg

is also of interest due to its avian taxa. A study by Spottiswoode *et al.* (2016) found that, while Serra Mecula hosts montane forests, these habitats are inhabited mostly by species associated with low- to mid-altitude riparian forest species. The only montane forest species recorded was Lemon Dove (*Aplopelia larvata* – LC). The authors suggest that the lack of montane avian taxa may be due to the remoteness of this mountain. This biogeographical pattern in bird species may parallel the suggestion by Timberlake *et al.* (2004) that Serra Mecula has greater botanical affinity with the Manica Highlands than with the closer Mulanje-Namuli-Ribáuè mountain arc. More research is recommended to elucidate these bio-geographical patterns.

Key ecosystem services

Serra Mecula makes a significant contribution to the tourism experience of Niassa Reserve, with trails maintained on the mountain to allow for visitors (Timberlake *et al.* 2004). The intact habitats here are likely to draw tourists, particularly for walking tours, providing an incentive for their protection.

The mountain also acts as an important source of fresh water, with drainage supplying rivers around Mecula town that probably provide water for both direct consumption and for agriculture. A number of these streams are tributaries for the Lugenda

River to the south. In addition, wetlands south of Serra Mecula provide important habitat for the Vulnerable fish species, *Nothobranchius niassa*, with the dambos within this IPA covering a large part of this species' highly limited range (Nagy & Watters 2019).

It is likely that the miombo within this IPA serves as a source of wood for timber and fuel used by local people. There was no evidence that trees with high market value, such as *Millettia stuhlmannii*, are being selectively harvested within the reserve as a whole (Timberlake *et al.* 2004). Honey collection also occurs on the foot slopes of Serra Mecula (T. Alves, pers. comm. 2021).

Ecosystem service categories

- Provisioning – Raw materials
- Provisioning – Fresh water
- Habitat or supporting services – Habitats for species
- Cultural services – Tourism

IPA assessment rationale

Serra Mecula and Mbatamila qualifies as an Important Plant Area under criterion A(i) due to the presence of the globally Vulnerable species *Justicia attenuifolia*. Although this species only occurs in the Mbatatmila area, Serra Mecula is included within this IPA as it is strongly suspected that the habitats in this area host more species of conservation importance. With further research, further trigger species will probably be uncovered, providing the potential to separate the lowland and montane areas as independent IPAs. Mid-altitude forest, a habitat of conservation importance in Mozambique, is present on Serra Mecula. However, with only 136 ha of this forest type in this IPA, none of the necessary thresholds for triggering sub-criterion C(iii) are met. Overall, only two endemic species are currently known from this IPA, although it is highly likely more will be recorded with further research.

Priority species (IPA Criteria A and B)

FAMILY	TAXON	IPA CRITERION A	IPA CRITERION B	≥ 1% OF GLOBAL POP'N	≥ 5% OF NATIONAL POP'N	IS 1 OF 5 BEST SITES NATIONALLY	ENTIRE GLOBAL POP'N	SPECIES OF SOCIO-ECONOMIC IMPORTANCE	ABUNDANCE AT SITE
Fabaceae	*Baphia massaiensis* subsp. *gomesii*		B(ii)	✓					unknown
Lamiaceae	*Rotheca luembensis* subsp. *niassensis*		B(ii)	✓	✓	✓			unknown
Rubiaceae	*Justicia attenuifolia*	A(i)		✓	✓	✓			unknown
		A(i): 1 ✓	B(ii): 2						

Inselberg vegetation (TB)

Peak of Serra Mecula (TB)

Threatened habitats (IPA Criterion C)

HABITAT TYPE	IPA CRITERION C	≥ 5% OF NATIONAL RESOURCE	≥ 10% OF NATIONAL RESOURCE	IS 1 OF 5 BEST SITES NATIONALLY	ESTIMATED AREA AT SITE (IF KNOWN)
Medium Altitude Moist Forest 900–1400 m [MOZ-02]	C(iii)				1.36

Protected areas and other conservation designations

CONSERVATION AREA TYPE	CONSERVATION AREA NAME	RELATIONSHIP OF IPA TO CONSERVATION AREA
National Reserve	Niassa Special Reserve	protected/conservation area encompasses IPA
Key Biodiversity Area	Niassa Special Reserve	protected/conservation area encompasses IPA

Threats

THREAT	SEVERITY	TIMING
Small-holder farming	medium	Ongoing – trend unknown
Roads & railroads	low	Past, not likely to return
Recreational activities	low	Ongoing – trend unknown
Increase in fire frequency/intensity	medium	Past, likely to return

MOUNT MASSANGULO

Assessors: Sophie Richards, Iain Darbyshire

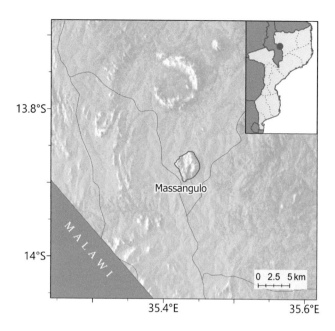

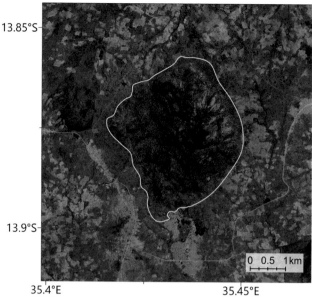

INTERNATIONAL SITE NAME		Mount Massangulo	
LOCAL SITE NAME (IF DIFFERENT)		Monte Massangulo	
SITE CODE	MOZTIPA039	PROVINCE	Niassa

LATITUDE	-13.87657	LONGITUDE	35.43433
ALTITUDE MINIMUM (m a.s.l.)	1,135	ALTITUDE MAXIMUM (m a.s.l.)	1,640
AREA (km²)	11	IPA CRITERIA MET	A(i)

Site description

The Mount Massangulo IPA is a mountain in the N'gauma District of Niassa Province, close to the Malawi border. The town of Massangulo, centred on one of the oldest Catholic missions in the region, lies at the foot of the mountain, while to the west is the main road from Lichinga running south to Mandimba.

The mountain itself reaches a peak of 1,640 m, with the IPA covering an area of 11 km². Much of the IPA is miombo woodland, however, there is some montane forest in the gullies of the mountain which hosts the only population in Mozambique of the globally Endangered species, *Streptocarpus erubescens*.

Botanical significance

This IPA is of botanical significance as the only site in Mozambique to host *Streptocarpus erubescens*, a globally Endangered, near-endemic species known also from a small number of sites across the border in Malawi. Most of the localities in Malawi are threatened by clearance of the montane forest on which this species depends; therefore, the intact patches of montane forest on Mount Massangulo are of great importance for the continued survival of this species (Darbyshire & Rokni 2020a). A survey of the *S. erubescens* population within this IPA is recommended to establish its size and health, as the last botanical record of the species at this site was taken in 1967 (*Torre* #10803).

An additional threatened species, *Oncella curviramea* (VU), is known to occur at this site. Massangulo represents one of only two localities in Mozambique for this parasitic species (Polhill & Wiens 1998).

Two Mozambican endemics are also known to occur at this IPA, *Pavetta gardeniifolia* var. *appendiculata*, known from only Massangulo in Niassa province and a small number of other localities in Zambezia province, and *Ceropegia cyperifolia* (LC), which has a range of just 3,826 km².

Numerous important timber species have been recorded from the miombo woodlands in this area, including *Albizia gummerifera*, *Brachystegia spiciformis*, *B. utilis* and *Newtonia buchananii* (GBIF. org 2021a) which are likely harvested and used by local people.

Habitat and geology

Massangulo reaches a peak of 1,640 m, with two smaller peaks, at 1,560 m and 1,610 m, to the south-east and south-west. The geology of the area is mostly sandy-clay soils underlain by granitoid rock (*Torre* #10773). Much of the drainage from the mountain appears to flow southwards towards the Chitape river.

Although the site has not been subjected to a formal inventory, a number of collections have been made on and around Mount Massangulo, particularly by Portuguese botanist António Gomes e Sousa (Exell 1937). Much of the lower and mid-altitude slopes of Mount Massangulo are covered in miombo woodland, however, some of the miombo on the flatter areas of the western slope of the mountain has been converted into machambas. Miombo is composed of *Brachystegia*, *Uapaca* and *Julbernardia* spp. on the southern slopes (*Torre* #10773). *Brachystegia boehmii* is the dominant species in the more open canopy areas (Exell 1937). In woodland clearings, trees of *Piliostigma thonningii* are common with the herb *Dolichos kilimandscharicus* frequent in the understorey (Exell 1937). At mid-altitudes the grass species *Eragrostis arenicola* is abundant in miombo clearings, occurring on the dry soils in these areas (*Gomes e Sousa* #1414).

Forested areas occur within gullies at mid to high altitudes, particularly on the south-facing slopes. The species composition of these areas has not yet been documented; however, *Newtonia buchananii* has been recorded within these forests (*Torre* #10826), and probably dominates as is the case for several montane forests on thin soils in Mozambique (Burrows *et al.* 2018). The montane forests on Mount Massangulo are important for the globally Endangered species *Streptocarpus erubescens*, which grows in the rocky understorey (Darbyshire & Rokni 2020a), while *Oncella curviramea* (VU) is known to parasitise at least one species of *Combretum* within these forests (*Torre* #11047).

Conservation issues

Mount Massangulo does not fall within a protected area, Key Biodiversity Area or Important Bird Area. However, many of the upper slopes are protected by the Environmental Act (Lei . 20/97 of 1997) which prohibits the cultivation of crops on the steeper slopes of Mozambique's mountains (Timberlake *et al.* 2007), although these areas of Mount Massangulo are largely undisturbed in any case, possibly due to their inaccessibility.

Since 2015 there has been significant loss of miombo on the lower western slopes of the mountain, including thinning of woodland (presumably through felling for timber or fuelwood) and clearing of land for agriculture (World Resources Institute 2020; Google Earth 2021). While clearings on the plains around Mount Massangulo have been present for some time, with botanical records from the 1930s documenting these (*Gomes e Sousa* #1339), until recently the mountain itself appeared largely undisturbed (Google Earth 2021).

Close to the IPA boundary are a number of pine and eucalyptus forestry plantations, owned by Green Resources S. A. (a Norwegian forestry company), totalling 3,322 ha in Ngaúma district. Following a number of land disputes with local people, some of whom previously farmed the land converted to forestry (Røhnebæk Bjergene 2015), the company agreed in 2020 to cede their land rights to local communities in several districts across Niassa province, including in Ngaúma where this IPA is situated (Agencia de Informacao de Mocambique 2020). If the land can be farmed by local people, this may relieve pressure in the Massangulo area which may, in turn, slow further agricultural expansion within the IPA.

The faunal taxa of Mount Massangulo has not yet been catalogued, however, with a significant area of intact woodlands and forest, it is likely there will be some taxa of interest within this IPA.

Key ecosystem services

Drainage from the mountain into the Chitape River to the south appears to be important for local land use, with a number of settlements, machambas and a forestry plantation all being situated by the streams that originate on the mountain. The thinning of miombo woodland in the west of the IPA suggests that this site is a source of wood, probably for timber or fuel.

Ecosystem service categories

- Provisioning – Raw materials
- Provisioning – Fresh water

IPA assessment rationale

Mount Massangulo qualifies as an IPA under sub-criterion A(i), with one Endangered species, *Streptocarpus erubescens,* and one Vulnerable species, *Oncella curviramea.*
As only two Mozambican endemic taxa (*Pavetta gardeniifolia* var. *appendiculata* and *Ceropegia cyperifolia*) and one near endemic (*S. erubescens*) are known from the site, Mount Massangulo does not meet the threshold for 3% of species of conservation importance under sub-criterion B(ii). It is recommended that further research be conducted into the botanical diversity of this site and to monitor the populations of the known priority species.

Priority species (IPA Criteria A and B)

FAMILY	TAXON	IPA CRITERION A	IPA CRITERION B	≥ 1% OF GLOBAL POP'N	≥ 5% OF NATIONAL POP'N	IS 1 OF 5 BEST SITES NATIONALLY	ENTIRE GLOBAL POP'N	SPECIES OF SOCIO-ECONOMIC IMPORTANCE	ABUNDANCE AT SITE
Apocynaceae	*Ceropegia cyperifolia*		B(ii)	✓	✓	✓			common
Gesneriaceae	*Streptocarpus erubescens*	A(i)	B(ii)	✓	✓	✓			unknown
Loranthaceae	*Oncella curviramea*	A(i)			✓	✓	✓		unknown
Rubiaceae	*Pavetta gardeniifolia* var. *appendiculata*		B(ii)	✓	✓	✓			unknown
		A(i): 2 ✓	B(ii): 3						

Threatened habitats (IPA Criterion C)

HABITAT TYPE	IPA CRITERION C	≥ 5% OF NATIONAL RESOURCE	≥ 10% OF NATIONAL RESOURCE	IS 1 OF 5 BEST SITES NATIONALLY	ESTIMATED AREA AT SITE (IF KNOWN)
Medium Altitude Moist Forest [MOZ-02]	C(iii)				1.0

Protected areas and other conservation designations

CONSERVATION AREA TYPE	CONSERVATION AREA NAME	RELATIONSHIP OF IPA TO CONSERVATION AREA
No formal protection	N/A	

Threats

THREAT	SEVERITY	TIMING
Small-holder farming	low	Ongoing – trend unknown
Logging & wood harvesting	medium	Ongoing – trend unknown

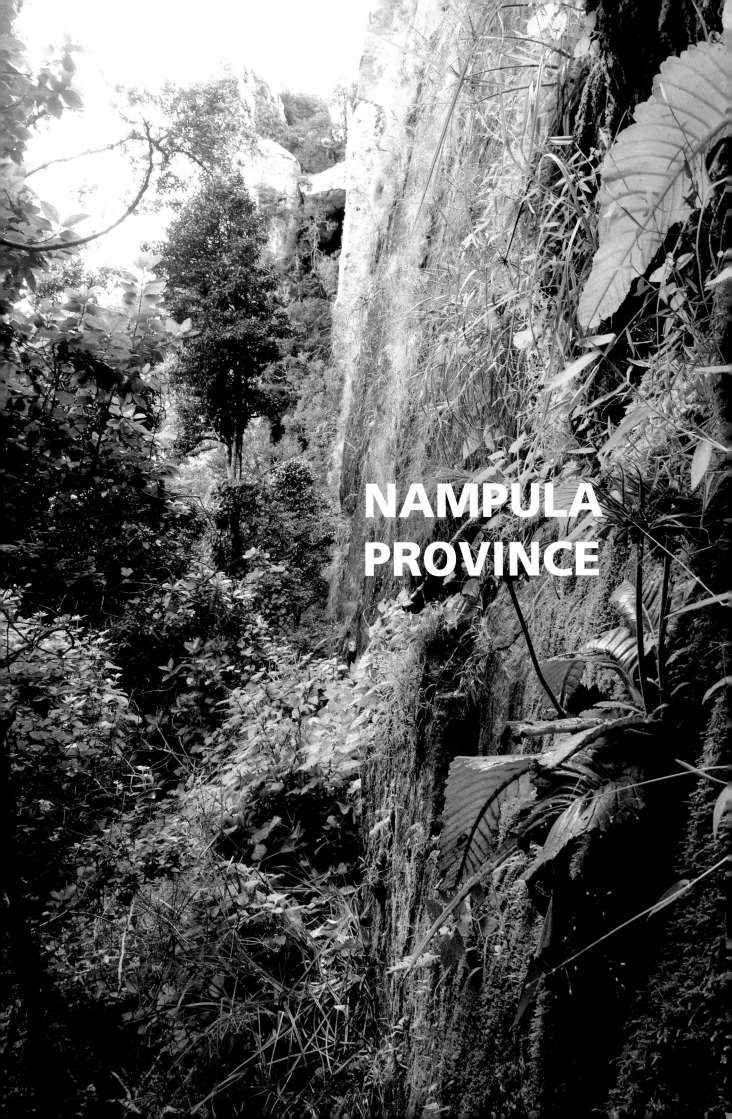

NAMPULA
PROVINCE

ERÁTI

Assessor: Iain Darbyshire

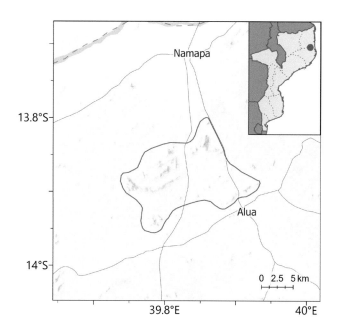

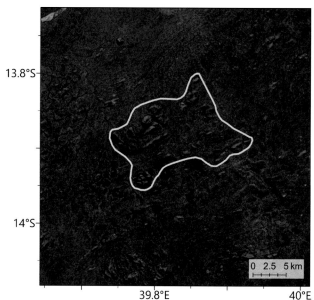

INTERNATIONAL SITE NAME		Eráti	
LOCAL SITE NAME (IF DIFFERENT)		–	
SITE CODE	MOZTIPA008	PROVINCE	Nampula

LATITUDE	-13.84000	LONGITUDE	39.85300
ALTITUDE MINIMUM (m a.s.l.)	262	ALTITUDE MAXIMUM (m a.s.l.)	765
AREA (km²)	174	IPA CRITERIA MET	A(i), A(iii), A(iv)

Site description

The Eráti IPA comprises a series of lowland granitic inselbergs in Eráti District of northern Nampula Province which support an interesting woodland flora with a number of highly localised plant species. The site is intersected by the main N1 road between the towns of Namapa and Alua, ca. 15–30 km south of the Lúrio River. As such, this is amongst the most accessible group of a diffuse belt of inselbergs across northern Nampula to the south of the Lúrio. Most of these peaks have not been botanised and further surveys of this highly under-explored region are likely to reveal other key inselberg sites in northern Nampula Province, which could either result in the expansion of the Eráti IPA or to the recognition of further IPAs in this region.

Botanical significance

Despite its proximity to the N1 road and ease of access, the Eráti hills are under-explored botanically. However, surveys by Antonio Rocha da Torre and Jorge Paiva in the 1960s, mainly on Mount Cheovi (Geovi) in the northeast of the IPA, revealed the presence of a number of range-restricted and threatened species amongst the woodland and rock flora for which this site is considered to be of high importance. Of particular note is the presence of *Allophylus torrei* (EN). This small tree or shrub is restricted to inselberg woodlands in Cabo Delgado and Nampula Provinces, and Eráti is one of only four known locations for this species (Darbyshire *et al.* 2019d). Other noteworthy inselberg woodland species here include the shrublet *Indigofera pseudomoniliformis* (VU), endemic to northern Mozambique; the wild "Ibo coffee" *Coffea zanguebariae* (VU); and the shrub *Croton kilwae* (EN) which was noted to be frequent on the small inselbergs around Alua in the southeast of the IPA during recent surveys there (Ernst Schmidt, pers. comm. 2020). This IPA is also the only known locality for *Rotheca sansibarensis* var. *eratensis*, a local variety of this widespread species. A species of *Cola* collected from along a watercourse on Mount Cheovi (*Torre & Paiva* #9874) has been provisionally identified as *C. discoglypremnophylla* (EN) but Lawrence & Cheek (2019) note that fertile material is needed from the Mozambique sites in order to confirm its presence there. Whatever its identity proves to be, it is likely that this *Cola* will be a further species of conservation concern. Elsewhere, the only known locality for the recently described miombo woodland liana *Momordica mossambica*

Croton kilwae (JEB)

(DD) falls within the eastern boundary of the site (Schaefer 2009), but its continued presence here requires confirmation given the widespread loss of miombo along the N1 road corridor. Similarly, the little-known shrub *Pavetta micropunctata* was recorded from dense woodland on damp black clay-humus soils at the foot of Mount Cheovi (*Torre & Paiva* #9887). This latter species has yet to be assessed on the IUCN Red List but is highly likely to be globally threatened.

Given the low level of botanical coverage to date, it is highly likely that other rare and range-restricted species will be uncovered within the Eráti IPA in the future. For example, the type collection of *Syncolostemon namapensis* (*Balsinhas & Marrime* #335) was made in 1961 from the base of "Serra Malala" which is understood to be a part of the inselberg chain to the west of Namapa, only ca. 15 km from the Eráti IPA. This species is otherwise known only from Tunduru in Tanzania, and is likely to be globally threatened.

Habitat and geology

This site comprises a series of low-lying inselbergs derived from deposits of the middle to upper Proterozoic, with surficial geology including granites and gneisses (Instituto Nacional de Geológia 1987). The climate of the region is classified as mainly semi-arid and dry sub-humid with average annual rainfall ranging from 800–1,200 mm (WCS *et al.* 2021). The vegetation of the IPA has not been documented in any detail, but the predominant habitat in the lowlands is miombo woodland of the Nampula Granite Escarpment Miombo type of Lötter *et al.* (in prep.), dominated by a number of *Brachystegia* spp. and *Julbernardia globiflora*. Sheltered areas amongst the inselberg cliffs and gulleys support a more densely wooded vegetation, corresponding to the Northern Inselberg Woodland and Forest of Lötter *et al.* (in prep.), with important species including *Sterculia* spp. and *Millettia stuhlmannii* as well as miombo taxa. The more open rocky slopes support a xerophytic plant community including a range of succulent taxa, but this has not been documented within the Eráti IPA.

Conservation issues

This site is not currently protected and no biodiversity management is in place. The majority of the IPA is included within the Eráti Key Biodiversity Area on the basis of the important population of *Allophylus torrei* (WCS *et al.* 2021); the IPA boundary is slightly larger to include the hills near Alua that have a sizable population of *Croton kilwae*.

The main general threat in this region is agricultural expansion, although this is less impactful on the rocky slopes than in the miombo woodlands of the intervening lowlands (WCS *et al.* 2021). Analysis of historical satellite imagery reveals some losses of woody vegetation around the inselbergs within the past 25 years (Darbyshire *et al.* 2019d), but considerable areas of natural vegetation remain that should support the rare and threatened species. A plausible threat to this habitat is an increased frequency of fires encroaching into the gulleys and slopes from neighbouring agricultural areas where it is used as a means of land clearance.

Key ecosystem services

This site provides important natural habitat for biodiversity in an area otherwise largely transformed by agriculture. The ecosystem services provided are otherwise undocumented, although it is likely that the woodlands provide a range of provisioning and regulatory services, including prevention of excessive erosion on the steep inselberg slopes.

Ecosystem service categories

- Regulating services – Erosion prevention and maintenance of soil fertility
- Habitat or supporting services – Habitats for species
- Habitat or supporting services – Maintenance of genetic diversity

IPA assessment rationale

The inselbergs of Eráti qualify as an IPA under criterion A, as they are inferred to contain important populations of four threatened species under criterion A(i): *Allophylus torrei* (EN), *Croton kilwae* (EN), *Coffea zanguebariae* (VU) and *Indigofera pseudomoniliformis* (VU); it is the only site within Mozambique's IPA network for the foremost and lattermost of these species. The site also qualifies under criterion A(iii) on the basis of being the only known locality for *Momordica mossambica* (DD) and *Rotheca sansibarensis* var. *eratensis*, and under criterion A(iv) as it is one of fewer than five sites known globally for *Pavetta micropunctata*.

Priority species (IPA Criteria A and B)

FAMILY	TAXON	IPA CRITERION A	IPA CRITERION B	≥ 1% OF GLOBAL POP'N	≥ 5% OF NATIONAL POP'N	IS 1 OF 5 BEST SITES NATIONALLY	ENTIRE GLOBAL POP'N	SPECIES OF SOCIO-ECONOMIC IMPORTANCE	ABUNDANCE AT SITE
Cucurbitaceae	*Momordica mosambica*	A(iii)	B(ii)	✓	✓	✓	✓		unknown
Euphorbiaceae	*Croton kilwae*	A(i)			✓	✓	✓		common
Fabaceae	*Indigofera pseudomoniliformis*	A(i)	B(ii)	✓	✓	✓			unknown
Lamiaceae	*Rotheca sansibarensis* var. *eratensis*	A(iii)	B(ii)	✓	✓	✓	✓		unknown
Rubiaceae	*Coffea zanguebariae*	A(i)				✓		✓	unknown
Rubiaceae	*Pavetta micropunctata*	A(iv)	B(ii)	✓	✓	✓			unknown
Sapindaceae	*Allophylus torrei*	A(i)	B(ii)	✓	✓	✓			unknown
		A(i): 4 ✓ A(iii): 2 ✓ A(iv): 1 ✓	B(ii): 5						

Protected areas and other conservation designations

CONSERVATION AREA TYPE	CONSERVATION AREA NAME	RELATIONSHIP OF IPA TO CONSERVATION AREA
No formal protection	N/A	
Key Biodiversity Area	Eráti	IPA encompasses protected/conservation area

Threats

THREAT	SEVERITY	TIMING
Shifting agriculture	medium	Ongoing – trend unknown
Small-holder farming	medium	Ongoing – trend unknown
Logging & wood harvesting	unknown	Ongoing – trend unknown
Increase in fire frequency/intensity	unknown	Ongoing – trend unknown

MATIBANE FOREST

Assessors: Jo Osborne, Iain Darbyshire, Hermenegildo Matimele

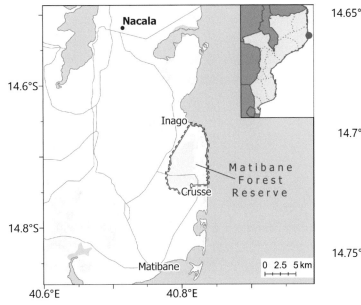
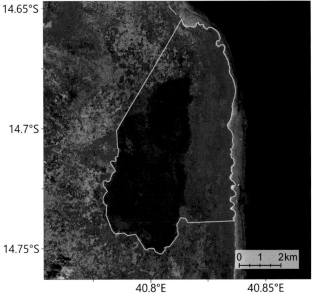

INTERNATIONAL SITE NAME		Matibane Forest	
LOCAL SITE NAME (IF DIFFERENT)		Floresta de Matibane	
SITE CODE	MOZTIPA005	PROVINCE	Nampula

LATITUDE	-14.71331	LONGITUDE	40.80208
ALTITUDE MINIMUM (m a.s.l.)	0	ALTITUDE MAXIMUM (m a.s.l.)	102
AREA (km²)	45.4	IPA CRITERIA MET	A(i), C(iii)

Androstachys johnsonii and *Icuria dunensis* coastal dry forest at Matibane (JO)

Site description

Matibane Forest is a coastal site in Mossuril District of Nampula Province, ca. 20 km to the south-east of Nacala town and 12 km north of Matibane village. The site consists of coastal forest to the west and a zone of coastal scrub and thicket, ca. 2 km wide, to the east between the forest edge and shore. Matibane was gazetted as a Forest Reserve in 1957 (Portaria No. 8459 of 22.7.57), originally for the protection and recovery of *Androstachys johnsonii*, an overexploited timber tree. The reserve is co-managed by government and local communities. Although there is no settlement within the core zone of the Forest Reserve, there are three communities living around the edge: Crusse to the south, and Inago and Namalasa to the north-west. Local people cultivate machambas (small-scale areas of cultivation) within the coastal zone. The intact Rovuma coastal dry forest at this site is of global conservation importance and the site is of particular interest for being the only locality globally where the threatened legume genera *Icuria* and *Micklethwaitia*, both endemic to Mozambique, occur together.

Botanical significance

Matibane Forest is of high botanical importance for the ca. 23 km² of the intact coastal dry forest within the southern portion of the proposed Rovuma Centre of Plant Endemism (Burrows & Timberlake 2011; Darbyshire *et al.* 2019a), this being one of the most highly threatened and fragmented habitat types in Mozambique. It is a critical site for *Icuria dunensis*, a globally Endangered tree species endemic to Mozambique for which Matibane Forest is the northernmost known locality (Darbyshire *et al.* 2019e). This tree is highly restricted and known from few sites along a ca. 360 km stretch of coastline, where it occurs in small and fragmented forest patches, many of which are under increasing pressure from human encroachment. Matibane Forest is currently the only protected area where this tree occurs. Here, *Icuria dunensis* grows with the important timber species *Androstachys johnsonii* (LC), with approximately 0.84 km² of *Icuria*-dominated forest recorded in the central portion of this site. Matibane has been identified as one of only three *Icuria* forests assessed to be in "very good condition" using a Forest Ecological

Condition Index (A. Massingue, pers. comm.), the others being Mogincual [MOZTIPA029] and Moebase [MOZTIPA032].

A second nationally endemic tree species of high importance at this site is *Micklethwaitia carvalhoi* (VU), which is here at the southernmost end if its range and is locally abundant with some dominant or co-dominant stands, particularly in the northern portion of the reserve.

Other threatened species of note at this site are *Hexalobus mossambicensis* (VU), *Monanthotaxis trichantha* (VU), *Pavetta dianeae* (EN), *Premna tanganyikensis* (VU) and *Tarenna pembensis* (EN). In total, Matibane Forest supports nine nationally endemic plant species and fourteen species that are threatened with extinction. Botanical inventory of this important site is currently incomplete, and a full survey is desirable as it may well reveal further rare and threatened plant species.

Habitat and geology

The principal habitat at Matibane is low-lying, semi-evergreen, dry coastal forest with flat or slightly undulating topography on deep sands (Müller *et al.* 2005). The forest canopy is dominated by *Androstachys johnsonii*, often forming almost pure stands or mixed with either *Icuria dunensis* or *Micklethwaitia carvalhoi*. Other tree species noted by Müller *et al.* (2005) include *Afzelia quanzensis*, *Albizia forbesii*, *A. glaberrima*, *Balanites maughamii*, *Fernandoa magnifica*, *Markhamia obtusifolia*, *Mimusops caffra*, *Schrebera trichoclada* and *Sclerocarya birrea*, but none of these are ever dominant. The forest understorey is dense and rich in small trees, shrubs and lianas, with a *Combretum* sp., several *Strychnos* spp. and a number of Rubiaceae including *Hyperacanthus*

microphyllus – here at the northernmost extent of its range – all common; *Hymenocardia ulmoides* was also noted to be common by Müller *et al.* (2005). To the north-east of the forest, between the forest edge and the shore, there is a zone of coastal scrub and thicket vegetation.

Matibane is noted to have a particularly high density of *Icuria* trees where it is dominant, although it is restricted here to less than 1 km² of the Forest Reserve (A. Massingue, pers. comm.). Seedling and sapling recruitment has been observed to be good.

Rainfall is rather low, with an average annual rainfall of 800 mm per year recorded at nearby Nacala, which is concentrated in the months December to March. Temperatures remain high throughout the year with average monthly high temperatures varying from 29–31°C at Nacala (weatherbase.com).

Conservation issues

Illegal logging and charcoal production pose a serious threat to Matibane Forest. *Androstachys johnsonii*, known as 'mecrusse', is selectively logged from within the core zone of the forest reserve. *Micklethwaitia carvalhoi*, known as 'ivate', is cut for construction locally, particularly around the north edge of the forest and is also used for charcoal making within the forest. Environmental officers (Fiscais) from each of the three local communities are employed part time (for example, three days per week) but have insufficient resources to control the illegal activity. They patrol on foot as they have no access to vehicles. Some illegal hunting of 'impala' takes place within the forest. A non-native *Opuntia* (cactus) species occurs along a disused railway line but does not appear to have spread through the forest.

Vitex carvalhoi (JB)

Seedlings of *Icuria dunensis* at Matibane Forest (JO)

Matibane Forest Reserve (JO)

The current extent of the Forest Reserve is smaller than that originally gazetted and most of the natural habitat in the original southwest extent of the reserve has now been cleared. Evidence from Google Earth (2021) imagery shows some notable declines in forest extent in the northern section of the current reserve buffer, but that this has stabilised since the mid-1990s. In these areas, fallow fields and regenerating areas dominated by *Hyparrhenia* spp. are frequent (Müller *et al.* 2005). These openings appear to be maintained by regular fires.

This site is also included within Mozambique's Key Biodiversity Areas network on the basis of its population of *Icuria dunensis* (WCS *et al.* 2021).

Key ecosystem services
The dry coastal forest at Matibane provides local timber for construction, in addition to medicinal plants that are used by the local communities. Timber species include *Androstachys johnsonii* 'mecrusse' and *Micklethwaitia carvalhoi* 'ivate'.

The coastal scrub and thicket vegetation provides timber, rope and m'siro, a cosmetic product extracted from the roots of *Olax dissitiflora*. Economically important timber species include *Afzelia quanzensis* 'chamfuta', *Dalbergia melanoxylon* 'pau-preto' and *Sterculia quinqueloba* 'metonha' while rope is made from the bark of *Sterculia africana*. Other species from coastal scrub and thicket vegetation are used locally for harvesting fruit (*Mimusops caffra*, *Sclerocarya birrea*, *Strychnos cocculoides* and *Strychnos spinosa*) and wood for implement handles (*Strychnos cocculoides*).

It is also likely that both the dry coastal forest and coastal scrub and thicket act as a source pool for insect pollinators, contributing to the diversity of pollinators available for local machamba agriculture.

Ecosystem service categories

- Provisioning – Food
- Provisioning – Raw materials
- Provisioning – Medicinal resources
- Regulating services – Pollination

IPA assessment rationale

Matibane Forest qualifies as an IPA under criteria A and C. Under sub-criterion A(i), the site supports important populations of 14 globally threatened plant species. Under criterion C(iii) the site includes a significant area of coastal dry forest of the Rovuma CoE, including patches dominated by *Icuria dunensis* (EN) and *Micklethwaitia carvalhoi* (VU), for which this is considered to be one of the five best sites nationally. Overall, the site is known to support 10 plant species of high conservation importance as defined under IPA criterion B(ii). Nine of these are nationally endemic species, while the tenth is a regional endemic with a restricted range of less than 10,000 km^2. As there are fewer than 16 species of high conservation importance, the site does not meet the threshold to qualify as an IPA under criterion B(ii) for Mozambique. However, the flora at this site has not been extensively surveyed to date and may support other qualifying species.

Priority species (IPA Criteria A and B)

FAMILY	TAXON	IPA CRITERION A	IPA CRITERION B	≥ 1% OF GLOBAL POP'N	≥ 5% OF NATIONAL POP'N	IS 1 OF 5 BEST SITES NATIONALLY	ENTIRE GLOBAL POP'N	SPECIES OF SOCIO-ECONOMIC IMPORTANCE	ABUNDANCE AT SITE
Annonaceae	*Hexalobus mossambicensis*	A(i)	B(ii)	✓	✓	✓		✓	unknown
Annonaceae	*Monanthotaxis trichantha*	A(i)		✓	✓	✓			unknown
Commelinaceae	*Aneilema mossambicense*		B(ii)						unknown
Fabaceae	*Icuria dunensis*	A(i)	B(ii)	✓	✓	✓		✓	abundant
Fabaceae	*Micklethwaitia carvalhoi*	A(i)	B(ii)	✓	✓	✓		✓	abundant
Fabaceae	*Millettia mossambicensis*		B(ii)					✓	unknown
Lamiaceae	*Premna tanganyikensis*	A(i)		✓	✓	✓			unknown
Lamiaceae	*Vitex carvalhi*	A(i)		✓	✓	✓			unknown
Loranthaceae	*Agelanthus longipes*	A(i)		✓	✓	✓			unknown
Rubiaceae	*Paracephaelis trichantha*	A(i)			✓	✓			unknown
Rubiaceae	*Pavetta curalicola*		B(ii)						unknown
Rubiaceae	*Pavetta dianeae*	A(i)	B(ii)	✓	✓	✓			occasional
Rubiaceae	*Pavetta mocambicensis*	A(i)	B(ii)	✓	✓	✓			unknown
Rubiaceae	*Psydrax micans*	A(i)				✓			unknown
Rubiaceae	*Tarenna pembensis*	A(i)	B(ii)	✓	✓	✓			occasional
Rutaceae	*Zanthoxylum tenuipedicellatum*	A(i)	B(ii)	✓	✓	✓			unknown
Sapotaceae	*Vitellariopsis kirkii*	A(i)			✓	✓			unknown
		A(i): 14 ✓	B(ii): 10						

Threatened habitats (IPA Criterion C)

HABITAT TYPE	IPA CRITERION C	≥ 5% OF NATIONAL RESOURCE	≥ 10% OF NATIONAL RESOURCE	IS 1 OF 5 BEST SITES NATIONALLY	ESTIMATED AREA AT SITE (IF KNOWN)
Rovuma *Micklethwaitia* Coastal Dry Forest [MOZ-12b]	C(iii)		✓	✓	
Rovuma *Icuria* Coastal Dry Forest [MOZ-12c]	C(iii)			✓	0.84

Protected areas and other conservation designations

CONSERVATION AREA TYPE	CONSERVATION AREA NAME	RELATIONSHIP OF IPA TO CONSERVATION AREA
Forest Reserve (conservation)	Reserva Florestal de Matibane	IPA encompasses protected/conservation area
Key Biodiversity Area	Reserva Florestal de Matibane	protected/conservation area encompasses IPA

Threats

THREAT	SEVERITY	TIMING
Small-holder farming	low	Ongoing – trend unknown
Logging & wood harvesting	medium	Ongoing – trend unknown

GOA AND SENA ISLANDS

Assessors: Iain Darbyshire, Papin Mucaleque

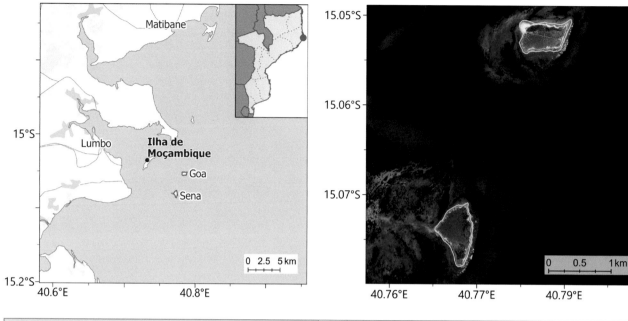

INTERNATIONAL SITE NAME		Goa and Sena Islands	
LOCAL SITE NAME (IF DIFFERENT)		Ilhas Goa e Sena	
SITE CODE	MOZTIPA010	PROVINCE	Nampula

LATITUDE	-15.05310	LONGITUDE	40.78440
ALTITUDE MINIMUM (m a.s.l.)	0	ALTITUDE MAXIMUM (m a.s.l.)	8
AREA (km²)	0.65	IPA CRITERIA MET	A(i)

Site description

Goa and Sena are two small, low-lying coral islets separated by ca. 2.6 km in the Indian Ocean in Ilha de Moçambique (Mozambique Island) District of Nampula Province, measuring ca. 0.3 km² and 0.35 km², respectively. These islands are located in the mouth of Mossuirl Bay, approximately 6.5–8 km offshore from the mainland. They lie east and southeast of the famous Ilha de Moçambique, a World Heritage Site which is one of the oldest European settlements in East Africa and an important early trading centre under Portuguese rule from the early sixteenth century until Independence. Ilha de Moçambique is now almost entirely built up away from the beaches, but the islands of Goa and Sena are largely undisturbed. Sena Island is also known locally as the Ilha das Cobras (Snake Island).

Botanical significance

These islands support important examples of coral rag thicket, a vegetation type that occurs only sporadically on the northern Mozambique coast – this is one of the most southern localities for this habitat nationally. The coral rag thicket supports one endemic species, *Euphorbia angularis* (VU), which is common throughout Goa Island but is absent on Sena Island (Mucaleque 2020a). The islands also support two near-endemic species of *Barleria* – *B. setosa* (EN) and *B. laceratiflora* (EN) (Darbyshire *et al.* 2015; Luke *et al.* 2015b; Darbyshire 2018). The former is common on both Goa and Sena (Mucaleque 2020a); it is also known historically from Ilha de Moçambique and the adjacent mainland Cabeceira Pequena near Mossuril; however, due to habitat loss at these latter two locations, this species may now be restricted to Goa and Sena Islands.

Barleria laceratiflora was recorded from Goa Island in 1947; it was not refound on either island during a brief botanical visit in 2020 (Mucaleque 2020a) but could easily be overlooked if not in flower. It is known elsewhere only from Lindi Bay in coastal southeast Tanzania where it is considered to be threatened by development and habitat loss (Luke *et al.* 2015b). To our knowledge, only Goa Island had previously been surveyed for its plant diversity before the brief botanical survey on both islands in 2020 for the TIPAs: Mozambique project.

Habitat and geology

The islands are surrounded by coral reefs and are formed from exposed coral rag deposits. Two main vegetation types are recorded: coral rag thicket and mangroves. On Goa Island, the coral rag thicket is common throughout and forms dense impenetrable stands with both deciduous and evergreen shrubby elements. The dominant species here are *Euphorbia angularis*, *Grewia glandulosa*, *Pemphis acidula* and *Suriana maritima*. The thicket on Sena is somewhat less dense and shorter, and *E. angularis* is absent, whilst *Salvadora persica* is more numerous. *Euphorbia tirucalli* and *Aloe* and *Sansevieria* (*Dracaena*) species are also noted in the Sena thickets (Mucaleque 2020a).

On Goa Island, the mangroves are restricted to the northern portion, whilst on Sena they are more widespread especially on the eastern side, and are associated with open saline pools. In both cases, *Rhizophora mucronata* is the dominant species, with *Pemphis acidula* occurring commonly along the mangrove margins (Mucaleque 2020a).

Conservation issues

Although the focus of the Ilha de Moçambique UNESCO World Heritage Site (WHS) is on the fortifications and architecture of the main island, and its association with the history of early navigation and trade in the Indian Ocean, Goa and Sena Islands fall within the proposed WHS buffer zone (UNESCO 2020). This IPA is not included within Mozambique's Key Biodiversity Areas network at present.

At present there is very limited disturbance on both islands. Goa Island is inhabited only by a lighthouse keeper on the eastern side and is otherwise only visited occasionally by fishermen. A previous occupant of the lighthouse grazed cattle on the island and this encouraged other local residents to bring their cattle, but this activity ceased in 2008 and has not caused lasting damage (Mucaleque 2020a). The coral rag thicket is largely intact except for two paths that cross the island north to south and east to west; *Barleria setosa* is quite frequent along the edges of these paths. An introduced *Opuntia* species (cactus) is present in small numbers but does not appear to be particularly invasive. Otherwise, the only problem is with litter and debris from the sea. On Sena Island, fishermen set up temporary camps with small huts built from harvested wood but these have only a very minor impact on the vegetation.

Tourism is expanding rapidly along the coast of Mossuril District, and the Ilha de Moçambique World Heritage Site is one of the fastest growing tourist destinations in Mozambique both for its historical

Goa Island (PM)

Coral rag thicket on Sena Island (PM)

interest and for its beautiful beaches. There is a concern that this growing tourist industry will expand to the nearby Goa and Sena Islands in the future, which could be damaging if not carefully controlled. However, previous plans to develop tourist lodges on Goa Island were rejected by government, in part due to the island's association with the World Heritage Site (Mucaleque 2020a). At present, only very few tourists visit these islands, and Sena remains difficult to access as there is no regular boat link.

Another potential future threat is from increased extreme weather events and flooding in light of human-induced climate change; these islands are so low-lying (mostly below 5 m) that they could be badly impacted by rising sea levels or increased storm events, although the intact vegetation may offer some resilience.

Key ecosystem services

The islands have some potential as an ecotourism destination but this would have to be carefully managed in order to prevent disturbance to the fragile habitats. They provide important habitat for marine species, including turtle species, and for migratory birds (Mucaleque 2020a).

Ecosystem service categories

- Regulating services – Erosion prevention and maintenance of soil fertility
- Habitat or supporting services – Habitats for species
- Habitat or supporting services – Maintenance of genetic diversity
- Cultural services – Tourism
- Cultural services – Aesthetic appreciation and inspiration for culture, art and design

IPA assessment rationale

Goa and Sena Islands qualify as an IPA under Criterion A(i), as they support critical populations of three globally threatened species: the endemic *Euphorbia angularis* which is considered to be Vulnerable, and the two Endangered, near-endemic *Barleria* species, *B. setosa* and *B. laceratiflora*; based on current knowledge, this IPA is considered to be the most important site globally for these three species.

Mangroves on Sena Island (PM)

Euphorbia angularis on Goa Island (PM)

Barleria setosa on Sena Island (PM)

Priority species (IPA Criteria A and B)

FAMILY	TAXON	IPA CRITERION A	IPA CRITERION B	≥ 1% OF GLOBAL POP'N	≥ 5% OF NATIONAL POP'N	IS 1 OF 5 BEST SITES NATIONALLY	ENTIRE GLOBAL POP'N	SPECIES OF SOCIO-ECONOMIC IMPORTANCE	ABUNDANCE AT SITE
Acanthaceae	*Barleria laceratiflora*	A(i)	B(ii)	✓	✓	✓			unknown
Acanthaceae	*Barleria setosa*	A(i)	B(ii)	✓	✓	✓			common
Euphorbiaceae	*Euphorbia angularis*	A(i)	B(ii)	✓	✓	✓	✓		common
		A(i): 3 ✓	B(ii): 3						

Protected areas and other conservation designations

CONSERVATION AREA TYPE	CONSERVATION AREA NAME	RELATIONSHIP OF IPA TO CONSERVATION AREA
UNESCO World Heritage Site	Island of Mozambique (Ilha de Moçambique): Buffer Zone	protected/conservation area encompasses IPA

Threats

THREAT	SEVERITY	TIMING
Tourism & recreation areas	unknown	Future – inferred threat
Logging & wood harvesting	low	Ongoing – stable
Climate change & severe weather – Storms & flooding	unknown	Future – inferred threat

MOGINCUAL

Assessors: Iain Darbyshire, Alice Massingue

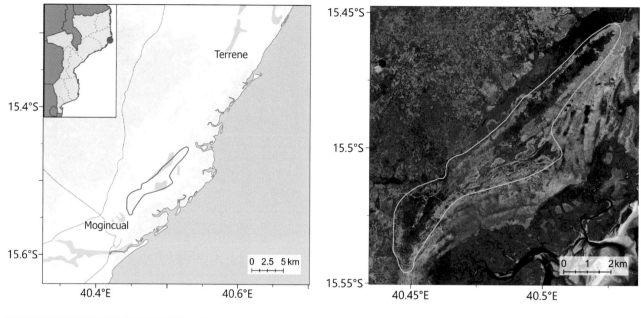

INTERNATIONAL SITE NAME		Mogincual	
LOCAL SITE NAME (IF DIFFERENT)		–	
SITE CODE	MOZTIPA029	PROVINCE	Nampula

LATITUDE	-15.49730	LONGITUDE	40.48840
ALTITUDE MINIMUM (m a.s.l.)	4	ALTITUDE MAXIMUM (m a.s.l.)	48
AREA (km²)	21.0	IPA CRITERIA MET	A(i), C(iii)

Site description

The Mogincual IPA is located northeast of the town of Mogincual in the District of the same name in coastal Nampula Province. The site covers an area of seasonally wet coastal sands with patches of coastal dry forest. This site is a part of the Rovuma Centre of Plant Endemism (Burrows & Timberlake 2011; Darbyshire *et al.* 2019a).

Botanical significance

The Mogincual IPA is of high botanical importance for holding the second-largest confirmed area of *Icuria*-dominated coastal dry forest globally. *Icuria dunensis* ('icuri' or 'ncuri') is an endemic species and genus of leguminous tree that forms mono-dominant or co-dominant stands in small and isolated patches along a ca. 360 km stretch of the Mozambique coastline (Darbyshire *et al.* 2019e).

The area of *Icuria*-dominated forest surveyed to date at Mognicual is approximately 3.25 km² but the total area is potentially larger as some forest patches that are believed to contain *Icuria* in the northeastern portion of the IPA have not yet been surveyed. It has been identified as one of only three *Icuria* forests assessed to be in "very good condition" using a Forest Ecological Condition Index (A. Massingue, unpubl. data).

Recent survey work in this forest revealed a small but significant population of the Endangered endemic tree species *Scorodophloeos torrei*. At least 10 mature individuals were found here, but a full population survey was not carried out (A. Massingue, pers. obs.). This species is otherwise known from only three subpopulations, two of which are highly threatened (Darbyshire & Rokni

2020b), hence Mogincual is a globally important site for this species. Other rare and threatened species are likely to be uncovered at this site following a more complete botanical inventory.

Habitat and geology

Icuria dunensis here forms moderately dense stands with observed seedling/sapling recruitment. *Icuria* tree density at this site is estimated at 280 per ha, significantly lower than at Matibane Forest Reserve where a density of almost 900 per ha has been estimated (A. Massingue, pers. obs.). As elsewhere within their range, the *Icuria* forests are associated with low-lying sands with a high water table. The forest patches lie within a matrix of more open dune vegetation, and with some intervening depressions that hold surface water in the wet season.

Species diversity within the IPA has not been well documented to date, and it is likely that this site will contain other plant species of conservation interest including Rovuma Centre endemics.

The climate is highly seasonal, with rainfall peaking in December to March; the average annual precipitation is ca. 1,050 mm.

Conservation issues

The Mogincual IPA is not currently protected. However, the forest patches are in good condition and do not appear to have been reduced in size by anthropogenic activities over recent decades, as evidenced by historical satellite imagery available on Google Earth (2021). As with most areas of coastal Mozambique, Mogincual District is experiencing a significant and sustained rise in human population, with an increase of 72% recorded between the 1997 and 2017 censuses (Instituto Nacional de Estatistica Moçambique 2021) and this may result in increasing pressure on the coastal habitats in the

future. Elsewhere within its range, *Icuria* trees are threatened by stripping of the bark for fishing boat construction and for making ropes (Darbyshire *et al.* 2019e), but it is not clear if this is the case at Mogincual.

Key ecosystem services

The coastal forest and scrub vegetation, including the stands of *Icuria*, help to stabilise and protect the coastal sand deposits and so prevent excessive erosion during extreme weather events. They are also likely to provide important habitat and supporting services for fauna.

Ecosystem service categories

- Regulating services – Moderation of extreme events
- Regulating services – Erosion prevention and maintenance of soil fertility
- Habitat or supporting services – Habitats for species
- Habitat or supporting services – Maintenance of genetic diversity

IPA assessment rationale

Mogincual qualifies as an Important Plant Area under criterion A(i) in view of its globally important populations of *Icuria dunensis* and *Scorodophloeos torrei*, both of which are assessed as Endangered on the IUCN Red List. It also qualifies under criterion C(iii) as the *Icuria* coastal dry forest is a nationally threatened and range-restricted habitat, and the Mogincual IPA is estimated to contain over 10% of the total area of remaining *Icuria* forest.

Priority species (IPA Criteria A and B)

FAMILY	TAXON	IPA CRITERION A	IPA CRITERION B	≥ 1% OF GLOBAL POP'N	≥ 5% OF NATIONAL POP'N	IS 1 OF 5 BEST SITES NATIONALLY	ENTIRE GLOBAL POP'N	SPECIES OF SOCIO-ECONOMIC IMPORTANCE	ABUNDANCE AT SITE
Fabaceae	*Icuria dunensis*	A(i)	B(ii)	✓	✓	✓			frequent
Fabaceae	*Scorodophloeus torrei*	A(i)	B(ii)	✓	✓	✓			occasional
		A(i): 2 ✓	B(ii): 2						

Threatened habitats (IPA Criterion C)

HABITAT TYPE	IPA CRITERION C	≥ 5% OF NATIONAL RESOURCE	≥ 10% OF NATIONAL RESOURCE	IS 1 OF 5 BEST SITES NATIONALLY	ESTIMATED AREA AT SITE (IF KNOWN)
Rovuma *Icuria* Coastal Dry Forest [MOZ-12c]	C(iii)		✓	✓	3.25

Protected areas and other conservation designations

CONSERVATION AREA TYPE	CONSERVATION AREA NAME	RELATIONSHIP OF IPA TO CONSERVATION AREA
No formal protection	N/A	

Threats

THREAT	SEVERITY	TIMING
Gathering terrestrial plants	unknown	Future – inferred threat

QUINGA

Assessors: Iain Darbyshire, Alice Massingue

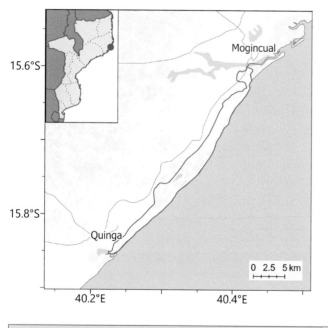
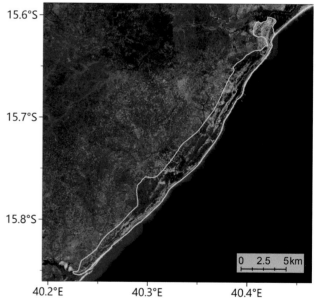

INTERNATIONAL SITE NAME		Quinga	
LOCAL SITE NAME (IF DIFFERENT)		–	
SITE CODE	MOZTIPA030	PROVINCE	Nampula

LATITUDE	-15.73550	LONGITUDE	40.34300
ALTITUDE MINIMUM(m a.s.l.)	0	ALTITUDE MAXIMUM (m a.s.l.)	85
AREA (km²)	63	IPA CRITERIA MET	A(i), C(iii)

Coastal habitats within the Quinga IPA (AM)

Site description

The Quinga IPA is located in the coastal Liupo District of Nampula Province in northern Mozambique, between the latitudes of −15.85° and −15.60°. It comprises a stretch of ca. 37 km of coastline between the village of Quinga in the southwest and the estuary of the Mogincual River in the northeast, and extends inland for up to 3.5 km. This site contains some of the most intact coastal dune vegetation systems in Nampula Province and includes significant patches of well-preserved coastal dry forest. Whilst this area has been highly under-botanised to date, it is known to contain globally important populations of several threatened species and is very likely to prove to be one of the most important remaining sites in the southern portion of the proposed Rovuma Centre of Plant Endemism (CoE).

Botanical significance

The coastal dune formations between Quinga and the Mogincual River mouth support a number of rare and threatened species and habitats of the proposed Rovuma CoE (Burrows & Timberlake 2011; Darbyshire et al. 2019a). Quinga is believed to be a critical site for Blepharis dunensis, an Endangered local endemic of coastal northern Mozambique. This species has been recorded here at Quinga Beach growing in open dry dune scrub (A.R. Torre & J. Paiva #11439), and was recently re-recorded at this site (A. Massingue, pers. obs.). Given that the other two known localities for B. dunensis – Angoche and Pebane – are both now highly disturbed and with little intact dune vegetation remaining, the Quinga IPA may be the prime locality globally for this species (Darbyshire et al. 2019f).

This is also a globally important site for the endangered Icuria dunensis ('icuri' or 'ncuri') which forms mono-dominant or co-dominant dry forest stands. Several patches of Icuria forest have been confirmed within this IPA, including well-preserved patches in the far north of the site along the margin of the Mogincual River Estuary which are reported to be in good condition (A. Massingue, pers. obs.; Darbyshire et al. 2019e). Whilst these forests have not been surveyed in full, a review of satellite imagery available on Google Earth (2021) imagery suggests that up to 6 km² of forest containing Icuria may be present within the IPA, second only in area to the Moebase Icuria forests [MOZTIPA032].

The Critically Endangered shrub Warneckea sessilicarpa has recently been discovered here, where it was found to be locally common to dominant along the dunes at Quinga beach (A. Massingue, pers. obs.). Elsewhere within its narrow range, this species is associated with Icuria forest and so it is likely to occur in and around the Icuria patches of the Quinga IPA.

Other interesting species recorded at this site to date include the Mozambican endemics *Dracaena* (formerly *Sansevieria*) *subspicata* (not assessed but likely to be LC) and *Chamaecrista paralias* (LC). The Vulnerable shrub or climbing shrub *Acacia* (*Senegalia*) *latistipulata* may also occur here as it has been recorded from just outside the IPA boundary, along the route between Mogincual and Quinga (*A.R. Torre & J. Paiva* #11496). Given the highly incomplete botanical survey at this site to date, the likelihood of finding other rare species is high. This may include other species endemic to the southern portion of the Rovuma CoE, such as *Scorodophloeus torrei* (EN) and *Ammannia moggii* (CR). The latter of these is known to date only from the Angoche area, ca. 50 km to the southwest of the Quinga IPA, and is Critically Endangered due to extensive sand mining operations there (Mucaleque 2020b). This species should be sought for in the seasonal wetlands of the coastal dune systems at Quinga.

Habitat and geology

The vegetation of this site is a mosaic of coastal thicket, woodland and dry forest on coastal dune formations, together with extensive areas of seasonal wetlands and damp grasslands in the inter-dunal slacks. The coastal sands are rich in heavy minerals including ilmenite (titanium ore) (Kenmare Resources 2018). The vegetation assemblages have not been studied in detail to date. As elsewhere within its range, the *Icuria* forests are associated with low-lying sands with a high water table. The coastal thicket vegetation is dominated by *Sideroxylon inerme*, *Flacourtia indica* and *Mimosops* cf. *obovata* (A. Massingue, pers. obs.); this is the "Dune Thicket-Forest [14b]" vegetation type of Burrows *et al.* (2018). Areas with better developed soils, e.g. along small rivers, support a woodland with *Afzelia quanzensis* and *Millettia stuhlmannii* amongst the dominant species (A. Massingue, pers. obs.). It is likely that the wooded areas on the inland side of the IPA include areas of miombo woodland.

The climate is highly seasonal, with a hot wet season from December to March/April, peaking in January. At nearby Mogincual to the north, annual rainfall is approximately 1,037 mm per year. Temperatures peak in December with an average high of 33°C (climatedata.eu).

Coastal thicket at Quinga (AM)

Interior of *Icuria*-dominated forest within the Quinga IPA (AM)

Conservation issues

There is no formal protection or management for biodiversity within this IPA at present. Some of the woodland habitats are subject to continuing encroachment, particularly on the landward side of the IPA where smallholder farms increase in density moving inland. There have been notable expansions of farmland clearly evident over the past 20 years from historical imagery available on Google Earth (2021). However, there are still significant areas of intact habitats along the coastal strip and within the dune systems. At present, some of these areas are not visited frequently except by fishing communities. In such areas, the most likely threat is from fire which can encroach into these coastal habitats from the neighbouring agricultural lands inland; evidence of recent fires was observed to be widespread during surveys in 2017 (A. Massingue, pers. obs.). In the vicinity of Quinga, there is greater footfall around the beach, with much fishing activity and recreation, and this may impact the quality of the habitat. There is an urgent need to protect the remaining intact coastal habitats at this site, given their high botanical importance.

A significant future threat lies in the fact that a large portion of the IPA falls within the Quinga North mining concession for which an exploration license is held by Kenmare Resources plc. who operate the Moma Titanium Mineral Mine to the southwest. The Quinga North concession is believed to hold commercially viable concentrations of heavy minerals, including ilmenite, rutile and zircon. Reconnaissance exploration of this concession began in 2018 (Kenmare Resources 2018).

Key ecosystem services

Although not surveyed in detail, the ecosystem services provided by this site are likely to be considerable. The dense thicket, woodland and forest vegetation help to stabilise the dune systems, which would otherwise be exposed to erosion from the Indian Ocean. These habitats are also likely to provide a range of provisioning services including building materials and fuelwood, although this requires more sustainable management. They also provide an important habitat for a range of fauna, whilst the rich mosaic of habitats including the seasonal coastal wetlands are likely to be of importance for migrating birds.

Despite the extensive beaches and beautiful coastline, tourism potential at this site is considered to be low at present in view of its isolation – the main access is the road from Liupo to Quinga, a distance of ca. 40 km.

Ecosystem service categories

- Provisioning – Food
- Provisioning – Raw materials
- Regulating Services – Carbon sequestration and storage
- Regulating services – Moderation of extreme events
- Regulating services – Erosion prevention and maintenance of soil fertility
- Habitat or supporting services – Habitats for species
- Habitat or supporting services – Maintenance of genetic diversity
- Cultural services – Recreation and mental and physical health

Warneckea sessilicarpa (AM)

IPA assessment rationale

Quinga qualifies as an IPA under criterion A(i), containing globally important populations of three threatened species: *Icuria dunensis* (EN), *Blepharis dunensis* (EN) and *Warneckea sessilicarpa* (CR). It is the only site within the current IPA network in Mozambique to contain *Blepharis dunensis*, and Quinga is considered likely to be a global stronghold for this species. Quinga also qualifies under criterion C(iii) as it contains up to 6 km² of Rovuma *Icuria*-dominated coastal dry forest.

Blepharis dunensis (AM)

Priority species (IPA Criteria A and B)

FAMILY	TAXON	IPA CRITERION A	IPA CRITERION B	≥ 1% OF GLOBAL POP'N	≥ 5% OF NATIONAL POP'N	IS 1 OF 5 BEST SITES NATIONALLY	ENTIRE GLOBAL POP'N	SPECIES OF SOCIO-ECONOMIC IMPORTANCE	ABUNDANCE AT SITE
Acanthaceae	*Blepharis dunensis*	A(i)	B(ii)	✓	✓	✓			unknown
Fabaceae	*Icuria dunensis*	A(i)	B(ii)	✓	✓	✓			occasional
Melastomataceae	*Warneckea sessilicarpa*	A(i)	B(ii)	✓	✓	✓			frequent
		A(i): 3 ✓	B(ii): 3						

Threatened habitats (IPA Criterion C)

HABITAT TYPE	IPA CRITERION C	≥ 5% OF NATIONAL RESOURCE	≥ 10% OF NATIONAL RESOURCE	IS 1 OF 5 BEST SITES NATIONALLY	ESTIMATED AREA AT SITE (IF KNOWN)
Rovuma *Icuria* Coastal Dry Forest [MOZ-12c]	C(iii)		✓	✓	5.8

Protected areas and other conservation designations

CONSERVATION AREA TYPE	CONSERVATION AREA NAME	RELATIONSHIP OF IPA TO CONSERVATION AREA
No formal protection	N/A	

Threats

THREAT	SEVERITY	TIMING
Small-holder farming	medium	Ongoing – increasing
Mining & quarrying	unknown	Future – inferred threat
Logging & wood harvesting	medium	Ongoing – trend unknown
Increase in fire frequency/intensity	unknown	Ongoing – increasing

MULIMONE FOREST

Assessors: Iain Darbyshire, Camila de Sousa, Tereza Alves, Jaime Rofasse Timóteo, Clayton Langa

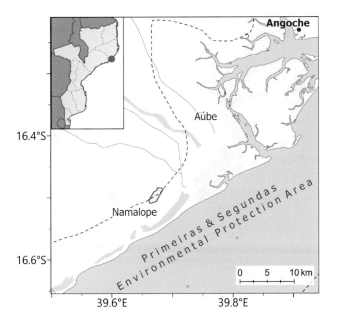

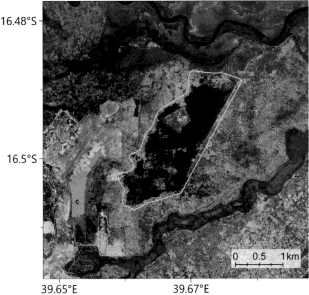

INTERNATIONAL SITE NAME		Mulimone Forest	
LOCAL SITE NAME (IF DIFFERENT)		Floresta de Mulimone	
SITE CODE	MOZTIPA031	PROVINCE	Nampula / Larde

LATITUDE	-16.49840	LONGITUDE	39.67120
ALTITUDE MINIMUM (m a.s.l.)	9	ALTITUDE MAXIMUM (m a.s.l.)	30
AREA (km²)	3.24	IPA CRITERIA MET	A(i), C(iii)

Site description

Mulimone Forest is situated in coastal Larde District of Nampula Province, approximately 55 km ENE of the town of Moma and 40 km SW of the town of Angoche. This small site of 3.24 km² is situated immediately adjacent to the Namalope heavy mineral sand operation of the Moma Titanium Minerals Mine owned by Kenmare Resources plc, one of the world's largest titanium mines. The IPA encompasses a patch of coastal dry forest of the proposed Rovuma Centre of Plant Endemism (Burrows & Timberlake 2011; Darbyshire *et al.* 2019a). It was identified as of biodiversity importance during the Environmental Impact Assessment ahead of commercial exploitation of the Namalope deposit, and has subsequently been protected from development by Kenmare.

Botanical significance

This site contains a globally important stand of *Icuria*-dominated coastal dry forest. *Icuria dunensis* ('icuri' or 'ncuri') is a leguminous tree endemic to Mozambique which forms mono-dominant or co-dominant stands in small and isolated patches along a ca. 360 km stretch of the Mozambique coastline and is assessed as globally Endangered (Darbyshire *et al.* 2019e). The area of *Icuria*-dominated forest at Mulimone is approximately 2.38 km². It is one of only five sites identified globally as being of high importance for *Icuria* forest. A recent vegetation survey of this site (J. Timóteo, unpubl. data) has also revealed the presence of two further globally threatened tree species endemic to Mozambique: *Brachystegia oblonga* (EN) and *Scorodophloeus torrei* (EN) both of which are locally frequent here. This site is of high importance for both these species given that most of the few other known sites are highly threatened. *Brachystegia oblonga*, in particular, is otherwise known only from highly disturbed woodland and dry forest remnants at Moma and at Gobene near Bajone, where it is severely threatened (Alves *et al.* 2014b).

The understorey of the *Icuria* forest has previously been reported to support a population of *Warneckea sessilicarpa* (Alves & Sousa 2007), but the recent surveys have not found that species here (C. de Sousa & J. Timóteo, pers. obs. 2021), so that record requires confirmation before being included in the IPA assessment. This species is noted to be locally common in *Icuria* stands at nearby Pilivili (A. Massingue, pers. comm. 2021).

Habitat and geology

In the areas of closed forest, *Icuria dunensis* forms dominant stands, with *Haplocoelum foliolosum* subsp. *mombasense*, *Brachystegia oblonga* and *Scorodophloeus torrei* also being frequent, and with occasional *Hymenaea verrucosa* amongst other species. Mature *Icuria* trees up to 30 m tall are recorded and substantial regeneration is observed in both *Icuria* and *Brachystegia oblonga* (Alves & Sousa 2007; J. Timóteo, unpubl. data). In areas of more open, disturbed forest, *Icuria* and *Haplocoeleum* are still present but with other species including *Mimusops obtusifolia* and *Olax dissitiflora* with *Strynchnos* sp. abundant (J. Timóteo, unpubl. data).

As elsewhere within its range, the *Icuria* forests are associated with low-lying ancient sand-dune deposits; at Mulimone the forest is on a slightly raised area of white sands. These dunes are rich in heavy minerals including high-grade ilmenite (titanium ore) (Kenmare Resources 2018). A layer of ca. 5 cm with leaf-litter covers the soil surface; this improves soil humidity and decreases soil temperature, allowing for forest regeneration. Immediately beyond the forest boundary, most of the vegetation has been substantially transformed, particularly to the west where the extensive mining operations are surrounded by infrastructure and settlements.

The climate at this site is highly seasonal, with ca. 90% of the rainfall occurring in December to March; the average annual precipitation is approximately 1,050 mm, but with ca. 1,521 mm evapotranspiration (Kassam *et al.* 1981).

Conservation issues

Kenmare Resources acquired the Congolone heavy mineral sand deposit concession in the late 1980s, and began construction of the Moma Titanium Mineral Mine on the Namalope deposit in 2004, with production from 2007 until the present day. Even prior to the development of the Namalope operation, the *Icuria* forest patch in Mulimone was small and clearly demarcated from the surrounding vegetation. Historical satellite imagery available on Google Earth shows that the forest extent in late 1984 was approximately 2.42 km². A portion of forest

Icuria-dominated forest at Mulimone (CS)

Mature *Icuria dunensis* tree (CS)

in the northwest of the site was destroyed in the early 1990s and this opening ("o buraco") expanded through that decade, destroying approximately 0.17 km² of forest. Some natural regeneration can now be observed in this area, although some machambas of cassava have been established there. Since the survey of this forest during the EIA of the Namalope deposit, and the formal description of *Icuria dunensis* in the late 1990s (Wieringa 1999), this site has been protected by Kenmare as part of their programme of environmental and social responsibilities at the Moma Mine site. However, the forest is increasingly threatened by agricultural encroachment from local communities, with areas in the south of the forest in particular having been opened up using fire to clear the land for cassava and cowpea cultivation. Although some of the larger trees are left standing, they are often killed by the fires. This encroachment has accelerated since 2017. Other threats include pit-sawing for timber, cutting of poles and the stripping of *Icuria* bark for use in boat-making (C. Sousa, pers. obs. 2018). It is estimated that ca. 30% of Mulimone Forest has been lost through these activities within the recent past (J. Timóteo *et al.*, unpubl. data). If protected against uncontrolled fires and agriculture encroachment, the forest may recover; enrichment planting in gaps may accelerate this natural regeneration process.

Approximately half of the area of this IPA falls within the extensive (>8,000 km²) Primeiras & Segundas Environmental Protection Area (APAIPS), gazetted in 2012, which extends along the coast south to Pebane and north to Angoche. The *Icuria* forests of the area between Moma and Angoche were highlighted as of high importance in the preliminary assessment of the coastal vegetation within the proposed reserve (Alves & Sousa 2007), but the northern part of the Mulimone Forest falls outside of the boundary. This whole area has also recently been recognised as a Key Biodiversity Area (WCS *et al.* 2021).

Key ecosystem services

The *Icuria* forest helps to stabilise and protect the coastal sand deposits and so prevent excessive erosion during extreme weather events. It also provides important habitat and supporting services in an area that is otherwise heavily transformed. The *Icuria* trees provide a provisioning service for local communities, supplying bark and wood, which could be managed sustainably.

Strip-barking of an *Icuria* tree for boat-making (JM)

Ecosystem service categories

- Provisioning – Raw materials
- Regulating services – Erosion prevention and maintenance of soil fertility
- Habitat or supporting services – Habitats for species
- Habitat or supporting services – Maintenance of genetic diversity

IPA assessment rationale

The Mulimone Forest qualifies as an Important Plant Area under criterion A(i) in view of its important populations of *Icuria dunensis* (EN), *Brachystegia oblonga* (CR) and *Scorodophloeus torrei* (EN). This site also qualifies under criterion C(iii) as the *Icuria* coastal dry forest is a nationally threatened and range-restricted habitat, and as the Mulimone IPA is estimated to contain approximately 10% of the total area of remaining *Icuria* forest and is one of the best five sites globally.

Saplings of *Scorodophloeus torrei* at Mulimone (CS)

Priority species (IPA Criteria A and B)

FAMILY	TAXON	IPA CRITERION A	IPA CRITERION B	≥ 1% OF GLOBAL POP'N	≥ 5% OF NATIONAL POP'N	IS 1 OF 5 BEST SITES NATIONALLY	ENTIRE GLOBAL POP'N	SPECIES OF SOCIO-ECONOMIC IMPORTANCE	ABUNDANCE AT SITE
Fabaceae	*Brachystegia oblonga*	A(i)	B(ii)	✓	✓	✓			frequent
Fabaceae	*Icuria dunensis*	A(i)	B(ii)	✓	✓	✓			abundant
Fabaceae	*Scorodophloeus torrei*	A(i)	B(ii)	✓	✓	✓			frequent
		A(i): 3 ✓	B(ii): 3						

Threatened habitats (IPA Criterion C)

HABITAT TYPE	IPA CRITERION C	≥ 5% OF NATIONAL RESOURCE	≥ 10% OF NATIONAL RESOURCE	IS 1 OF 5 BEST SITES NATIONALLY	ESTIMATED AREA AT SITE (IF KNOWN)
Rovuma *Icuria* Coastal Dry Forest [MOZ-12c]	C(iii)		✓	✓	2.28

Icuria trees left standing in machambas within the Mulimone Forest (CS)

Protected areas and other conservation designations

CONSERVATION AREA TYPE	CONSERVATION AREA NAME	RELATIONSHIP OF IPA TO CONSERVATION AREA
Environmental Protection Area	Primeiras & Segundas Environmental Protection Area (APAIPS)	protected/conservation area overlaps with IPA
Important Bird Area	Primeiras & Segundas Environmental Protection Area (APAIPS)	protected/conservation area overlaps with IPA
Key Biodiversity Area	Primeiras & Segundas Environmental Protection Area (APAIPS)	protected/conservation area overlaps with IPA

Threats

THREAT	SEVERITY	TIMING
Commercial & industrial areas	high	Ongoing – stable
Housing & urban areas	medium	Ongoing – stable
Mining & quarrying	high	Ongoing – stable
Small-holder farming	high	Ongoing – increasing
Gathering terrestrial plants	medium	Ongoing – increasing

MOUNT INAGO AND SERRA MERRIPA

Assessors: Sophie Richards, Iain Darbyshire

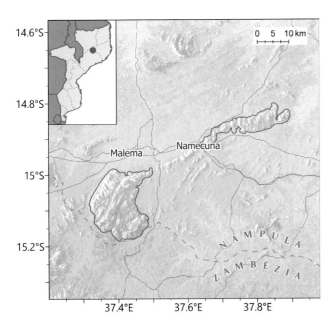

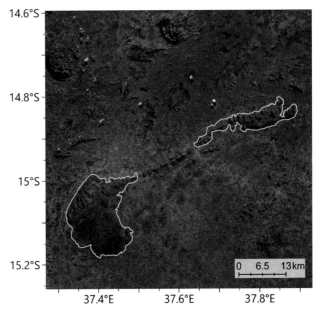

INTERNATIONAL SITE NAME		Mount Inago and Serra Merripa	
LOCAL SITE NAME (IF DIFFERENT)		Monte Inago e Serra Merripa	
SITE CODE	MOZTIPA048	PROVINCE	Nampula

Section of Mount Inago with agricultural land in the foreground (JB)

Steep slopes of Inago (JB)

LATITUDE	-14.97373	LONGITUDE	37.58455
ALTITUDE MINIMUM (m a.s.l.)	655	ALTITUDE MAXIMUM (m a.s.l.)	1,870
AREA (km²)	379	IPA CRITERIA MET	A(i)

Site description

The Inago-Merripa Important Plant Area falls within Malema District in western Nampula Province. This IPA is 379 km² in area consisting of two separate polygons: one encompassing Mount Inago (–15.07°, 37.40°) and another around Serra Merripa or Murripa (–14.85°, 37.80°). Geographically, these mountains are connected through a spine of inselbergs, but there are densely populated areas and agricultural land intervening, so this has been excluded from the boundary.

Inago-Merripa falls within the proposed Mulanje-Namuli-Ribáuè Centre of Plant Endemism (Darbyshire et al. 2019a). The IPA is towards the northern edge of this mountain chain, with Mount Naumli to the south-west and the Ribáuè Mountains to the north-east. Although Inago-Merripa has been delineated to avoid much of the cultivated areas surrounding, agriculture is commonplace on both mountains, and woodland and montane forests are heavily fragmented as a result. Despite the degradation of habitats, there are a number of rare and threatened species known from these mountains that merit the recognition of the site as an IPA.

Botanical significance

Inago-Merripa hosts a number of rare and threatened species. The globally Endangered species *Rytigynia torrei* is of particular importance as it is only known from Serra Merripa and a single other location, Taratibu in Cabo Delgado Province (see The Quirimbas Inselbergs [MOZTIPA022]). *R. torrei* occurs in the ecotone between savanna and mid-altitude forest, both of which are highly disturbed habitats due to the expansion of machambas in this IPA (Darbyshire et al. 2019g). This species was last observed on Serra Merripa in 1967 (*Torre & Correia* #16555) but there has been little botanical collecting in this part of the IPA and so this species may well continue to exist there, particularly in less accessible areas of the mountain. Research should be undertaken to establish whether this species is still extant at this site and, if so, conservation would be required to prevent this species becoming Critically Endangered.

There are also five Vulnerable species present within this IPA. Both *Ammannia parvula* and *Cynanchum oresbium* are only known from Nampula Province, both occurring at only a single other locality outside this IPA. A further Vulnerable species, the cycad *Encephalartos gratus* is known in this IPA from Serra Merripa, although it may yet be found to occur on Mount Inago. This species is native to Mozambique and Malawi but is threatened by habitat clearance, which is estimated to be reducing the area of occupancy by 30% per generation (Donaldson 2010a).

Another Vulnerable species native to both Mozambique and Malawi, *Gladiolus zambesiacus*, was collected on Mount Inago (*Matimele* #2533). This species is currently known from fewer than 11 locations in montane and submontane areas of southern Malawi and northern Mozambique (Darbyshire *et al.* 2018a). There is only a single other confirmed locality for this species in Mozambique, nearby Mount Namuli [MOZTIPA004]. The Inago-Namuli specimens represent a distinct form, different from the Malawian specimens, with narrower leaves and thickened leaf margins and midribs (Goldblatt 1993). Further research may find that the Inago-Namuli form represents a distinct species.

The final Vulnerable species present at this IPA, *Khaya anthotheca*, is widespread, found from Mozambique to tropical West Africa, but is threatened by overharvesting as a valuable hardwood species (Hawthorne 1998). There are nine Mozambican endemic species in total occurring within this IPA. The threatened species *Ammannia parvula* and *Cynanchum oresbium* are the most range-limited of these endemics. *Encephalartos turneri* (LC) also has a narrow range, being endemic to Nampula and Niassa provinces, but is abundant within this limited area (Bösenberg 2010). Alongside these endemic species is the near-endemic *Euphorbia decliviticola* (LC), known from a range of only 5,744 km², predominantly in Mozambique with one location in Malawi (Osborne *et al.* 2019d).

Justicia asystasioides, collected on Mount Inago in 2017 (*Matimele* #2537), is a further species of note. Although also native to Tanzania and Malawi, this IPA and Mount Mabu [MOZTIPA012] are the only known locations for this species in Mozambique. There are probably further species of conservation interest within the remaining forest fragments of Inago and Merripa. For example, *Memecylon nubigenum* (EN) and *Pyrostria chapmanii* (EN) occur

Upland valley on Mount Inago showing rocky areas and woodlands (JB)   Rocky area of Mount Inago with cycad species (JB)

on both Mount Namuli and Ribáuè, either side of this IPA, and so are likely to occur within Inago-Merripa. Further botanical collecting at this site, particularly on Serra Merripa, is needed with great urgency as habitat continues to be lost before its conservation value can be fully established.

Habitat and geology

Inago-Merripa IPA encompasses Mount Inago (1,870 m) and Serra Merripa (1,023 m), excluding the spine of inselbergs that run between these two peaks. The mountains are granite-porphyrite intrusions, dating from around 630–550 million years ago, in the surrounding 1,100–850 million year old Nampula and Namarroi series migmatites (Bayliss *et al.* 2010; Macey *et al.* 2010). Average temperatures for Malema district range from 15–22°C in June and July to 21–29°C in October. Rainfall can vary year on year but averages at 1,300 mm annually, with much of this falling between December and March (Bayliss *et al.* 2010; World Weather Online 2021). The mountains, however, also receive moisture from frequent mists (*Torre & Correia* #16564).

Biodiversity surveys of a range of taxa were conducted in 2009 by Bayliss *et al.* (2010), and the plant species that characterise the different ecosystems on Mount Inago were recorded as part of this work. By contrast, there has been little botanical collecting on Serra Merripa, with many of the specimens dating from A. R. Torre's visits in the 1960s. Much of the description below is based on the Inago survey work by Bayliss *et al.* (2010). It is highly likely that the species compositions of ecosystems on Serra Merripa are similar to those on Inago due to the short distance between them, although Serra Merripa peaks at a lower altitude so montane elements are likely to be absent or very limited.

Below 1,000 m, much of this IPA is covered by woodland vegetation, although this is highly fragmented by agriculture (Google Earth 2021). Woodland on the foothills and lower slopes of Mount Inago was described by Bayliss *et al.* (2010) as having distinct strata, with a 10–15 m canopy comprised of *Parinari curatellifolia*, *Brachystegia* sp., *Albizia adianthifolia*, *Burkea africana* and *Syzygium cordatum*. Although the *Brachystegia* species that is present in these woodlands was not identified by Bayliss *et al.* (2010), Lötter *et al.*

(in prep.) state that *B. torrei*, and sometimes *B. spiciformis* and *B. bussei*, are present within this habitat type (categorised by Lötter *et al.* (in prep.) as Northern Inselberg Woodland). The mid-canopy is composed of species such as *Protea petiolaris*, *Cussonia arborea*, *Vitex doniana* and *Strychnos* sp., while herbs and grasses, such as *Themeda triandra* and *Hyparrhenia* spp., cover the ground layer.

At lower altitudes, this woodland transitions into miombo. It is likely that this miombo has a similar composition as the woodland of the footslopes, although *Brachystegia* is probably more dominant on the surrounding plain. It is clear from satellite imagery that this habitat, like the inselberg woodland, is also highly fragmented by agriculture (Google Earth 2021).

Bayliss *et al.* (2010) also made collections in the gallery forests on Mount Inago, with canopy trees reaching 15–30 m and including species such as *Khaya anthotheca*, *Breonadia salicina*, *Englerophytum magalismontanum* and *Newtonia buchananii*. The mid-canopy, at 8–15 m tall, consists of species such as *Sterculia africana*, *Cussonia spicata*, *Trema orientalis* and *Anthocleista grandiflora*, with smaller shrubs and tree ferns occurring below. Collection data from Serra Merripa describe riverine soils as red clay (*Torre & Paiva* #10439), the habitat within these areas is otherwise largely undocumented, but probably bears similarities to those of Mount Inago.

At higher altitudes, of 1,000–1,600 m, there are patches of forest categorised by Bayliss *et al.* (2010) as "semi-deciduous, wet forest". Canopy trees are relatively large, with diameter at breast height (DBH) over 50 cm and height of 20–30 m. Species include *Drypetes natalensis*, *Schefflera umbellifera* and *Newtonia buchananii*. Forest on Serra Merripa is likely to be very similar, with this habitat described as dense fog forest including *Newtonia buchananii*, *Ekebergia capensis* and *Khaya anthotheca* (*Torre & Correia* #16583), although the latter species is probably associated more with riverine forest (Burrows *et al.* 2018). Smaller trees such as *Myrianthus holstii* occur in the forest understorey on Mount Inago, while the herbaceous species in this forest are yet to be documented in any detail. Although mid-altitude forest was thought to have been extensive on these mountains, patches of forest are limited to around 1–10 ha in size and are now largely confined to inaccessible areas, below granite domes and in steep ravines (Bayliss *et al.* 2010). Owing to extensive conversion of habitat on Serra Merripa, there appears to be little or even no montane forest remaining on this inselberg, whereas it was previously extensive on the south-facing plateau of this inselberg (Google Earth 2021; Lötter *et al.*, in prep.).

In more exposed rocky areas, particularly on rocky slopes and granite domes above 1,500 m, the landscape is open (Google Earth 2021). Succulent *Euphorbia*, including *E. corniculata* and *E. decliviticola*, and *Aloe* species, including the endemic subspecies *A. menyharthii* subsp. *ensifolia*, occur in these areas. On high altitude plateaux, tussocks of the sedge *Coleochloa setifera* are frequent. *Exacum zombense* (LC), a near-endemic herbaceous species typical of seasonal seepages in the shallow soils of rock faces and outcrops, was recorded from high-altitude wooded grassland on Mount Inago (*Matimele* #2529). Other rock-loving flora, such as *Aeollanthus serpiculoides*, *Cyanotis lanata* and *Linderniella gracilis* (*Matimele* #2528, #2525, #2526) have also been collected from these high-altitude areas. The Vulnerable species *Ammannia parvula* occurs on granite outcrops of Mount Inago, probably occurring in seasonal pools in the rocky landscape. Cycads are also present on exposed outcrops, with *Encephalartos turneri* (LC) occurring on both mountains. *E. gratus* (VU) has also been recorded from the rocky slopes of Serra Merripa, and further investigation may find that this species is also present on Mount Inago.

Machambas occur throughout the Inago-Merripa IPA, forming a patchwork within mid-altitude forest and lowland woodland habitats. On Mount Inago, these plots are mainly used to grow maize; however, other crops such as tomatoes, onions, green peppers, and beans are also grown. In these areas, irrigation systems are used to channel watercourses onto productive land (Bayliss *et al.* 2010).

Conservation issues

The habitats within this IPA are highly fragmented by agricultural expansion. Mid-altitude to montane forest is now largely restricted to small patches in inaccessible areas such as steep slopes below granite domes and in ravines. On Mount Inago forest patches are generally 1–5 ha in size but would have previously covered large areas of the mid-altitude plateaux (Bayliss *et al.* 2010). The forested areas of Inago became occupied during the Mozambican Civil War, as local people sought refuge in the relative security of this mountainous landscape. Following settlement by refugees, forests began to be cleared to make way for machambas (Bayliss *et al.* 2010). It is likely that the Serra Merripa also had areas of montane forest, particularly on the southern plateau below the rockier peaks (Lötter *et al.*, in prep.), however, much of this has now been cleared for agriculture (Google Earth 2021).

Clearing of montane habitats for machambas is of particular conservation concern, given the rare and threatened species that reside within these habitats, but also because a loss of forested areas may reduce the provision of ecosystem services, such as water regulation and soil stabilisation. Cultivation on the mountain also involves the irrigation of crops through redirecting water (Bayliss *et al.* 2010), which may reduce water availability for riverine habitats as well as for communities who live downstream, particularly in the dry season. On the other hand, the riverine forests are thought to regulate the flow of water during the wet season (Bayliss *et al.* 2010), so degradation of these areas could expose the mountain slopes to erosion and could cause flooding downstream. The use of fire to clear land in preparation for farming is an additional threat as it brings the risk of unintentional burns of larger swathes of land (Bayliss *et al.* 2010).

Bayliss *et al.* (2010) recommend that local authorities should encourage people to move agricultural land off the mountain and instead farm fertile areas of the surrounding plain, with support from NGOs to develop more sustainable agriculture. However, much of the land surrounding the mountains is or has been used for farming already (Google Earth 2021), while the frequent mists and irrigation systems on the mountain are likely to support a more reliable harvest than those on the plains. People who farm on the mountains may, therefore, be resistant towards the idea of moving. A great deal of support and cooperation would be required to facilitate alternative, sustainable production systems for local people.

One project on nearby Mount Namuli could offer some guidance on balancing conservation with sustainable production and community development. NGOs Legado and Nitidae are collaborating on a project, running between 2018 and 2022, to support communities around Namuli by securing land rights for local people, improving healthcare provision and supporting greater market access. Alongside this, the project has been supporting communities in developing a long-term natural resource management plan while also working towards establishing the site as a Community Conservation Area (Nitidae 2021). Some of the strategies employed by this project could also be applied within this IPA to support local environmental restoration and economic development.

Although this IPA does not fall within a protected area, Mount Inago has been recognised as a KBA. Amongst the triggers for this KBA are a species of frog, *Nothophryne inagoensis* (EN), and chameleon, *Rhampholeon bruessoworum* (CR), that are both known only from the forest fragments on Mount Inago. Both are severely threatened by the continued encroachment of agriculture into these areas (IUCN SSC Amphibian Specialist Group 2019; Tolley *et al.* 2019b). The presence of these species would also allow the KBA to qualify as an Alliance of Zero Extinction site. In addition, the butterfly taxon *Cymothoe baylissi* subsp. *monicae* is known only from Mount Inago and Mount Mabu. As a sedentary species, which generally does not move between forest patches, *C. baylissi* subsp. *monicae* is at high risk of becoming locally extirpated on Mount Inago (Van Velzen *et al.* 2016). Protection of the remaining forest patches and the restoration of areas to reconnect these forest fragments is of great importance in preventing the extinction of these species.

Avian taxa recorded on Mount Inago include Thyolo Alethe (*Chamaetylas choloensis* – VU), with this IPA being the most easterly locality for this species, and East Coast Akalat (*Sheppardia gunningi* – NT) (BirdLife International 2017a, 2018). Both of these species were observed by Bayliss *et al.* (2010) in forest patches on the mountain, further highlighting the importance of these forests for their conservation value.

The landslide hazard, a measure of risk and impact of landslides on people and assets, was assessed as high to very high in this IPA (World Bank 2019).

A geological analysis by Mizuno *et al.* (2018) found that Mount Inago is constantly displacing a few millimetres each year but is also at risk of a deep-seated landslide following an earthquake or heavy rainfall. Such an event is predicted to have the potential to displace up to 200 million m³ of debris, which would be catastrophic in scale. The authors state that the last such event occurred in 887 AD following an earthquake and moved an estimated 350 million m³ of debris. There may be little that can be done to prevent deep-seated landslides here, as the depth of the presumed slip surface is far beyond the roots of vegetation on this mountain, although changes in hydrology through habitat clearance may well have some impact. Further research to confirm the risk of landslide on this mountain is recommended by Mizuno *et al.* (2018). This risk should be assessed primarily for consideration of the safety of the people who live on and around Inago, for which such a disaster would cause huge loss of life and suffering, but would also be of importance for conservation planning, particularly if habitat protection or restoration could play any role in mitigating landslide risk or magnitude. *Ex situ* conservation measures may also need to be considered for the highly threatened or site endemic species should there be a high risk of landslide in the near future.

Key ecosystem services

The mountains of this IPA are important sources of water, not only for people who live and farm on their slopes but for those in nearby towns and villages. Parts of Malema town depend on water that originates from Mount Inago while agriculture in the valleys that fall between the inselbergs is strongly associated with streams and rivers originating from the mountains (Bayliss *et al.* 2010; Google Earth 2021). The water provision from these inselbergs is likely of great importance during the dry season, while farming on the mountain itself may provide greater food security than farming the surrounding plains, particularly as the mid-altitude plateaux receive frequent mists. However, the continued degradation of forest in these areas will likely increase evapotranspiration, lowering moisture availability, and increase soil erosion which in turn may reduce the viability of agriculture on the mountains and could reduce water availability for downstream communities.

Bayliss *et al.* (2010) observed evidence of large volumes of water moving through riverine forests during the wet season. It is likely that these forests slow water flow during this season, mitigating against flooding in the valleys below.

Timber, including commercial hardwoods, is extracted from this IPA (Bayliss *et al.* 2010). Timber species may include *Khaya anthotheca*, a globally Vulnerable species threatened by over-harvesting for hardwood (Hawthorne 1998), known from the riverine forests at this site. Overharvesting of this and other forest species may reduce water availability, which is probably of greater value as an ecosystem service to local communities than the value of the timber extracted (Bayliss *et al.* 2010).

Ecosystem service categories

- Provisioning – Raw materials
- Provisioning – Fresh water
- Regulating services – Moderation of extreme events
- Regulating services – Erosion prevention and maintenance of soil fertility
- Habitat or supporting services – Habitats for species

IPA assessment rationale

Inago-Merripa qualifies as an IPA under sub-criterion A(i), as it contains important populations of one Endangered species, *Rytigynia torrei*, and four Vulnerable species, *Ammannia parvula*, *Cynanchum oresbium*, *Gladiolus zambesiacus* and *Encephalartos gratus*. Although an additional Vulnerable species, *Khaya anthotheca*, is present at this site, this species has a widespread distribution and it is unlikely that 1% of the global or 5% of the national population is found here. Overall, there are 10 endemic or near endemic species under sub-criterion B(ii), but this represents only 2% of the national list of species of B(ii) qualifying species, lower than the 3% threshold required. It is highly likely that more endemic and near endemic species will be found at this site with further investigation, and it may well qualify under B(ii) in future. Although there is mid-altitude montane forest present at this site, a habitat type of conservation importance nationally, it is heavily degraded and the remaining area does not trigger sub-criterion C(iii). However, this habitat should still be conserved, and restored where possible, given its high value for biodiversity and ecosystem services.

Agriculture surrounding Mount Inago (JB)

Priority species (IPA Criteria A and B)

FAMILY	TAXON	IPA CRITERION A	IPA CRITERION B	≥ 1% OF GLOBAL POP'N	≥ 5% OF NATIONAL POP'N	IS 1 OF 5 BEST SITES NATIONALLY	ENTIRE GLOBAL POP'N	SPECIES OF SOCIO-ECONOMIC IMPORTANCE	ABUNDANCE AT SITE
Apocynaceae	Cynanchum oresbium	A(i)	B(ii)	✓	✓	✓			frequent
Apocynaceae	Huernia erectiloba		B(ii)	✓	✓				unknown
Asphodelaceae	Aloe menyharthii subsp. ensifolia		B(ii)	✓	✓				unknown
Asteraceae	Bothriocline steetziana		B(ii)	✓	✓				unknown
Crassulaceae	Kalanchoe hametiorum		B(ii)	✓	✓				unknown
Euphorbiaceae	Euphorbia corniculata		B(ii)	✓	✓				frequent
Euphorbiaceae	Euphorbia decliviticola		B(ii)	✓	✓	✓			unknown
Iridaceae	Gladiolus zambesiacus	A(i)		✓	✓	✓			unknown
Lythraceae	Ammannia parvula	A(i)	B(ii)	✓	✓	✓			unknown
Meliaceae	Khaya anthotheca	A(i)						✓	frequent
Rubiaceae	Rytigynia torrei	A(i)	B(ii)	✓	✓	✓			unknown
Zamiaceae	Encephalartos gratus	A(i)		✓	✓				unknown
Zamiaceae	Encephalartos turneri		B(ii)	✓					frequent
		A(i): 6 ✓	B(ii): 10						

Threatened habitats (IPA Criterion C)

HABITAT TYPE	IPA CRITERION C	≥ 5% OF NATIONAL RESOURCE	≥ 10% OF NATIONAL RESOURCE	IS 1 OF 5 BEST SITES NATIONALLY	ESTIMATED AREA AT SITE (IF KNOWN)
Medium Altitude Moist Forest 900–1,400 m [MOZ-02]	C(iii)				1.6

Protected areas and other conservation designations

CONSERVATION AREA TYPE	CONSERVATION AREA NAME	RELATIONSHIP OF IPA TO CONSERVATION AREA
Key Biodiversity Area	Mount Inago	IPA encompasses protected/conservation area

Threats

THREAT	SEVERITY	TIMING
Small-holder farming	high	Ongoing – trend unknown
Logging & wood harvesting	low	Ongoing – trend unknown
Housing & urban areas	low	Ongoing – trend unknown
Increase in fire frequency/intensity	unknown	
Roads & railroads	low	Past, not likely to return
Abstraction of surface water (agricultural use)	medium	Ongoing – trend unknown
Avalanches/landslides	unknown	Future – inferred threat

RIBÁUÈ-M'PALUWE

Assessors: Jo Osborne, Iain Darbyshire, Hermenegildo Matimele, Camila de Sousa, Tereza Alves

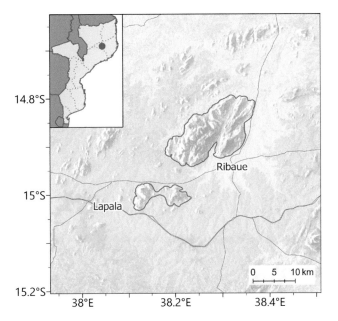

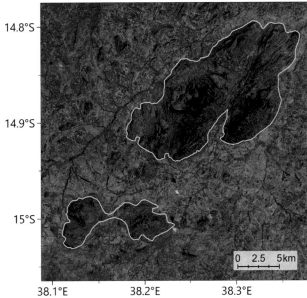

INTERNATIONAL SITE NAME		Ribáuè-M'paluwe	
LOCAL SITE NAME (IF DIFFERENT)		–	
SITE CODE	MOZTIPA001	PROVINCE	Nampula

LATITUDE	-14.87444	LONGITUDE	38.27750
ALTITUDE MINIMUM (m a.s.l.)	480	ALTITUDE MAXIMUM (m a.s.l.)	1,675
AREA (km²)	221	IPA CRITERIA MET	A(i), A(iii), B(ii), C(iii)

Serra de Ribáuè (ID)

Aloe rulkensii (TR)

Medium altitude moist forest in the Serra de Ribáuè (JO)

Site description

The Ribáuè-M'paluwe IPA comprises a series of granite inselbergs in Nampula Province of northern Mozambique near the town of Ribáuè in the district of the same name. The main area of the IPA is made up of the Serra de Ribáuè to the west and the Serra de M'paluwe to the east, separated by a wide valley. Outlying Serra Nametere and Mount Matharia to the south of the Ribáuè-Mutúali road are also included within the site. The inselbergs rise from a relatively flat landscape at ca. 500–600 m altitude up to 1,675 m at the peak of Monte M'paluwe. This massif forms a part of the belt of granitic inselbergs and massifs running NE-SW across southern Malawi and Nampula and the Zambezia provinces of Mozambique, which together comprise the proposed Mulanje-Namuli-Ribáuè Centre of Plant Endemism [CoE] (Darbyshire et al. 2019a).

Botanical significance

The Ribáuè-M'paluwe massif is one of Mozambique's most important sites for plant diversity and endemism. The site supports significant areas of granite outcrop flora and medium-altitude moist forest, both of which are restricted habitat types with a high species diversity. Five plant species are only found at this site: *Aloe rulkensii* (CR), *Baptorhachis foliacea* (DD), *Coleus cucullatus* (VU), *Dombeya leachii* (EN) and *Polysphaeria ribauensis* (EN). *Aloe rulkensii* was only recently described and was observed growing on shaded, vertical, granite cliff faces on the edges of moist forest in close association with the spectacular orange-red-flowered herb

Streptocarpus myoporoides (EN), which is otherwise known only from nearly Mount Nállume (McCoy & Baptista 2016); both of these species are scarce at Ribáuè. *Baptorhachis foliacea*, the only member of a genus endemic to Mozambique (Darbyshire et al. 2019a), is a small annual grass from rocky hillslopes, known only from a historical collection from Serra Nametere (*M.R. Carvalho* #508). Attempts to re-find this species in October 2017 were unsuccessful, but a visit to that site at the end of the rainy season may be more productive. *Coleus cucullatus*, a succulent shrub, is locally common on the open rock slopes of the massif, whilst the large-flowered shrub *Dombeya leachii* is occasional in scrub vegetation along forest margins and riverine thickets (J. Osborne et al., pers. obs.). *Polysphaeria ribauensis* is an understorey forest shrub which is locally frequent at Ribáuè but was only very recently described (Darbyshire et al. 2019h).

Other scarce and threatened species include the forest margin shrub or treelet *Vepris macedoi* (EN), again only found on the Ribáuè-M'paluwe massif and nearby Mount Nállume, and the locally abundant succulent *Aloe ribauensis* (EN), which is otherwise known only from the southern end of the Mueda Plateau (McCoy et al. 2014). Overall, the site supports 15 national endemic plant taxa, 12 near-endemics and 15 globally threatened taxa on the IUCN Red List. The moist forests are important for nationally rare species, such as *Calycosiphonia spathicalyx*, *Trichoscypha ulugurensis* and *Olea* aff. *madagascariensis*, the latter at its only known site in Mozambique (I. Darbyshire, pers. obs.).

In addition, several taxa that are potentially new to science have been recorded in the Ribáuè-M'paluwe IPA, including one potential new genus of Asparagaceae (T. Rulkens, pers. comm.), and the Critically Endangered shrub *Rytigynia* sp. C of *Flora Zambesiaca* (Bridson 1998).

Habitat and geology

Steeply sloping granitic rock outcrops, mid-altitude moist forest and miombo woodland are the dominant natural habitat types at the Ribáuè-M'paluwe massif. The site also includes smaller areas of gallery forest, marsh, seasonal stream gullies, seepage on granite rock, and shaded granite cliffs. The large, domed peaks comprise Pre-Cambrian granite-syenites of the Nampula group, dating to ca. 1,100–850 mya (Instituto Nacional de Geológia 1987).

Using remote sensing analyses, Montfort (2019) recorded 17.08 km^2 of extant moist evergreen forest, of which the majority (11.85 km^2) is on Serra de Ribáuè, with a more limited extent (4.73 km^2) on Serra de M'paluwe. The forest composition changes with elevation and soil depth and moisture availability. The lower elevation forests are dominated by *Newtonia buchananii*, with *Maranthes goetzeniana* also common, and have a canopy up to 25 m in height and with emergents to 30 m. Frequent understorey trees and shrubs include a range of Rubiaceae species, together with e.g. *Drypetes* spp., *Garcinia* spp., *Filicium decipiens*, *Funtumia africana*, *Olax* aff. *madagascariensis* and *Rinorea ferruginea*. Frequent lianas, particularly at forest margins and along riverine fringes, include *Agelaea pentagyna* and *Millettia lasiantha*. Dominant understorey herbs include *Mellera lobulata* and *Pseuderanthemum subviscosum*. Higher up on the granite slopes over thin soils, shorter and denser forest assemblages occur. In some areas, these are dominated by *Syzygium cordatum*, whilst in others there is a more mixed assemblage, with *Garcinia kingaensis* noted as abundant, along with *Aphloia theiformis*, *Gambeya gorungosana*, *Pyrostria chapmanii* and *Synsepalum muelleri*. Some riverine fringing forests persist at lower elevations; Müller *et al.* (2005) noted the presence of *Breonadia salicina*, *Milicia excelsa* and *Syzygium owariense* amongst other species in this habitat.

The granite rock outcrops have a high diversity of micro-habitats according to the slope, aspect, soil depth and moisture availability. These outcrops support a diverse flora of herbs, shrubs, geophytes and succulents including abundant *Aloe ribauensis*, *Aloe chabaudii*, *Euphorbia mlanjeana*, *Xerophyta* spp. including the range-restricted *X. pseudopinifolia*, and the cycad *Encephalartos turneri*. *Coleochloa setifera* provides the dominant cover and there are also areas of bare granite rock. In areas of seepage over rocks, a rich herb community develops, with abundant *Exacum zombense* and other typical seepage plants such as *Drosera*, *Utricularia* and *Xyris* spp.

Miombo woodland is extensive at lower elevations, although much has now been removed. Dominant species include *Brachystegia spiciformis*, *Uapaca*

Cynanchium oresbium (JO)

Pyrostria chapmanii (ID)

nitida and *Uapaca kirkiana*, with *Pterocarpus angolensis* and *Stereospermum kunthianum* also frequent; the suffruticose perennial *Cryptosepalum maraviense* can be conspicuous in the ground layer. Müller *et al.* (2005) also note the presence of stands of the bamboo *Oxytenanthera abyssinica.*

Large areas of the site are now given over to subsistence agriculture or are in various stages of fallow and degraded former forest; the invasive shrub *Vernonanthura polyanthes* can be abundant in such areas at altitudes below ca. 1,200 m. Montfort (2019) recorded ca. 70% of the land cover on the two main mountains of the massif to be given over to agriculture, fallow or secondary vegetation.

Conservation issues

There are two Forest Reserves within the IPA, the Ribáuè Forest Reserve and the M'paluwe Forest Reserve. The reserves were established in 1957 with the objectives to protect the catchment area and to study the restricted moist forest and gallery forest. Currently, the reserves are not being managed for their biodiversity and there is no control of agricultural expansion within the reserve boundaries.

Expansion of slash and burn agriculture on the slopes of the Ribáuè-M'paluwe massif is a serious threat to the forest and woodland habitats. The main crops grown are maize as a cash and subsistence foodcrop, and tomatoes as a cashcrop (Nitidae, pers. comm. 2021). Fire is used to clear forest and woodland and also spreads uncontrolled into the adjacent granite rock vegetation, causing significant damage. Where forest and woodland has been cleared, unsustainable agricultural practices lead to rapid soil erosion driving further forest clearance for access to fertile forest soil. In addition, the invasive South American shrub *Vernonanthura polyanthes* forms dense stands on fallow land, inhibiting forest and woodland regeneration and probably outcompeting forest margin species such as *Dombeya leachii.*

Using satellite imagery and analyses, Montfort (2020) estimated that 37% of forest and miombo on Serra de Ribáuè and 47% on Serra de M'paluwe has been lost during the period 2000–2020, and that the rate of deforestation is accelerating. Unless interventions are taken, she estimates that the forest resources will be exhausted within the next 35 years. In response to this severe threat, Nitidae and Legado have initiated a programme of community engagement in more sustainable agricultural practices and diversified livelihood options in order to balance community needs with biodiversity conservation.

The northern portion of the IPA has recently been designated as the Monte Ribáuè-Mphaluwe Key Biodiversity Area (KBA), on the basis of both its flora and its fauna, the latter including the endemic

Habitat destruction in the Serra de Ribáuè caused by burning (JO)

Ribáuè Mongrel Frog (*Nothophryne ribauensis*, EN). The site would also qualify as an Alliance for Zero Extinction (AZE) site. The IPA is larger in extent than the KBA in order to accommodate the only known site for *Baptorhachis foliacea* at Serra Nametere.

Key ecosystem services

The Ribáuè-M'paluwe massif protects the water catchment for the local area and the water supply for the town of Ribáuè and a commercial water bottling plant, Aguas de Ribaue. The natural vegetation also has a key role in protecting the steep slopes from soil erosion, and acts as a carbon sink. Local communities use botanical resources for a range of purposes. Interviews with local inhabitants conducted by Nitidae as part of an ongoing study of agrarian dynamics within the massif recorded the following uses of the forest resources: agriculture (55% of inhabitants interviewed), sourcing of bamboo for construction (45%), mushroom harvesting (28%), sourcing of construction wood other than bamboo (18%), hunting (18%) and sourcing of wood for cooking (9%) (Nitidae, pers. comm. 2021).

Shaded cliffs with *Streptocarpus myoporoides* (TR)

Ecosystem service categories

- Provisioning – Raw materials
- Provisioning – Fresh water
- Regulating services – Local climate and air quality
- Regulating services – Carbon sequestration and storage
- Regulating services – Erosion prevention and maintenance of soil fertility
- Habitat or supporting services – Habitats for species

IPA assessment rationale

The Ribáuè-M'paluwe massif qualifies as an IPA under all three criteria. Under Criterion A(i), the site supports 15 globally threatened taxa, for five of which this site contains the entire known global population. Under Criterion A(iii), the site supports one highly restricted endemic taxon, *Baptorhachis foliacea*, which is currently considered to be Data Deficient. Overall, the site supports 22 plant taxa of high conservation importance, in excess of the threshold of 3% under sub-criterion B(ii). Under Criterion C(iii), the site includes a significant area of Medium Altitude Moist Forest, one of Mozambique's national priority habitats recognised during the first Mozambique TIPAs workshop in Maputo in January 2018.

Serra de M'paluwe (ID)

Priority species (IPA Criteria A and B)

FAMILY	TAXON	IPA CRITERION A	IPA CRITERION B	≥ 1% OF GLOBAL POP'N	≥ 5% OF NATIONAL POP'N	IS 1 OF 5 BEST SITES NATIONALLY	ENTIRE GLOBAL POP'N	SPECIES OF SOCIO-ECONOMIC IMPORTANCE	ABUNDANCE AT SITE
Apocynaceae	Cynanchum oresbium	A(i)	B(ii)	✓	✓	✓			scarce
Apocynaceae	Huernia erectiloba		B(ii)						occasional
Apocynaceae	Stomatostemma pendulina	A(i)	B(ii)	✓	✓	✓			scarce
Apocynaceae	Strophanthus hypoleucos	A(i)		✓	✓	✓			frequent
Asphodelaceae	Aloe ribauensis	A(i)	B(ii)	✓	✓	✓			common
Asphodelaceae	Aloe rulkensii	A(i)	B(ii)	✓	✓	✓	✓		scarce
Asteraceae	Bothriocline moramballae		B(ii)						unknown
Crassulaceae	Kalanchoe hametiorum		B(ii)						unknown
Euphorbiaceae	Euphorbia decliviticola		B(ii)						unknown
Fabaceae	Baphia massaiensis subsp. gomesii		B(ii)						unknown
Gesneriaceae	Streptocarpus myoporoides	A(i)	B(ii)	✓	✓	✓			scarce
Lamiaceae	Coleus cucullatus	A(i)	B(ii)	✓	✓	✓	✓		common
Lamiaceae	Plectranthus mandalensis	A(i)	B(ii)	✓	✓	✓			scarce
Malvaceae	Dombeya leachii	A(i)	B(ii)	✓	✓	✓	✓		occasional
Melastomataceae	Memecylon nubigenum	A(i)	B(ii)	✓	✓	✓			scarce
Orchidaceae	Polystachya songaniensis		B(ii)						frequent
Poaceae	Baptorhachis foliacea	A(iii)	B(ii)	✓	✓	✓	✓		unknown
Polygalaceae	Polygala adamsonii		B(ii)						scarce
Rubiaceae	Polysphaeria ribauensis	A(i)	B(ii)	✓	✓	✓	✓		occasional
Rubiaceae	Pyrostria chapmanii	A(i)	B(ii)	✓	✓	✓			occasional
Rubiaceae	Rytigynia sp. C of F.Z.	A(i)	B(ii)	✓	✓	✓	✓		unknown
Rutaceae	Vepris macedoi	A(i)	B(ii)	✓	✓	✓			unknown
Vitaceae	Cissus aristolochiifolia	A(i)		✓	✓	✓			unknown
Zamiaceae	Encephalartos turneri		B(ii)						common
		A(i): 15 ✓ A(iii): 1 ✓	B(ii): 22 ✓						

Threatened habitats (IPA Criterion C)

HABITAT TYPE	IPA CRITERION C	≥ 5% OF NATIONAL RESOURCE	≥ 10% OF NATIONAL RESOURCE	IS 1 OF 5 BEST SITES NATIONALLY	ESTIMATED AREA AT SITE (IF KNOWN)
Medium Altitude Moist Forest [MOZ-02]	C(iii)			✓	15.5

Protected areas and other conservation designations

CONSERVATION AREA TYPE	CONSERVATION AREA NAME	RELATIONSHIP OF IPA TO CONSERVATION AREA
Forest Reserve (conservation)	Ribáuè Forest Reserve	IPA encompasses protected/conservation area
Forest Reserve (conservation)	Mepalué Forest Reserve	IPA encompasses protected/conservation area
Key Biodiversity Area	Monte Ribaue-Mphaluwe	IPA encompasses protected/conservation area

Threats

THREAT	SEVERITY	TIMING
Small-holder farming	high	Ongoing – increasing
Shifting agriculture	high	Ongoing – increasing
Increase in fire frequency/intensity	high	Ongoing – increasing
Invasive non-native/alien species	high	Ongoing – increasing

Steep granite slopes with *Encephalartos turneri* and *Euphorbia mlanjeana* (JO)

MOUNT NÁLLUME

Assessors: Jo Osborne, Iain Darbyshire

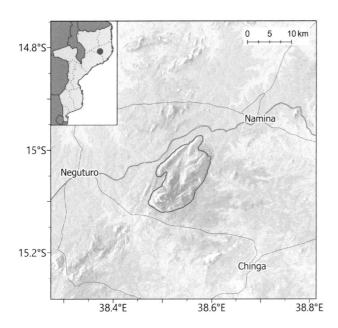

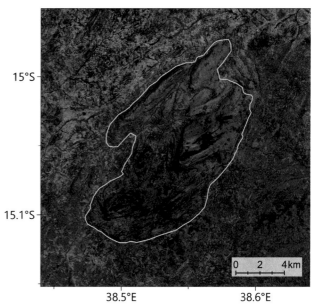

INTERNATIONAL SITE NAME		Mount Nállume	
LOCAL SITE NAME (IF DIFFERENT)		Monte Nállume	
SITE CODE	MOZTIPA018	PROVINCE	Nampula

LATITUDE	-15.05650	LONGITUDE	38.54674
ALTITUDE MINIMUM (m a.s.l.)	558	ALTITUDE MAXIMUM (m a.s.l.)	1,420
AREA (km²)	120	IPA CRITERIA MET	A(i)

Site description

Mount Nállume, also known as Serra Chinga, is a large granite inselberg in Ribáuè and Murrupula Districts of Nampula Province, ca. 25 km south-east of the town of Ribáuè. It forms part of a band of inselbergs in northern Mozambique running north-east from Mount Namuli, which together form the eastern part of the proposed Mulanje-Namuli-Ribáuè Centre of Plant Endemism (Darbyshire *et al.* 2019a). The IPA includes a series of irregular granite rock outcrops, partially covered by forest and reaching an elevation of ca. 1,420 m. The site covers an area of approximately 115 km² and is not formally protected at present.

Botanical significance

Significant areas of both medium-altitude moist forest and granite inselberg habitat can be found at Mount Nállume. Together, these habitats support three globally threatened endemic plant taxa, of which two, the striking orange-red-flowered herb *Streptocarpus myoporoides* (EN) and the small tree *Vepris macedoi* (EN), are only found on Mount Nállume and the nearby Ribáuè Massif (Osborne *et al.* 2019e, Darbyshire & Rokni 2019). The *Streptocarpus* species is restricted to damp shaded rocks and tree trunks in upland forest, whilst the *Vepris* occurs primarily along the forest margins. The third taxa, *Euphorbia grandicornis* subsp. *sejuncta* (EN), a spiny succulent of exposed granite slopes, is known from only Mount Nállume and two sites to the east of Nampula city (Osborne *et al.* 2019f). These records are all based on historical botanical expeditions dating from the Portuguese colonial era in the late 1960s, and so the continued presence of these species at Nállume needs to be confirmed; no recent botanical collections are known from here.

This site also holds an important population of the large forest tree *Maranthes goetzeniana*, which is widespread in the region but sparsely distributed and assessed as Near Threatened (Timberlake *et al.* 2018); it has been noted as abundant at Nállume in past botanical surveys (*J.M.A. Macêdo* #3258, 3265). Other notable species that occur at Mount Nállume include two further Mozambique endemics, the cycad *Encephalartos turneri* (LC) and the suffrutescent herb *Bothriocline moramballae* (LC). The site is not well-studied botanically and other notable plant species are likely to occur here.

Habitat and geology

The landscape at Mount Nállume consists predominantly of granite inselberg slopes ranging from curved granite domes to steep cliffs. The migmatoid granites are of the Nampula Group from the Proterozoic eon, dating from 1,100 to 850 mya (Instituto Nacional de Geológia 1987). These granite slopes support an interesting and diverse flora of herbs and shrubs, typically including many succulents that can survive the harsh water-scarce environment. Crevices and gullies in the rock provide numerous microhabitats that support plant diversity.

Moist forest patches cover a significant area of the inselberg, with a canopy mostly 15–20 m tall, though reaching over 40 m in places (P. Platts, pers. comm. 2020). Although not well inventoried to date, collections made in the late 1960s by A.R. Torre and M.F. Correia, and by J.M. de Aguiar

Macêdo note the presence of *Newtonia buchananii* and *Maranthes goetzeniana* as important trees, this being in keeping with the forests of the nearby Ribáuè Massif. Other forest taxa noted by past collectors include *Bersama abyssinica*, *Dracaena* sp., likely to be *D. mannii*, *Myrianthus holstii*, *Filicium decipiens* (highly localized in Mozambique but also present on Ribáuè), *Rauvolifia caffra* and *Xylopia* sp., likely to be *X. aethiopica* (*A.R. Torre & M.F. Correia* #16387, 16470). However, a large part of the forest has been cleared recently through logging and for subsistence agriculture, particularly at the base of the granite slopes.

On top of the inselbergs, water from the moist forest forms swamps and drains into frequent streams. At the base of the granite slopes meandering streams are conspicuous on satellite imagery (Google Earth 2021), supporting narrow bands of riparian forest within a mosaic of agricultural land, secondary scrub or grassland, and fragments of miombo woodland.

Although the plant diversity of Nállume has not been documented in any detail, it is likely to be comparable to the Ribáuè Massif [MOZTIPA001] given their proximity and similar altitudinal range and habitat types, as observed on satellite imagery (Google Earth 2021).

Annual rainfall at nearby Ribáuè town averages 799 mm per year, and is concentrated in a short rainy season from December to March; the dry season is prolonged (weather-atlas.com). However, the moist forests are likely to receive additional moisture throughout the year from

Mount Nállume peak (PP)

Forest on Mount Nállume, including recent destruction (JB)

low cloud cover, which results from the high air humidity that remains at over 60% for most of the year, and often over 80%.

Conservation issues

The forest at Mount Nállume is under increasing threat due to logging and forest clearance for subsistence agriculture. Biologists who visited the site in 2019 estimate forest loss of more than 30% over the past 10 years and suggest that all of the forest could be lost within 15 years at the current rates of deforestation (Njagi 2019). As soil fertility and the reliability of water supply diminish in the surrounding lowlands, local communities are increasingly moving into the higher elevation, forested areas to farm, but the clearance of forest is exasercbating the reduction in water supply and the erosion of soils. Fire presents another threat to the forest, both unintentional spread of fires used by local people to clear agricultural fields surrounding the inselbergs and fires set intentionally by hunters to drive animals into traps in the forest (Njagi 2019). These fires are damaging the forest edge but are a secondary threat when compared to the current rate of forest clearance for subsistence agriculture (P. Platts, pers. comm. 2019). The site is not formally protected at present and there is an urgent need for conservation action. Community-led conservation efforts, together with diversification of local livelihood options and adoption of "conservation agriculture" techniques, are most likely to be effective in increasing the sustainability of land use practices and in reducing biodiversity loss.

In terms of other biodiversity, the site is noteworthy for a range of faunal taxa, including a potential new species of pygmy chameleon (*Rhampholeon*) and a potential endemic butterfly species (Njagi 2019). A new species of *Leptomyrina* butterfly, *L. congdonii*, was recently described from the upper elevations of Mounts Nállume, Inago, Mabu, Mecula and Namuli, where it feeds on Crassulaceae (Bayliss *et al.* 2019).

Key ecosystem services

Mount Nállume has a high plant diversity value, providing moist forest and granite inselberg habitats for flora and fauna within a predominantly agricultural plain. Timber and medicinal plants from the forest are used by local people, some of whom also depend on wildlife hunting for food (Njagi 2019). The inselberg forests have cultural and spiritual value to the local community, being considered sacred and used as sites for performing traditional rituals (Njagi 2019). The forested inselbergs provide a watershed for the local area and the vegetation contributes to carbon sequestration and storage.

Rock flora on Nállume, with *Encephalartos turneri* (JB)

Streptocarpus myoporoides (photographed on Serra de M'paluwe) (TR)

Ecosystem service categories

- Provisioning – Food
- Provisioning – Raw materials
- Provisioning – Fresh water
- Provisioning – Medicinal resources
- Regulating services – Carbon sequestration and storage
- Habitat or supporting services – Habitats for species
- Cultural services – Spiritual experience and sense of place
- Cultural services – Cultural heritage

IPA assessment rationale

Mount Nállume qualifies as an IPA under criteria A. Under criterion A(i), the site supports populations of three globally threatened taxa: *Euphorbia grandicornis* subsp. *sejuncta* (EN), *Streptocarpus myoporoides* (EN) and *Vepris macedoi* (EN) – it is the only site within the current Mozambique IPA network to contain the first of these taxa. The site nearly qualifies under criterion C(iii), having significant areas of Medium Altitude Moist Forest, a restricted and nationally threatened habitat, but it is not considered to be among the five best sites nationally for that habitat.

Priority species (IPA Criteria A and B)

FAMILY	TAXON	IPA CRITERION A	IPA CRITERION B	≥ 1% OF GLOBAL POP'N	≥ 5% OF NATIONAL POP'N	IS 1 OF 5 BEST SITES NATIONALLY	ENTIRE GLOBAL POP'N	SPECIES OF SOCIO-ECONOMIC IMPORTANCE	ABUNDANCE AT SITE
Asteraceae	*Bothriocline moramballae*		B(ii)						unknown
Euphorbiaceae	*Euphorbia grandicornis* subsp. *sejuncta*	A(i)	B(ii)	✓	✓	✓			unknown
Gesneriaceae	*Streptocarpus myoporoides*	A(i)	B(ii)	✓	✓	✓			unknown
Rutaceae	*Vepris macedoi*	A(i)	B(ii)	✓	✓	✓			unknown
Zamiaceae	*Encephalartos turneri*		B(ii)						frequent
		A(i): 3 ✓	B(ii): 5						

Threatened habitats (IPA Criterion C)

HABITAT TYPE	IPA CRITERION C	≥ 5% OF NATIONAL RESOURCE	≥ 10% OF NATIONAL RESOURCE	IS 1 OF 5 BEST SITES NATIONALLY	ESTIMATED AREA AT SITE (IF KNOWN)
Medium Altitude Moist Forest [MOZ-02]	C(iii)				10.0

Protected areas and other conservation designations

CONSERVATION AREA TYPE	CONSERVATION AREA NAME	RELATIONSHIP OF IPA TO CONSERVATION AREA
No formal protection	N/A	

Threats

THREAT	SEVERITY	TIMING
Shifting agriculture	high	Ongoing – increasing
Small-holder farming	high	Ongoing – increasing
Hunting & collecting terrestrial animals	unknown	Ongoing – trend unknown
Logging & wood harvesting	high	Ongoing – trend unknown
Increase in fire frequency/intensity	unknown	Ongoing – increasing

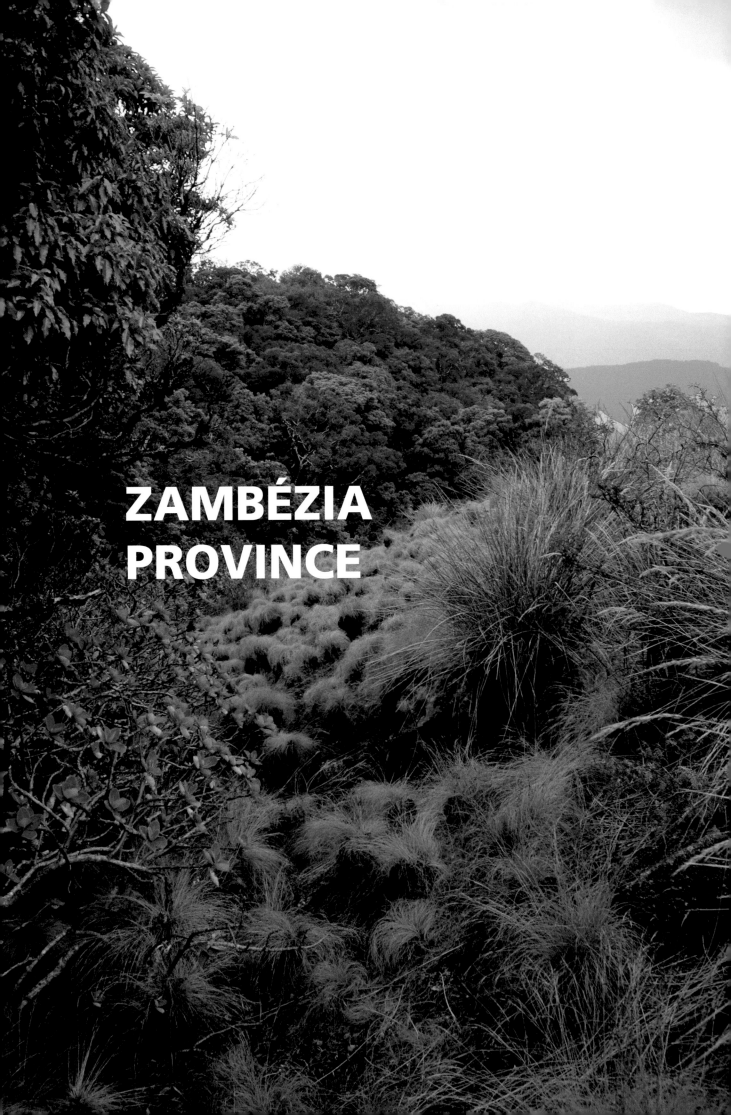

ZAMBÉZIA
PROVINCE

MOUNT NAMULI

Assessors: Iain Darbyshire, Jonathan Timberlake

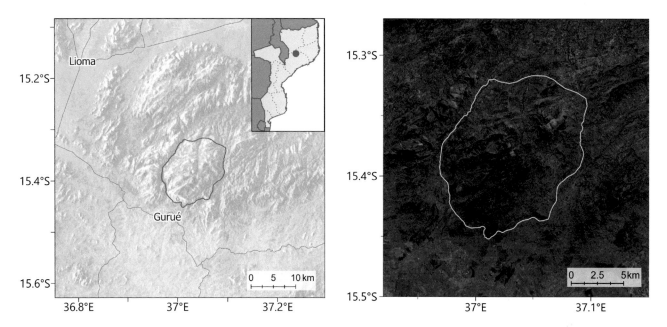

INTERNATIONAL SITE NAME		Mount Namuli	
LOCAL SITE NAME (IF DIFFERENT)		Monte Namúli / Serra de Gurué	
SITE CODE	MOZTIPA004	PROVINCE	Zambézia

LATITUDE	-15.37861	LONGITUDE	37.03167
ALTITUDE MINIMUM (m a.s.l.)	750	ALTITUDE MAXIMUM (m a.s.l.)	2,419
AREA (km²)	146	IPA CRITERIA MET	A(i), A(iii), B(ii), C(iii)

Site description

The Mount Namuli IPA is situated in Gurué District of Zambézia Province in northern Mozambique, ca. 250 km inland from the Indian Ocean coastline. It comprises a series of granitic inselbergs linked by a high plateau. With one of its three main peaks rising to 2,419 m a.s.l., Mount Namuli is the second highest point in Mozambique after Mount Binga in the Chimanimani Mountains. Immediately to the south of the massif is the small town of Gurué, a major centre of tea production in Mozambique particularly during the colonial era.

Several rivers arise on the massif, the main ones being the Malema River east of the main plateau, which flows north to join the Lúrio, one of the major rivers of northern Mozambique, and the Licungo River to the west, which flows southwards to the

Indian Ocean near Quelimane. The northern flanks of Namuli are drained by the Namparro River, which joins the Malema further north (Timberlake *et al.* 2009). The mountain has a number of impressive waterfalls, including the Cascata de Namuli on the Licungo which falls ca. 100 m down a rock face.

Mount Namuli is one of the major components of the proposed montane Mulanje-Namuli-Ribáuè Centre of Plant Endemism (CoE) in southern Malawi and northern Mozambique (Darbyshire *et al.* 2019a). It supports a rich mosaic of habitats, from lowland riverine forest and succulent-rich rocky slopes, through mid-elevation to montane forests, to montane grasslands and shrublands, and extensive areas of rock outcrops. This area has been subject to a range of biodiversity studies, dating as far back as the early exploration of Joseph Last in

1886, but with most of the more intensive surveys conducted within the past 25 years (Timberlake 2021a). However, despite being well documented as a site of global biodiversity importance, Namuli is not formally protected and some of its key habitats are highly threatened.

Botanical significance

In view of its high number of endemic and range-restricted plant species, Mount Namuli is one of the most important botanical sites in Mozambique, and indeed in montane southern tropical Africa. Nineteen strict endemic taxa have been described to date (this figure includes *Buchnera namuliensis*, as a record from Dondo in Sofala Province is considered to be erroneous). A significant proportion of these endemics have only recently been documented, with eight described since 2010 including the montane herbs *Coleus namuliensis*, *Crepidorhopalon namuliensis* and *Crotalaria namuliensis* (Harris *et al.* 2011; Downes & Darbyshire 2018; Darbyshire *et al.* 2019i). The number of endemics is likely to continue to increase as further survey work uncovers additional new species and records; a wet-season survey of the montane geophyte flora may be particularly productive as this has not been surveyed extensively to date (Darbyshire *et al.* 2019i; Timberlake 2021a). Within Mozambique, only the Chimanimani Mountains hold higher numbers of point-endemic species.

In addition to its endemics, this massif supports a high number of range-restricted montane species, including several taxa that are otherwise known from or were previously thought to be endemic to Mount Mulanje in Malawi, for example *Gnidia chapmanii*, *Pimpinella mulanjensis*, *Senecio peltophorus* and *Xyris makuensis* (Harris *et al.* 2011). In total, Timberlake (2021a) recorded a further 20 taxa known from Namuli and three or fewer mountains in the Mulanje-Namuli-Ribáuè CoE. A significant proportion of the endemic and near-endemic flora is found in the montane grasslands and rocky areas which, fortunately, are amongst the least severely impacted habitats on Namuli (see Threats). Several of these species, including all those listed above, are assessed to be Least Concern on the IUCN Red List despite their narrow ranges. However, a number of rare forest species are also recorded, and most of these are assessed as globally threatened, including the strict-endemic herb *Isoglossa namuliensis* (CR) and several near-endemics, including two mistletoes *Agelanthus patelii* (EN) and *Helixanthera schizocalyx* (EN) and three woody species, *Memecylon nubigenum* (EN), *Pyrostria chapmanii* (EN) and *Faurea racemosa* (EN). The lattermost is commercially exploited as an important local timber source for carpentry (Darbyshire *et al.* 2018b). A small number of endemics are found in woodland and riverine habitats at lower elevations, including the

Mount Namuli (AMR)

riverine herb *Plectranthus guruensis* (EN) and the charismatic shrub *Dombeya lastii* (EN).

In terms of habitats, Namuli holds some of the most extensive areas of montane grassland in Mozambique, a habitat that is particularly species-rich. The remnants of medium-altitude and montane forest are also of national importance given that these habitats are range-restricted and threatened in Mozambique. However, these forests are diminishing at an alarming rate and urgent conservation action is needed if they are to be safeguarded.

A preliminary checklist for the massif above 1,200 m elevation (descending to 1,000 m on the western flanks) recorded 603 taxa of vascular plants (Timberlake 2021a). However, this is only a preliminary list and more comprehensive surveys would undoubtedly result in significant increases, in particular the montane herbaceous/grass flora, the remnants of woodland on the lower slopes, and the drier, rocky northern slopes. The latter areas have not been surveyed in detail to date and would likely add a number of succulent species to the list (Timberlake 2021a).

Habitat and geology

The Namuli massif is a batholith, a complex of inselbergs or intrusions linked by a high plateau, exposed by millions of years of subsequent erosion (Timberlake *et al.* 2009). The extensive plateau at ± 1,800 m rises from a pediplain at ± 800 m elevation in the south and west. The large inselbergs are essentially granitic; the peaks and ridges comprise Precambrian granite-porphyrite intrusions into migmatites of the Nampula and Namarroi series, dating from ca. 1,100–850 mya (Instituto Nacional de Geológia 1987; Timberlake 2021a). Namuli experiences a rainy season between November and March and a dry season from May to October. Mean annual rainfall at Gurué town at the foot of the massif is 1,995 mm. However, rainfall on the high plateau will be considerably higher than this and may reach 3,000 mm, and higher elevations also receive significant moisture year-round from mists. Temperatures peak in October, just before the onset of the rains, and are at their lowest in the middle of the dry season in July; overnight mild frosts are likely to occur at higher elevations between June and August (Timberlake 2021a).

The habitats of Mount Namuli have been documented in some detail in previous studies (Dowsett-Lemaire 2008; Timberlake *et al.* 2009; Timberlake 2021a); a summary is given below. Six main vegetation types have been recorded: forest, woodland, montane scrub, montane grassland, rocky slopes and outcrops, and cultivated areas. The lattermost is covered under Conservation issues.

Montane grassland on Mount Namuli (AMR)

Impatiens psychadelphoides (BW)

Crepidorhopalon namuliensis (BW)

Timberlake *et al.* (2009) estimated a total forest cover on Namuli of ca. 12.5 km², of which the large majority was montane, over 1,600 m elevation. However, rapid forest losses have occurred in the intervening period (see Threats) and total forest cover is now less than 10 km². Darbyshire *et al.* (2021) estimate a total of ca. 7 km² of forest remaining over 1,400 m elevation. The largest blocks of forest are found in the broad valleys and on less steep slopes of the high plateau, although smaller patches extend into the deeper valleys and the steeper slopes of the peaks. The montane forest, from ca. 1,600–2,200 m elevation, is characterized by a canopy layer typically 18–25 m though lower in some areas, with emergents to 30(–40) m tall. Common emergent species are *Cryptocarya liebertiana*, *Ekebergia capensis*, *Faurea racemosa* and *Olea capensis*. The canopy has a mixed species composition, varying somewhat between patches, with many characteristic Afromontane species such as *Albizia gummifera*, *Aphloia theiformis*, *Cassipourea malosana*, *Podocarpus milanjianus*, *Prunus africana* and *Syzygium afromontanum*, in addition to the emergent species which are all frequent; *Garcinia kingaensis* can also be particularly common. Frequent understorey shrubs include *Alchornea laxiflora*, *Anisotes pubinervis* and *Lasianthus kilimandscharicus*. The herbaceous layer is rich in ferns, reflecting the high moisture availability from frequent mists and rains.

At lower elevations, below 1,600 m, small areas of mid-altitude forest remain, estimated at only 1.35 km² by Timberlake *et al.* (2009). These differ from the montane forests in having a higher canopy and a greater presence of *Albizia gummifera*, *Ficus* spp., *Newtonia buchananii*, *Parinari excelsa*, *Syzygium cordatum* and Sapotaceae species: *Englerophytum magalismontanum*, *Gambeya* (formerly *Chrysophyllum*) *gorungosana* and *Synsepalum muelleri*. Riverine fringes can also support forest to lower elevations, with the typical species composition similar to that of the mid-elevation forest along with characteristic riverine species such as *Breonadia salicina*.

Woodland is not extensive on Namuli and miombo is essentially absent, though was probably previously present at lower elevations. The most extensive woodland type is found on montane forest margins, where a fire-impacted *Erica benguellensis*-dominated woodland to 20 m tall can be common. Elsewhere, secondary woodland dominated by *Syzygium cordatum* was noted by Dowsett-Lemaire (2008).

Montane scrub, found mainly at elevations of over 1,700 m, is extensive in fertile, well-drained areas and typically comprises stands of the bracken fern, *Pteridium aquilinum*, together with dense shrubs to 2.5 m tall, including *Kotschya recurvifolia* and *Tetradenia riparia*. Regular burning of this habitat was noted by Timberlake *et al.* (2009), and this is almost certainly a secondary habitat that is increasing due to forest loss and increased frequency of fire.

Montane grassland, mainly at 1,850–2,000 m elevation, is extensive with an estimated extent of ca. 2.3 km² (Timberlake 2017, 2021a), but this may be an under-estimate given that this habitat forms a mosaic with both the vegetation on rocky outcrops and the montane scrub. The largest montane grassland area is in the east of the massif on the Muretha Plateau above the Malema valley at ca. 1,850 m, with a second extensive area on the Nachone Plateau on the west slopes of Mounts Pilani and Pesse. The grassland is tussocky, typically growing on deep peats with seasonal waterlogging. The dominant grass species are *Loudetia simplex*, with *Themeda triandra* and *Eragrostis* spp. on better-drained areas. A range of herbs and geophytes are common within these grasslands, with characteristic species including *Euphorbia depauperata*, *Helichrysum* spp., *Kniphofia splendida* and a range of terrestrial orchids. The grasslands and associated communities on the drier northern slopes have not been well surveyed to date, and some of them are extensively grazed by cattle. Grasses here are less tussocky; dominant species include *Setaria sphacelata* in grazed areas.

The montane rocky slopes and outcrops are the most extensive habitats at this site. *Coleochloa setifera* dominates on the thin soils between exposed rock faces, together with a range of xerophytic taxa such as *Aloe mawii*, *Crassula globularioides* and *Xerophyta kirkii*. On the drier northern slopes, the endemic succulent *Euphorbia namuliensis* is also frequent alongside other succulent taxa. Areas of seasonal seepage are frequent in the rocky areas, with the thin damp mats supporting a rich flora including geophytes such as *Hypoxis nyasica*, *Merwilla plumbea* and terrestrial orchids, together

with typical wet grassland herbs such as *Drosera* and *Xyris* spp.; the endemic *Crepidorhopalon namuliensis* is confined to this habitat.

Fast-flowing streams and waterfalls are an under-explored habitat on Mount Namuli that may be of interest. The endemic rheophyte *Inversodicraea torrei* (VU) has not been recorded since the 1940s but has probably been overlooked. Other rheophytes may also be present, although only the widespread *Hydrostachys polymorpha* has been noted to date (Timberlake 2021a).

Conservation issues

Mount Namuli should be considered one of the greatest conservation priorities in Mozambique (Timberlake 2021a). Despite this site being internationally renowned as a site of high importance for a range of biodiversity, the entirety of this IPA is unprotected at present, and is one of the most highly threatened montane regions in Mozambique (Timberlake 2017; 2021a). Recent decades have seen a significant expansion of agricultural practices on the mountain. The mid-elevation and lower montane forests are being cleared for subsistence and small cash-crop agriculture, with a notable recent expansion in potato cultivation being particularly problematic (Timberlake *et al.* 2009; Timberlake 2017, 2021a). Potato yields are reasonably high in the first year, but quickly diminish in subsequent cycles, and one plot can be farmed for a maximum of five years before the soil fertility is reduced and new areas need to be farmed (Legado, pers. comm.; Darbyshire *et al.* 2018b; Timberlake 2021a). Comparisons of satellite image available on Google Earth between September 2013 and November 2015 indicate an estimated forest loss of 10–30% over this short time period, and this clearance is ongoing, with clear-felling of many patches of moist forest on the Muretha plateau and upper Nivolo valley noted during recent surveys (Timberlake 2017, 2021a). These losses are particularly severe at lower elevations, along forest margins and in smaller patches. Most of the threatened plant species on Namuli are associated with forest and forest margins.

Selective timber extraction is also problematic, specifically impacting the range-restricted and threatened large tree species *Faurea racemosa* or 'tchetchere' which, whilst common at Namuli, was clearly being logged unsustainably in the mid

2000s for use in local carpentry and construction (Darbyshire *et al.* 2018b; Timberlake 2021a). This problem is believed to be ongoing.

Within the montane grasslands, Ryan *et al.* (1999) reported grazing by goats and feral pigs to be a significant problem, and the impact of the pigs in digging up the delicate, species-rich seepage areas over rock was also noted by Timberlake *et al.* (2009). However, more recent visits indicate that the pigs and goats have been removed – perhaps because of their damaging impact on the newly established potato plots – and they do not appear to have caused lasting damage to the grasslands and seepages (Timberlake 2017, 2021a). The drier northern slopes were used for cattle production during the colonial era, and a number of cattle owners remain in this area today (Timberlake *et al.* 2009; Timberlake 2021a). More problematic is the increase in uncontrolled dry-season wildfires, deliberately set for land clearance or to aid hunting. Frequent burning of the grassland and cleared areas between the forests is likely to be causing forest margins to further recede. Such fires also prevent forest regeneration in fallow areas. However, the montane grasslands and rocky areas, where most of the endemic and near-endemic species are located, are likely to be adapted to fire to at least some extent and so the threats to these species are limited.

At lower elevations, very little is left of the natural vegetation, except for narrow woodland and forest fringes along rivers. Much of the lower elevation forests and woodlands (up to ca. 1,200 m) around Gurué and in the Licungo Valley were cleared for tea plantations during the colonial era in the early 20th century (Timberlake *et al.* 2009; Timberlake 2021a). Beyond the tea plantations, there are extensive areas of subsistence agriculture and fallow lands, with many of the tea workers holding small subsistence plots for cassava, maize and sweet potatoes. The fallow areas are particularly susceptible to dry season fires, which suppress the regeneration of woodland and forest. The extensive losses and degradation of low elevation habitats threaten some of the rare woodland species, such as *Dombeya lastii* and *Gymnosporia guruensis*. Neither of these species has been recorded since the 1970s and their continued presence here requires confirmation given the extent of habitat transformation.

To address some of these conservation issues, the international NGO Legado, in partnership with Nitidae and local NGO LUPA, have been working since 2014 with local communities around Namuli to balance livelihoods and sustainable stewardship of the montane ecosystems under their "Thriving Futures" programme (Legado 2021). Work to date has included securing community land rights, increasing access to health care, improving market access for products, and developing sustainable agricultural practices and improved management of natural resources through community leadership (Legado 2021). Nitidae are now working in partnership with the Rainforest Trust to establish a 56 km² Community Conservation Area (CCA) in the core area of the mountain above 1,200 m elevation, with the reduction of slash-and-burn agriculture a primary aim (Rainforest Trust 2021), and with forest guards established to help reduce deforestation and wildfires (Timberlake 2021a).

Beyond its importance for plants, Mount Namuli is also known to be an important site for a range of fauna. It is an Important Bird Area, with the forests being particularly important for a range of bird species including the Namuli Apalis (*Apalis lynesi*, EN), which were thought to be endemic until they were discovered on nearby Mount Mabu, as well as Spotted Ground Thrush (*Geokichla guttata*, EN), Thyolo Alethe (*Chamaetylas choloensis*, VU) and Dapple-throat (*Arcanator orostruthus*, VU) (Ryan *et al.* 1999; Dowsett-Lemaire 2008; BirdLife International 2021a). Ryan *et al.* (1999) stated that this site is arguably the most critical IBA in Mozambique, with the remaining forests a particularly high priority for conservation action. Butterfly surveys between 2005 and 2008 recorded 126 species above 1,200 m, including five new species and two new subspecies to science, as well as the first records of three species previously thought to be endemic to Mount Mulanje (Timberlake *et al.* 2009). Namuli is also an Alliance for Zero Extinction (AZE) site, triggered by the presence of the endemic Vincent's Bush Squirrel (*Paraxerus vincenti*, EN) and Mount Namuli Pygmy Chameleon (*Rhampholeon tilburyi*, EN), although the latter has since been recorded from the Ribáuè Mountains and Mount Socone in Nampula Province. Amongst the plant species on Namuli, *Dombeya lastii*, *Isoglossa namuliensis* and *Plectranthus guruensis* would also trigger AZE status. Most recently, Namuli has been designated as a Key Biodiversity Area, largely on

Montane grassland and forest patches on Namuli (AMR)

Inside montane forest in the mist (AMR)

the basis of the proposed CCA area, with plants comprising the majority of the qualifying taxa on which the assessment is based. The IPA covers a larger area than that proposed for the KBA, because we include areas outside of the proposed CCA that support threatened and endemic plant species, notably the lower altitude riverine fringes and the drier northern slopes of the mountain.

Key ecosystem services

As well as providing habitat for a wealth of biodiversity, this site provides important ecosystem services for the ca. 13,500 people who reside on and around the mountain. In particular, this massif is a major regional water source, with the Malema River being a significant tributary of the Lúrio River, a critical water resource in northern Mozambique. Locally, the rivers supply water to local communities and the combination of moist climate and readily available river water on the south side feeds Gurué's commercial tea industry. The intact vegetation of the steep slopes of the massif helps to maintain soil stability and prevent excessive erosion. These habitats are also an important carbon sink. They provide important provisioning services, including wood for fuel and carpentry, fibres, and fodder for domestic animals. Timberlake (2021a) noted that some level of use of natural resources, such as limited cattle grazing and collection of fibres from *Kniphofia splendida*, is not incompatible with the conservation of the important habitats and species, but that current land-use practices in the forests in particular are unsustainable.

Mount Namuli is of cultural importance as a sacred site for the Lomwe people. It also has significant potential for ecotourism in view of its spectacular scenery, unique flora and fauna, and hiking and rock-climbing potential. However, the difficulties of access to this site would have to be overcome before this tourism potential could be realised (Timberlake 2021a), and any improvement in road access would probably bring with it other challenges and development threats.

Ecosystem service categories

- Provisioning – Raw materials
- Provisioning – Fresh water
- Regulating services – Local climate and air quality
- Regulating services – Carbon sequestration and storage
- Regulating services – Erosion prevention and maintenance of soil fertility
- Habitat or supporting services – Habitats for species
- Habitat or supporting services – Maintenance of genetic diversity
- Cultural services – Aesthetic appreciation and inspiration for culture, art and design
- Cultural services – Spiritual experience and sense of place

IPA assessment rationale

Mount Namuli qualifies as an IPA under all three criteria. Under criterion A(i), it supports important populations of 22 globally threatened plant taxa, two of which are Critically Endangered endemics – *Isoglossa namuliensis* and *Tephrosia whyteana* subsp. *gemina* – whilst 10 are Endangered and 11 are Vulnerable. It also supports the entire global population of three species that are currently assessed as Data Deficient and one that has not yet been evaluated on the IUCN Red List, hence these species trigger criterion A(iii). Under criterion B(ii), Mount Namuli supports 40 (ca. 8.4%) of the species on the national list and so is well in excess of the 3% threshold for this criterion; 19 of these taxa are endemic to this IPA. Under criterion C(iii), this site supports nationally important areas of Montane Grassland and Montane Moist Forest, both of which are nationally range-restricted habitats, the latter also being nationally threatened.

Cyanotis sp. nov. *"namuliensis"* (AMR)

Priority species (IPA Criteria A and B)

FAMILY	TAXON	IPA CRITERION A	IPA CRITERION B	≥ 1% OF GLOBAL POP'N	≥ 5% OF NATIONAL POP'N	IS 1 OF 5 BEST SITES NATIONALLY	ENTIRE GLOBAL POP'N	SPECIES OF SOCIO-ECONOMIC IMPORTANCE	ABUNDANCE AT SITE
Acanthaceae	*Asystasia malawiana*	A(i)		✓	✓	✓			occasional
Acanthaceae	*Isoglossa namuliensis*	A(i)	B(ii)	✓	✓	✓	✓		unknown
Acanthaceae	*Sclerochiton hirsutus*	A(i)	B(ii)	✓	✓	✓			scarce
Apiaceae	*Pimpinella mulanjensis*		B(ii)						occasional
Apocynaceae	*Ceropegia nutans*	A(i)	B(ii)	✓	✓	✓	✓		scarce
Asphodelaceae	*Aloe torrei*	A(iii)	B(ii)	✓	✓	✓	✓		unknown
Asteraceae	*Bothriocline moramballae*		B(ii)						unknown
Asteraceae	*Helichrysum lastii*		B(ii)						scarce
Asteraceae	*Senecio peltophorus*		B(ii)						occasional
Balsaminaceae	*Impatiens psychadelphoides*	A(i)		✓	✓	✓			unknown
Campanulaceae	*Lobelia blantyrensis*		B(ii)						unknown
Celastraceae	*Gymnosporia gurueensis*	A(i)	B(ii)	✓	✓	✓			unknown
Commelinaceae	*Cyanotis "namuliensis"* ined.		B(ii)				✓		unknown

Priority species (IPA Criteria A and B)

FAMILY	TAXON	IPA CRITERION A	IPA CRITERION B	≥ 1% OF GLOBAL POP'N	≥ 5% OF NATIONAL POP'N	IS 1 OF 5 BEST SITES NATIONALLY	ENTIRE GLOBAL POP'N	SPECIES OF SOCIO-ECONOMIC IMPORTANCE	ABUNDANCE AT SITE
Crassulaceae	Crassula zombensis		B(ii)						unknown
Euphorbiaceae	Euphorbia namuliensis		B(ii)				✓		frequent
Fabaceae	Crotalaria namuliensis		B(ii)				✓		frequent
Fabaceae	Crotalaria torrei		B(ii)				✓		frequent
Fabaceae	Indigofera namuliensis	A(iii)	B(ii)	✓	✓	✓	✓		unknown
Fabaceae	Rhynchosia clivorum subsp. gurueensis	A(iii)	B(ii)	✓	✓	✓	✓		unknown
Fabaceae	Rhynchosia torrei		B(ii)				✓		common
Fabaceae	Tephrosia whyteana subsp. gemina	A(i)	B(ii)	✓	✓	✓	✓		scarce
Iridaceae	Gladiolus zambesiacus	A(i)		✓	✓	✓			unknown
Lamiaceae	Coleus namuliensis		B(ii)				✓		occasional
Lamiaceae	Plectranthus guruensis	A(i)	B(ii)	✓	✓	✓	✓		unknown
Lamiaceae	Plectranthus mandalensis	A(i)	B(ii)	✓	✓	✓			frequent
Lamiaceae	Stachys didymantha		B(ii)						unknown
Linderniaceae	Crepidorhopalon namuliensis		B(ii)				✓		occasional
Loranthaceae	Agelanthus patelii	A(i)	B(ii)	✓	✓	✓			unknown
Loranthaceae	Helixanthera schizocalyx	A(i)	B(ii)	✓	✓	✓			occasional
Malvaceae	Dombeya lastii	A(i)	B(ii)	✓	✓	✓	✓		scarce
Melastomataceae	Dissotis johnstoniana var. johnstoniana		B(ii)						unknown
Melastomataceae	Memecylon nubigenum	A(i)	B(ii)	✓	✓	✓			unknown
Orobanchaceae	Buchnera namuliensis		B(ii)				✓		scarce
Poaceae	Alloeochaete namuliensis	A(i)	B(ii)	✓	✓	✓	✓		frequent
Poaceae	Digitaria appropinquata	A(iii)	B(ii)				✓		unknown
Poaceae	Digitaria megasthenes	A(i)	B(ii)	✓	✓	✓			unknown
Podostemaceae	Inversodicraea torrei	A(i)	B(ii)	✓	✓	✓	✓		scarce
Polygalaceae	Polygala adamsonii		B(ii)						occasional
Proteaceae	Faurea racemosa	A(i)		✓	✓	✓		✓	common
Rosaceae	Prunus africana	A(i)						✓	unknown
Rubiaceae	Pavetta gurueensis	A(i)	B(ii)	✓	✓	✓			unknown
Rubiaceae	Pyrostria chapmanii	A(i)	B(ii)	✓	✓	✓			unknown

Priority species (IPA Criteria A and B)

FAMILY	TAXON	IPA CRITERION A	IPA CRITERION B	≥ 1% OF GLOBAL POP'N	≥ 5% OF NATIONAL POP'N	IS 1 OF 5 BEST SITES NATIONALLY	ENTIRE GLOBAL POP'N	SPECIES OF SOCIO-ECONOMIC IMPORTANCE	ABUNDANCE AT SITE
Thymelaeaceae	*Gnidia chapmanii*		B(ii)						occasional
Velloziaceae	*Xerophyta splendens*		B(ii)						scarce
Vitaceae	*Cissus aristolochiifolia*	A(i)		✓	✓	✓			unknown
Xyridaceae	*Xyris makuensis*		B(ii)						occasional
Zamiaceae	*Encephalartos gratus*	A(i)		✓	✓	✓			unknown
		A(i): 22 ✓ A(iii): 4 ✓	B(ii): 40 ✓						

Threatened habitats (IPA Criterion C)

HABITAT TYPE	IPA CRITERION C	≥ 5% OF NATIONAL RESOURCE	≥ 10% OF NATIONAL RESOURCE	IS 1 OF 5 BEST SITES NATIONALLY	ESTIMATED AREA AT SITE (IF KNOWN)
Montane Moist Forest [MOZ-01]	C(iii)		✓	✓	7
Medium Altitude Moist Forest [MOZ-02]	C(iii)				1.3
Montane Grassland [MOZ-09]	C(iii)		✓	✓	2.3

Protected areas and other conservation designations

CONSERVATION AREA TYPE	CONSERVATION AREA NAME	RELATIONSHIP OF IPA TO CONSERVATION AREA
No formal protection	N/A	
Important Bird Area	Mount Namuli	protected/conservation area overlaps with IPA
Key Biodiversity Area	Monte Namuli	IPA encompasses protected/conservation area
Alliance for Zero Extinction Site	Mount Namuli	protected/conservation area overlaps with IPA

Threats

THREAT	SEVERITY	TIMING
Shifting agriculture	high	Ongoing – increasing
Small-holder farming	high	Ongoing – increasing
Agro-industry farming	low	Past – not likely to return
Nomadic grazing	low	Ongoing – stable
Logging & wood harvesting	medium	Ongoing – stable
Increase in fire frequency/intensity	medium	Ongoing – increasing

SERRA TUMBINE

Assessors: Sophie Richards, Iain Darbyshire

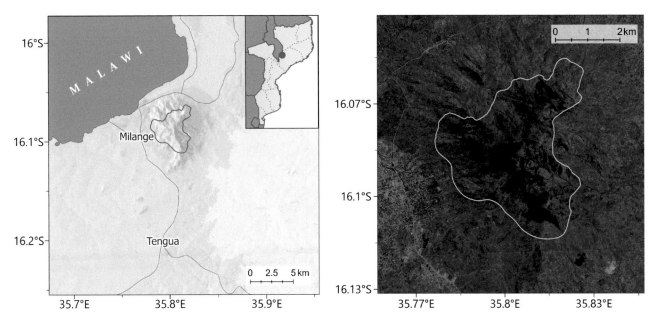

INTERNATIONAL SITE NAME		Serra Tumbine	
LOCAL SITE NAME (IF DIFFERENT)		Machinjiri	
SITE CODE	MOZTIPA036	PROVINCE	Zambézia

LATITUDE	-16.08730	LONGITUDE	35.80350
ALTITUDE MINIMUM (m a.s.l.)	810	ALTITUDE MAXIMUM (m a.s.l.)	1,548
AREA (km²)	13.7	IPA CRITERIA MET	A(i)

Site description

Serra Tumbine is a mountain in Milange District, Zambézia Province, Mozambique. The peak reaches 1,548 m and the entire mountain is approximately 8 km in diameter (Woolley 1987). Serra Tumbine is considered to be an outlier of the Mulanje massif, which is located a short distance across the border in Malawi, and is also part of the Mulanje-Namuli-Ribáuè centre of endemism (Darbyshire *et al.* 2019a). Milange town lies at the western foot of the mountain and there are a small number of residents who live along the 474 road that runs from the east of Serra Tumbine to Milange. Serra Tumbine has a number of aquifers, and it appears from satellite imagery that there is agricultural land associated with the streams that originate on the mountain, particularly around Milange town.

The area of this IPA is 13.7 km². While the dense montane forest is of conservation significance, only remnants of this habitat remain due to extensive conversion to agriculture. During the Mozambican Civil War, people fleeing the conflict settled on Serra Tumbine and began cultivating crops on the mountain, particularly the north-eastern slopes (Manuel 2007). It is thought that the loss of forested areas from the mountain may have contributed to a catastrophic landslide in 1998 (World Bank 2019). At lower altitudes, the habitat is characterised by miombo woodland, although much of this area has also been converted to agriculture and has been excluded from this IPA.

Botanical significance

Three threatened species have been recorded from Serra Tumbine. There is one Endangered species,

Streptocarpus leptopus, which is only known from this IPA and neighboring Mount Mulanje in Malawi and which is threatened by clearing of its forest habitat at both sites (Richards 2021a). Two Vulnerable species are also known from this site: *Encephalartos gratus* (VU), a cycad mainly threatened in Mozambique by annual burning and the resulting decrease in seedling recruitment (Strugnell 2002; Donaldson 2010a; Burrows *et al.* 2018), and *Pavetta chapmanii* (VU), which is only known from six locations within the Mulanje-Namuli-Ribáuè chain of mountains (Timberlake 2020). In addition, a Near Threatened species, *Cola mossambicensis*, has also been collected from Serra Tumbine. Many of the collections for these species are historical, so it would be highly desirable to confirm the continued presence of each at this site and to establish how large the populations are within this IPA.

There has not yet been a full botanical inventory of Serra Tumbine and it is possible that more species of conservation significance are present but have yet to be documented from the site or assessed for the IUCN Red List. For instance, nearby Mount Mulanje is known for its high number of endemic and near endemic species (Strugnell 2002) and, although many of these species are found in high-altitude grasslands and rocky outcrops that are largely absent from Serra Tumbine, we could still expect some species found on Mulanje to be shared with Serra Tumbine.

Although the remaining area of mid-altitude forest is not extensive enough for Serra Tumbine to meet C(iii) of the IPA criteria, it has been suggested that the high root density of the forest stabilises the soil and so prevents erosion and landslides (Manuel 2007). Although much of this forest has been cleared for agriculture (World Bank 2019), it is likely that the remaining fragments still provide this important ecosystem service.

Habitat and geology

The remaining mid-altitude montane forest on Serra Tumbine is of great conservation importance. Although only very limited botanical survey work has been conducted here, in a botanical collection at this site by Correia (*MF* #510), the forest was described as dense, with species including *Albizia*, *Newtonia* (described on the specimen voucher as *Piptadenia* by Correira, but this almost certainly a synonym), *Chrysophyllum* and *Macaranga*. In the nearby Chisongeli forest on Mount Mulanje, *Newtonia buchananii* has been recorded as a dominant species (Dowsett-Lemaire 1988) and it is therefore also likely to be dominant in the forests of Serra Tumbine. There is around 2–4 km^2 of forest remaining on Tumbine, mostly located at altitudes above 1,000 m, with the largest patch overlooking Milange town and some smaller patches in steep gullies and on the northerly peak.

On all of the lower slopes of the mountain (< 1,000 m), as well as at higher altitudes on the western side (up to around 1,300 m), subsistence agriculture dominates, with crops including maize, beans, banana, manioc and sorghum (Manuel 2007). Amongst the agricultural clearings below 1,000 m, there are small patches of miombo woodland remaining, likely dominated by *Brachystegia*, probably *B. spiciformis* as is the case on the foothills of Mount Mulanje (Dowsett-Lemaire 1988). It has been suggested that these miombo woodlands play an important role as a buffer vegetation below mid-montane forests (Timberlake *et al.* 2007); however, very little remains on Serra Tumbine.

On Mount Chiperone, forest clearance has been observed to promote the establishment of edge species, such as *Albizia gummifera*, which in turn prevent the re-establishment of forest species (Timberlake *et al.* 2007). It is possible that the same process has also occurred on Serra Tumbine where forest has been cleared by anthropogenic disturbance or possibly due to landslides.

In terms of geology, Serra Tumbine is a Late Cretaceous to Early Jurassic syenite intrusion within the surrounding Pre-Cambrian metamorphic granulites and gneisses (Woolley 1987; Manuel 2007). The soils derived from the syenites are dark brown with a humic top layer, while the granulites and gneisses form lateritic soils. Both soil types are deep with an overlying layer of colluvial materials varying in size from fine sediments to large boulders (Manuel 2007). Rainfall on the mountain is around 1,200 to 2,000 mm per year, peaking between January and March (Manuel 2007). There are no temperature data for the mountain itself, although nearby Milange town experiences its highest average temperature of 27°C in October and November and an average low of 19°C in June and July (World Weather Online 2021), although it is likely cooler on the upper slopes of the mountain which may possibly experience mists, as observed on Mount Chiperone to the south (Timberlake *et al.* 2007).

Conservation issues

This IPA does not fall within a protected area, Important Bird Area (IBA) or Key Biodiversity Area. There has been little scientific research into the animals of Serra Tumbine, however, the nearby Mount Mulanje is an IBA and it is possible that some of the important bird species from Mulanje occupy Serra Tumbine, at least transiently.

As a result of the Mozambican Civil War, people from various places fleeing conflict settled on Serra Tumbine and began occupying the slopes of the mountain and cultivating crops. The relatively low root density of these crops, compared to that of the forest that previously occupied these slopes, provides less soil stabilisation and is believed to have contributed to a catastrophic landslide at this site in 1993 (Manuel 2007). These agricultural practices on the slopes of the mountain have continued since the last landslide (Achar 2012).

Other causes of the 1998 landslide are also linked to the clearing of vegetation, including high levels of tree felling for charcoal production and fires (Manuel 2007). It is unclear whether the fires have increased in frequency due to anthropogenic burning. Fires are used on Mount Mulanje by hunters to clear the bush (Wisborg & Jumbe 2010) and on Mount Chiperone both to hunt and to clear areas for small-scale cultivation (Timberlake *et al.* 2007); we may therefore expect that at least some of the fires on Serra Tumbine have been related to human activities.

The landslides themselves, of which there have been four between 1940 and 2000, have probably caused massive disturbance to vegetation, as records of the 1993 landslide suggest that high volumes of debris were carried down the mountain (World Bank 2019). This most recent landslide had a huge impact on the local landscape – it is reported that 1,000 hectares of crops were destroyed. The extent of damage to the mid-altitude forest stands is not well-documented, but it is known that tree trunks came down the mountain in the debris and that large landslides have the potential to clear forests, remove topsoil and make land less productive (Forbes & Broadhead 2013), the latter consequence

possibly exacerbating the problem of conversion of forest to agriculture on the mountain.

Support for sustainable food and timber production is required to prevent further landslide catastrophes while also enabling local people to meet their consumption needs. A project run by the NGOs Legado and Nitidae on Mount Namuli has been working with communities to establish a community protected area, secure land rights for local people and promote sustainable economic development for these communities. Part of this work includes a process of delineating a core zone, where conservation is a priority, while using agroecological research to increase production of crops outside this core zone (Nitidae 2021). A similar approach on Serra Tumbine could help protect and regenerate forests on the mountain, which would benefit both local communities and the threatened species residing in this habitat.

Key ecosystem services

The forested areas of Serra Tumbine are important for stabilising the soils, due to the high root density provided by the forest vegetation. It is thought that deforestation, combined with heavy rains and the geology of the mountain, make landslides more likely on Serra Tumbine (World Bank 2019). As a result of the 1998 landslide, 200 people died, 4,000 were displaced and disease outbreaks quickly followed. In addition, 1,000 hectares of crops were lost, much of the area became flooded and there was damage to homes, roads, bridges and water supplies (Manuel 2007). It is therefore clear that forest stands on the mountain play an important role in protecting the lives and well-being of the people living nearby.

It is likely that the many aquifers under Serra Tumbine are an important source of water, at the very least providing moisture required for farming and possibly also a source of clean drinking water, especially when refugees began occupying the mountain in the 1990s.

Timber is extracted from the mountain for the production of charcoal (Manuel 2007). It has not been documented whether hunting occurs here, but hunting is reported on other mountains in the area, Mulanje and Chiperone (Timberlake *et al.* 2007; Wisborg & Jumbe 2010), so it is likely that some hunting also occurs on Serra Tumbine.

However, with very little forest left, the numbers of game available would have likely decreased. Food production on the mountain is largely from the small-scale cultivation of crops such as maize, beans, banana trees, manioc and sorghum. While the continuation of these agricultural practices makes future landslides more likely (Achar 2012), and encroachment will lead to the further loss of the already nationally threatened mid-altitude forests, local people are reliant on these farms for food and so sustainable alternatives should be found.

Ecosystem service categories

- Provisioning – Raw materials
- Provisioning – Food
- Provisioning – Fresh water
- Regulating services – Moderation of extreme events
- Regulating services – Erosion prevention and maintenance of soil fertility

IPA assessment rationale

Serra Tumbine qualifies as an IPA under criterion A. Three sub-criterion A(i) species are recorded from this IPA: *Encephalartos gratus* (VU), *Pavetta chapmanii* (VU) and *Streptocarpus leptopus* (EN). Serra Tumbine is particularly important for the lattermost species as one of only two sites globally and the only Mozambican IPA from which *S. leptopus* is known.

Priority species (IPA Criteria A and B)

FAMILY	TAXON	IPA CRITERION A	IPA CRITERION B	≥ 1% OF GLOBAL POP'N	≥ 5% OF NATIONAL POP'N	IS 1 OF 5 BEST SITES NATIONALLY	ENTIRE GLOBAL POP'N	SPECIES OF SOCIO-ECONOMIC IMPORTANCE	ABUNDANCE AT SITE
Gesneriaceae	*Streptocarpus leptopus*	A(i)		✓	✓	✓			unknown
Rubiaceae	*Pavetta chapmanii*	A(i)			✓	✓			unknown
Zamiaceae	*Encephalartos gratus*	A(i)			✓	✓			unknown
		A(i): 3 ✓							

Threatened habitats (IPA Criterion C)

HABITAT TYPE	IPA CRITERION C	≥ 5% OF NATIONAL RESOURCE	≥ 10% OF NATIONAL RESOURCE	IS 1 OF 5 BEST SITES NATIONALLY	ESTIMATED AREA AT SITE (IF KNOWN)
Medium Altitude Moist Forest [MOZ-02]					1.8

Protected areas and other conservation designations

CONSERVATION AREA TYPE	CONSERVATION AREA NAME	RELATIONSHIP OF IPA TO CONSERVATION AREA
No formal protection	N/A	

Threats

THREAT	SEVERITY	TIMING
Small-holder farming	high	Ongoing – trend unknown
Increase in fire frequency/intensity	unknown	Ongoing – trend unknown
Logging & wood harvesting	low	Ongoing – trend unknown

MOUNT MABU

Assessor: Iain Darbyshire

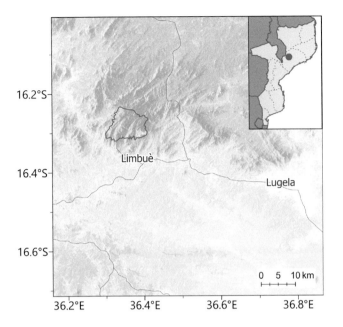

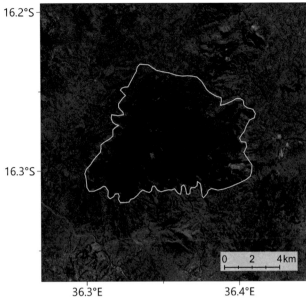

INTERNATIONAL SITE NAME		Mount Mabu	
LOCAL SITE NAME (IF DIFFERENT)		Monte Mabu	
SITE CODE	MOZTIPA012	PROVINCE	Zambézia

LATITUDE	-16.27430	LONGITUDE	36.35680
ALTITUDE MINIMUM (m a.s.l.)	640	ALTITUDE MAXIMUM (m a.s.l.)	1,650
AREA (km²)	75	IPA CRITERIA MET	A(i), C(iii)

Early morning view of Mount Mabu (TH)

Streamside forest on Mount Mabu (TH)

Transition from Medium Altitude Forest to *Coleochloa*-dominated rock outcrops on Mabu (TH)

Site description

The Mount Mabu IPA covers an area of 75 km² in Lugela District of central Zambézia Province. It is located ca. 120 km to the southwest of Mount Namuli and 80 km east-southeast of Mount Mulanje in Malawi. The site is rather isolated, the nearest town being Lugela, ca. 45 km to the east. Mabu is one of a series of inselbergs and massifs that form a broad archipelago-like chain running from southern Malawi through Zambézia and Nampula Provinces of northern Mozambique, which have together been proposed as the Mulanje-Namuli-Ribáuè Centre of Plant Endemism (Darbyshire *et al.* 2019a). This site was the focus of significant publicity in the late 2000s when a series of biodiversity surveys revealed the biological importance of its forests; it is sometimes labelled the "Google Forest" as it was identified from Google Earth imagery as a key site of potentially high biodiversity importance during a review of important montane sites in northern Mozambique.

Botanical significance

Of primary botanical importance at Mount Mabu is the presence of extensive areas of intact moist forest, estimated to cover 78.8 km². The large majority of this forest (ca. 53 km²) is at elevations of 950–1,400 m and it is posited that Mabu holds the largest continuous tract of medium altitude moist forest in southern Africa (Bayliss *et al.* 2014). It also contains a smaller area (ca. 10 km²) of montane moist forest. Together, these forests support a varied flora with a number of rare and threatened species. To date, only one potential endemic species has been discovered, an as yet undescribed species of *Vepris* related to *V. trichocarpa* which is locally common in the forests at 980–1,600 m. However, Mabu is also considered to be the most important locality globally for a number of range-restricted and threatened forest species including *Helixanthera schizocalyx* (EN), *Pavetta gurueensis* (VU) and *Polysphaeria harrisii* (EN), all of which are globally threatened (Darbyshire *et al.* 2019h). It is

Asystasia malawiana (TH)

Pavetta gurueensis (TH)

also a critical site for the scarce Afromontane forest tree *Faurea racemosa* (EN), a species that is exploited for its timber elsewhere in its range, but is a major component of the montane forest canopy at Mabu where it is considered to be secure (Darbyshire *et al.* 2018b). A second canopy species of conservation importance here is *Maranthes goetzeniana* (NT) which is fairly common at elevations up to 1,400 m (Dowsett-Lemaire & Dowsett 2009).

The site is also notable for several outlier populations of plant species. Species previously thought to be endemic to Mount Mulanje have recently been discovered here, notably the rock-dwelling taxa *Senecio peltophorus* (LC) and *Streptocarpus milanjianus* (VU), the former also now known from Mount Namuli. *Cryptostephanus vansonii* (LC) is otherwise known only from the Chimanimani-Nyanga Mountains on the Mozambique-Zimbabwe border. Several other taxa have their southern limit at Mabu, such as *Justicia asystasioides* and *Crotonogynopsis australis*, the latter otherwise known only from the Udzungwa and Mahenge Mountains of southern Tanzania. A similar situation is noted in the fauna of Mabu (Bayliss *et al.* 2014). Other nationally rare species recorded here include the bamboo *Oreobambos buchwaldii*, otherwise known in Mozambique only from Moribane Forest Reserve in the Chimanimani foothills, and the diminutive orchid *Polystachya songaniensis*, otherwise known only from Mulanje and Zomba in Malawi and the Ribáuè Massif in Nampula Province; the Mozambique populations of this latter species may prove to be a distinct subspecies (A. Schuiteman, pers. comm. 2018).

During botanical surveys on the mountain in 2008, 249 plant species were recorded (Timberlake *et al.* 2012). However, only a small portion of the site on the eastern side has so far been surveyed botanically and there is a high likelihood that further discoveries of rare and potentially new species will be made if future surveys are conducted, particularly targeting other portions of the site.

Habitat and geology

Mount Mabu is a granitic massif, formed from an igneous intrusion of granite-syenite of the Namarroi Group (ca. 1,100–850 mya), surrounded by migmatites of the same series (Instituto Nacional de Geológia 1987; Timberlake *et al.* 2012). The massif is intersected by a series of steep valleys. Most of the site is densely forested, but with small areas of exposed granitic outcrops on the higher peaks. Climate data are not available for the massif, but limited data from the nearby Madal tea estates at ca. 400 m elevation record a high mean annual rainfall of 2,119 mm, with the main rainy season from November to March. Mean annual temperature was 23.7°C and always exceeding 20°C; the occurrence of frost on the mountain is likely to be rare (Timberlake *et al.* 2012).

Detailed accounts of the vegetation and species assemblages of Mount Mabu are provided by Dowsett-Lemaire & Dowsett (2009) and Timberlake *et al.* (2012) and are summarized here. The lower slopes, primarily below 1,000 m but rising to higher elevations on the drier northern side, are low diversity woodlands dominated by *Pterocarpus angolensis* lower down and *Syzygium cordatum* as elevation rises. The understorey of this transitional woodland is dominated by *Aframomum* sp. with some patches of the bamboo *Oxytenanthera abyssinca*. Some small areas of lowland forest occur, particularly along streams, where *Albizia adianthifolia* can dominate together with *Macaranga capensis*. The medium altitude (950–1,400 m) moist forests are tall, with the canopy up to 40–45(–50) m tall, frequently comprising trees of *Cryptocarya liebertiana*, *Drypetes gerrardii*, *Gambeya* (formerly *Chrysophyllum*) *gorongosana*, *Maranthes goetzeniana*, *Newtonia buchananii* and *Strombosia scheffleri*, the lattermost often being dominant. The substrata are diverse, with frequent small trees including *Drypetes natalensis*, *Pavetta gurueensis*, *Rawsonia lucida*, *Rinorea ferruginea* and *Synsepalum muelleri*. Lianas are numerous, with *Millettia lasiantha* particularly common. At higher elevations (1,350–1,650 m), an Afromontane forest with a canopy of up to 25 m tall dominates. *Newtonia buchananii* drops out above 1,400 m, whilst *Albizia adianthifolia* is replaced by *A. gummifera*. Common tree species include *Olea capensis* and *Podocarpus milanjianus*, with *Aphloia theiformis*, *Faurea racemosa*, *Macaranga capensis*, *Myrsine* (formerly *Rapanea*) *melanophloeos*, *Prunus africana* and *Syzygium afromontanum* at higher elevations. Exposed granite-migmatite slopes and peaks support a lithophytic flora dominated by *Coleochloa setifera*. Patches of montane shrubland are frequent towards the peaks, where *Aeollanthus buchnerianus*, *Aeschynomene nodulosa*, *Kotschya recurvifolia* and *Tetradenia riparia* are among the common shrubs and *Myrsine melanophloeos* is frequent as a small tree.

Much of the habitat is intact and undisturbed. However, there are areas of abandoned tea plantations on the south-eastern slopes of the mountain which are in the process of converting to secondary woodland and forest, with *Albizia adianthifolia* as a common overtopping tree (Dowsett-Lemaire & Dowsett 2009). There are areas of subsistence machamba agriculture in this area.

Conservation issues

Although not formally protected at present, Mount Mabu is currently in the process of being established as a conservation area. The Mount Mabu Conservation Project, led by Flora and Fauna International and Justiça Ambiental and supported by CEPF, ran between 2013 and 2016 with the aim to establish community-based conservation efforts and education as steps towards establishing a Community Conservation Area (CCA), and to delineate areas for forest conservation and ecotourism; a draft management plan was developed as part of this work (CEPF 2021). A consortium of conservation organisations has now been tasked with establishing the protected area and implementing a revised management plan, under the PROMOVE Biodiversidade project (Biofund 2021).

Significant threats at Mount Mabu are minimal, in part because of the relative inaccessibility of most of the forest due to the steep rocky terrain, and in part because of the small human population in the vicinity of this site (Bayliss *et al.* 2014). The spiritual significance of the site for local communities has also contributed to its protection. A number of low-level or potential threats to the biodiversity and ecological integrity of Mabu have been noted by Timberlake *et al.* (2012). These include potential expansion of agriculture into the lower

Helixanthera schizocalyx (CC)

parts of medium-altitude forest if the local human population increases; increased frequency of fires burning right up to the forest boundary which may inhibit forest regeneration at the margins; a slight danger of logging for timber, particularly in the surrounding moist woodland and in lower-elevation parts of the forest; and the unsustainable level of bush meat hunting in the forest by the local population. Some recent (post-2010) encroachment into the forest on the southwestern slopes is observable on satellite imagery available via Google Earth (2021).

In addition to its significance for plants, Mount Mabu is an important site for a range of faunal groups. Surveys in the 2000s revealed several new species of reptiles and butterflies, such as the bush viper *Atheris mabuensis* (EN), known only from Mabu and Namuli, and the endemic Mount Mabu Pygmy Chameleon (*Rhampholeon maspictus*, NT). The avifauna is particularly important and Mount Mabu is an Important Bird Area (BirdLife International 2021b), with seven globally threatened or near-threatened species recorded including the Namuli Apalis (*Apalis lynesi*, EN), previously thought to be endemic to Mount Namuli (Dowsett-Lemaire & Dowsett 2009). The spectacle of butterfly hill-topping can be observed on the summit of Mabu in October and November (Timberlake *et al.* 2012; Bayliss *et al.* 2014). This site has recently been designated as a Key Biodiversity Area (WCS *et al.* 2021).

Key ecosystem services

Bayliss *et al.* (2014) note that the extensive forests at Mabu are important for their carbon storage, estimating that the total carbon storage value of the forest area is 3.6 Tg, of which 2.7 Tg would be released if the forest were converted to bushland and cropland. The forests also provide other important regulatory services, notably in the prevention of soil erosion on the steep montane slopes and flooding events in the adjacent lowlands.

Mount Mabu and its rivers hold spiritual significance for the local communities. The site has some potential for community-led ecotourism, particularly for trekking and wildlife enthusiasts, although its remoteness and steep terrain may limit the scale of ecotourism options. Bayliss *et al.* (2014) also suggested that bottled mineral water could be a viable commercial venture at Mabu.

Ecosystem service categories

- Provisioning – Fresh water
- Regulating services – Local climate and air quality
- Regulating services – Carbon sequestration and storage
- Regulating services – Moderation of extreme events
- Regulating services – Erosion prevention and maintenance of soil fertility
- Habitat or supporting services – Habitats for species
- Habitat or supporting services – Maintenance of genetic diversity
- Cultural services – Tourism
- Cultural services – Spiritual experience and sense of place

IPA assessment rationale

Mount Mabu qualifies as an IPA under criterion A(i) as it contains important populations of three globally Endangered species and four globally Vulnerable species. Mabu is considered likely to be the global stronghold for most of these species in view of the excellent habitat quality at this site and the high levels of threat at their other known localities. In addition, it contains a nationally important population of the widespread but globally Vulnerable Afromontane species *Prunus africanus*, which is exploited in parts of its range for its medicinal bark. Mabu also qualifies under criterion C(iii) on the basis of containing the largest extent of continuous Medium Altitude Moist Forest in Mozambique. It also contains a smaller extents of Montane Moist Forest, but this site is not considered to meet the thresholds for that habitat under criterion C(iii). To date, 10 of the priority species under criterion B(ii) have been found at Mount Mabu, equating to 2% of the national B(ii) species list, and so falling below the 3% threshold for this sub-criterion. However, this figure is likely to rise as further surveys of this site are conducted in the future.

Priority species (IPA Criteria A and B)

FAMILY	TAXON	IPA CRITERION A	IPA CRITERION B	≥ 1% OF GLOBAL POP'N	≥ 5% OF NATIONAL POP'N	IS 1 OF 5 BEST SITES NATIONALLY	ENTIRE GLOBAL POP'N	SPECIES OF SOCIO-ECONOMIC IMPORTANCE	ABUNDANCE AT SITE
Acanthaceae	*Asystasia malawiana*	A(i)		✓	✓	✓			occasional
Acanthaceae	*Sclerochiton hirsutus*	A(i)	B(ii)	✓	✓	✓			unknown
Asteraceae	*Senecio peltophorus*		B(ii)						scarce
Euphorbiaceae	*Crotonogynopsis australis*		B(ii)						unknown
Gesneriaceae	*Streptocarpus milanjianus*	A(i)	B(ii)	✓	✓	✓			scarce
Iridaceae	*Freesia grandiflora* subsp. *occulta*		B(ii)						unknown
Loranthaceae	*Helixanthera schizocalyx*	A(i)	B(ii)	✓	✓	✓			occasional
Orchidaceae	*Polystachya songaniensis*		B(ii)						frequent
Proteaceae	*Faurea racemosa*	A(i)		✓	✓	✓		✓	frequent
Rosaceae	*Prunus africana*	A(i)			✓	✓		✓	frequent
Rubiaceae	*Pavetta gurueensis*	A(i)	B(ii)	✓	✓	✓			common
Rubiaceae	*Polysphaeria harrisii*	A(i)	B(ii)	✓	✓	✓			common
Rutaceae	*Vepris* sp. nov.		B(ii)				✓		common
		A(i): 8 ✓	B(ii): 10						

Threatened habitats (IPA Criterion C)

HABITAT TYPE	IPA CRITERION C	≥ 5% OF NATIONAL RESOURCE	≥ 10% OF NATIONAL RESOURCE	IS 1 OF 5 BEST SITES NATIONALLY	ESTIMATED AREA AT SITE (IF KNOWN)
Montane Moist Forest [MOZ-01]	C(iii)				10.1
Medium Altitude Moist Forest [MOZ-02]	C(iii)		✓	✓	53.0

Protected areas and other conservation designations

CONSERVATION AREA TYPE	CONSERVATION AREA NAME	RELATIONSHIP OF IPA TO CONSERVATION AREA
No formal protection (but gazettement in progress)	N/A	
Important Bird Area	Mount Mabu	IPA encompasses protected/conservation area
Key Biodiversity Area	Mount Mabu	IPA encompasses protected/conservation area

Threats

THREAT	SEVERITY	TIMING
Small-holder farming	low	Ongoing – trend unknown
Increase in fire frequency/intensity	low	Ongoing – trend unknown
Hunting & collecting terrestrial animals	low	Ongoing – trend unknown

MOUNT CHIPERONE

Assessors: Sophie Richards, Iain Darbyshire, Hermenegildo Matimele

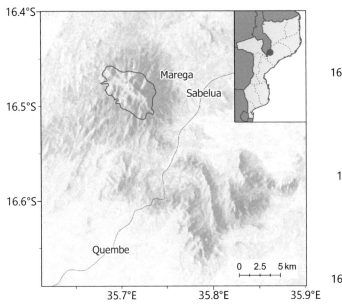

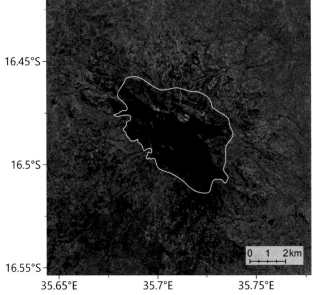

INTERNATIONAL SITE NAME		Mount Chiperone	
LOCAL SITE NAME (IF DIFFERENT)		Monte Chiperone	
SITE CODE	MOZTIPA035	PROVINCE	Zambézia

LATITUDE	-16.48144	LONGITUDE	35.71007
ALTITUDE MINIMUM (m a.s.l.)	850	ALTITUDE MAXIMUM (m a.s.l.)	2,054
AREA (km²)	24.2	IPA CRITERIA MET	A(i), C(iii)

View of Mount Chiperone peaks (TH)

Mid-altitude montane forests on Mount Chiperone (TH)

Site description

Mount Chiperone is an isolated peak located in Milange District, Zambézia Province, Mozambique. This inselberg falls within the Mulanje-Namuli-Ribáuè centre of endemism - a chain of mountains from southern Malawi to northern Mozambique where a high number of endemic or near endemic plant species are located (Darbyshire *et al.* 2019a). Two villages are situated to the south of the mountain, Sabelua and Marega, each with associated agricultural land in the surrounding area. Drainage from the mountain gives rise to the Rio Macololo and the Rio Muse in the south-west, both of which are tributaries of the Shire River in Malawi.

The IPA is 24.2 km² in area and encompasses the montane forests on the mid- to high-altitude slopes of the mountain. These forests are of conservation importance due to their large extent of mostly undisturbed vegetation across an altitudinal gradient. Much of this area is above 900 m altitude, peaking at 2,054 m above sea level. The miombo woodland that surrounds the moist forest has largely been excluded. This area is used by local people for subsistence agriculture, particularly on the southern and western slopes and in some valleys to the north. However, the woodland fragments that remain provide an important ecological buffer for the forest in the IPA above (Timberlake *et al.* 2007).

Botanical significance

The main botanical significance of Mount Chiperone is the expansive area of undisturbed mid- to high-altitude humid forest, providing an uninterrupted forest transition from around 1,000 to 2,000 m above sea level. As many of the forests at this site are regarded as sacred by local people, they are largely in pristine condition. Chiperone represents one of the five best nationally for montane forest, with an extent of around 1.7 km² of forest above 1,600 m altitude, while it is also only slightly below the threshold for qualifying as one of the five best sites for mid-altitude forest, with an estimated 9.4 km² of habitat.

Although there are currently no plant species known from only this site, it is home to some globally threatened species. Two of these threatened species are used directly by people: *Prunus africana* (VU) and *Coffea salvatrix* (EN). The medicinal bark of *P. africana* was previously exported at high volumes to Europe and the United States to treat benign prostate hyperplasia before an export ban was implemented. This species has a long history of use in traditional healing and as a timber tree throughout its native range, although it is not currently thought to be under threat from commercial harvesting at this site (Jimu 2011). *Coffea salvatrix*, or Mukofi, is a crop wild relative of commercial Arabica and Robusta coffees and may have some potential as a specialist coffee, particularly as it is a drought-tolerant species (O'Sullivan & Davis 2017). A third threatened species, *Pavetta chapmanii* (VU), is only known from six locations within the Mulanje-Namuli-Ribáuè chain of mountains (Timberlake 2020). All three of these forest species are threatened by the conversion of habitat to agriculture.

In addition, Chiperone represents the only confirmed Mozambican populations of three species as yet unassessed for the IUCN Red List, *Erica microdonta*, *Cyperus amauropus* and *Pollia condensata*, representing the southernmost collections within their respective ranges, while collections of *Abrus melanospermus* subsp. *suffruticosus* and *Cyrtorchis arcuata* subsp. *whytei* on Chiperone both represent range extensions for these subspecies (Harris *et al.* 2011). A total of 229 species were included in a checklist of plants above 800 m on Chiperone compiled by Timberlake *et al.* (2007), after a scientific expedition in 2006. Previous scientific expeditions to Chiperone have been focused on avifauna, with four visits recorded between 1950 and 2005. There are still large areas of this mountain, particularly to the north-west, where botanical inventory has yet to be undertaken,

it is therefore likely that the checklist will continue to grow and will include more threatened and rare plant species.

Habitat and geology

Inventory work was undertaken in 2006 by Timberlake *et al.* (2007), who documented many of the key species in ecosystems on the mountain. The following habitat description is based upon this account.

The vegetation on Chiperone varies with altitude. Mid-altitude forest, between 1,000–1,600 m, is dominated by tree species such as *Newtonia buchananii, Strombosia scheffleri, Rinorea convallarioides* and *Gambeya* (formerly *Chrysophyllum*) *gorungosana*. The understorey is open and dominated by *Dracaena fragrans* and *Pseuderanthemum subviscosum*. The mid-slopes have shallow soils because of their steepness, which may explain the observed dominance of trees with low trunk diameters.

At higher altitudes, above 1,600 m, the forest composition changes, with shorter and more sclerophyllous tree species dominating

Montane forest at high altitudes, with fern and lichen epiphytes in the understorey (TH)

Clearing of land surrounding the IPA for subsistence agriculture (TH)

including *Peddiea africana, Diospyros whyteana, Maytenus undata, M. acuminata, Myrsine africana, Ochna holstii, Vepris bachmannii* and *Olea capensis* subsp. *macrocarpa*. There is a greater incidence of fruticose lichens (thought to be *Usnea* spp.) and understorey ferns here, probably due to the lower temperatures and frequent moist air derived from low cloud cover.

At the exposed peak, between 1,900 and 2,000 m, there is a dense thicket vegetation, primarily of *Erica microdonta*, with *Aloe arborescens* forming decumbent masses. Fruticose lichens are also common at this altitude, again these are thought to be *Usnea*. Other exposed outcrops are rare, as even the most inaccessible ridges are covered by woody vegetation. Where rock is exposed, the plant community consists of *Aloe*, grasses, sedges (particularly *Coleochloa*, probably *C. setifera*) and low shrubs. There is a small patch of *Acacia abyssinica* woodland, which is confined to one south-easterly ridge as far as is understood. This woodland is of particular note due to the high number of ephiphytic ferns, orchids and lichens. The root parasite *Sarcophyte sanguinea* subsp. *piriei* is also common.

At the very edges of this IPA, and surrounding its boundaries, miombo woodland dominates. Much of this vegetation consists of *Brachystegia* species, and while this plant community is common throughout southern Africa, it has an important role as a buffer to the forested areas above. Where soils are fertile, this woodland has been cleared for cultivation. Extensive clearing near the settlements on the southern and eastern slopes has been observed, but much of this has been excluded from the IPA. As soils become exhausted, there may be expansion into the forested areas at higher altitudes, which are largely undisturbed at present.

The soils throughout the IPA are quite shallow. Chiperone is comprised of synetite (Jurassic/ Cretaceous period, ca. 150 Mya) and is an igneous intrusion into the surrounding country rock, migmatites of the Namarroi Series (850– 1,100 Mya). This geology is unlike most of the massifs and hills in northern Mozambique, which are comprised mostly of granites and migmatites. That said, this geology is shared with the nearby Mount Tembe, Morrumbala and Serra Tumbine at Milange, along with sections of Mount

Mulanje, across the border in Malawi, and Mount Gorongosa in central Mozambique.

Climate data from the Chiperone itself are not available. Temperatures in nearby Milange town range from 17.1°C to 28.9°C, with annual rainfall averaging 1,734 mm, although annual precipitation on the lower slopes of Chiperone is thought to be much lower, around 1,000 mm.

Conservation issues

This IPA does not fall within a protected area and is under no formal conservation measures. The forest and its wildlife fall under the Forest and Wildlife Act (Lei. 10/99 of 1999) and, in most areas, The Environmental Act (Lei. 20/97 of 1997), the latter prohibiting the cultivation of annual crops on slopes steeper than 7° and perennial crops on slopes steeper than 14° (Timberlake *et al.* 2007; ECOLEX.org 2020).

Mount Chiperone has been designated as an Important Bird Area (IBA), with the IPA described here falling completely within the boundaries of the IBA. In terms of bird diversity, this site hosts the only known population of the White-winged Apalis (*Apalis chariessa* – NT) in Mozambique (BirdLife International 2020a). In addition, the largest population of Thyolo Alethe (*Chamaetylas choloensis* – VU) globally is known from Chiperone. Elsewhere this species is greatly threatened by deforestation, particularly in Malawi (BirdLife International 2018).

This site is also recognised as a Key Biodiversity Area (KBA) with trigger species including *C. choloensis* and three reptile species that are only known from Mount Chiperone. One of these species, Mount Chiperone Pygmy Chameleon (*Rhampholeon nebulauctor*), has been assessed as globally Vulnerable (Tolley *et al.* 2019c). The remaining two trigger species are the recently described *Nadzikambia chiperone* and *Lygodactylus chiperone* (Tolley 2017, 2018). The presence of these unique reptile species allows the site to qualify as a KBA under the irreplaceability criterion (Tolley 2017). A number of small mammals have been recorded from the massif, and Leopard (*Panthera pardus* – VU) are reported to be common (Timberlake *et al.* 2007).

At lower altitudes, below 800 m, much of the woodland has been cleared for cultivation. However, wood extraction above ca. 1,300 m is limited due to the steep inaccessible terrain and a local reluctance to enter the dense forest due to spiritual beliefs in the area (Spottiswoode *et al.* 2008). At the margins of the forest, however, there has been some minor tree felling but a greater threat is posed by uncontrolled fires, particularly in the gullies. Fires are used to clear areas for subsistence agriculture at lower altitudes. However, these fires have been known to spread up the mountain, with evidence of burning 50 m in from the margins of the moist forest (Timberlake *et al.* 2007). These uncontrolled fires result in the loss of humus and moisture in the soils, and in the clearing of understorey plants. Edge and gap species, such as *Trema orientalis* and *Albizia gummifera*, can then establish and prevent forest regeneration by outcompeting forest tree species (Timberlake *et al.* 2007).

Despite the shallow and rocky soils, local people establish machambas on steep slopes of the mountain for cultivating maize and beans. With good rainfall, the larger fields surrounding the mountain produce higher yields than the small clearings above. However, when there is little rainfall, the machambas on the slopes will still yield some food, due to the moister air and reduced evapotranspiration on the mountain slopes, while the crops in the surrounding peneplain are liable to fail in these conditions (Timberlake *et al.* 2007). There could therefore be an increasing threat of encroachment from agriculture, particularly if drought stress becomes more common in the area. As Chiperone is linked to both weather patterns in the local area (see "Key ecosystem services") and to the drainage of clean water into nearby rivers, encroachment of agriculture could further exacerbate water shortages, and therefore the environmental pressures, in the local and wider area.

Key ecosystem services

Local villages to the south and east, with a total population of 1,000 to 2,000 residents, depend on the streams coming off the mountain as their only source of clean water for several kilometres (Timberlake *et al.* 2007). Local people also depend on this water source to support subsistence agriculture, particularly on machambas on steeper slopes that are an important food source during drier years. In Malawi, "chiperone" refers to a misty weather system that forms over Mount Chiperone and crosses into Malawi. It is thought that this weather system is an important source of moisture for the tea plantations in Malawi during the dry season (Timberlake *et al.* 2007).

Owing to locally held spiritual beliefs, people from the surrounding villages are generally reluctant to enter the forested areas above ca. 1,300 m (Spottiswoode *et al.* 2008). The humid forest on Chiperone may therefore be regarded as a sacred forest using the criteria suggested by Virtanen (2002).

Timber is rarely harvested from within the IPA boundary, and the collection of plants for medicinal properties occurs mostly in the surrounding woodland areas. Bushmeat hunting, however, does take place on higher slopes, although this practice is limited to a small group of local residents. Bushbuck (*Tragelaphus scriptus*), Bushpig (*Potamochoerus larvatus*), and Duiker (*Cephalophus* spp.) are caught using gin traps (Timberlake *et al.* 2007). Honey is also reported to be collected in the forest (Spottiswoode *et al.* 2008).

Deforestation on nearby Serra Tumbine has resulted in catastrophic landslips above Milange town, suggesting that the forest on Chiperone may also play an important role in stabilising the substrate on the mountain (Timberlake *et al.* 2007).

There is no known tourism in this area.

Ecosystem service categories

- Provisioning – Food
- Provisioning – Fresh water
- Provisioning – Medicinal resources
- Regulating services – Local climate and air quality
- Regulating services – Erosion prevention and maintenance of soil fertility
- Cultural services – Spiritual experience and sense of place

IPA assessment rationale

Mount Chiperone qualifies as an IPA under criteria A and C. The massif supports three sub-criterion A(i) species: *Coffea salvatrix* (EN), *Prunus africana* (VU) and Pavetta *chapmanii* (VU). Much of the Timberlake *et al.* (2007) checklist of plants is yet to be assessed for the IUCN Red List, and so it is likely that more species meet criterion A(i) but are in need of assessment. This site also qualifies under sub-criterion C(iii) as one of the five best sites for montane moist forest nationally. Although the median altitude forests at this site do not trigger sub-criterion C(iii), the habitat here is of high-quality and still of conservation importance.

Priority species (IPA Criteria A and B)

FAMILY	TAXON	IPA CRITERION A	IPA CRITERION B	≥ 1% OF GLOBAL POP'N	≥ 5% OF NATIONAL POP'N	IS 1 OF 5 BEST SITES NATIONALLY	ENTIRE GLOBAL POP'N	SPECIES OF SOCIO-ECONOMIC IMPORTANCE	ABUNDANCE AT SITE
Rosaceae	*Prunus africana*	A(i)				✓		✓	unknown
Rubiaceae	*Coffea salvatrix*	A(i)			✓	✓		✓	frequent
Rubiaceae	*Pavetta chapmanii*	A(i)			✓	✓			unknown
		A(i): 3 ✓							

Threatened habitats (IPA Criterion C)

HABITAT TYPE	IPA CRITERION C	≥ 5% OF NATIONAL RESOURCE	≥ 10% OF NATIONAL RESOURCE	IS 1 OF 5 BEST SITES NATIONALLY	ESTIMATED AREA AT SITE (IF KNOWN)
Montane Moist Forest [MOZ-01]	C(iii)			✓	1.7
Medium Altitude Moist Forest [MOZ-02]	C(iii)				9.4

Protected areas and other conservation designations

CONSERVATION AREA TYPE	CONSERVATION AREA NAME	RELATIONSHIP OF IPA TO CONSERVATION AREA
Important Bird Area	Mount Chiperone	protected/conservation area encompasses IPA
Key Biodiversity Area	Mount Chiperone	protected/conservation area encompasses IPA

Threats

THREAT	SEVERITY	TIMING
Small-holder farming	low	Ongoing – trend unknown
Increase in fire frequency/intensity	medium	Ongoing – trend unknown

MOUNT MORRUMBALA

Assessors: Sophie Richards, Iain Darbyshire

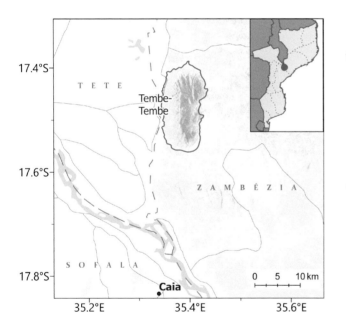
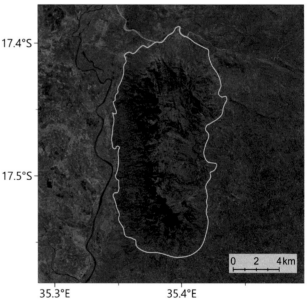

INTERNATIONAL SITE NAME		Mount Morrumbala	
LOCAL SITE NAME (IF DIFFERENT)		Monte Morrumbala	
SITE CODE	MOZTIPA052	PROVINCE	Zambézia

LATITUDE	-17.48000	LONGITUDE	35.38500
ALTITUDE MINIMUM (m a.s.l.)	45	ALTITUDE MAXIMUM (m a.s.l.)	1,172
AREA (km²)	135	IPA CRITERIA MET	A(i), C(iii)

Site description

Mount Morrumbala, known sometimes as Mount Tembe, is an IPA in Morrumbala District of Zambézia Province, adjacent to its border with Tete Province and 30 km south of the Mozambique-Malawi border. The name "Morrumbala" probably means barrier, referring to the position of the mountain separating the district from the Shire River to the west (Inguaggiato *et al.* 2009). The IPA covers an area of 135 km² with the Shire River running to the west and the Nambuur River to the north of this site. Morrumbala town, the district centre, is a short distance away from the IPA, 20 km to the north-east, and there are a number of communities that live along the Shire River. Tembe-Tembe is the closest village and, while it mostly falls outside this IPA, a small number of homesteads on the lower slopes are inside the site boundary.

Little botanical research has taken place in recent decades on Mount Morrumbala, but from previous botanical collections, it is clear that there are plant species of conservation interest at this site.

Botanical significance

Mount Morrumbala is the only known locality for Critically Endangered species *Crassula morrumbalensis*. This species is known from a single collection made in 1942 by A.R. Torre. Although little is known about *C. morrumbalensis*, suitable habitat is described as "moist savanna", presumably the escarpment miombo that dominates the eastern slopes. This habitat is currently experiencing continued degradation through the opening of machambas (Google Earth 2021; World Resources Institute 2021), and so urgent research is required to confirm the continued presence of this species and to establish the impact of this habitat loss on its population size.

One additional threatened species has been recorded within this IPA. The Vulnerable species *Celosia pandurata* is endemic to central Mozambique, occurring in lowland forests of Mount Morrumbala, and is threatened at multiple locations by the expansion of agriculture (Richards, in press [a]). There is also a doubtful record of *Coffea zanguebariae* (VU), collected in a steep gully on the western side of the mountain (GBIF. org 2021b). The voucher states that this specimen

is a multi-stemmed 12 m tall tree with "pale, longitudinally scaly bark", whereas by contrast *C. zanguebariae* is usually up to 6 m and tends to have smooth bark (A. Davis, pers. comm. 2021). This specimen could potentially be *C. salvatrix* (EN), which occurs in similar moist montane forests, including Mount Chiperone [MOZTIPA035], another mountain of western Zambézia Province. *C. salvatrix* is known to have buff-coloured, cracked bark similar to the specimen description, although, like *C. zanguebariae*, it is a bush or small tree (Bridson & Verdcourt *et al.* 2003). Further collections of *Coffea* on Morrumbala are required to identify the taxon as there is a strong possibility that it may be a threatened species.

In addition to these threatened species, there are two Near Threatened species, *Searsia* (formerly *Rhus*) *acuminatissima* and *Cola mossambicensis*, that have been recorded from this IPA. Although not endemic, the ranges of both these species fall predominantly within Mozambique.

Overall, there are four Mozambican endemics known from this IPA, including the two threatened species *C. morrumbalensis* and *Celosia pandurata*, alongside *Bothriocline moramballae* (LC) and *Pavetta gardeniifolia* var. *appendiculata*. Mount Morrumbala is the southernmost location for these latter two taxa, both of which are known from fewer than five locations globally.

There has been limited botanical collecting within this IPA in recent decades, with many of the collections made in the 1940s (F.A. Mendonça and A.R. Torre) and in the 1970s (T. Müller and G. Pope). Further botanical surveying is needed to fully characterise the habitats and to establish the continued presence and population sizes of threatened species such as *Crassula morrumbalensis*. Further investigation could also reveal additional threatened or rare species.

There are two nationally threatened and restricted habitat types present at this IPA, low altitude moist forest and medium altitude moist forest. It is unlikely that there is enough medium altitude forest to trigger C(iii) as Morrumbala only reaches a peak of 1,172 m and so this habitat type is limited to a small area below the peak ridge. There is a much larger area of low altitude moist forest within this IPA, probably of 10–15 km².

Overall, Morrumbala is one of the five best sites for low altitude forest nationally and so triggers sub-criterion C(iii) of the IPA criteria, although further research is required to accurately delineate and measure this fragmented habitat at this site. To the north of this IPA, across the Nambuur River, there is additional intact, lowland forest. This nearby area is currently understudied; however, it could be included in the IPA if found to be botanically interesting.

Habitat and geology

Mount Morrumbala is the result of multiple, predominantly syenite, intrusions in the surrounding plain (*Coelho* 1959; *Araújo et al.* 1973). The mountain reaches a peak of 1,172 m and is 15 km across from north to south and 5 km from east to west. Soils have not been fully categorised, but the lower slopes are known to have thin, rocky soils (*Andrada* #1570) while the soils towards the west of the IPA, by the Shire River, are clayey (*Dungo* #185). It is likely that the soils in gullies are deep and with greater fertility and moisture than elsewhere in this IPA.

The site experiences a winter dry season, between April and October, with temperatures recorded at nearby Morrumbala town ranging from an average low of 17.2°C to a high of 29.8°C in summer (MAE 2005a). Average monthly precipitation in the district is 1,017 mm, but given the topology of this IPA, there is probably a stark difference in precipitation and temperature between the IPA and the surrounding plain, with the mountain experiencing lower temperatures and higher precipitation – some of which probably occurs through frequent mists.

According to analysis by Lötter *et al.* (in prep.), the northern and western lowlands of Mount Morrumabala can be classified as "Central Lowland Moist Forest". Although the species composition at this site is yet to be documented, some of the collections made in this area include species typical of this vegetation type (GBIF.org 2021b). Canopy trees include *Albizia adianthifolia*, *Bersama abyssinica*, *Newtonia buchananii*, *Macaranga capensis* and *Terminalia* (formerly *Pteleopsis*) *myrtifolia*, while trees and shrubs in the understorey include *Cola greenwayi* and *Vangueria esculenta* alongside the herbaceous *Celosia pandurata* (VU). Lianas and climbers such as *Landolphia buchananii*, *Gouania longispicata* and *Tiliacora funifera* have also been recorded in these forests. At higher altitudes on the northern and western slopes, Lötter *et al.* (in prep.) delineate a small strip of Central Mid-elevation Moist Forest vegetation below the peak ridge. It is unclear how these forests may differ from the lowland forests at this site, but *Newtonia buchananii* may be more dominant in these forests, as is the case in other moist mid-elevation forests.

On the eastward slopes of the mountain, the vegetation is largely moist miombo, similar to that found on the southern escarpments of Mount Gorongosa (Lötter *et al.*, in prep.). Little is known of this vegetation on Mount Morrumbala. One collection describes woodland on the mountain as dominated by *Brachystegia tamarindoides* subsp. *microphylla* (*Muller & Pope* #1973), a species typical of areas with thin soils which also dominates the escarpment miombo of Mount Gorongosa. Although this collection was made on the western slopes of the mountain, it is highly likely that the eastern miombo is also dominated by *B. tamarindoides* subsp. *microphylla* and that this vegetation type also occurs in a mosaic within the forests on the western slopes. The understorey of this woodland has not been documented; however, it is known to host the only known population of the Critically Endangered species *Crassula morrumbalensis*, a perennial, succulent herb. As *Crassula* species are often associated with rocky areas, it is highly likely that this species occurs in areas of rocky miombo at this site.

There are a number of riverine forests in deep gorges on the mountain and it is very possible that the Vulnerable species *Khaya anthotheca* occurs in these areas, as is the case in several other montane gallery forests. The herbaceous species *Impatiens oreocallis* was collected by a waterfall on the mountain, growing within the spray zone. A species native to Malawi, Mozambique and Tanzania, this is probably one of the southernmost collections of *I. oreocallis*.

Summit vegetation has not yet been described. *Decorsea schlechteri* is known to be associated with rocky outcrops on the mountain although this species probably occurs on the slopes below the summit.

Conservation issues

Mount Morrumbala does not fall within a protected area, Key Biodiversity Area or Important Bird Area. However, with the presence of the entire global population of Critically Endangered species *Crassula morrumbalensis*, the site would qualify as an Alliance of Zero Extinction site and KBA under sub-criterion A1e.

Many of the slopes of this site are theoretically protected by the Environment Act (Lei . 20/97 of 1997) which prohibits the cultivation of annual crops on slopes steeper than 7° and perennial crops on slopes steeper than 14° (Timberlake *et al.* 2007). However, in practice, this law appears to have little impact on preventing cultivation on this mountain, with a continued increase in agricultural expansion on Morrumbala. Agricultural expansion particularly impacts the miombo on the eastern slopes and the lowland forest on the western slopes, this is probably due to the flatter terrain in these places. Both the forest and miombo in these accessible areas are heavily fragmented, with the rate of tree cover loss in this IPA accelerating since 2001 (World Resources Institute 2021). Only the less accessible areas, including forests on steep slopes and in gullies, remain completely intact.

This loss of key habitats will inevitably be a major threat to the rare and range-limited species that occur on Mount Morrumbala. Urgent research and conservation action is needed to protect these habitats and the species that reside within. One reason for the increased pressure on land within this IPA may be land disputes elsewhere, which are particularly pronounced in the Shire River valley and around nearby Morrumbala town (MAE 2005a). Work with local communities to solve these issues could go a long way towards minimising agricultural expansion onto Morrumbala. The land within Morrumbala District is otherwise highly favourable and the most productive across Zambézia Province. There is evidence that solving land disputes could be an effective strategy and this is one of the key actions implemented by the conservation and development project, Legado: Namuli, on Mount Namuli led by Legado and Nitidae (Nitidae 2021). This project has helped secure land rights for local people as a key action towards protecting valuable montane habitats from agricultural expansion. Applying this approach to Mount Morrumbala could similarly relieve pressures on the montane habitats of this site.

Slowing the rate of agricultural expansion on Mount Morrumbala, alongside restoration of habitats, may also be of great importance to local communities as the loss of woodland and forest, particularly at higher altitudes, may increase the risk of landslides. Loss of substrate stabilising forest on Serra Tumbine, a mountain 150 km to the north-east of this IPA with a similar syenite geology, led to a catastrophic landslide following heavy rains in 1998 (see MOZTIPA036). The World Bank categorise the rainfall-triggered landslide hazard as "very high" within this IPA (World Bank 2019). Further research is urgently required to understand whether the maintenance of complex forest and woodland ecosystems on this mountain could help mitigate the risk of landslide.

Given the number of threats, including catastrophic events such as landslides, *ex situ* conservation should be considered for *Crassula morrumbalensis alongside in situ* actions. *C. morrumbalensis* is predicted to have orthodox seed storage behaviour (Wyse & Dickie 2018) and so collection of seeds for seed banking is highly recommended.

Much like the plant taxa of this site, the animal taxa of Morrumbala are yet to be inventoried. Inventory work for avian taxa has been described as urgent by Spottiswoode *et al.* (2008) due to the presence of evergreen forest, which is known to provide habitat for rare and threatened birds elsewhere in Mozambique.

Key ecosystem services

There are a number of streams originating on Mount Morrumbala which are tributaries of the Shire River. Agriculture in the Shire River valley has been commercially important, including sugar and cotton plantations (Inguaggiato *et al.* 2009), and continues to be important to livelihoods. However, agriculture should be undertaken sustainably to prevent disturbance of montane habitats which could, in turn, increase evapotranspiration, decrease water availability and increase erosion risk.

The forests and woodlands on this mountain are probably valued locally for timber and fuel –

wood is the primary source of fuel for cooking in Morrumbala District (MAE 2005a).

This IPA may also host important gene resources due to the potential presence of a *Coffea* crop wild relative which could support the breeding of commercial coffee species. This species may also have commercial value as a beverage itself.

Ecosystem service categories

- Provisioning – Food
- Provisioning – Raw materials
- Provisioning – Fresh water
- Regulating services – Moderation of extreme events
- Regulating services – Erosion prevention and maintenance of soil fertility
- Habitat or supporting services – Maintenance of genetic diversity

IPA assessment rationale

Mount Morrumbala qualifies as an Important Plant Area under subcriterion A(i), with two threatened species meeting this threshold. The site hosts the only known population of the Critically Endangered species *Crassula morrumbalensis*. Urgent conservation research and planning are required to protect this species from extinction. In addition, one Vulnerable taxon, *Celosia pandurata*, also triggers A(i) of the IPA criteria and it is highly likely that more threatened species will be documented from this site. Mount Morrumbala also qualifies under C(iii) as it is one of the top five sites nationally for low altitude moist forest. Although another threatened and restricted habitat type, mid montane moist forest, is also present within this IPA, there is only a very limited area of this habitat and so it is unlikely to trigger C(iii).

Priority species (IPA Criteria A and B)

FAMILY	TAXON	IPA CRITERION A	IPA CRITERION B	≥ 1% OF GLOBAL POP'N	≥ 5% OF NATIONAL POP'N	IS 1 OF 5 BEST SITES NATIONALLY	ENTIRE GLOBAL POP'N	SPECIES OF SOCIO-ECONOMIC IMPORTANCE	ABUNDANCE AT SITE
Amaranthaceae	*Celosia pandurata*	A(i)	B(ii)	✓	✓				unknown
Asteraceae	*Bothriocline moramballae*		B(ii)	✓	✓	✓			unknown
Crassulaceae	*Crassula morrumbalensis*	A(i)	B(ii)	✓	✓	✓	✓		unknown
Rubiaceae	*Pavetta gardeniifolia* var. *appendiculata*		B(ii)	✓	✓	✓			unknown
		A(i): 2 ✓	B(ii): 4						

Threatened habitats (IPA Criterion C)

HABITAT TYPE	IPA CRITERION C	≥ 5% OF NATIONAL RESOURCE	≥ 10% OF NATIONAL RESOURCE	IS 1 OF 5 BEST SITES NATIONALLY	ESTIMATED AREA AT SITE (IF KNOWN)
Low Altitude Moist Forest [MOZ-03]	C(iii)			✓	10.0
Medium Altitude Moist Forest 900–1,400 m [MOZ-02]	C(iii)				2.0

Protected areas and other conservation designations

CONSERVATION AREA TYPE	CONSERVATION AREA NAME	RELATIONSHIP OF IPA TO CONSERVATION AREA
No formal protection	N/A	

Threats

THREAT	SEVERITY	TIMING
Small-holder farming	high	Ongoing – increasing
Housing & urban areas	low	Ongoing – trend unknown
Increase in fire frequency/intensity	unknown	Ongoing – trend unknown
Logging & wood harvesting	unknown	Ongoing – trend unknown
Avalanches/landslides	high	Future – inferred threat

MOEBASE

Assessors: Iain Darbyshire, Alice Massingue

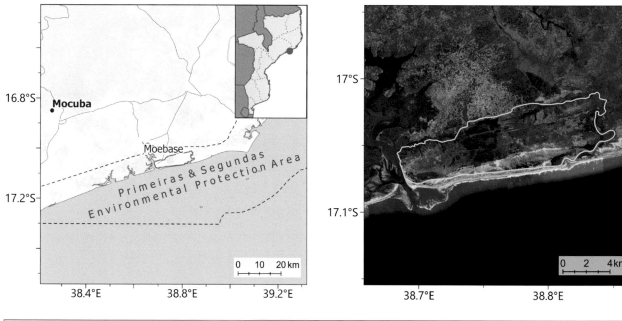

INTERNATIONAL SITE NAME		Moebase	
LOCAL SITE NAME (IF DIFFERENT)		–	
SITE CODE	MOZTIPA032	PROVINCE	Zambézia

LATITUDE	-17.04510	LONGITUDE	38.75080
ALTITUDE MINIMUM (m a.s.l.)	0	ALTITUDE MAXIMUM (m a.s.l.)	60
AREA (km²)	71	IPA CRITERIA MET	A(i), C(iii)

Site description

The Moebase IPA is located in Pebane District of Zambézia Province. It lies immediately south of the coastal village of Moebase, ca. 215 km ENE of the seaport of Quelimane. The site falls within the southern extension of the proposed Rovuma Centre of Plant Endemism [CoE] (Burrows & Timberlake 2011; Darbyshire *et al.* 2019a) and comprises a small area of dune systems with coastal dry forest patches and interdunal slacks on heavy mineral-rich coastal sands.

Botanical significance

This site is of high botanical importance because of the presence of three rare and threatened species of the Rovuma CoE. First, it is the southern-most locality for the globally Endangered endemic tree *Icuria dunensis* ('icuri' or 'ncuri'). The Moebase IPA holds the largest confirmed patches of *Icuria*-dominated forest, totalling ca. 9 km^2, and it is one of only three *Icuria* forests assessed to be in "very good condition" using a Forest Ecological Condition Index (A. Massingue, unpubl. data). This is also the first site at which *Icuria* was recognised as a distinct forest tree during surveying in August 1995 (Lubke *et al.* 2018). Second, the seasonal wetlands in the dune slacks at Moebase hold the only known extant population of the diminutive wetland herb *Triceratella drummondii*, which is assessed as Critically Endangered (S. Richards, in press [b]). This species was previously known also from the Gwanda area of Zimbabwe, over 1,000 km inland from Moebase, but it has not been observed at that locality since the 1960s despite three searches undertaken between 1996 and 2001 to no avail (Barker *et al.* 2001). Around 20 individuals were observed within a single population during an Environmental Impact Assessment at Moebase in 1997 (Barker *et al.* 2001). Finally, evidence has recently come to light that the Critically Endangered endemic shrub *Warneckea sessilicarpa* occurs within the *Icuria* forests – this is derived from a specimen (*Boana* #154) collected in 1997 which had previously been misidentified as *W. sousae* but is confirmed to be *W. sessilicarpa* by the global expert on this group, R.D. Stone (pers. comm.). Further discoveries of rare species are likely to be made with further exploration of these forest patches.

Habitat and geology

The *Icuria* forest is highly impressive here, forming dense, mono-dominant dry forest stands with many mature individuals, some reaching up to 40 m in height (A. Msssingue, pers. obs.), and with substantial recruitment evident. These forests occur on low-lying, seasonally damp, ancient sand dune systems behind the foredunes, between 1 km and 4 km from the shoreline (Lubke *et al.* 2018). The sands of these dunes are rich in heavy minerals including ilmenite (titanium ore), which may be of interest for mining (URS/Scott Wilson 2011). *Icuria* patches are surrounded by open dune scrub and woodland with typical species including *Garcinia livingstonei* and *Strychnos spinosa* (Barker *et al.* 2001). The interdunal slacks have a high water table and include areas of free-standing water in the wet season. Dominant species recorded in these wetlands include *Eragrostis ciliaris*, *Xyris anceps* and *Utricularia* sp.; *Triceretalla drummondii* was found in these slacks growing on open wet sands together with *Digitaria eriantha* and *Bulbostylis hispidula* (Barker *et al.* 2001).

Extensive areas of miombo woodland were previously found on the raised, freely drained decksand deposits to the east of the village and north of the *Icuria* patches, but this woodland has been seriously depleted in recent decades as the village and associated subsistence agriculture has expanded – these miombo woodlands are excluded from the IPA delineation. To the west, the Moebase Estuary has extensive stands of mangroves, and there are also mangroves along the eastern boundary.

A habitat survey was conducted as part of an initial Environmental Impact Assessment conducted on the Moebase mining exploration concession (see Conservation issues) but a full species inventory would be desirable and may well uncover further rare and threatened species.

The climate at Moebase is highly seasonal, with rainfall peaking in December to March; the average annual precipitation is approximately 1,350 mm.

Conservation issues

This IPA falls within the extensive (>8,000 km^2) Primeiras & Segundas Environmental Protection Area (APAIPS) which extends along the coast south to Pebane and north to Angoche. However, the emphasis here is on offshore and marine protection, and there is little evidence of conservation action at Moebase at present.

The surrounding area is subject to high population pressure from the expanding settlement of Moebase and to the encroachment of subsistence agriculture into natural habitats. A previously large area of miombo woodland on the eastern edge of the village has been largely destroyed over the past 30–40 years and it is feared that, now that this wood resource has been largely exhausted, the local population may target the *Icuria* patches more frequently (Darbyshire *et al.* 2019e). However, at present the *Icuria* patches are largely intact. The primary threats currently are clearance for access routes to the beach, uncontrolled burning, and the stripping of *Icuria* bark for making ropes (A. Massingue, pers. obs.). Areas of fixed dunes in both the west and east of the IPA have been converted to machambas.

A significant future threat to this IPA is that it falls within a mining concession (License 4623C, Moebase and Naburi deposits, currently owned by Pathfinder Minerals plc), and has commercially viable concentrations of heavy minerals (URS/Scott Wilson 2011). There is continued interest in the exploitation of these deposits.

The Moebase Region was previously recognised as an Important Bird Area (IBA) centering on the Moebase Estuary (Parker 2001), but this has since been expanded to the extensive Primeiras & Segundas Environmental Protection Area (APAIPS) IBA and Key Biodiversity Area (BirdLife International 2021c). The area is also of interest for other faunal groups including reptiles. As Moebase is the only known extant locality for the Critically Endangered species *Triceratella drummondii*, it would also qualify as an Alliance for Zero Extinction site.

Key ecosystem services

This IPA is considered to provide a range of ecosystem services. The stands of *Icuria* in particular help to stabilise and protect the coastal sand deposits and so prevent excessive erosion during extreme weather events. The forest also provides provisioning services for the local community, which could be managed at sustainable levels. It also provides important habitat and supporting services for a range of fauna.

Ecosystem service categories

- Provisioning – Raw materials
- Regulating services – Carbon sequestration and storage
- Regulating services – Moderation of extreme events
- Regulating services – Erosion prevention and maintenance of soil fertility
- Habitat or supporting services – Habitats for species
- Habitat or supporting services – Maintenance of genetic diversity
- Cultural services – Recreation and mental and physical health

IPA assessment rationale

Moebase qualifies as an Important Plant Area under criterion A(i) as it is the only known extant site for *Triceratella drummondii* (CR), and holds a globally important population of *Icuria dunensis* (EN). It is also deemed to contain over 5% of the global population of *Warneckea sessilicarpa* (CR) which is known from only four sites, one of which may no longer be extant – Moebase is likely to be a critical site for the survival of this species. It also qualifies under criterion C(iii) because the *Icuria*-dominated coastal dry forest of the Rovuma Centre of Endemism is a nationally threatened and range-restricted habitat, and the Moebase IPA contains the largest known example of this forest type.

Priority species (IPA Criteria A and B)

FAMILY	TAXON	IPA CRITERION A	IPA CRITERION B	≥ 1% OF GLOBAL POP'N	≥ 5% OF NATIONAL POP'N	IS 1 OF 5 BEST SITES NATIONALLY	ENTIRE GLOBAL POP'N	SPECIES OF SOCIO-ECONOMIC IMPORTANCE	ABUNDANCE AT SITE
Commelinaceae	*Triceratella drummondii*	A(i)	B(ii)	✓	✓	✓	✓		scarce
Fabaceae	*Icuria dunensis*	A(i)	B(ii)	✓	✓	✓			abundant
Melastomataceae	*Warneckea sessilicarpa*	A(i)	B(ii)	✓	✓	✓			unknown
		A(i): 3 ✓	B(ii): 3						

Icuria dunensis forest at Moebase (AM)

Threatened habitats (IPA Criterion C)

HABITAT TYPE	IPA CRITERION C	≥ 5% OF NATIONAL RESOURCE	≥ 10% OF NATIONAL RESOURCE	IS 1 OF 5 BEST SITES NATIONALLY	ESTIMATED AREA AT SITE (IF KNOWN)
Rovuma *Icuria* Coastal Dry Forest [MOZ-12c]	C(iii)	No	Yes	Yes	8.95

Protected areas and other conservation designations

CONSERVATION AREA TYPE	CONSERVATION AREA NAME	RELATIONSHIP OF IPA TO CONSERVATION AREA
Environmental Protection Area	Primeiras & Segundas	protected/conservation area encompasses IPA
Important Bird Area	Primeiras & Segundas Environmental Protection Area (APAIPS)	protected/conservation area encompasses IPA
Key Biodiversity Area	Primeiras & Segundas Environmental Protection Area (APAIPS)	protected/conservation area encompasses IPA

Threats

THREAT	SEVERITY	TIMING
Small-holder farming	medium	Ongoing – trend unknown
Gathering terrestrial plants	medium	Ongoing – increasing
Mining & quarrying	unknown	Future – inferred threat

SOFALA
PROVINCE

CATAPÚ

Assessors: Sophie Richards, Iain Darbyshire

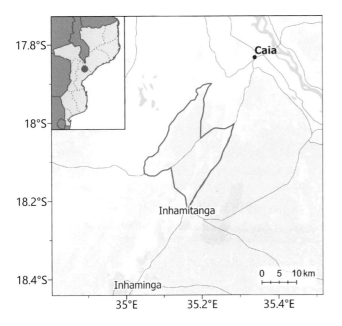

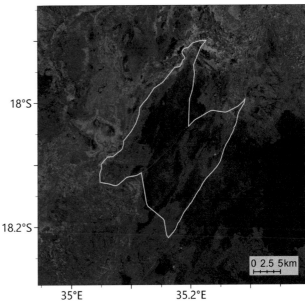

INTERNATIONAL SITE NAME		Catapú	
LOCAL SITE NAME (IF DIFFERENT)		–	
SITE CODE	MOZTIPA033	PROVINCE	Sofala

LATITUDE	-18.03630	LONGITUDE	35.17030
ALTITUDE MINIMUM (m a.s.l.)	30	ALTITUDE MAXIMUM (m a.s.l.)	190
AREA (km²)	352	IPA CRITERIA MET	A(i), C(iii)

Site description

Catapú is a timber concession in Cheringoma District, Sofala Province, at the far north of the Cheringoma Plateau. The site has been run by TCT-Dalmann Furniture since 1996 and was one of the first concessions in Africa to gain Forest Stewardship Council certification (Remane & Therrell 2019). Catapú is around 20 km south-west of the Zambezi River, with the Zangue River and flood plain to the west. To the east is the old Caia – Dondo railway, and to the south is Inhamitanga village. The local villages Mutondo, Pungue and Santove are involved with the activities of the concession, with the latter two falling partially within the concession (Catapú. net 2020).

The IPA is 352 km² in area and follows the boundary of the forestry concession. Altitude ranges from 30 m, towards Chirimadzi Valley in the north, to 190 m in the dry forests towards the southern boundary. Within Catapú, the Tissadze River flows from north to south, running south-east of the EN-1 road which itself bisects the site. Although the site is a timber concession, logging is targeted towards selected species that can be coppiced, and is complemented by a replanting scheme, hence it is considered to be a sustainable operation. The vegetation of Catapú is a mosaic of sand forest, dry deciduous thicket and woodland.

Botanical significance

The Catapú concession is of botanical importance due to the extensive area of high quality Inhamitanga Sand Forest, a restricted forest type of Mozambique. This forest is part of the wider Coastal Forests of Eastern Africa Biodiversity Hotspot, recognised for being both highly biodiverse and highly threatened (Burgess *et al.* 2004b).

In keeping with its hotspot status, the site itself is botanically rich. A survey of woody vegetation here found 238 species and infraspecific taxa from 167 genera and 59 families (Coates Palgrave *et al.* 2007). Although the majority (64.5%) of species recorded from this site are shared with the wider southern African flora region, ten species recorded are Mozambican endemics. These endemics include an as yet undescribed species of *Dovyalis*, *Dovyalis* sp. A of *Trees and Shrubs Mozambique* (Burrows *et al.* 2018), which is known only from this site and neighbouring Inhamitanga Forest Reserve; this species was previously confused with the Tanzanian species *D. xanthocarpa*.

In addition, a number of globally threatened species occur within Catapú, with ten species recorded to date. For some of these species, with threats from agricultural conversion elsewhere within their restricted ranges, Catapú is of major importance for the prevention of further declines and extinction. This IPA covers much of the range of two Vulnerable species, *Cordia megiae* and *Dorstenia zambesiaca*, both known only from the Cheringoma District (Coates Palgrave *et al.* 2014a; Mynard & Rokni 2019). By contrast, *Khaya anthotheca* (East African Mahogany – VU) is found across tropical Africa but is threatened by harvesting for timber (Hawthorne 1998).

Coffea racemosa (Inhambane coffee – NT), a crop wild relative of commercial coffee, is recorded from this IPA and is itself roasted and ground locally to make coffee (Rodrigues *et al.* 1975). A small number of hardwood species are harvested for timber, including *Millettia stuhlmannii*, *Afzelia quanzensis* (LC) and *Cordyla africana* (LC) (Coates Palgrave *et al.* 2007).

Habitat and geology

The substrate of this site is mostly sandy soils, underlaid by sublittoral sands, with outbreaks of sandstone and calcareous conglomerates. The sublittoral sands are of great importance as they accumulate the water necessary to support tall forest trees (Coates Palgrave *et al.* 2007). Around the pans and floodplains, particularly to the west along the Zangue River, are black clay alluvial soils which are seasonally wet (Coates Palgrave *et al.* 2014b).

The below habitat description is based on surveys of the concession completed by Coates Palgrave *et al.* (2007) – species lists can be viewed in that paper. The vegetation of Catapú is described as a mosaic of forests, woodland and thicket. Woodland can be further subdivided into miombo, which occurs towards Inhamitanga village in the south-eastern corner of the concession, and undifferentiated woodland, which lacks the mycorrhizal associations and dominant species that define miombo (B. Wursten, pers. comm. 2020), the latter vegetation covering a greater area of the IPA. The variation in plant communities may be related to moisture and nutrient levels in the substrate. In the nearby Inhamitanga Forest Reserve, for example, it is reported that floral composition varies with soil clay content, which has a greater capacity for water storage (Müller *et al.* 2005).

Mildbraedia carpinifolia (BW)

Dorstenia zambesiaca (MH)

This site contains a significant area of the restricted forest type, Inhamitanga Sand Forest. This forest is often patchy within the mosaic but is more dominant along the Via Pungue road (−18.021°, 35.171°), and in the southern portion of the IPA between the Tissadze River and EN-1 (−18.125°, 35.150°). Dominant canopy species include: *Afzelia quanzensis, Balanites maughamii, Cordyla africana, Fernandoa magnifica, Terminalia sambesiaca* and *Xylia torreana*. Emergent trees above the canopy include *Adansonia digitata* and *Millettia stuhlmannii*, while the understorey is sparse with almost no ferns, herbs or grasses. This absence of a herbaceous understorey is typical of Inhamitanga Sand Forest (B. Wursten, pers. comm. 2020). There are, however, understorey shrubs, including the Mozambican endemic *Millettia mossambicensis, Drypetes reticulata* and many species of liana (as listed in Coates Palgrave *et al.* 2007). In neighbouring Inhamitanga it has been reported that, in areas with higher moisture availability, there are areas with evergreen elements, where trees such as *Celtis mildbraedii* and *Drypetes gerrardii* are more prevalent (Müller *et al.* 2005). Such a pattern is likely also reflected in the Inhamitanga Sand Forest within this IPA.

To the east of the Tissadze river, variation in the mosaic is more prominent with species composition and vegetation density varying between forest, thicket and woodland. Much of the woodland patches are undifferentiated, rather than miombo, with emergent trees including *Rhodognaphalon mossambicense, Newtonia hildebrandtii* and *Millettia stuhlmannii*. Forest species, such as *Dalbergia boehmii, Drypetes reticulata* and *Strychnos madagascariensis* form dense stands. Thicket vegetation, which is similarly dense but with a lower canopy, also features *D. reticulata* and *S. madagascariensis* along with species such as *Albizia anthelmintica, Diospyros loureiriana* and *D. senensis* that are more typical of scrubby areas.

Towards the Tissadze River bridge (−18.184°, 35.149°), in the southernmost corner of the IPA, the most well-defined area of miombo woodland is recorded. Vegetation is sparse with little grass cover and few shrubs, suggesting poorer soils in this area (Coates Palgrave *et al.* 2007). *Brachystegia spiciformis* is the dominant miombo species. Trees are widely spaced, although some vegetation is concentrated around termite mounds, pans or on riverbanks where

typical species include *Cleistochlamys kirkii, Dovyalis hispidula, Flueggea virosa* and *Strychnos potatorum*. Pans are numerous within the forest/thicket/woodland mosaic. They appear as grass-covered depressions of 0.5 hectares or more and are often bordered by trees such as *Combretum imberbe* and *Acacia robusta* subsp. *usambarensis*.

Mean annual rainfall at the site is between 700 and 1,400 mm, with the rainy season occurring between November and March, while temperatures at nearby Inhamitanga town reach an average high of 28°C between October and December and a low of 21°C in June and July (Coates Palgrave *et al.* 2007; World Weather Online 2021). For several years in the 2000s, below average rainfall was recorded at the site and, during this particular period of water scarcity, *A. robusta* subsp. *usambarensis* was observed to be dying as the pans remained dry (Coates Palgrave *et al.* 2007).

To the west of the site is the Zangue river and floodplain. The grassland on alluvial soil here is of great importance as it is one of the few sites from which the highly range-restricted *Acacia torrei* (LC) has been recorded (Coates Palgrave *et al.* 2014b). Although much of the habitat lies outside of the IPA boundary, and there are no collections here, it is possible that *Acacia torrei* occurs towards the eastern boundary.

Habenaria stylites (MH)

Conservation issues

This IPA does not overlap with any protected areas. However, it falls within the vast Gorongosa-Marromeu Key Biodiversity Area, and trigger species for this site include four priority species for this IPA (*Acacia torrei*, *Cordia stuhlmannii*, *Dorstenia zambesiaca* and *Tarenna longipedicellata*). The site is also adjacent to the Coutada 12 former hunting concession, which has recently been acquired by a partnership between Gorongosa National Park and Entroposto with a view to integrating the area into Gorongosa National Park in future (Mozambique News Agency 2016). Catapú would then be part of a wide-spanning network of forestry concessions and Gorongosa Project areas, providing secure habitat corridors and ecological resilience across a large proportion of Sofala Province (Parque Nacional da Gorongosa 2020).

TCT Dalmann, the owners of Catapú, describe their forestry activities as being focussed on "sustainable and responsible management of the concession and the use of… resources" (TCT Dalmann 2020). Felling is limited to three species: *Millettia stuhlmannii*, *Afzelia quanzensis* and *Cordyla africana* (Coates Palgrave *et al.* 2007). *Millettia stuhlmannii*, which is the prime timber species at this site in terms of volumes extracted, has not been assessed for the IUCN Red List and, although *M. stuhlmannii* is widespread and frequent in south-east Africa, much of the population falls within Mozambique where there are concerns of overexploitation due to high market demand (Remane & Therrell 2019). However, the Catapú concession is one of the first in Africa to gain Forest Stewardship Council certification and there is a regime of replanting of *M. stuhlmannii* and other indigenous hardwoods (Catapú.net 2020; Coates Palgrave *et al.* 2007; Remane & Therrell 2019).

The forestry practice at the site follows a 27-year rotational cycle with 2,400 m³ of timber extracted per year. Post felling clean-ups are employed to minimise disturbance and coppiced trees are subject to labour-intensive management to ensure that regrowth produces usable timber. As a result, 55% of felled trees are coppiced successfully, with all three major tree species extracted having coppicing ability (TCT Dalmann 2020).

In an effort to diversify their activities and promote sustainability, TCT Dalmann has partnered with Premier African Minerals to begin exploration into lime mining towards the north of the concession. As mentioned in the habitat section, Catapú has a number of calcareous conglomerates in its geology, some of which have acceptable grades of calcium carbonate to allow for the production of cement, agricultural lime and aggregate (Premier African Minerals 2020). At present, it is not clear to what extent mining activity may threaten the plant communities at the site, and as of 2017, scoping work was ongoing.

As a major local employer, TCT Dalmann collaborates closely with the local communities and as such conservation efforts here have been largely successful, with little anthropogenic disturbance (beyond the forestry management) at the site. This is in contrast to the surrounding areas where encroachment from agriculture often leads to degradation of habitat (Cheek *et al.* 2019).

One major threat to the Catapú site is an increase in fire intensity and frequency. For instance, a devastating forest fire in 1994 resulted in the aggressive proliferation of the scrambling shrub *Acacia adenocalyx*, which forms dense thickets and prevents the regeneration of forest species. Some of the *A. adenocalyx* stands have since been cleared, allowing pioneer species such as *Fernandoa magnifica* to establish (Coates Palgrave *et al.* 2007). A firebreak area has also been established around the sawmill, within which a 4,000 ha sensitive forest is protected. Although some elements of miombo and woodland vegetation are adapted to fire, it has been reported that the plant community within this sensitive forest area is intolerant to burning (Coates Palgrave *et al.* 2007). A similar fire intolerance has been observed in another IPA, Licuáti Forest Reserve [MOZTIPA009], where fire frequency has increased due to slash and burn agriculture (Coates Palgrave *et al.* 2007). The firebreak at Catapú may therefore be protecting plant communities that would otherwise be lost.

In addition to high plant biodiversity, there is a game farm inside the concession of 9,960 ha. It is not clear if animals are being specifically reared for hunting; however, TCT Dalmann state that wildlife is still recovering towards previous levels (TCT Dalmann 2020) and so it is unlikely that the area is overpopulated with game such that damage is caused to plant communities. A wide range of mammal species have been reported from the site, including some threatened taxa such as the Samango Monkey (*Cercopithecus mitis* ssp. *labiatus* – VU) and, very rarely, Leopard (*Panthera pardus* – EN) and African Wild Dog (*Lycaon pictus* – VU) (TCT Dalmann 2020). In terms of bird species, Catapú is one of Sasol's top 200 birding sites in southern Africa (Hardaker & Sinclair 2001). TCT Dalmann have recorded over 120 bird species within Catapú, including observations of the Martial Eagle (*Polemaetus bellicosus* – VU), the White-headed Vulture (*Trigonoceps occipitalis* – CR) and the white-backed Vulture (*Gyps africanus* – CR) (Riddell *et al.* n.d.).

Given the high conservation value of this site, for both flora and fauna, it has received a number of visits from researchers conducting scientific studies (for example, Symes 2012; Remane & Therrell 2019) while a herbarium, the TCT-Catapu Cheringoma Herbarium, was set up nearby to house specimens from the site and surrounding areas.

Key ecosystem services
This IPA is primarily valued for the provisioning of hardwood timber, some of which is crafted on-site into bespoke furniture and wooden artefacts for which there is a market in the UK and Germany (Premier African Minerals 2020). In addition, Catapú wood is used to construct beehives for on-site apiculture with the honey produced sold nationally (TCT Dalmann 2020). The forest itself, along with others patches of forests on the Cheringoma plateau, have been reported to be important for water catchment (Timberlake & Chidumayo 2011).

Alongside these provisioning services, TCT Dalmann run a game farm of 9,960 hectares which is centred on the Tissadze river south of the EN-1 road (TCT Dalmann 2020). The wide range of animals on the game farm attracts both ecotourism and hunting safaris to the reserve. The income generated from both tourism and forestry are of importance for the livelihoods of local people, providing employment and the funding required to build schools in nearby villages (Catapú.net 2020). Along with the support of local people, the various income streams allow Catapú to be a self-sustaining business (Premier African Minerals 2020) that incorporates important conservation work, recognised by the Forest Stewardship Council, and presents a successful model of sustainable forest management.

There are some culturally important areas within the IPA, such as burial grounds. Efforts have been made by TCT Dalmann to map these sites to ensure that they are protected from destructive or culturally insensitive activities (TCT Dalmann 2020).

Ecosystem service categories

- Provisioning – Food
- Provisioning – Raw materials
- Provisioning – Fresh water
- Habitat or supporting services – Habitats for species
- Cultural services – Tourism
- Cultural services – Cultural heritage

IPA assessment rationale

Catapú qualifies as an IPA under both criterion A and criterion C. It was suggested by both Coates Palgrave *et al.* (2007) and Smith (2005) that the whole of the Cheringoma Plateau be designated as an IPA. However, based on the different management of Catapú and Inhamitanga, the latter being a forest reserve recently brought under the management of the Goronogsa National Park, these two sites have been recognised as separate IPAs.

Ten species meet sub-criterion A(i) with three Endangered species (*Cola clavata*, *Cordia torrei* and *Vepris myrei*) and six Vulnerable species (*Cordia stuhlmannii*, *Cordia megiae*, *Dorstenia zambesiaca*, *Habenaria stylites*, *Monodora stenopetala* and *Tarenna longipedicellata*). An additional Vulnerable species, *Khaya anthotheca*, occurs within this IPA but does not meet the thresholds required to trigger sub-criterion A(i) due to its widespread distribution. Catapú contains high-quality areas of the nationally important Inhamitanga Sand Forest, meeting the threshold for sub-criterion C(iii). Ten of the sub-criterion B(ii) qualifying species for Mozambique are recorded at this site, which does not meet the threshold of encompassing 3% of species of conservation importance, but their presence is still of note, particularly as many of these species have restricted ranges.

Priority species (IPA Criteria A and B)

FAMILY	TAXON	IPA CRITERION A	IPA CRITERION B	≥ 1% OF GLOBAL POP'N	≥ 5% OF NATIONAL POP'N	IS 1 OF 5 BEST SITES NATIONALLY	ENTIRE GLOBAL POP'N	SPECIES OF SOCIO-ECONOMIC IMPORTANCE	ABUNDANCE AT SITE
Acanthaceae	*Justicia gorongozana*		B(ii)	✓	✓	✓			occasional
Annonaceae	*Monodora stenopetala*	A(i)		✓	✓	✓			unknown
Boraginaceae	*Cordia megiae*	A(i)	B(ii)	✓	✓	✓			occasional
Boraginaceae	*Cordia stuhlmannii*	A(i)	B(ii)	✓	✓	✓			occasional
Boraginaceae	*Cordia torrei*	A(i)		✓	✓	✓			scarce
Capparaceae	*Maerua brunnescens*		B(ii)	✓	✓				unknown
Fabaceae	*Millettia mossambicensis*		B(ii)	✓	✓	✓			common
Malvaceae	*Cola clavata*	A(i)	B(ii)	✓	✓	✓			common
Meliaceae	*Khaya anthotheca*	A(i)						✓	unknown
Moraceae	*Dorstenia zambesiaca*	A(i)	B(ii)	✓	✓	✓			frequent
Orchidaceae	*Habenaria stylites*	A(i)		✓	✓	✓			unknown
Polygalaceae	*Carpolobia suaveolens*		B(ii)	✓	✓				unknown
Rubiaceae	*Tarenna longipedicellata*	A(i)	B(ii)	✓	✓	✓			scarce
Rutaceae	*Vepris myrei*	A(i)		✓	✓	✓			unknown
Salicaceae	*Dovyalis* sp. A of T.S.M.		B(ii)	✓	✓	✓			occasional
		A(i): 9 ✓	**B(ii): 10**						

Threatened habitats (IPA Criterion C)

HABITAT TYPE	IPA CRITERION C	≥ 5% OF NATIONAL RESOURCE	≥ 10% OF NATIONAL RESOURCE	IS 1 OF 5 BEST SITES NATIONALLY	ESTIMATED AREA AT SITE (IF KNOWN)
Inhamitanga Sand Forest [MOZ-04]	C(iii)		✓	✓	

Protected areas and other conservation designations

CONSERVATION AREA TYPE	CONSERVATION AREA NAME	RELATIONSHIP OF IPA TO CONSERVATION AREA
Forest Reserve (conservation)	Inhamitanga	protected/conservation area is adjacent to IPA
Important Bird Area	Zambezi Delta	protected/conservation area is adjacent to IPA
Key Biodiversity Area	Gorongosa-Marromeu	protected/conservation area encompasses IPA
Ramsar	Zambezi Delta	protected/conservation area is adjacent to IPA

Threats

THREAT	SEVERITY	TIMING
Increase in fire frequency/intensity	medium	Past, likely to return

INHAMITANGA FOREST

Assessors: Sophie Richards, Iain Darbyshire

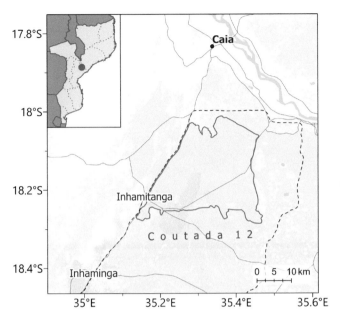

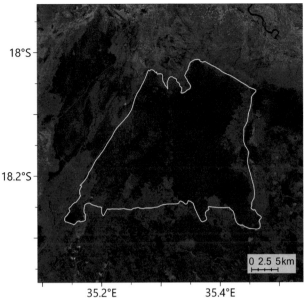

INTERNATIONAL SITE NAME		Inhamitanga Forest	
LOCAL SITE NAME (IF DIFFERENT)		Floresta de Inhamitanga	
SITE CODE	MOZTIPA034	PROVINCE	Sofala

LATITUDE	-18.12590	LONGITUDE	35.40260
ALTITUDE MINIMUM (m a.s.l.)	65	ALTITUDE MAXIMUM (m a.s.l.)	290
AREA (km²)	622	IPA CRITERIA MET	A(i), C(iii)

Site description

The Inhamitanga Important Plant Area is an area of forest and woodland that spans the border between the Cheringoma and Marromeu Districts of Sofala Province. The IPA covers an area of 622 km², falling within the former Coutada 12 hunting concession. The northern boundary is 15–20 km south of the Zambezi River. To the west is the Catapú timber concession, a separate IPA [MOZTIPA033], while the southern boundary is just south of the Inhamitanga-Chupanga (R1002) road and the railway line running east from Inhamitanga village. To the east, the IPA is bounded by wetlands of the Zambezi Delta, including Lago Nharica, and Camacho village. East of this site (–18.124°, 35.612°), running south of the 215 road, is another area of high-quality Inhamitanga Sand Forest, as delineated by Lötter *et al.* (in prep.). Further investigation is recommended in this latter area as, if it proves to be of botanical significance, it may warrant inclusion in this IPA or delineation of a separate IPA.

Inhamitanga is dominated by the Cheringoma Forest-Woodland Mosaic, a highly restricted habitat type in Mozambique. It falls within the Zambezi River Delta Ramsar site and Important Bird Area, and the Gorongosa-Marromeu Key Biodiversity Area. The site has long been recognised for its biological importance, with the establishment of a forest reserve, Inhamitanga Forest Reserve (following the 213 road), over 50 years ago (Müller *et al.* 2005), while the forest and the entire Coutada 12 is now managed by Gorongosa National Park (Parque Nacional da Gorongosa 2020).

Botanical significance

The Inhamitanga Forest is of high importance for the presence of a range of globally threatened species, with 11 recorded to date, including ten Vulnerable species and one Endangered species. This site is particularly important for *Tarenna longipedicellata* (VU) and *Dorstenia zambesiaca* (VU); both species are largely confined to the Cheringoma Plateau and the latter is restricted to a global range of only 50 km². In addition, this IPA is one of only two locations from which *Cephalophis lukei* (EN) is known in Mozambique. One further Endangered species, *Cola clavata*, is endemic to Sofala and Zambezia Provinces and threatened by expansion of agriculture and the burning of land associated with agriculture (Cheek & Lawrence 2019). Inhamitanga is a relatively secure area where threats from clearing of woodland are considerably less severe than in many neighbouring areas of Mozambique at present, with extensive areas of intact vegetation remaining (Darbyshire *et al.* 2019j), and therefore it is an important site for the continued existence of these threatened species.

In addition to a number of threatened species occurring within this IPA, a query of the Flora of Mozambique database states that there have been 177 different plant species collected at this site (Hyde *et al.* 2021), and at least ten species are endemic to Mozambique. This includes the Near Threatened species *Ochna angustata* and one as yet undescribed species, *Dicliptera* sp. B of *Flora Zambesiaca* (Darbyshire *et al.* 2015). The latter has been recorded within Inhamitanga forest and is only known from the area between the villages of Inhamitanga and Lacérdonia, with much of the area between these two villages falling within this IPA.

The Inhamitanga Sand Forest that covers much of this IPA is of national importance, as this vegetation type is limited only to the Cheringoma Plateau. The IPA also falls within the wider Coastal Forests of East Africa biodiversity hotspot, so defined because of the combination of high biodiversity and high threat levels (Burgess *et al.* 2004b). In particular, the forest community within this IPA is of note due to the unusual richness in woody plants, with a mixture of moist forest species and species that are more associated with drier habitats (Müller *et al.* 2005).

Habitat and geology

A survey of the Inhamitanga Forest Reserve within this IPA was carried out by Müller *et al.* (2005), and there has since been botanical collecting associated with the local TCT-Catapú Cheringoma Herbarium.

Inhamitanga Sand Forest is the dominant vegetation within this IPA. This forest type has a closed canopy and has variously been described as "dry deciduous forest" throughout (Lötter *et al.*, in prep.) to a mixture of "moist evergreen" and "dry deciduous" forest (Müller *et al.* 2005). The forest within this IPA is a mosaic of different elements, and although there is a long dry season and species composition is mostly deciduous, there are markedly evergreen components to this forest; for example, the evergreen tree *Celtis mildbraedii* is common at this site (B. Wursten, pers. comm. 2020). The dry season occurs between May and October, with annual rainfall around 1,000 mm, while temperatures in the area peak at 21–36°C in November and drop to 15–26°C in July (Burrows *et al.* 2018; World Weather Online 2021).

The IPA has a broadly flat topography, and the geology varies between sands and clayey loams. It has been suggested that the spatial variation in clay content in the soils, resulting in varying moisture availability, may be partially responsible for the diversity of species at this site, with more evergreen species found on soils with a higher clay content. In areas where soils retain more moisture, species such as *Celtis mildbraedii* and *Drypetes gerrardii* are common, while *Khaya anthotheca* (VU) occurs on the wettest patches of soil (Müller *et al.* 2005). In these areas, there is still a large contingent of deciduous species that dominate across all the forests in this area, including *Millettia stuhlmannii* and *Terminalia (Pteleopsis) myrtifolia*, while *Afzelia quanzensis* is common.

The Mozambican endemic *Millettia mossambicensis* (LC) is frequent within the shrub layer, while *Monodora stenopetala* (VU) also features. The shrub layer of this forest also varies in deciduous tendencies depending on moisture availability, with the semi-deciduous shrub *Rinorea elliptica* common in areas with higher moisture alongside lianas such as *Tiliacora funifera* and *Landolphia kirkii* (Müller *et al.* 2005). What may be a defining character of Inhamitanga Sand Forest is the scarcity of herbaceous plants in the understorey

Mildbraedia carpinifolia (BW)

Monodora stenopetala (BW)

and epiphytes (B. Wursten, pers. comm. 2020). Groundcover is sparse with almost no grasses and, instead, leaf litter covers the substrate (Muller *et al.* 2005).

More open areas of woodland occur towards the eastern margin of the forest. This vegetation is dominated by similar tree species as the forest, including *Millettia stuhlmannii* and *Afzelia quanzensis,* with widely scattered scrubs and a dense grass layer of mostly *Panicum* species (Müller *et al.* 2005). Towards the south-west of the IPA and Inhamitanga village, there are around 50 km² of degraded, open woodland with trees isolated by up to 100 m and some clusters of denser vegetation (Müller *et al.* 2005). It is likely that, before disturbance, the western-most section of woodland was formerly miombo, as suggested by the presence of occasional trees of *Brachystegia spicifoirmis* and the nearby miombo woodland in the south-eastern corner of the neighbouring Catapú concession (Coates Palgrave *et al.* 2007; Müller *et al.* 2005).

While this site falls within the Zambezi Delta Ramsar site, the wetlands are largely to the south and south-east of the site boundary.

Conservation issues

The importance of Inhamitanga Forest has long been recognised, with Inhamitanga Forest Reserve established over 50 years ago, covering only 18 km² of this IPA, following the 213 road from Inhamitanga village. However, few people knew of its existence and little formal protection has been afforded to the site (Coates Palgrave *et al.* 2007). As a result, the south-western portion of the reserve and surrounding woodland have been heavily degraded through intense and frequent burning, with some trees isolated by up to 100 m (Müller *et al.* 2005). In neighbouring Catapú, there is an area of sensitive forest to the east of the concession that is reported to be fire-intolerant and so is protected within a firebreak (Coates Palgrave *et al.* 2007). There may, therefore, be similarities in ecology between the fire-intolerant vegetation within Catapú and the woodlands in the south-west of Inhamitanga, which may explain the intense degradation in this part of the IPA.

Despite past disturbances from fire, there is low population pressure on the area as a whole, with anthropogenic activities mostly limited to Inhamitanga village, to the south-west corner, and to agricultural land outside the north-west corner of the IPA. The forested areas within the centre of the reserve have been subjected to some logging while the Inhamitanga-Chupanga road, which runs through the forest reserve in the south, may increase the risk of disturbance from fire, cyclones and extreme winds. However, the majority of the forest within the reserve is in good condition (Müller *et al.* 2005). It appears that much of the vegetation within the Inhamitanga IPA is largely undisturbed, as is suggested by satellite imagery from Google Earth and the general inaccessibility of much of the forest (Google Earth 2021).

In 2017, Gorongosa National Park, in partnership with Entreposto, formally took on the former Coutada 12 hunting concession as a Gorongosa Project (Parque Nacional da Gorongosa 2020).

These partners are undertaking ecological assessments, community engagement and analysis of tourism potential with a view to proposing to government that the site become part of Gorongosa National Park (Mozambique News Agency 2016). Conservation activities here are also undertaken in collaboration with neighboring Catapú (M. Stalmans, pers. comm. 2021), providing a landscape-scale approach to conserving the important Inhamitanga Sand Forest habitat.

Inhamitanga IPA also falls within the vast Gorongosa-Marromeu Key Biodiversity Area (KBA), with three trigger species for this KBA (*Cordia stuhlmannii, Dorstenia zambesiaca* and *Tarenna longipedicellata*) also recognised as priority species for this IPA.

Key ecosystem services

The forests on the Cheringoma plateau, including Inhamitanga, have been noted for their importance in water catchment (Timberlake & Chidumayo 2011). Although logging within the forest reserve is illegal, there is some evidence of tree extraction in and around the area towards Inhamitanga village (Müller *et al.* 2005). The IPA has a number of hardwood trees, including *Millettia stuhlmannii* and *Afzelia quanzensis*, with the latter in particular known to be overexploited in Mozambique due to high market demand (Remane & Therrell 2019). A successful sustainable forestry scheme has been developed in the neighbouring Catapú concession (Coates Palgrave *et al.*, 2007) and the sustainable extraction methods could be replicated with care in parts of Inhamitanga, particularly towards Inhamitanga village where land is already thought to be utilised. It is unclear how management under Gorongosa National Park will impact wood extraction at the site.

There have been previous attempts to monetise the carbon sequestration services provided by the forests in the form of carbon credits sold by Envirotrade. However, it is not thought that the aims of this project were realised (Kill, 2013).

There may be tourism potential at the site, particularly as nearby Catapú hosts wildlife safaris, and with Coutada 12 now under the management of Gorongosa National Park, the expansion of tourism here may become more feasible. The south of the IPA is particularly accessible due to the Inhamitanga-Chupanga road.

Ecosystem service categories

- Provisioning – Raw materials
- Provisioning – Fresh water
- Regulating services – Carbon sequestration and storage

Tarenna longipedicellata (BW)

IPA assessment rationale

Inhamitanga qualifies as an IPA under sub-criterion A(i), with ten Vulnerable species and one Endangered species recorded from this IPA meeting the A(i) threshold. Overall, there are ten endemic and range-restricted species present at this site. However, this total represents only 2% of endemic and range-limited species and, as such, this site does not qualify as an IPA under criterion B at present. As much of this site has not been extensively inventoried, there may be more A(i) or B(ii) qualifying taxa found in the future. An additional Vulnerable species, *Khaya anthotheca*, occurs within this IPA but does not meet the thresholds required to trigger sub-criterion A(i) due to its widespread distribution. With an extensive and largely undisturbed area of Inhamitanga Sand Forest habitat, Inhamitanga also qualifies under IPA sub-criterion C(iii) and is the most important site both nationally and globally for this habitat type.

Cephalophis lukei (MH)

Priority species (IPA Criteria A and B)

FAMILY	TAXON	IPA CRITERION A	IPA CRITERION B	≥ 1% OF GLOBAL POP'N	≥ 5% OF NATIONAL POP'N	IS 1 OF 5 BEST SITES NATIONALLY	ENTIRE GLOBAL POP'N	SPECIES OF SOCIO-ECONOMIC IMPORTANCE	ABUNDANCE AT SITE
Acanthaceae	*Cephalophis lukei*	A(i)		✓	✓	✓			unknown
Acanthaceae	*Dicliptera* sp. B of F.Z.		B(ii)	✓	✓	✓			unknown
Amaranthaceae	*Celosia pandurata*	A(i)	B(ii)	✓	✓	✓			unknown
Annonaceae	*Monodora stenopetala*	A(i)		✓	✓	✓			unknown
Apocynaceae	*Pleioceras orientale*	A(i)			✓	✓			unknown
Boraginaceae	*Cordia megiae*	A(i)	B(ii)	✓	✓	✓			scarce
Boraginaceae	*Cordia stuhlmannii*	A(i)	B(ii)	✓	✓	✓			occasional
Capparaceae	*Maerua brunnescens*		B(ii)	✓					unknown
Euphorbiaceae	*Mildbraedia carpinifolia*	A(i)			✓	✓			common
Fabaceae	*Millettia mossambicensis*		B(ii)	✓	✓				frequent
Malvaceae	*Cola clavata*	A(i)	B(ii)	✓	✓	✓			frequent
Meliaceae	*Khaya anthotheca*	A(i)						✓	unknown
Moraceae	*Dorstenia zambesiaca*	A(i)	B(ii)	✓	✓	✓			occasional
Ochnaceae	*Ochna angustata*		B(ii)	✓					unknown
Rubiaceae	*Tarenna longipedicellata*	A(i)	B(ii)	✓	✓	✓			scarce
		A(i): 11 ✓	B(ii): 10						

Threatened habitats (IPA Criterion C)

HABITAT TYPE	IPA CRITERION C	≥ 5% OF NATIONAL RESOURCE	≥ 10% OF NATIONAL RESOURCE	IS 1 OF 5 BEST SITES NATIONALLY	ESTIMATED AREA AT SITE (IF KNOWN)
Inhamitanga Sand Forest [MOZ-04]	C(iii)		✓	✓	350.0

Protected areas and other conservation designations

CONSERVATION AREA TYPE	CONSERVATION AREA NAME	RELATIONSHIP OF IPA TO CONSERVATION AREA
Game Reserve	Coutada 12	protected/conservation area encompasses IPA
Forest Reserve (conservation)	Inhamitanga	IPA encompasses protected/conservation area
Key Biodiversity Area	Gorongosa-Marromeu	protected/conservation area encompasses IPA
Ramsar	Zambezi Delta	protected/conservation area encompasses IPA
Important Bird Area	Zambezi Delta	protected/conservation area encompasses IPA

Threats

THREAT	SEVERITY	TIMING
Increase in fire frequency/intensity	medium	Ongoing – trend unknown
Logging & wood harvesting	low	Ongoing – trend unknown

MOUNT GORONGOSA

Assessors: Iain Darbyshire, Jo Osborne, Bart Wursten

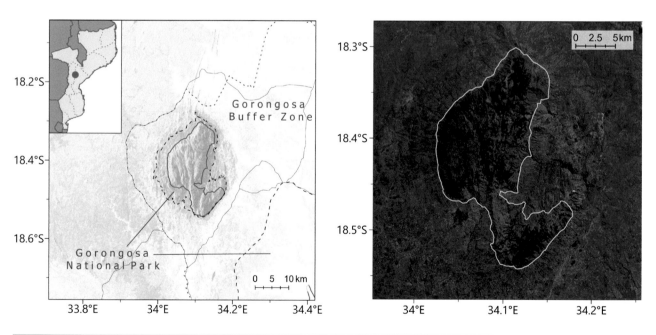

INTERNATIONAL SITE NAME		Mount Gorongosa	
LOCAL SITE NAME (IF DIFFERENT)		Monte Gorongosa	
SITE CODE	MOZTIPA002	PROVINCE	Sofala

View of Mount Gorongosa from the southeast (BW)

LATITUDE	-18.39833	LONGITUDE	34.08556
ALTITUDE MINIMUM (m a.s.l.)	580	ALTITUDE MAXIMUM (m a.s.l.)	1,863
AREA (km²)	216	IPA CRITERIA MET	A(i), B(ii), C(iii)

Site description

Mount Gorongosa is an isolated massif in Gorongosa District of Sofala Province in central Mozambique, approximately 180 km inland from the Indian Ocean coastline. It rises from ca. 400–450 m elevation at the base to over 1,850 m at the summit. There are three main peaks, the highest of which is Mount Gogogo in the southwest of the massif at 1,863 m elevation. Between the peaks is an extensive upland area with three main river valleys and with a varied, undulating terrain, sometimes referred to as the "plateau". Mount Gorongosa has extensively forested slopes, whilst the summit comprises a mosaic of montane grassland, bushland and rock outcrops with forest patches in gulleys and sheltered areas; the lower slopes previously supported extensive miombo woodlands but have now been heavily denuded of woody vegetation. This site is considered to be an outlier of the proposed Chimanimani-Nyanga Centre of Plant Endemism (Darbyshire *et al.* 2019a).

Botanical significance

Mount Gorongosa supports two endemic plant species, both of which are considered to be threatened: *Impatiens wuerstenii* (VU) and *Streptocarpus brachynema* (EN). A currently undescribed species of *Justicia* – *Justicia* sp. B of *Flora Zambesiaca* (EN) – is known with certainty only from the forests of Mount Gorongosa but is tentatively also recorded from Mount Namuli. Several other globally threatened plant species are recorded from this site, including *Aloe rhodesiana* (VU) and *Rhinacanthus submontanus* (VU), for which Mount Gorongosa is the only known site in Mozambique.

The mountain has phytogeographical links with the Chimanimani-Nyanga (Manica) Highlands that run along the Zimbabwe-Mozambique border over 100 km to the west, and it supports outlier populations of several Chimanimani-Nyanga endemics, including *Cineraria pulchra*, *Cynorkis anisoloba* (LC), *Euphorbia citrina*, *Jamesbrittenia carvalhoi* (LC), *Lysimachia gracilipes*, *Pavetta comostyla* subsp. *comostyla* var. *inyangensis*, *Pelargonium mossambicense*, *Polystachya subumbellata* (LC), *Protea caffra* subsp. *gazensis*, *Tephrosia montana* and *Vernonia calvoana* subsp. *meridionalis*. Some of these, such as the two orchid species, have large and important populations on Gorongosa. A number of these species have not yet been evaluated on the IUCN Red List and may well prove to be globally threatened – these include the recently described twining herb *Vincetoxicum monticolum*, known only from Bvumba in Zimbabwe and Tsetserra and Gorongosa in Mozambique (Goyder *et al.* 2020).

A botanical survey of the mountain in 2007 recorded 605 species of vascular plant at elevations of 700 m and above (Müller *et al.* 2012), which is significantly smaller than the flora of some other montane areas of southern tropical Africa, but the authors note that this list is likely to increase significantly when further surveys are conducted in different seasons. It is highly likely that additions to the plant list will include further outlier populations of Chimanimani-Nyanga highland endemics and/or additional new, endemic species.

Mount Gorongosa is also important for its extensive montane and sub-montane forests (but see Conservation issues) and for the areas of rocky montane grassland and shrubland on the summit plateau, two habitats that are highly restricted in Mozambique and support rare and threatened species.

Habitat and geology

The Mount Gorongosa massif mainly comprises Late Jurassic granites but with intrusions of gabbro on the western and southern sides. The gabbros form gentle-sloped, undulating terrain whereas the granite forms steeper slopes, with some extensive sheer faces and much exposed rock. The rainfall pattern at lower altitudes is markedly seasonal with peak rainfall in the austral summer (December-March) derived from moist air from the Indian Ocean, and with a prolonged dry season. However, orographic rainfall and mists at higher elevations maintain more constant moisture availability year-round (Müller *et al.* 2012).

The lower, seasonally dry slopes would originally have supported extensive areas of *Brachystegia*-dominated miombo woodland with forest confined to river valleys and gulleys, but much of this habitat has now been transformed to farmland.

The dominant habitat on the upper slopes with higher rainfall and mist is moist evergreen forest. The forest composition is variable, with three main altitudinal belts recognisable – these are documented in detail by Müller *et al.* (2012), from which the following summary is derived. Montane forest (over 1,600 m) is characterised by the dominance of *Syzygium afromontanum*, with *Aphloia theiformis*, *Macaranga mellifera*, *Maesa lanceolata*, *Olea capensis* subsp. *hochstetteri*, *Podocarpus milanjianus* and *Myrsine melanophloeos* also important to co-dominant. Mixed Submontane Forest (1,300–1,600 m) is typically dominated by *Craibia brevicaudata*, with several of the montane forest species also common, together with *Cassipourea malosana*, *Gambeya gorungosana* and *Strombosia scheffleri*. Medium Altitude Forest (900–1,300 m) is characterised by the presence of *Newtonia buchananii*, often as the dominant species, with other frequent species including *Albizia gummifera*, *Drypetes gerrardii*, *Ficus* spp. and *Trichilia dregeana*, as well as several species of the submontane forest.

At altitudes of over 1,700 m, montane grassland is extensive which supports a rich grassland and geophytic flora. Rocky slopes can support areas of shrubland and wooded grassland, sometimes dominated by *Erica hexandra* and *Widdringtonia nodiflora*.

Conservation issues

In 2010, Mount Gorongosa was incorporated into the Gorongosa National Park, one of Mozambique's

Upper Nhanda Valley with *Kniphofia splendida* (BW)

Within elfin montane forest on Gorongosa (BW)

Streptocarpus brachynema (BW)

flagship protected areas. Prior to this, the mountain had not been formally protected. The lower slopes (below 1,100 m and particularly below 700 m) have been settled extensively and most of the natural vegetation, miombo woodlands and lowland forest, has long since been cleared. In recent decades, there has been an increasing threat to the moist forests on the mid-slopes due to encroachment of small-scale subsistence and cash-crop agriculture, using burning to clear the forest. This is particularly impacting the western slopes of the mountain which have experienced severe losses (see Müller *et al.* 2012, figure 19). This loss has continued at an increased rate in the past 10 years, with particular problems during the conflict between the government and the Renamo opposition which flared up in 2014–2015. Renamo forces used the mountain as their base and the resultant fighting on the mountain led to burning of large areas of forest that have subsequently been cleared and used for cultivation. Such clearance has led to the loss, for example, of the forest patch within which *Justicia* sp. B has previously been recorded on Gorongosa (Darbyshire *et al.* 2019k), and has also led to the loss of the globally southern-most population of the forest herb *Brachystephanus africanus* (B. Wursten, pers. obs.).

The upper slopes are considered to be a sacred site by some local communities and so afforded some protection, but recent satellite images show that even on top of the plateau and in the inner valleys the forests are now fragmented where they were still near-pristine in 2007 when the last vegetation surveys were carried out. Increased frequency of fire is also a concern, and this may penetrate more of the high plateau as the forest fragmentation continues.

It is hoped that its inclusion within this flagship National Park – and its ambitious conservation, scientific research and education programmes under the umbrella of the Gorongosa Restoration Project – will result in the long-term protection and rehabilitation of Mount Gorongosa. As part of these efforts, an agroforestry scheme has recently been launched with shade-cropping of arabica coffee, intercropped with native trees such as *Albizia adianthifolia*, *Khaya anthotheca* and *Millettia stuhlmannii*, to help reforest some of the lower slopes whilst also providing a significant source of income for local communities, through the sale of "Gorongosa Coffee" to national and international markets.

Gorongosa National Park is also an Important Bird Area (BirdLife International 2021d), with Mount Gorongosa being cited as of highest importance within the IBA because it supports rare forest species such as the Chirinda Apalis (*Apalis chirindensis*, LC), Plain-backed Sunbird (*Anthreptes reichenowi*, NT) and Swynnerton's Robin (*Swynnertonia swynnertoni*, VU), as well as having a disjunct population of Green-headed Oriole

(*Oriolus chlorocephalus*). The pygymy chameleon *Rhampholeon gorongosae* (EN) is endemic to the mountain; this, together with *Streptocarpus brachynema*, would qualify the mountain as an Alliance for Zero Extinction site. This site is included within the vast Gorongosa-Marromeu Key Biodiversity Area.

Key ecosystem services

Mount Gorongosa, and the natural habitats it supports, is of importance for a range of ecosystem services including: protection of soil resources and local climate; providing most of the water flow to the surrounding lowlands and so maintaining regional ecological functioning; provisioning of resources for local communities; and cultural heritage. There is strong potential for ecotourism on Mount Gorongosa once safe access can be guaranteed; walking tours of the mountain combined with safari tours in the lowland portion of the National Park could be an ideal ecotourism package.

Justicia sp. A of *Flora Zambesiaca* (BW)

Impatiens wuerstenii (BW)

Ecosystem service categories

- Provisioning – Raw materials
- Provisioning – Fresh water
- Regulating services – Local climate and air quality
- Regulating services – Carbon sequestration and storage
- Regulating services – Erosion prevention and maintenance of soil fertility
- Habitat or supporting services – Habitats for species
- Habitat or supporting services – Maintenance of genetic diversity
- Cultural services – Tourism
- Cultural services – Spiritual experience and sense of place
- Cultural services – Cultural heritage

IPA assessment rationale

Mount Gorongosa qualifies as an IPA under all three criteria. Under IPA criterion A(i), it contains ten globally threatened plant species, four of which are only represented at this site within the Mozambique IPA network, including two that are endemic to this massif: *Impatiens wuerstenii* (VU) and *Streptocarpus brachynema* (EN). It qualifies under IPA criterion B as it supports just over 3% (16 species) of the national list of priority species under criterion B(ii). As noted above, this massif is under-botanised and further surveys are likely to yield further endemic and range-restricted species and so the number of criterion B(ii) species at this site is likely to increase, as is the number of threatened species. This site also qualifies under IPA criterion C(iii) as it supports approximately 80 km² of moist evergreen forest, with both Medium Altitude and Montane Forest assemblages, both of which are threatened and highly range-restricted in Mozambique. It also holds extensive areas of Montane Grassland. It is considered to be one of the five best sites nationally for these habitats.

Priority species (IPA Criteria A and B)

FAMILY	TAXON	IPA CRITERION A	IPA CRITERION B	≥ 1% OF GLOBAL POP'N	≥ 5% OF NATIONAL POP'N	IS 1 OF 5 BEST SITES NATIONALLY	ENTIRE GLOBAL POP'N	SPECIES OF SOCIO-ECONOMIC IMPORTANCE	ABUNDANCE AT SITE
Acanthaceae	*Justicia* sp. A of F.Z.	A(i)	B(ii)	✓	✓	✓			scarce
Acanthaceae	*Rhinacanthus submontanus*	A(i)		✓	✓	✓			unknown
Apocynaceae	*Vincetoxicum monticola*		B(ii)						unknown
Asphodelaceae	*Aloe rhodesiana*	A(i)		✓	✓	✓			frequent
Asteraceae	*Cineraria pulchra*		B(ii)						unknown
Asteraceae	*Vernonia calvoana* subsp. *meridionalis*		B(ii)						unknown
Balsaminaceae	*Impatiens wuerstenii*	A(i)	B(ii)	✓	✓	✓	✓		frequent
Dioscoreaceae	*Dioscorea sylvatica*	A(i)			✓	✓			unknown
Euphorbiaceae	*Euphorbia citrina*		B(ii)						unknown
Euphorbiaceae	*Tannodia swynnertonii*	A(I)			✓	✓			unknown
Fabaceae	*Lotus wildii*		B(ii)						frequent
Fabaceae	*Tephrosia montana*		B(ii)						unknown
Geraniaceae	*Pelargonium mossambicense*		B(ii)						unknown
Gesneriaceae	*Streptocarpus brachynema*	A(i)	B(ii)	✓	✓	✓	✓		occasional
Lauraceae	*Ocotea kenyensis*	A(i)			✓	✓		✓	scarce
Meliaceae	*Khaya anthotheca*	A(i)			✓	✓		✓	occasional
Orchidaceae	*Cynorkis anisoloba*		B(ii)						frequent
Orchidaceae	*Polystachya subumbellata*		B(ii)						frequent
Primulaceae	*Lysimachia gracilipes*		B(ii)						unknown
Proteaceae	*Protea caffra* subsp. *gazensis*		B(ii)						frequent
Rubiaceae	*Pavetta comostyla* var. *inyangensis*		B(ii)						frequent
Sapindaceae	*Allophylus chirindensis*	A(i)		✓	✓	✓			unknown
Scrophulariaceae	*Jamesbrittenia carvalhoi*		B(ii)						frequent
		A(i): 10 ✓	B(ii): 16 ✓						

Threatened habitats (IPA Criterion C)

HABITAT TYPE	IPA CRITERION C	≥ 5% OF NATIONAL RESOURCE	≥ 10% OF NATIONAL RESOURCE	IS 1 OF 5 BEST SITES NATIONALLY	ESTIMATED AREA AT SITE (IF KNOWN)
Montane Moist Forest [MOZ-01]	C(iii)		✓	✓	9
Medium Altitude Moist Forest [MOZ-02]	C(iii)		✓	✓	71
Montane Grassland [MOZ-09]	C(iii)			✓	

Protected areas and other conservation designations

CONSERVATION AREA TYPE	CONSERVATION AREA NAME	RELATIONSHIP OF IPA TO CONSERVATION AREA
National Park	Gorongosa National Park	protected/conservation area encompasses IPA
Important Bird Area	Gorongosa Mountain and National Park	protected/conservation area encompasses IPA
Key Biodiversity Area	Gorongosa-Marromeu	protected/conservation area encompasses IPA

Threats

THREAT	SEVERITY	TIMING
Shifting agriculture	high	Ongoing – increasing
Small-holder farming	high	Ongoing – increasing
Increase in fire frequency/intensity	medium	Ongoing – increasing

Montane habitats on Mount Gorongosa (BW)

UREMA VALLEY AND SANGARASSA FOREST

Assessors: Sophie Richards, Iain Darbyshire

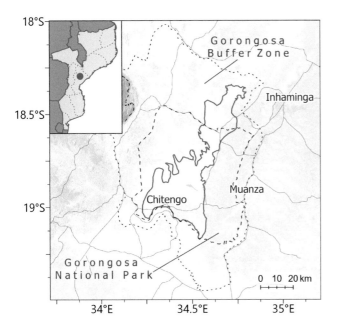

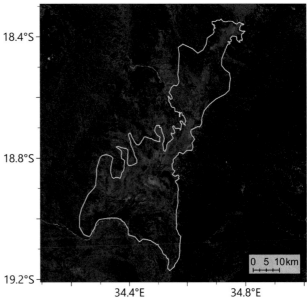

INTERNATIONAL SITE NAME		Urema Valley and Sangarassa Forest	
LOCAL SITE NAME (IF DIFFERENT)		Vale do Urema e Floresta de Sangarrassa	
SITE CODE	MOZTIPA038	PROVINCE	Sofala

LATITUDE	-18.57350	LONGITUDE	34.83650
ALTITUDE MINIMUM (m a.s.l.)	18	ALTITUDE MAXIMUM (m a.s.l.)	110
AREA (km²)	1594	IPA CRITERIA MET	A(i), B(ii), C(iii)

Site description

The Urema Valley and Sangarassa Forest IPA falls within Gorongosa National Park and Buffer Zone. Spanning four districts of Sofala Province, Maringue in the north-west, Cheringoma in the north-east, Muanza in the east and the vast majority within Gorongosa District, this site is centred on the far south of the African Great Rift Valley. With highly seasonal water levels, the vegetation is a complex mixture of open floodplain woodland-grassland to closed woodland and dry forest (Stalmans & Beilfuss 2008; Parque Nacional da Gorongosa 2019). The most distinct patch of dry forest within this IPA is Sangarassa Forest (-18.97°, 34.33°), 2 km north of the Pungue River. There are a number of species of conservation importance recorded from this small patch of dry forest and, as such, it is given particular prominence within this report.

The IPA is 1,594 km² in area and has been delineated to encompass much of the valley floodplain within Gorongosa National Park. The northern boundary is adjacent to Chipanha village, 7 km south of the Gorongosa Buffer Zone boundary, and the southern boundary follows the Pungue River along the southern boundary of the national park core zone. While there is floodplain habitat both to the north and south of this IPA, in the national park buffer zone, this area is more heavily populated and so has been excluded, however, this IPA could be expanded if species of conservation interest are found in these areas in future.

Botanical significance

Urema Valley and Sangarassa Forest is home to a number of endemic and threatened species. For instance, *Vepris myrei* (EN) is known from this site, occurring in the dry forest patches north-east of Chitengo, where it is described as common (*Tinley #2777*), and in Sangarassa Forest. This species is threatened elsewhere by habitat conversion, and its presence at this relatively secure location is important in preventing the extinction of *V. myrei*. Although there are unconfirmed records of *V. myrei* in both Malawi and Zimbabwe, it is thought that these may be specimens of a closely related species, *V. rogersii*, and so *V. myrei* may well be endemic to Mozambique, although further investigation is required to confirm this (Timberlake 2021b).

Two Vulnerable species have also been recorded from this site. *Erythrococca zambesiaca* (VU) is of particular importance. Although it is also native to Malawi, *E. zambesiaca* is a range-restricted species (under sub-criterion B(ii) of the IPA criteria), with an EOO (extent of occurrence) of 788 km². Threatened elsewhere by the conversion of habitat to agriculture, the presence of *E. zambesiaca* within Sangarassa Forest not only is important for the continued survival of this species but is the only known location within Mozambique and represents the southern edge of its range (Timberlake 2019).

Celosia pandurata (VU), an endemic species, is also recorded from Sangarassa forest. A total of four Mozambican endemics occur within this forest patch and, hence, Sangarassa Forest is of particular importance within this IPA.

Within the IPA as a whole, 12 Mozambican endemics have been recorded. One of these species, *Acacia torrei* (LC) is limited only to a range of ca. 1,700 km² on the black alluvial clay soils of the Urema and Zangue valleys. This species is locally common in the north of the IPA and is currently assessed as Least Concern; the continued protection that *A. torrei* receives within Gorongosa National Park is central to preventing it from becoming globally threatened with extinction (Coates Palgrave *et al.* 2014b). The Urema Valley also hosts the largest known population of another endemic species, *Gyrodoma hispida* (LC). *G. hispida* has been described as common within this IPA and so this is a key locality for preventing this species from becoming threatened with extinction (Richards 2021b).

The over 700 km² of seasonally inundated grasslands within this IPA represent a habitat type of conservation interest for Mozambique (Stalmans & Beilfuss 2008). As well as hosting significant populations of the two endemic species mentioned above, this habitat type has a limited range across Mozambique. The seasonally inundated grasslands of this site represent one of the largest and highest quality examples of this habitat nationally and therefore trigger sub-criterion C(iii) of the IPA criteria for this site.

Habitat and geology

The plant communities within this IPA are highly variable, which probably reflects the underlying soil structure and moisture availability within the landscape (Stalmans & Beilfuss 2008). Much of the IPA has sandy soils, with a black clay colluvial fan to the north (Steinbruch 2010). *Acacia torrei* in particular is reliant on these areas of black clay and is restricted to this substrate within Sofala Province (Coates Palgrave *et al.* 2014b).

Lake Urema, just south of the centre of the valley, is supplied with drainage from both Mount Gorongosa and the surrounding plateau to the west and the Cheringoma Plateau to the east, with overflow joining the Pungue River at the southern

Edge of Sangarassa Forest, featuring *Newtonia hildebrandtii* (BW)

boundary of this IPA (Stalmans & Beilfuss 2008). During the wet season, December to March, the water levels of Lake Urema and associated rivers increase to cover up to 40% of Gorongosa's area (Stahl 2020), with much of the central stretch of this IPA, along with the southern boundary by the Pungue River, being inundated with water (Parque Nacional da Gorongosa 2019).

The vegetation types of Gorongosa National Park are categorised in Stalmans & Beilfuss (2008), and a summary of the relevant vegetation types is provided below.

The floodplain region is a largely open landscape dominated by seasonally inundated grasslands of various types including: *Echinochloa – Chrysopogon, Setaria* and *Cynodon dactylon – Digitaria didactyla* assemblages; the latter community is concentrated around Lake Urema and has almost no woody plants. On the lower slopes and drainage lines south of Lake Urema are areas of palm savanna consisting of open to closed *Hyphaene* stands with a grassy understorey. Stands of *Acacia xanthophloea*, mixed *Acacia–Combretum* and *Faidherbia albida* also form open to closed areas of woodland within the floodplain and alluvial fan.

West of Chitengo Camp is Sangarassa Forest, a 1.6 km² area of vegetation that is described on specimen vouchers as dense sand forest (e.g., *Wursten* #911). The forest is dominated by species such as *Newtonia hildebrandtii* and *Xylia torreana* (*Tinley* #2331). The understorey includes the Mozambican endemic *Millettia mossambicensis*, while some species are associated with the termite mounds that border seasonal pans, such as *Cola mossambicensis* (NT).

To the south of the IPA, following the Pungue River, is closed woodland/dry forest dominated by *Piliostigma thonningii* and, in seasonally flooded areas, *Borassus aethiopium* (Stalmans & Beilfuss 2008; Hyde et al. 2021).

Conservation issues

The entirety of the Urema Valley and Sangarassa Forest falls within Gorongosa National Park and Buffer Zone, with only the most northerly 220 km² of the IPA falling within the buffer zone. This IPA is also encompassed within Gorongosa Mountain and National Park Important Bird Area and Gorongosa-Marromeu Key Biodiversity Area. The Urema Valley wetlands are of particular importance for bird species; Grey Crowned Crane (*Balearica regulorum* – EN) has been recorded here, while the area is possibly an important over-wintering ground for Great Snipe (*Gallinago media* – NT) (BirdLife International 2021d). In addition, a 2014 count found that the population sizes of two avian taxa meet Ramsar site criteria; Yellow-billed Stork (*Mycteria ibis* – LC) exceeded the threshold of 1% of the population in sub-Saharan Africa (870 individuals), and the population of African Darter (*Anhinga rufa* – LC) exceeded 1% of individuals of this species in southern and eastern Africa (1,000 individuals) (Stalmans et al. 2014), although the area is not currently listed as a Ramsar site.

Mimosa pigra, a species that features in IUCN's "100 of the World's Worst Invasive Alien Species" (van der Weijden et al. 2004), is a major threat to the wetlands within this IPA and has established on the floodplain (Stalmans & Beilfuss 2008). This species forms dense thickets, excluding other species and converting floodplains into scrubland (Beilfuss 2007). However, it is thought that the

Savanna grasslands in the Urema floodplain (MS)

Grasslands surrounding outflow of Lake Urema (MS)

Acacia torrei (BW)   *Orbea halipedicola* (BW)

re-introduction of ruminant grazers is helping to contain shrub encroachment (Guyton *et al.* 2020).

The management strategy for the site includes controlled burning of the valley early in the dry season to reduce the fuel load for the late season fires (Stahl 2020). Research into fire and herbivory dynamics has been undertaken (see Stahl 2020) towards improving the use of fire for the continued restoration of the park following the Mozambican Civil War.

During the civil war, large herbivore populations declined by over 90% within Gorongosa National Park (Stalmans *et al.* 2019). Today, as populations continue to recover, a number of species within the national park are centred around the Urema Valley region. Hippo were released around Lake Urema in 2008 and African wild dogs were released in 2018, while the recovering Sungwe lion pride is centred around the streams south-west of Lake Urema. The vegetation within this IPA, as an important component of the ecosystem, therefore, making an important contribution to the conservation of mammals within the national park, as well as to the tourism that these mammal species attract.

As the number of large herbivores within the park has increased, a change in the dominant species has been recorded. Elephant, hippo and African buffalo previously dominated the pre-war large herbivore biomass, whereas in 2018 over 74% of large herbivore biomass recorded was waterbuck (Stalmans *et al.* 2019). It would be informative to monitor how these changing herbivory dynamics may be impacting plant communities, particularly habitats vital for rare or threatened plants species.

Much of this IPA has been unaffected by the conversion of habitat to agriculture, a major threat to plant species across Mozambique (Darbyshire *et al.* 2019a), probably because the vast majority of the IPA area falls within Gorongosa National Park. A total of ca. 200,000 people live within the buffer zone, and Gorongosa National Park partners with these communities to build sustainable livelihood opportunities (Parque Nacional da Gorongosa 2019). To this end, GNP are working towards having a large Community Conservation Area proclaimed at the north-east boundary, which would include part of this IPA (M. Stalmans, pers. comm. 2021). Monitoring of populations of *Acacia torrei* in this area could be considered within conservation actions, to safeguard against threats to a key area of habitat for this range-restricted endemic.

Key ecosystem services

The Urema Valley region is regularly inundated with water (Stalmans & Beilfuss 2008). However, following Cyclone Idai in 2019, one of the worst weather events recorded in the southern hemisphere (Warren 2019), much of Gorongosa National Park was submerged. Although the cyclone made landfall near Beira, the Gorongosa area was impacted by heavy rain and extensive flooding, with communities living south of the national park severely impacted (Parque Nacional da Gorongosa 2019). However, the vegetation within the park is thought to have mitigated some of the impacts of the cyclone on local communities, with water being gradually released for over five months after the event, due to the complexity of the landscape in the Urema Valley area and beyond (Parque Nacional da Gorongosa 2019).

The landscapes of the Urema Valley, particularly the grasslands surrounding Lake Urema, have the greatest suitability for supporting grazers across the entirety of Goronogosa National Park and Buffer Zone (Stalmans & Beilfuss 2008). Therefore, through supporting some of the more charismatic mammals, the vegetation within this IPA makes a major contribution to attracting tourism to the area.

Gyrodoma hispida (BW)

Ecosystem service categories

- Regulating services – Moderation of extreme events
- Habitat or supporting services – Habitats for species
- Cultural services – Tourism

IPA assessment rationale

Urema Valley and Sangarassa Forest qualifies as an Important Plant Area under criterion A(i), due to the presence of one Endangered species, *Vepris myrei*, and two Vulnerable species *Erythrococca zambesiaca* and *Celosia pandurata*. In addition, there are 13 endemic and near-endemic species within this IPA and the site therefore qualifies under sub-criterion B(ii) as one of the top 15 sites nationally for range-restricted and endemic species. The presence of a large, high-quality expanse of seasonally inundated grassland, a nationally restricted habitat type associated with endemic species, at the site triggers sub-criterion C(iii).

Priority species (IPA Criteria A and B)

FAMILY	TAXON	IPA CRITERION A	IPA CRITERION B	≥ 1% OF GLOBAL POP'N	≥ 5% OF NATIONAL POP'N	IS 1 OF 5 BEST SITES NATIONALLY	ENTIRE GLOBAL POP'N	SPECIES OF SOCIO-ECONOMIC IMPORTANCE	ABUNDANCE AT SITE
Amaranthaceae	*Celosia pandurata*	A(i)	B(ii)	✓	✓				unknown
Apocynaceae	*Orbea halipedicola*		B(ii)	✓	✓	✓			frequent
Asparagaceae	*Dracaena subspicata*		B(ii)	✓	✓				unknown
Asteraceae	*Gyrodoma hispida*		B(ii)	✓	✓	✓			common
Capparaceae	*Maerua brunnescens*		B(ii)	✓	✓				unknown
Euphorbiaceae	*Erythrococca zambesiaca*	A(i)	B(ii)	✓	✓	✓			unknown
Euphorbiaceae	*Euphorbia ambroseae* var. *ambrosae*		B(ii)	✓	✓				unknown
Euphorbiaceae	*Jatropha scaposa*		B(ii)	✓	✓				unknown
Fabaceae	*Acacia torrei*		B(ii)	✓	✓	✓			common
Fabaceae	*Millettia mossambicensis*		B(ii)	✓	✓				unknown
Malvaceae	*Grewia transzambesica*		B(ii)	✓					unknown
Polygalaceae	*Carpolobia suaveolens*		B(ii)	✓					scarce
Rubiaceae	*Psydrax moggii*		B(ii)	✓	✓				unknown
Rutaceae	*Vepris myrei*	A(i)		✓	✓	✓			frequent
		A(i): 3 ✓	B(ii): 13						

Threatened habitats (IPA Criterion C)

HABITAT TYPE	IPA CRITERION C	≥ 5% OF NATIONAL RESOURCE	≥ 10% OF NATIONAL RESOURCE	IS 1 OF 5 BEST SITES NATIONALLY	ESTIMATED AREA AT SITE (IF KNOWN)
Seasonally Inundated Grassland [MOZ-10]	C(iii)			✓	780.0

Protected areas and other conservation designations

CONSERVATION AREA TYPE	CONSERVATION AREA NAME	RELATIONSHIP OF IPA TO CONSERVATION AREA
National Park	Gorongosa National Park and Buffer Zone	protected/conservation area encompasses IPA
Important Bird Area	Gorongosa Mountain and National Park	protected/conservation area overlaps with IPA
Key Biodiversity Area	Gorongosa-Marromeu	protected/conservation area encompasses IPA

Threats

THREAT	SEVERITY	TIMING
Small-holder farming	low	Ongoing – trend unknown
Storms & flooding	high	Past, likely to return
Fire & fire suppression – Trend unknown/unrecorded	unknown	Ongoing – stable

CHERINGOMA LIMESTONE GORGES

Assessors: Sophie Richards, Iain Darbyshire

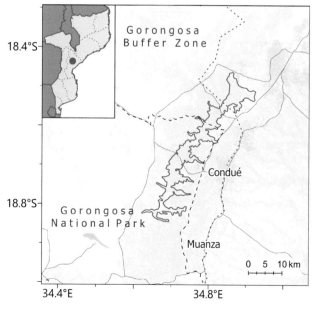

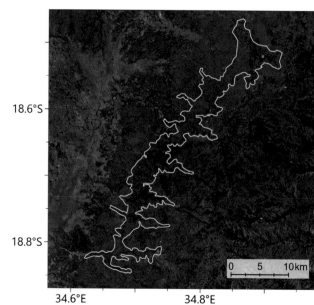

INTERNATIONAL SITE NAME		Cheringoma Limestone Gorges	
LOCAL SITE NAME (IF DIFFERENT)		Desfiladeiros de calcário de Cheringoma	
SITE CODE	MOZTIPA037	PROVINCE	Sofala

LATITUDE	-18.64661	LONGITUDE	34.76241
ALTITUDE MINIMUM (m a.s.l.)	30	ALTITUDE MAXIMUM (m a.s.l.)	326
AREA (km²)	182	IPA CRITERIA MET	A(i), C(iii)

Site description

The Cheringoma Limestone Gorges span the border between Muanza and Cheringoma Districts of Sofala Province. Falling mostly within Gorongosa National Park and Buffer Zone, this IPA encompasses the gorge system where the Cheringoma Plateau meets the southern end of the African Great Rift Valley. The humid conditions within these gorges host forest that, although not yet extensively studied, is known to be biodiverse and of great conservation value (Byrne 2013). This IPA is unique, covering most of the only limestone forest known with certainty from Mozambique. This habitat is rare and threatened across tropical Africa, where it is only known to occur in Kenya, Tanzania and Mozambique (Cheek *et al.* 2019).

The site boundary closely follows the gorge habitat, delineated by Stalmans & Beilfuss (2008), excluding some of the more degraded patches in the far north but encompassing 182 km² in area. The north of the IPA extends just beyond the Khodzhue Gorge system (-18.52°, 34.90°) and the southern boundary is just beyond Muanza Gorge (-18.82°, 34.68°). Less than 10 km west is a separate IPA, Urema Valley [MOZTIPA038], that encompasses the wetlands of Gorongosa National Park.

Botanical significance

Limestone forests are rare in tropical Africa, with known locations limited only to Kenya, Tanzania and Mozambique, while the Cheringoma Limestone Gorges are the only area known to support forest on limestone in Mozambique (Cheek *et al.* 2019). Globally, limestone is associated with narrowly endemic plant species, as the physiological challenges of the substrate provide a selection pressure through which adaptations may develop, some of which restrict these species to a single or a small number of limestone patches (Cheek *et al.* 2019). One example of a narrow range endemic within this IPA is *Cola cheringoma*. A globally Endangered species, *C. cheringoma* is described as locally common but is restricted only to Cheringoma limestone forests, with at least one locality known from this IPA. It is likely that the population within

View overlooking forests of Nhamfisse Gorge in the centre of the IPA (JEB)

View of limestone gorge and the overlooking Cheringoma Plateau (MS)

Gorge lip displaying the overlying sandstone on the plateau and the limestone beneath (MS)

the gorges is larger than is currently known, with a number of *Cola mossambicensis* specimens collected here possibly representing misidentified individuals of *C. cheringoma* (Cheek *et al.* 2019). In addition to this narrow endemic, an as yet undescribed species of *Justicia*, *Justicia* sp. B of *Flora Zambesiaca* (LC), is also thought to be endemic to the limestone outcrops of the Cheringoma gorges and, although common at the two localities from which it is known, there has been no collections made outside this IPA (Darbyshire *et al.* 2019k). With further investigation, it is possible that more Cheringoma limestone endemics will be documented from these gorges.

In total, six national endemic taxa have been recorded from this IPA and, including *C. cheringoma* (EN), two threatened species. *Khaya anthotheca*, the second threatened species, is known from gorge margin habitat and, despite being assessed as globally Vulnerable, does not qualify under sub-criterion A(i) of the IPA criteria due to its extensive range covering parts of west, central and southern Africa. An additional Vulnerable species, *Diplocyclos tenuis*, has been recorded at this site (Wursten s.n.). However, the IUCN Red List assessment does not take this locality into account, with the range considered including only Tanzania, Kenya and one locality in Cabo Delgado province of Mozambique.

We therefore could expect this Red List assessment to be downgraded should this additional locality be considered.

Although limited botanical studies of the site have been undertaken to date, two surveys of the limestone forests of Cheringoma have been completed since 2004, with an unpublished checklist from Burrows *et al.* (2012) numbering around 320 species. This checklist records a number of interesting plant taxa within the gorges, including *Mondia whitei*, a medicinal species categorised as Endangered in both South Africa and eastern Africa (Aremu *et al.* 2011). *Antiaris toxicaria* subsp. *usambarensis* var. *welwitschia*, a taxon native to Mozambique, Tanzania, Kenya, Zambia and the Democratic Republic of Congo, is also recorded from these gorges. This species is used to make bark cloth in West Africa, while the latex is used for arrow poison (Burrows *et al.* 2018). Although frequent within the Cheringoma Gorges, *A. toxicaria* subsp. *usambarensis* var. *welwitschia* is rare in Mozambique, known from only two other localities in Cabo Delgado province (Burrows *et al.* 2018). These northerly localities are threatened, with some areas already transformed and, therefore, this IPA is of great importance to the national population, representing the only legal

Forest at the base of Khodzue Gorge in the north of the IPA (JEB)

protection for this species in the *Flora Zambesiaca* area (J. Burrows, pers. comm. 2021).

In addition, two fern species, *Thelypteris opulenta* (*Amblovenatum opulentum*) and *T. unita* (*Sphaerostephanos unitus*), are present within the Cheringoma Limestone Gorges, representing the only collections for these species within the *Flora Zambesiaca* region (Burrows *et al.* 2012).

Habitat and geology

Geology is a major contributor to the plant diversity of this IPA. Drainage from the Cheringoma Plateau eroded the overlying sandstone, carving out the deep gorges and revealing the underlying Eocene limestone (Cheek *et al.* 2019). The gorges, some of which are 70 m in depth, provide a sheltered environment in which conditions are hot and humid (Byrne 2013; Burrows *et al.* 2018). The mean annual temperature is 34°C and the mean annual rainfall, of over 1,100 mm, are relatively high for the area (Stalmans & Beilfuss 2008; Burrows *et al.* 2018). These conditions produce a flora distinct from the *Androstachys johnsonii*-dominated woodland on the gorge lip (Burrows *et al.* 2018). The limestone gorge forest is of particular conservation interest due to the association of narrow range endemics, notably *Cola cheringoma*. Tree species within the

gorge forest include *Albizia glaberrima* (LC), *Celtis philippensis* (LC) and *Khaya anthotheca* (VU), while large shrubs include *Pavetta klotzschiana*, *Grandidiera boivinii* (LC) and *Combretum pisoniiflorum* (Burrows *et al.* 2018).

The small areas of plateau included in this IPA are populated by miombo woodland, dominated by *Brachystegia spiciformis* and *Julbernardia globifora,* and are also underlain by limestone (Lötter *et al.*, in prep.). Relatively low levels of habitat transformation have been observed on the plateau, although there is some agricultural land near Muanza Gorge towards the south-east of the IPA (Stalmans & Beilfuss 2008).

Conservation issues

Much of this IPA falls within Gorongosa National Park (GNP) and Buffer Zone. The focus of Gorongosa has been set out in the 2020–2050 Strategic Plan and involves improving the capacity of the national park to "preserve, protect and manage the diverse ecosystems within the Park" while also working with communities within the buffer zone, making a particular effort to reach women in these communities, to improve sustainable economic opportunities (Parque Nacional da Gorongosa 2019). This IPA also falls with the Gorongosa-

Marromeu Key Biodiversity Area and Gorongosa Mountain and National Park Important Bird Area.

One of the major threats to biodiversity within GNP is the expansion of agriculture and settlements (Biofund 2013), but relative to areas outside the national park, the site management offers greater security for plant communities. Outside the national park and buffer zone, north of this IPA, gorge vegetation has been degraded by agriculture. However, GNP are working towards having a large Community Conservation Area declared in this area (M. Stalmans, pers. comm. 2021). This additional protection may therefore offer greater protection to the most northerly gorge habitat in future.

The inaccessibility of the larger gorges probably affords them additional protection from anthropogenic disturbance. There is, however, some transformation around Muanza Gorge (-18.82, 34.68) within the GNP (Stalmans & Beilfuss 2008). Much of this agricultural land surrounds the road from Muanza town, although the rate of habitat conversion here appears to have peaked in the 2000s and has since slowed (World Resources Institute 2019; Google Earth 2021).

In the buffer zone, the focus of GNP is on economic and social development within communities, with around 200,000 people living within the entire buffer zone (Parque Nacional da Gorongosa 2019). The development of sustainable livelihoods contributes to the reduction of pressure on land and natural resources around the communities in the area.

While the geology of this site is strongly linked to its unique biodiversity, limestone is also a valuable material that is extracted in the area. Limestone extraction has been observed near Condué, upstream of Antiaris Gorge, (-18.71°, 34.83°) with clearing of vegetation for mining and associated infrastructure such as access roads (Cheek *et al.* 2019). It appears that similar extraction is occurring to the east of Muanza Gorge (-18.815°, 34.735°) (Google Earth 2021).

The fauna of the Cheringoma Limestone Gorges has not yet been studied extensively, but some interesting vertebrate taxa have been recorded, including several species of bat and an undescribed species of frog in the genus *Kassina* (Conneely 2013; Parque Nacional da Gorongosa 2016).

Key ecosystem services

Despite being largely inaccessible, the aesthetic value of the gorges contributes to the tourism experience at Gorongosa National Park. The Cheringoma Plateau is known to be an important water catchment area in the region. In addition, the regulation of water and stabilisation of soils provided by riverine forest in the gorges probably regulates the hydrology and prevents sediment build-up downstream in the Urema Valley. The valley is home to a number of charismatic mammal species, which are both of conservation importance and major tourist attractions. The continued regulation of water on the Cheringoma plateau and in the gorges may be important to maintaining the quality of this habitat downstream.

Ecosystem service categories

- Provisioning – Fresh water
- Regulating services – Erosion prevention and maintenance of soil fertility
- Cultural services – Tourism

IPA assessment rationale

The Cheringoma Limestone Gorges qualify as an Important Plant Area under sub-criterion A(i) due to the presence of, *Cola cheringoma* (EN), a species only known from the Cheringoma area. A total of six Mozambican endemic species in total are known from this IPA, falling below the threshold of 3% of Mozambican species (equivalent to 16 species) of high conservation importance within the site required to trigger sub-criterion B(ii). However, as this area has not yet been extensively studied, it is possible that more species of conservation importance will be recorded within the gorges with further investigation. This IPA also covers the only known limestone forest in Mozambique (Cheek *et al.* 2019). Given the uniqueness of this habitat and its association with narrow range endemic species, the limestone gorge forest qualifies under sub-criterion C(iii) of the IPA criteria.

Justicia sp. B of Flora Zambesiaca (BW)

Euphorbia ambroseae var. ambroseae (JEB)

Priority species (IPA Criteria A and B)

FAMILY	TAXON	IPA CRITERION A	IPA CRITERION B	≥ 1% OF GLOBAL POP'N	≥ 5% OF NATIONAL POP'N	IS 1 OF 5 BEST SITES NATIONALLY	≥ 10% OF GLOBAL POP'N	ENTIRE GLOBAL POP'N	SPECIES OF SOCIO-ECONOMIC IMPORTANCE	ABUNDANCE AT SITE
Acanthaceae	Justicia gorongozana		B(ii)	✓	✓	✓				occasional
Acanthaceae	Justicia sp. B of F.Z.		B(ii)	✓	✓	✓	✓	✓		frequent
Asparagaceae	Dracaena subspicata		B(ii)	✓	✓					unknown
Euphorbiaceae	Euphorbia ambroseae var. ambrosae		B(ii)	✓	✓	✓	✓			occasional
Euphorbiaceae	Euphorbia bougheyi		B(ii)	✓	✓	✓	✓			occasional
Malvaceae	Cola cheringoma	A(i)	B(ii)	✓	✓	✓	✓			frequent
Meliaceae	Khaya anthotheca	A(i)								unknown
		A(i): 1 ✓	B(ii): 6							

Threatened habitats (IPA Criterion C)

HABITAT TYPE	IPA CRITERION C	≥ 5% OF NATIONAL RESOURCE	≥ 10% OF NATIONAL RESOURCE	IS 1 OF 5 BEST SITES NATIONALLY	ESTIMATED AREA AT SITE (IF KNOWN)
Cheringoma Limestone Forest [MOZ-13]	C(iii)	✓	✓	✓	150.0

Protected areas and other conservation designations

CONSERVATION AREA TYPE	CONSERVATION AREA NAME	RELATIONSHIP OF IPA TO CONSERVATION AREA
National Park	Gorongosa National Park and Buffer Zone	protected/conservation area overlaps with IPA
Important Bird Area	Gorongosa Mountain and National Park	protected/conservation area overlaps with IPA
Key Biodiversity Area	Gorongosa-Marromeu	protected/conservation area overlaps with IPA

Threats

THREAT	SEVERITY	TIMING
Small-holder grazing, ranching or farming	low	Ongoing – trend unknown
Mining & quarrying	low	Ongoing – trend unknown

MANICA
PROVINCE

MOUNT MURUWERE-BOSSA

Assessors: Sophie Richards, Iain Darbyshire

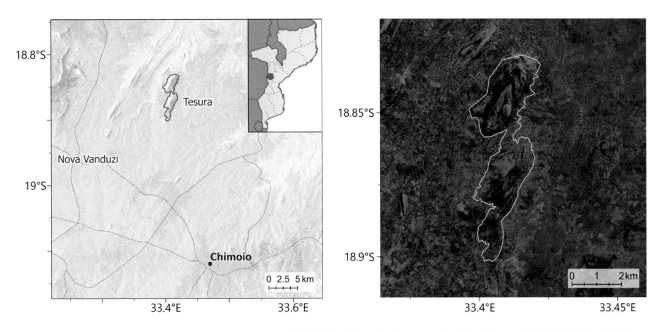

INTERNATIONAL SITE NAME		Mount Muruwere-Bossa	
LOCAL SITE NAME (IF DIFFERENT)		Monte Urueri e Monte Bossa	
SITE CODE	MOZTIPA040	PROVINCE	Manica

LATITUDE	-18.85969	LONGITUDE	33.40989
ALTITUDE MINIMUM (m a.s.l.)	490	ALTITUDE MAXIMUM (m a.s.l.)	1,030
AREA (km²)	10	IPA CRITERIA MET	A(i)

View of Mount Muruwere, surrounded by farmland (TR)

Site description

Mount Muruwere-Bossa IPA comprises two inselbergs 30 km north-northwest of the city of Chimoio in Manica District of Manica Province. The site includes Mount Muruwere in the north, peaking at 1,030 m, and Mount Bossa to the south-east, peaking at 870 m. The IPA is 10 km² in area and to the east of the boundary is the village of Tesura, with small number of houses and agricultural land surrounding these inselbergs.

The plant diversity of the IPA has not yet been inventoried comprehensively. However, the exposed gneiss outcrops of Mount Muruwere are known to be a highly important habitat, hosting the only confirmed population of the Critically Endangered cycad, *Encephalartos pterogonus*.

Botanical significance

Mount Muruwere is of great conservation importance as the only locality from which *Encephalartos pterogonus* (CR) is known with certainty. *E. pterogonus* inhabits the granite outcrops of the inselberg. Although the steep sides make the inselberg difficult to access, this species is threatened due to over-collection. As an attractive and rare cycad, it is sought after by collectors as an ornamental and is listed, along with all other *Encephalartos* species, on Appendix I of CITES (UNEP 2021). As of 2006, there were reported to be at least 246 mature plants and reasonable recruitment of seedlings (300–400 out of 1,000 planted) following re-introduction efforts by the "Plantas de Moçambique" project in 2004 (Capela 2006). There are reports that this species may also occur on Mount Dengalenga (Donaldson 2010b), 5 km west of this IPA, but this is yet to be confirmed.

Some authors (see Capela 2006) argue that *E. pterogonus* does not show enough consistency of characters to justify recognition as a species, and is instead a form within the *E. manikensis* complex. However, as this species is currently accepted by most authorities, we accept the species here. If, however, *E. pterogonus* is in the future accepted as a form of *E. manikensis*, this site could still qualify as an IPA, as *E. manikensis* is globally Vulnerable, although the site would likely then be of lower priority.

Another species of interest, *Euphorbia graniticola* (LC), was observed on Mount Muruwere growing between crevices in the gneiss rock (T. Rulkens,

pers. comm. 2021). A Mozambican endemic, *E. graniticola* is known only from the inselbergs of Manica Province (Darbyshire *et al.* 2019l) and this IPA is one of only two, alongside Mount Zembe [MOZTIPA011], known to host this species. In addition, an as yet undescribed species of *Jatropha* was collected slightly east of this IPA and may represent an additional Mozambican endemic (T. Rulkens, pers. comm. 2021). This species requires further investigation; if it does prove to be new to science, it is likely to be globally threatened as this area is under pressure from expanding agriculture. In this case, it should be established whether this species occurs within this IPA, and if it does not, a boundary change could be considered to encompass this species and to highlight it as a conservation priority.

Dracaena (formerly *Sansevieria*) *pedicellata* (LC) occurs in large colonies in the seasonally moist forests of Mount Muruwere. This is a species of medicinal importance; the leaves and rhizome of *D. pedicellata* are used in parts of Manica Province to treat a number of illnesses in both people and poultry (Rulkens & Baptista 2009).

Habitat and geology

The geology of the site is described as gneiss derived from granite. Both of the inselbergs comprise exposed rock outcrops which are an important habitat for *Encephalartos pterogonous*, particularly those at forest margins on Mount Muruwere (Donaldson 2010b).

There is yet to be a comprehensive survey of the botanical diversity at this site and the most recent visits were made in the late 2000s. The following description of the vegetation on Mount Muruwere is based on personal communication with T. Rulkens (2021).

Forested areas occur mostly along rivers within gullies and on plateaux where soils are sufficiently deep. The species composition of the forest canopy has not been described; however, large groups of *Dracaena pedicellata* (LC) occur in the understorey. *D. pedicellata* is known to occur with tree species such as *Albizia gummifera* and *Millettia stuhlmannii*, the latter species has been observed on Mount Muruwere and may be dominant in places. While the Chimoio area experiences a dry season between May and October, it is thought that areas where *D. pedicellata* grows tend to be moist due to frequent mists and drizzles on mountains in this area and runoff from the rocky outcrops (Rulkens & Baptista 2009). The large tree aloe *Aloidendron barberae* has also been observed within these forests. In more exposed areas of the mountain, dry forest-thicket areas and miombo woodland inhabit the slopes, suggesting that a mosaic of vegetation types has arisen on the mountain according to moisture availability.

The steeper slopes of Mount Muruwere retain a shallow layer of soil, with short grasses (the species composition of which has not yet been documented) and the lithophytic sedge *Coleochloa setifera* inhabiting these areas. In the grasslands, species such as *Aloe chabaudii, A. cameronii* (LC), *Euphorbia graniticola* (LC) and a *Drimia* species, possibly *Drimia intricata* var. *intricata*, also occur. *Bulbine latifolia* grows at higher altitudes near the summit. Where soils are deeper, the grass communities are taller and numerous *Aloe excelsa* may be observed.

There are few streams on the inselberg; however, there are small pools of water on rocky plateaux at mid altitudes, which are likely to be seasonal.

The surrounding plain has deep and sandy soils and was previously miombo, dominated by *Brachystegia boehmii*, which is reported to form stands in the area (*Rulkens* #2010/1) (GBIF.org 2021c). However, much of this land is now given over to agriculture, with cultivation of *Sorghum* and other crops (*Rulkens* #2010/1), and so has been largely excluded from this IPA.

Conservation issues

Mount Muruwere does not fall within a protected area, Key Biodiversity Area or Important Bird Area. However, given that the entire confirmed population of *Encephalartos pterogonus* occurs within this IPA, and that this species is Critically Endangered, Muruwere could qualify as an Alliance of Zero Extinction site and also as a KBA under sub-criterion Ae1.

Due to the steep sides of these inselbergs, many parts of Muruwere are difficult to access, probably providing some passive protection to most areas of vegetation. Although there is extensive conversion to agriculture in the surrounding plain, with a particular expansion of farmland between 2004 and 2009 (Google Earth 2021), the slopes of the mountains remain undisturbed. However, miombo woodland is known to act as a buffer to the denser vegetation on mountains elsewhere in Mozambique (Timberlake *et al.* 2007), and the loss of this vegetation around Mount Muruwere may result in greater exposure of forests on the mountain, for example with greater evapotranspiration, and in changes in the plant community along the forest margins.

There is a species-specific threat to *Encephelartos pterogonus* of over-collection. As an attractive ornamental with a highly limited range, the threat of over-collection is high and there is a risk of reproductive failure through decreasing population size (Donaldson 2010b). A reintroduction programme was initiated in 2003 when 1,000 seedlings were planted by the "Plantas de Moçambique" project as a conservation measure,

Euphorbia graniticola (TR)

Aloidendron barberae on the slopes of Mount Muruwere (TR)

Dracena subspicata, a medicinal plant, common in the forest understorey of Mount Muruwere (TR)

with 300–400 seedlings observed to have survived following this reintroduction and the overall population consisting of at least 246 mature individuals when surveyed in the mid-2000s (Capela 2006). There has been no record of how the seedlings and population have progressed since this time. However, this is due to be surveyed in the near future as part of a project led by the University of Kent and supported by the Mohamed Bin Zayed Fund (D. Roberts, pers. comm. 2021).

Key ecosystem services

The medicinal species *Dracaena pedicellata* has been recorded from within in the IPA, and is used to treat a variety of diseases in humans and poultry (Rulkens & Baptista 2009).

Although there are few running streams on Mount Muruwere (T. Rulkens, pers. comm. 2021), there may be some regulation of soil moisture provided by the vegetation on these inselbergs that benefits the agricultural land below, while the forests probably stabilise the soil on the slopes and prevent soil erosion.

Old wall structures on Mount Muruwere may be of archaeological and anthropological interest (T. Rulkens, per. comm. 2021) and, therefore, this site may also be of cultural significance.

Ecosystem service categories

- Provisioning – Medicinal resources
- Regulating services – Erosion prevention and maintenance of soil fertility
- Cultural services – Cultural heritage

IPA assessment rationale

The Mount Muruwere IPA qualifies as an Important Plant Area under sub-criterion A(i) due to the presence of the only confirmed population of the Critically Endangered species *Encephalartos pterogonus*. As this area has not been extensively inventoried for plant species, it is possible that additional A(i) species will be recorded.

Priority species (IPA Criteria A and B)

FAMILY	TAXON	IPA CRITERION A	IPA CRITERION B	≥ 1% OF GLOBAL POP'N	≥ 5% OF NATIONAL POP'N	IS 1 OF 5 BEST SITES NATIONALLY	ENTIRE GLOBAL POP'N	SPECIES OF SOCIO-ECONOMIC IMPORTANCE	ABUNDANCE AT SITE
Euphorbiaceae	*Euphorbia graniticola*		B(ii)	✓	✓			✓	frequent
Zamiaceae	*Encephalartos pterogonus*	A(i)	B(ii)	✓	✓	✓	✓	✓	frequent
		A(i): 1 ✓	B(ii): 2						

Protected areas and other conservation designations

CONSERVATION AREA TYPE	CONSERVATION AREA NAME	RELATIONSHIP OF IPA TO CONSERVATION AREA
No formal protection	N/A	

Threats

THREAT	SEVERITY	TIMING
Small-holder farming	low	Ongoing – trend unknown
Gathering terrestrial plants	medium	Ongoing – trend unknown

SERRA GARUZO

Assessors: Jo Osborne, Iain Darbyshire

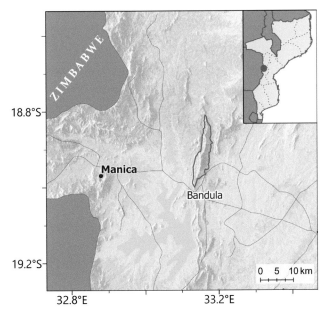

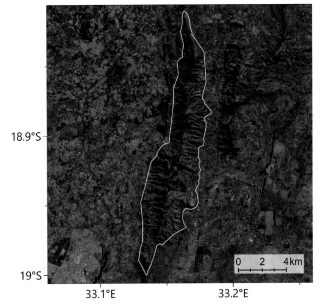

INTERNATIONAL SITE NAME		Serra Garuzo	
LOCAL SITE NAME (IF DIFFERENT)		–	
SITE CODE	MOZTIPA014	PROVINCE	Manica

Serra Garuzo (JO)

LATITUDE	-18.91859	LONGITUDE	33.15576
ALTITUDE MINIMUM (m a.s.l.)	700	ALTITUDE MAXIMUM (m a.s.l.)	1,486
AREA (km²)	51	IPA CRITERIA MET	A(i)

Site description

Serra Garuzo is a mountain ridge in Manica Province, to the east of Manica town and north of Bandula, straddling Manica and Vanduzi Districts. The ridge runs north-south for approximately 20 km, arising from a plain at ca. 700 m elevation and reaching a peak of just under 1,500 m. Numerous stream gullies run-off the ridge to the east and west. The IPA includes the whole length of the mountain ridge, up to 4 km wide and covering an area of approximately 50 km². The site is not formally protected and the slopes on either side of the ridge are managed by different communities.

Botanical significance

The Serra Garuzo IPA lies within the Chimanimani-Nyanga Highlands (sub-)Centre of Plant Endemism (Darbyshire *et al.* 2019a). Significant areas of medium altitude moist forest can be found at this site, though this habitat has become fragmented, particularly on the east side of the ridge. Medium altitude moist forest is a species-rich but restricted

and threatened habitat in Mozambique. At Garuzo, the forest supports the threatened tree species *Tannodia swynnertonii* (VU) at one of only two known sites nationally, and the near-threatened understorey herb *Cyathula divulsa* (NT) at its only known site in Mozambique. Species that are near-endemic to Mozambique, notably the forest tree *Maranthes goetzeniana* (NT; Timberlake *et al.* 2018) and the understorey shrub *Pavetta comostyla* var. *inyangensis*, also occur. The forest is not well-studied botanically and other interesting species are likely to occur here; one taxon of note is an undescribed species of *Dracaena* (formerly *Sansevieria*) (T. Rulkens, pers. comm. 2019). Garuzo is particularly noteworthy for supporting a healthy population of the threatened cycad *Encephalartos manikensis* (VU), which is found in open rocky sites and at forest margins. This cycad is known from only a few sites in Mozambique and Zimbabwe and is threatened by the illegal collection of wild plants for use in horticulture (Donaldson 2010c). Other species of note in rocky areas and associated woodland at Garuzo are *Aeschynomene* sp. B of *Flora Zambesiaca* at its only known site (Verdcourt 2000), *Huernia volkartii* var. *repens* for which Garuzo is one of only four known sites globally, and the Chimanimani-Nyanga Highlands endemic *Plectranthus chimanimanensis* (LC).

Habitat and geology

The north-south montane ridge of Serra Garuzo is a part of the Frontier (Fronteira) Formation of the Precambrian Gairezi Group, comprising mica schists and quartzites; this formation has not been dated precisely (Instituto Nacional de Geológia 1987; Manhica 2012).

The medium altitude moist forest on the mountain slopes reaches a canopy height of 25 –30 m, with large trees including *Strombosia scheffleri*, *Myrianthus holstii*, *Vepris bachmannii* and *Zanthoxylum gilletii*. The forest understorey includes both bare patches and areas of dense shrubs with Acanthaceae and Rubiaceae species often dominating. On the top of the ridge,

Forest margin at Serra Garuzo (JO)

Ridgetop miombo woodland (JO)

montane grassland and rocky areas provide habitat for diverse herbs, shrubs, succulents and grasses; *Loudetia simplex* is locally dominant in the grassland whilst the more open slopes hold patches of the sedge *Coleochloa setifera*. Conspicuous succulents include *Aloe cameronii* and *A. chabaudii*. Patches of low miombo (*Brachystegia*) woodland are also found here and extend down the western slopes of the ridge. There is a greater extent of woody vegetation remaining on the western slopes than on the eastern slopes, including both forest and woodland. Stream gullies in the mountain slopes provide further habitat diversity at this site, along gradients of slope and soil moisture.

Conservation issues

The Serra Garuzo IPA is not currently protected or included within any conservation schemes. The following information on threats was gathered during fieldwork on the east side of the ridge in 2018 (Osborne & Matimele 2018). Here, much of the forest has been cleared for subsistence agriculture including the cultivation of maize and yams (*Colocasia esculenta*). Remaining forest patches are healthy with only low levels of disturbance from selective timber cutting for local use. Some previously cultivated areas on the slopes have been recently abandoned following the prohibition of cultivation there by the government administrator (Chefe do Posto) in Vanduzi. Several dense patches of the invasive shrub *Vernonanthura polyanthes* were recorded on a rocky slope at ca. 1,350 m elevation. This shrub was originally introduced into Mozambique from South America as a nectar source for bees. It is a potential threat to the montane grassland and shrubland vegetation as it can form dense stands on disturbed and open ground.

Key ecosystem services

Serra Garuzo has a high plant biodiversity value and provides an island of natural habitat for flora and fauna within the surrounding agricultural plain. The mountain ridge provides a watershed for the local area and the vegetation contributes to carbon sequestration and storage. The remaining forest has spiritual and cultural value for local communities and provides timber for local use. In addition, the site supports populations of at least two socio-economically important plants, *Encephalartos manikensis*, a cycad valued in the horticultural trade, and *Coffea mufindiensis* subsp. *australis*, a wild relative of coffee.

Ecosystem service categories

- Provisioning – Raw materials
- Provisioning – Fresh water
- Regulating services – Carbon sequestration and storage
- Habitat or supporting services – Habitats for species
- Habitat or supporting services – Maintenance of genetic diversity
- Cultural services – Spiritual experience and sense of place
- Cultural services – Cultural heritage

Rock outcrops on Garuzo with *Aloe chabaudii* (JO)

IPA assessment rationale

Serra Garuzo qualifies as an IPA under criterion A(i), supporting two globally threatened species: *Tannodia swynnertonii* (VU) and *Encephalartos manikensis* (VU). There is a large and healthy population of the latter species for which Garuzo is the only site within Mozambique's IPA network. The site also supports significant areas of Medium Altitude Moist Forest, a restricted and nationally threatened habitat, but is not considered to be among the best five sites nationally for that habitat and so Serra Garuzo does not qualify under criterion C(iii).

Encephalartos manikensis on Garuzo (JO)

Priority species (IPA Criteria A and B)

FAMILY	TAXON	IPA CRITERION A	IPA CRITERION B	≥ 1% OF GLOBAL POP'N	≥ 5% OF NATIONAL POP'N	IS 1 OF 5 BEST SITES NATIONALLY	ENTIRE GLOBAL POP'N	SPECIES OF SOCIO-ECONOMIC IMPORTANCE	ABUNDANCE AT SITE
Euphorbiaceae	*Tannodia swynnertonii*	A(i)			✓	✓			unknown
Rubiaceae	*Pavetta comostyla* var. *inyangensis*		B(ii)						occasional
Zamiaceae	*Encephalartos manikensis*	A(i)		✓	✓	✓		✓	frequent
		A(i): 2 ✓	B(ii): 1						

Threatened habitats (IPA Criterion C)

HABITAT TYPE	IPA CRITERION C	≥ 5% OF NATIONAL RESOURCE	≥ 10% OF NATIONAL RESOURCE	IS 1 OF 5 BEST SITES NATIONALLY	ESTIMATED AREA AT SITE (IF KNOWN)
Medium Altitude Moist Forest [MOZ-02]	C(iii)				

Protected areas and other conservation designations

CONSERVATION AREA TYPE	CONSERVATION AREA NAME	RELATIONSHIP OF IPA TO CONSERVATION AREA
No formal protection	N/A	

Threats

THREAT	SEVERITY	TIMING
Small-holder farming	high	Ongoing – trend unknown
Logging & wood harvesting	medium	Ongoing – trend unknown
Invasive non-native/alien species	medium	Ongoing – increasing

TSETSERRA

Assessors: Jo Osborne, Iain Darbyshire

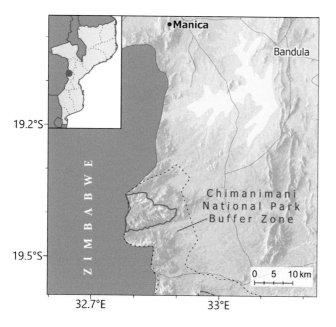
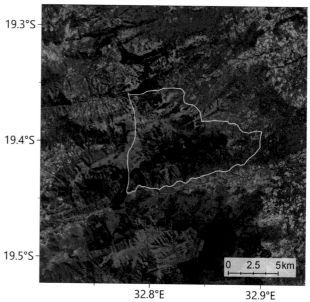

INTERNATIONAL SITE NAME		Tsetserra	
LOCAL SITE NAME (IF DIFFERENT)		–	
SITE CODE	MOZTIPA007	PROVINCE	Manica

LATITUDE	-19.39326	LONGITUDE	32.79879
ALTITUDE MINIMUM (m a.s.l.)	890	ALTITUDE MAXIMUM(m a.s.l.)	2,274
AREA (km²)	77	IPA CRITERIA MET	A(i), A(iii), A(iv), B(ii), C(iii)

Montane shrubland on Tsetserra Plateau, with old farm buildings and pine plantations in the background (JO)

Site description

Tsetserra (or Tsetsera) is a montane plateau in the Manica Highlands to the north of the Chimanimani Mountains, in Sussendenga District of Manica Province. It is situated ca. 70 km to the WSW of Chimoio town. It forms part of a cross-border plateau extending into Mozambique from Zimbabwe's Manicaland Province, where is it known as Himalaya. In Mozambique, the plateau reaches an elevation of over 2,200 m and has a history of use during the colonial period. The ruins of several buildings remain along with remnants of pine plantations. There is access via a single road that runs up onto the plateau from the east. The site includes both the montane plateau and the steep, forested slopes below. It lies within the buffer zone of the Chimanimani National Park, also known as the Chimanimani Trans-Frontier Conservation Area (TFCA).

Botanical significance

Tsetserra is a highly important site for plant diversity and endemism in Mozambique, being an important component of the Chimanimani-Nyanga (sub-) Centre of Plant Endemism [CoE] (Darbyshire *et al.* 2019a). It supports nationally significant areas of montane grassland and scrubland on the plateau and montane moist forest on the slopes below,

which are highly restricted habitats in Mozambique. Four plant species are only known from this site: the rhizomatous perennial grass *Digitaria fuscopilosa* (DD), the herbs *Phyllanthus manicaensis* (VU) and *Phyllanthus tsetserrae* (CR), and the subshrub *Pterocephalus centennii* (CR). Two further taxa, *Euphorbia depauperata* var. *tsetserrensis* and *Geranium exellii* (EN), are endemic to the Tsetserra-Himalaya plateau. Further globally Endangered species that occur here include the herb *Dierama inyangense*, for which this is the only known site in Mozambique, and the shrub *Myrica* (formerly *Morella*) *chimanimaniana* at its only known site away from the Chimanimani Mountains (Osborne & Matimele 2018). Overall, the site supports important populations of 15 species that are globally threatened and many range-restricted species of the Chimanimani-Nyanga CoE. A majority of these taxa occur in the plateau grasslands, scrublands and upper forest margins. However, a number of range-restricted montane forest species are also noteworthy, including the woody Rubiaceous taxa *Pavetta comostyla* var. *inyangensis*, *Pavetta umtalensis* (LC) and *Tricalysia ignota*, as well as the recently described forest climber *Vincetoxicum monticolum*, which is likely to be globally threatened (Darbyshire *et al.* 2019a; Goyder *et al.* 2020).

The plant diversity of Tsetserra has only been partially explored to date and further rare and threatened species may be found by future expeditions. One currently undescribed species is noted from this site: *Sericanthe* sp. A (Nyanga taxon) of *Flora Zambesiaca* (Bridson 1998) which occurs on rock outcrops on the edges of forest.

Habitat and geology

The high-altitude plateau at Tsetserra is underlain by red sandy clay soils derived from schist bedrock. Surficial geology is Precambrian in age. The plateau is dominated by montane grassland and shrubland vegetation with occasional rocky outcrops and poorly drained areas that increase the plant diversity. The grasslands support a varied flora with many herbs and geophytes. Frequent shrubby species include *Helichrysum* spp., with *Hypericum revolutum* also plentiful and *Erica hexandra* occasional.

On the slopes below the plateau there are large areas of intact evergreen montane moist forest with stream gullies and rocky areas on the slopes providing habitat diversity. The forest above 1,600 m is of the Central Montane Forest vegetation unit of Lötter *et al.* (in prep.). The forest composition at Tsetserra has not been fully inventoried to date. A forest plot surveyed at 1,794 m elevation (J. Osborne *et al.*, unpubl. data 2018) recorded *Macaranga mellifera* and *Vepris bachmannii* as the dominant species, with trees of *Kiggelaria africana*, *Tabernaemontana stapfiana* and *Erythroxylum emarginatum* also recorded. Other tree species that are noted to be

of importance during recent surveys and/or by past botanical collectors include *Aphloia theiformis*, *Myrsine* (formerly *Rapanea*) *melanophloeos*, *Pittosporum viridiflorum*, *Podocarpus milanjianus*, *Rauvolfia caffra* and *Syzygium afromontanum*. Common components of the understorey include *Peddiea africana* and *Psychotria zombamontana*, whilst *Dracaena* sp., *Halleria lucida*, *Nuxia congesta* and *Polyscias fulva* are amongst the species of forest margins and clearings. The ground layer is dominated by pteridophytes, with *Selaginella kraussiana* often abundant. Stream gullies support populations of *Ensete ventricosum*, an important food plant in Ethiopia. *Strelitzia caudata* is noted from rocky slopes (J. Osborne *et al.*, pers. obs.).

Conservation issues

Tsetserra falls within the extensive buffer zone of the Chimanimani National Park and TFCA, and both the core and buffer zones of this protected area have recently been designated as the large Parque Nacional de Chimanimani Key Biodiversity Area. The TFCA buffer zone is not considered to be well protected or managed for biodiversity at present, and Tsetserra faces a number of ongoing and potential future threats. The vegetation on the plateau is highly disturbed in places. Invasive *Pinus patula*, a Mexican pine species planted commercially for timber from the 1950s, is regenerating across large areas. Some previous efforts have been made to clear areas of pine plantation here, and Ghiurghi *et al.* (2010) noted the positive recolonisation by *Chironia gratissima*, a range-restricted herb, in areas where pine had been cleared, but this clearance does not appear to be ongoing. Around

Forested slopes at Tsetserra (JO)

The Tsetserra Plateau (JO)

Jamesbrittenia carvalhoi (JO)

Gladiolus zimbabweensis (BW)

the derelict buildings there are *Eucalyptus* trees and several non-native ornamental species including *Fuchsia* and *Hydrangea*. Cattle and goat grazing are heavily impacting some areas and the invasive European weed species *Hypochaeris radicata* is abundant. Previous fire events were evident during field surveys on the plateau in 2018 (Osborne & Matimele 2018) and it is possible that increased fire frequency may also be impacting the grassland and scrubland vegetation, although data on fire frequency and management are not available at present. Despite these high levels of disturbance, good examples of montane grassland and shrubland habitat remain.

In a management plan for the Chimanimani National Reserve, Ghiurghi *et al.* (2010) note that the sandy-clay soil and the isolation of the Tsetserra Plateau grasslands provide a unique potential for disease-free seed potato production, and that plans were being developed by the Ministry for Agriculture and Rural Development to use ca. 50 ha of the current grassland. They added that if this were to take place, agriculture on the slopes would need to be forbidden to maintain the isolation, thereby having a positive impact on the conservation of vegetation on the slopes. While no evidence of cultivation was observed during fieldwork on Tsetserra in 2018 (J. Osborne *et al.*, pers. obs.) and the proposed area was small, seed potato

production remains a potential future threat to the grassland habitat on the plateau. Commercial farming was previously established on Tsetserra prior to Mozambican independence (Timberlake *et al.* 2016a), and Ghiurghi *et al.* (2010) describe the site as having been heavily transformed in the past by human intervention.

Along the roadside on the lower slopes to 1,500 m elevation there are scattered individuals of the invasive shrub *Vernonanthura polyanthes*, a plant from South America originally introduced into Mozambique as nectar source for bees. This shrub is a potential threat to the montane grassland and shrubland vegetation as it can form dense stands on disturbed ground (Timberlake *et al.* 2016b).

On the slopes below the plateau, the montane forests are extensive and in good condition, with only low levels of disturbance. Local people with packs of donkeys follow tracks through the forest to cross the Zimbabwe border for trading and there is some hunting of wildlife within the forest (Osborne & Matimele 2018). Recently, plans have been mooted to cultivate coffee as a shade crop on the forested slopes of Tsetserra as part of a habitat restoration plan for the Chimanimani TFCA under the draft "Plano de Restauração paisagem de Chimanimani" (C. de Sousa, pers. comm. 2021). Such a scheme would need to be carefully managed and focused on degraded forest areas in order to prevent damage to the intact forest ecosystem. Ghiurghi *et al.* (2010) report on some issues with increased wildfire frequency impacting the forest margins at this site, and they also noted some issues with the clearance of forest from some slopes for agriculture. They recommended that land use agreements with the Tsetserra communities be treated as a priority for management of this site, with the ultimate aim to create a "Tsetserra Community Reserve" that includes both the montane forests and high plateau.

Key ecosystem services

Tsetserra provides an essential ecosystem service to the local area by protecting a part of the watershed supplying water to the valleys below the plateau. The intact natural habitats, in particular the forested slopes, protect the soils from erosion. In addition, the vegetation contributes to carbon sequestration and storage and provides habitat for montane flora and fauna. The site has a high potential for tourism, as it has the highest road access of any point in Mozambique, providing ready access to a wide range of habitats and associated wildlife, as well as the scenic appeal and hiking potential (Ghiurghi *et al.* 2010).

Ecosystem service categories

- Provisioning – Fresh water
- Regulating services – Carbon sequestration and storage
- Regulating services – Erosion prevention and maintenance of soil fertility
- Habitat or supporting services – Habitats for species
- Cultural services – Tourism

IPA assessment rationale

Tsetserra qualifies as an Important Plant Area under all three criteria. Under criterion A(i), the site supports populations of fourteen globally threatened plant species that are inferred to meet the population threshold; the globally threatened *Prunus africana* is also recorded here but it is not clear whether this species meets any of the criterion A(i) thresholds at this site. In addition, four potentially threatened endemics occur here, one being highly restricted (having range of <100 km²) and three range-restricted (range >100 km² but <5,000 km²), thus qualifying the site under sub-criteria A(iii) and A(iv), respectively. Tsetserra is a botanically rich site supporting an exceptional number of species of high conservation importance, qualifying under criterion B(ii). The site supports 36 plant taxa of high conservation importance, including four nationally endemic species and 32 regional endemics with a restricted range of less than 10,000 km². Under criterion C(iii) the site includes significant areas of moist montane forest and montane grassland, two of Mozambique's national priority habitats recognised during the first Mozambique TIPAs workshop in Maputo in January 2018.

Roadside rocky habitats on the slopes of Tsetserra (JO)

Priority species (IPA Criteria A and B)

FAMILY	TAXON	IPA CRITERION A	IPA CRITERION B	≥ 1% OF GLOBAL POP'N	≥ 5% OF NATIONAL POP'N	IS 1 OF 5 BEST SITES NATIONALLY	ENTIRE GLOBAL POP'N	SPECIES OF SOCIO-ECONOMIC IMPORTANCE	ABUNDANCE AT SITE
Apocynaceae	*Asclepias cucullata* subsp. *scabrifolia*		B(ii)						unknown
Apocynaceae	*Vincetoxicum monticola*		B(ii)						unknown
Asphodelaceae	*Aloe inyangensis* var. *kimberleyana*		B(ii)						unknown
Asteraceae	*Cineraria pulchra*		B(ii)						unknown
Asteraceae	*Helichrysum acervatum*	A(iv)	B(ii)						unknown
Asteraceae	*Helichrysum chasei*	A(iv)	B(ii)						unknown
Asteraceae	*Lopholaena brickellioides*	A(iv)	B(ii)						unknown
Asteraceae	*Schistostephium oxylobum*	A(i)	B(ii)	✓	✓	✓			unknown
Caprifoliaceae	*Pterocephalus centennii*	A(i)	B(ii)	✓	✓	✓	✓		unknown
Euphorbiaceae	*Euphorbia citrina*		B(ii)						unknown
Euphorbiaceae	*Euphorbia depauperata* var. *tsetserrensis*	A(iii)	B(ii)	✓	✓	✓	✓		unknown
Fabaceae	*Crotalaria insignis*	A(i)	B(ii)	✓	✓	✓			unknown
Fabaceae	*Indigofera vicioides* subsp. *excelsa*		B(ii)						unknown
Fabaceae	*Tephrosia praecana*	A(i)	B(ii)	✓	✓	✓			unknown
Geraniaceae	*Geranium exellii*	A(i)	B(ii)	✓	✓	✓			unknown
Geraniaceae	*Pelargonium mossambicense*		B(ii)						unknown
Gesneriaceae	*Streptocarpus michelmorei*		B(ii)						unknown
Gesneriaceae	*Streptocarpus umtaliensis*		B(ii)						unknown
Iridaceae	*Dierama inyangense*	A(i)	B(ii)	✓	✓	✓			unknown
Iridaceae	*Gladiolus zimbabweensis*	A(i)	B(ii)	✓	✓	✓			unknown
Lamiaceae	*Coleus sessilifolius*		B(ii)						unknown
Loranthaceae	*Englerina oedostemon*		B(ii)						unknown
Myricaceae	*Myrica chimanimaniana*	A(i)	B(ii)	✓	✓	✓			unknown
Orchidaceae	*Disa zimbabweensis*	A(i)	B(ii)	✓	✓	✓			unknown
Orchidaceae	*Schizochilus lepidus*	A(i)	B(ii)	✓	✓	✓			unknown
Phyllanthaceae	*Phyllanthus manicaensis*	A(i)	B(ii)	✓	✓	✓	✓		unknown
Phyllanthaceae	*Phyllanthus tsetserrae*	A(i)	B(ii)	✓	✓	✓	✓		unknown
Poaceae	*Digitaria fuscopilosa*		B(ii)				✓		unknown
Polygalaceae	*Polygala zambesiaca*	A(i)	B(ii)	✓	✓	✓			unknown
Proteaceae	*Faurea rubriflora*		B(ii)						unknown

Priority species (IPA Criteria A and B)

FAMILY	TAXON	IPA CRITERION A	IPA CRITERION B	≥ 1% OF GLOBAL POP'N	≥ 5% OF NATIONAL POP'N	IS 1 OF 5 BEST SITES NATIONALLY	ENTIRE GLOBAL POP'N	SPECIES OF SOCIO-ECONOMIC IMPORTANCE	ABUNDANCE AT SITE
Rosaceae	*Prunus africana*	A(i)						✓	unknown
Rubiaceae	*Anthospermum zimbabwense*		B(ii)						unknown
Rubiaceae	*Otiophora inyangana* subsp. *inyangana*		B(ii)						unknown
Rubiaceae	*Pavetta comostyla* var. *inyangensis*		B(ii)						unknown
Rubiaceae	*Pavetta umtalensis*		B(ii)						unknown
Rubiaceae	*Tricalysia ignota*		B(ii)						unknown
Sapindaceae	*Allophylus chirindensis*	A(i)		✓	✓				occasional
Scrophulariaceae	*Jamesbrittenia carvalhoi*		B(ii)						occasional
		A(i): 14 ✓ A(iii): 1 ✓ A(iv): 3 ✓	B(ii): 36 ✓						

Threatened habitats (IPA Criterion C)

HABITAT TYPE	IPA CRITERION C	≥ 5% OF NATIONAL RESOURCE	≥ 10% OF NATIONAL RESOURCE	IS 1 OF 5 BEST SITES NATIONALLY	ESTIMATED AREA AT SITE (IF KNOWN)
Montane Moist Forest [MOZ-01]	C(iii)		✓	✓	11.5
Medium Altitude Moist Forest [MOZ-02]	C(iii)				
Montane Grassland [MOZ-09]	C(iii)			✓	

Protected areas and other conservation designations

CONSERVATION AREA TYPE	CONSERVATION AREA NAME	RELATIONSHIP OF IPA TO CONSERVATION AREA
National Park (buffer zone)	Chimanimani	protected/conservation area encompasses IPA
Key Biodiversity Area	Parque Nacional de Chimanimani	protected/conservation area encompasses IPA

Threats

THREAT	SEVERITY	TIMING
Agro-industry farming	unknown	Future – inferred threat
Agro-industry plantations	low	Past, not likely to return
Small-holder grazing, ranching or farming	medium	Ongoing – stable
Increase in fire frequency/intensity	medium	Ongoing – trend unknown
Invasive non-native/alien species	high	Ongoing – trend unknown

MOUNT ZEMBE

Assessors: Jo Osborne, Iain Darbyshire

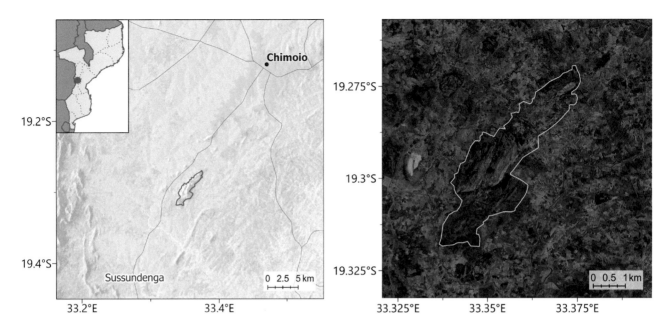

INTERNATIONAL SITE NAME		Mount Zembe	
LOCAL SITE NAME (IF DIFFERENT)		Monte Zembe	
SITE CODE	MOZTIPA011	PROVINCE	Manica

LATITUDE	-19.29845	LONGITUDE	33.35312
ALTITUDE MINIMUM (m a.s.l.)	575	ALTITUDE MAXIMUM (m a.s.l.)	1,203
AREA (km²)	7.6	IPA CRITERIA MET	A(i)

Site description

Mount Zembe is a granite inselberg in Macate District of Manica Province, 22 km south-west of Chimoio. It reaches 1,200 m in elevation, rising from the surrounding plains at ca. 600 m elevation. The site is approximately 6 km long by 2 km wide and comprises a series of granite rocks running north-east to south-west. This site is of importance for its interesting xerophytic flora on the exposed rock outcrops, as well as for the pockets of moist forest in gulleys.

Botanical significance

Mount Zembe is significant as it is the only known site for two plant species, the cycad *Encephalartos munchii* and the rosette-forming, low-growing aloe, *Aloe decurva*. These two endemic plants are assessed as Critically Endangered on the IUCN Red List (Donaldson 2010d; Osborne *et al.* 2019g). Both are largely confined to the summit of Mount Zembe, the cycad growing in bushland by streams and amongst rocks and boulders, the aloe being found only on exposed steep rocky slopes. Other interesting succulent species of note include the Mozambican endemic shrub or small tree euphorbia, *Euphorbia graniticola* (LC), and the scarce near-endemic stapeliad, *Huernia leachii* (LC). The site also supports a population of the Endangered wild coffee species *Coffea salvatrix*, or "mukofi" (O'Sullivan & Davis 2017), which occurs in the small patches of moist forest. These forest patches also hold a population of the Mozambique near-endemic *Dracaena* (formerly *Sansevieria*) *pedicellata* (LC) in the ground layer. A more complete botanical inventory of this site may reveal further species of conservation concern.

Encephalartos munchii (OB)

Aloe decurva (TR)

Habitat and geology

Mount Zembe is a granite inselberg that provides a range of different habitats according to slope, aspect, soil depth and moisture availability. Rock crevices and shallow soils over granite rock form the dominant habitat, supporting a range of herbs, including the tussock-forming sedge *Coleochloa setifera*, geophytes such as *Drimia intricata* and *Ledebouria* spp., and succulents including *Euphorbia* spp. and *Huernia leachii*. Open grassland covers flatter areas where deeper soils have formed. Woody vegetation, including small pockets of moist forest, is found where sufficient moisture is available in deeper rock crevices and stream gullies. The species composition of the different habitats on Mount Zembe has not been fully documented to date, and this should be considered a priority as a baseline for future monitoring.

Conservation issues

Mount Zembe is not currently protected and is not included within any other conservation prioritization schemes, except that it is listed as an Alliance for Zero Extinction site based on the presence of *Encephalartos munchii* (AZE 2018). An increase in fire frequency on Mount Zembe presents a serious threat to the vegetation, particularly damaging immature plants. There has been some quarrying of rock for construction materials at the foot of Mount Zembe, and whilst this is not considered to be a major threat at present, it may expand in the future and threaten this site (Osborne *et al.* 2019g). There is also a potential threat of over-harvesting by plant collectors for private collections and for the horticultural trade, particularly in the case of the cycad *Encephalartos munchii* and the aloe *Aloe decurva*, both of which are striking plants with the added appeal of their rarity. Other attractive succulents such as *Euphorbia graniticola* and *Huernia leachii* may also be targeted.

A reintroduction programme for *E. munchii* was initiated in 2003 when 1,000–1,300 seedlings were established by the Plantas de Moçambique project as a conservation measure (Capela 2006). There has been no record of how the seedlings and population have progressed since this time. However, this is due to be surveyed in the near future as part of a project led by the University of Kent and supported by the Mohamed Bin Zayed Fund (D. Roberts, pers. comm.).

Key ecosystem services

Mount Zembe has a high plant biodiversity value and provides an island of natural habitat for flora and fauna within an agricultural plain. A wild relative of coffee, *Coffea salvatrix*, occurs here and the site therefore contributes to maintenance of crop genetic diversity. The inselberg provides a watershed for the local area and the vegetation contributes to carbon sequestration and storage. In addition, Mount Zembe has spiritual significance and cultural value to local people.

Ecosystem service categories

- Provisioning – Fresh water
- Regulating services – Carbon sequestration and storage
- Habitat or supporting services – Habitats for species
- Habitat or supporting services – Maintenance of genetic diversity
- Cultural services – Spiritual experience and sense of place
- Cultural services – Cultural heritage

View of Mount Zembe (TR)

IPA assessment rationale

Mount Zembe qualifies as an Important Plant Area under criterion A(i), supporting populations of three globally threatened plant species: *Encephalartos munchii* (CR), *Aloe decurva* (CR) and *Coffea salvatrix* (EN). The only known populations of *Encephalartos munchii* and *Aloe decurva* occur here.

Lithophytic flora on the slopes of Mount Zembe (TR)

Priority species (IPA Criteria A and B)

FAMILY	TAXON	IPA CRITERION A	IPA CRITERION B	≥ 1% OF GLOBAL POP'N	≥ 5% OF NATIONAL POP'N	IS 1 OF 5 BEST SITES NATIONALLY	ENTIRE GLOBAL POP'N	SPECIES OF SOCIO-ECONOMIC IMPORTANCE	ABUNDANCE AT SITE
Asphodelaceae	*Aloe decurva*	A(i)	B(ii)	✓	✓	✓	✓	✓	scarce
Euphorbiaceae	*Euphorbia graniticola*		B(ii)						unknown
Rubiaceae	*Coffea salvatrix*	A(i)		✓	✓	✓		✓	unknown
Zamiaceae	*Encephalartos munchii*	A(i)	B(ii)	✓	✓	✓	✓	✓	frequent
		A(i): 3 ✓	B(ii): 3						

Protected areas and other conservation designations

CONSERVATION AREA TYPE	CONSERVATION AREA NAME	RELATIONSHIP OF IPA TO CONSERVATION AREA
No formal protection	N/A	
Alliance for Zero Extinction Site	Mount Zembe	protected/conservation area matches IPA

Threats

THREAT	SEVERITY	TIMING
Gathering terrestrial plants	unknown	Future – inferred threat
Mining & quarrying	low	Future – inferred threat
Increase in fire frequency/intensity	high	Ongoing – increasing

SERRA MOCUTA

Assessors: Jo Osborne, Iain Darbyshire

 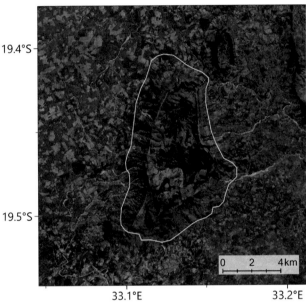

INTERNATIONAL SITE NAME		Serra Mocuta	
LOCAL SITE NAME (IF DIFFERENT)		–	
SITE CODE	MOZTIPA015	PROVINCE	Manica

LATITUDE	-19.46191	LONGITUDE	33.13061
ALTITUDE MINIMUM (m a.s.l.)	700	ALTITUDE MAXIMUM (m a.s.l.)	1,573
AREA (km²)	62	IPA CRITERIA MET	A(i), C(iii)

Site description

Serra Mocuta is a mountain in Sussundenga District of Manica Province, situated ca. 20 km north of the Chimanimani Mountains and 15 km west of Sussundenga town. It lies at the southernmost end of a series of mountain ridges extending north to Garuzo Forest [MOZTIPA014] and is part of the wider Chimanimani-Nyanga (sub-)Centre of Plant Endemism (Darbyshire *et al.* 2019a) that straddles the border between Mozambique and Zimbabwe. The mountain forms an oval crater ca. 7 km long by 4 km wide, reaching an elevation of 1,573 m on the southwest edge. The IPA includes the whole of the crater and parts of the outer slopes down to ca.

Varied habitats on Serra Mocuta (OB)

700 m elevation, covering an area of approximately 62 km². The site is not formally protected at present but supports a range of intact habitats.

Botanical significance

Significant areas of medium altitude moist forest, montane grassland and montane scrubland vegetation can be found at Serra Mocuta; these habitats are restricted and threatened in Mozambique and support a number of rare plant taxa. This site is likely to be of global importance for the globally threatened *Raphionacme pulchella*, which is assessed as Endangered (Osborne *et al.* 2019h). Quartzite rock habitat at Serra Mocuta supports the threatened perennial herb *Gutenbergia westii* (VU) in one of its few known localities away from the Chimanimani Mountains, and the shrub *Tephrosia chimanimaniana* (LC) which is otherwise known only from the high Chimanimani. A highly range-restricted endemic aloe, *Aloe cannellii* (LC), also occurs on the steep quartzite cliffs here. Although not currently considered to be threatened, *A. cannellii* is known from only two sites globally and the continued integrity of its favoured habitat at this IPA site

is critical for this species' survival (Osborne *et al.* 2019i). Other Chimanimani quartzite near-endemics recorded here are *Asclepias cucullata* subsp. *scabrifolia* and *Wahlenbergia subaphylla* subsp. *scoparia*. The range-restricted epiphytic or lithophytic herb *Streptocarpus michelmorei*, provisionally assessed as Near Threatened (I. Darbyshire, unpubl. data), is recorded here, whilst the areas of moist forest support two currently undescribed and highly range-restricted species: *Diospyros* sp. 2 of *Flora Zambesiaca* and *Rytigynia* sp. E of *Flora Zambesiaca*, the latter of which is known only from a single collection (*T. Müller & T. Gordon* #1785) from this site (White 1983b; Bridson 1998). Serra Mocuta is not well-studied botanically and other interesting plant species are likely to occur here, potentially including other species that are otherwise restricted to the Chimanimani quartzites. Botanical collections from this site date mainly from the 1960s and 1970s and few recent expeditions are known to have occurred. There is therefore an urgent need to conduct a thorough botanical survey of the site in order to confirm the continued presence of the species of high conservation interest and to fully inventory the flora.

Habitat and geology

The landscape at Serra Mocuta consists predominantly of rugged quartzite outcrops and undulating land on a red soil, probably derived from schist (Google Earth 2021), as found in the Chimanimani Mountains (Timberlake *et al.* 2016b). The Chimanimani quartzites are derived from sandstones of the Frontier Series of the Umkondo Group, dating to the Proterozoic eon ca. 1,875 mya (Timberlake *et al.* 2016b) of which Serra Mocuta can be considered an outlier. Quartzite outcrops occur around the edge of the crater and throughout the more rugged, southern part of the site. On shallow soils on the quartzite, open woodland, montane scrub vegetation and montane grassland occur. Based on notes from botanical collections made in the 1960s and 1970s (e.g., *A. Marques & A. Pereira* #995, 1006; *A. Sarmento et al.* #1219; *G.V. Pope & W.M. Biegel* #3521), the woodlands are dominated by *Brachystegia* spp., including *B. tamarindoides* subsp. *microphylla*, and/or *Uapaca sansibarica*, with other tree species of note including *Cussonia* sp. and *Parinari curatellifolia*.

Within the crater there is an undulating landscape of moist forest, dense scrub, and grassland, with grassland occurring mostly on raised areas and moist forest occurring in depressions and along drainage gullies. The species composition of these habitats has not been documented to our knowledge, but they are likely to be similar to those of the Chimanimani Mountains [MOZTIPA003 & 006] at comparable elevations. The outer slopes of the crater are covered mostly by woodland vegetation though some moist forest occurs in sheltered gullies.

Conservation issues

Little information is available concerning the current conservation issues faced at Serra Mocuta. Some grazing has been recorded (Osborne *et al.* 2019h) and previous fires in the grasslands are visible on satellite imagery (Google Earth 2021). On the outer slopes towards the base of the crater, there has been some considerable clearance for agriculture, and fires have spread into the woodland from the agricultural land below. However, vegetation on much of the upper slopes and crater appears to be largely intact. An emerging threat is the continuing spread of the invasive shrub *Vernonanthura polyanthes* in Sussendenga District (J. Massunde, pers. comm. 2021); in the nearby Chimanimani foothills this species has invaded large swathes of land following burning, outcompeting the natural vegetation.

Key ecosystem services

The mountain crater at Serra Mocuta has a high plant diversity value and provides important areas of intact natural habitat for flora and fauna within an agricultural plain. The mountain provides a watershed for the local area and the vegetation contributes to carbon sequestration and storage.

Ecosystem service categories

- Provisioning – Fresh water
- Regulating services – Carbon sequestration and storage
- Habitat or supporting services – Habitats for species

Aloe cannellii on Serra Mocuta (OB)

Quartzite outcrops on Serra Mocuta (OB)

IPA assessment rationale

Serra Mocuta qualifies as an Important Plant Area under criterion A(i), supporting populations of two globally threatened species, *Raphionacme pulchella* (EN) and *Gutenbergia westii* (VU). The site also qualifies under criterion C(iii), having significant areas of medium altitude moist forest, which is a nationally restricted and threatened habitat. It also contains areas of montane grassland, a further criterion C(iii) habitat, but is not considered to be among the best five sites nationally for that habitat.

Gutenbergia westii (MC)

Priority species (IPA Criteria A and B)

FAMILY	TAXON	IPA CRITERION A	IPA CRITERION B	≥ 1% OF GLOBAL POP'N	≥ 5% OF NATIONAL POP'N	IS 1 OF 5 BEST SITES NATIONALLY	ENTIRE GLOBAL POP'N	SPECIES OF SOCIO-ECONOMIC IMPORTANCE	ABUNDANCE AT SITE
Apocynaceae	*Raphionacme pulchella*	A(i)		✓	✓	✓			unknown
Asphodelaceae	*Aloe cannellii*		B(ii)						unknown
Asteraceae	*Gutenbergia westii*	A(i)	B(ii)	✓	✓	✓			unknown
Fabaceae	*Tephrosia chimanimaniana*		B(ii)						unknown
Gesneriaceae	*Streptocarpus michelmorei*		B(ii)						unknown
		A(i): 2 ✓	B(ii): 4						

Threatened habitats (IPA Criterion C)

HABITAT TYPE	IPA CRITERION C	≥ 5% OF NATIONAL RESOURCE	≥ 10% OF NATIONAL RESOURCE	IS 1 OF 5 BEST SITES NATIONALLY	ESTIMATED AREA AT SITE (IF KNOWN)
Medium Altitude Moist Forest [MOZ-02]	C(iii)			✓	
Montane Grassland [MOZ-09]	C(iii)				

Protected areas and other conservation designations

CONSERVATION AREA TYPE	CONSERVATION AREA NAME	RELATIONSHIP OF IPA TO CONSERVATION AREA
No formal protection	N/A	

Threats

THREAT	SEVERITY	TIMING
Small-holder grazing, ranching or farming	unknown	Ongoing – trend unknown
Small-holder farming	low	Ongoing – trend unknown
Fire & fire suppression – Trend Unknown/Unrecorded	unknown	Ongoing – trend unknown
Invasive non-native/alien species	unknown	Ongoing – trend unknown

CHIMANIMANI LOWLANDS

Assessors: Iain Darbyshire, Jo Osborne, Camila de Sousa

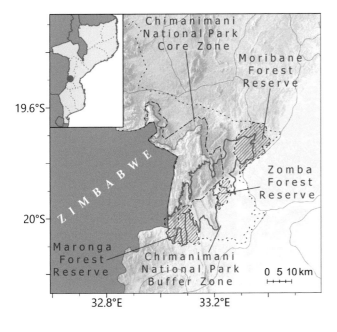
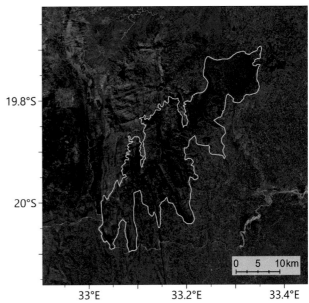

Forested slopes of the Chimanimani foothills, Maronga (ID)

Riverine forest with quartzite outcrops at Maronga (ID)

INTERNATIONAL SITE NAME		Chimanimani Lowlands	
LOCAL SITE NAME (IF DIFFERENT)		–	
SITE CODE	MOZTIPA003	PROVINCE	Manica

LATITUDE	-19.90778	LONGITUDE	33.16472
ALTITUDE MINIMUM (m a.s.l.)	147	ALTITUDE MAXIMUM (m a.s.l.)	1200
AREA (km²)	514	IPA CRITERIA MET	A(i), A(iii), A(iv), B(ii), C(iii)

Site description

This IPA encompasses the lowlands and foothills associated with the southern and eastern flanks of the Chimanimani Mountains in Sussundenga District of Manica Province, Mozambique. The site starts in the west on the Zimbabwe border at Makurupini Falls and the Lucite River, then extends northeast through portions of the Maronga, Zomba and Moribane Forest Reserves and also into the lower-elevation southern portion of the core Chimanimani National Park (CNP). Much of this area would originally have been covered in lowland moist forest, interspersed with miombo woodland and with natural wetlands and rock outcrops. There has been significant transformation and fragmentation of natural habitats outside of the core CNP but significant remnants of the key habitats are still intact. It is immediately abutted to the north and west by the Chimanimani Highlands IPA with the 1,200 m elevation contour forming the boundary; they are separated here to draw attention to their differing floras and management issues (Rokni *et al.* 2019), but could be treated together as the Chimanimani IPA.

Botanical significance

The Chimanimani Lowlands are of botanical importance primarily for the extensive areas of lowland moist evergreen and semi-deciduous forest, the largest extent of this highly threatened forest type in Mozambique (Timberlake *et al.* 2016b; Rokni *et al.* 2019). Although low in species richness, these forests contain a significant number of both regionally and globally rare plant species, several of which are restricted to these forests together with the contiguous Haroni-Rusitu forests of Zimbabwe. These include the spectacular herb *Streptocarpus acicularis* (CR), which is only known from the type locality within this IPA, the understorey shrub *Vepris drummondii* (VU) and the recently described understorey tree *Synsepalum chimanimani* (EN) (Rokni *et al.* 2019). They also support a number of outlier and edge-of-range populations. For example, the fig tree *Ficus mucuso* (LC) and the forest climber *Raphidiocystis chrysocoma* are both West African species that are known in the *Flora Zambesiaca* region only from these forests, whilst *Phyllanthus myrtaceus* is primarily a South African species but with an outlier population in the

Swampy clearing in Moribane Forest, with invasive *Vernonanthura polyanthes* around the swamp margins (MC)

Chimanimani Lowlands (Timberlake *et al.* 2016b; Rokni *et al.* 2019). The forested areas between the Zimbabwe border and the Thekeza area of the Zomba community in the western portion of the IPA are particularly important for most of these scarce species.

Also of high botanical significance are low-elevation quartzite outcrops which support an interesting rock flora, differing from the high elevation quartzites that are so famous for their plant endemism in the Chimanimani Highlands. Particularly notable species include the endemic *Ficus muelleriana* (EN), a tiny fig that climbs on the rock faces, and *Otiophora lanceolata* (VU), a locally abundant shrublet. Other scarce species of this habitat include *Gutenbergia westii* (VU) and a disjunct population of *Sclerochiton caeruleus* (NT). Where quartzites outcrop along shaded river valleys, the Chimanimani endemic grass *Danthoniopsis chimanimaniensis* (EN) can be frequent, together with the shrub *Vernonia muelleri* subsp. *muelleri* (NT).

Small areas of seasonally wet grassland on sands support an interesting, though not diverse, herb flora including *Crepidorhopalon flavus* (VU) whose range is centred on the southern Chimanimani foothills, and *Mesanthemum africanum* (LC), a Chimanimani endemic that is mainly found in the high mountains but occurs at much lower abundance in these lowland wet grasslands.

In total, 532 plant taxa were recorded during recent surveys of these lowland habitats, although these were far from exhaustive (Timberlake *et al.* 2016b), and further interesting species are likely to be discovered in the future. A more thorough survey of the lowland quartzite outcrops after the main rainy season might be particularly productive given that Chimanimani quartzites are so rich in endemics at higher elevations.

Habitat and geology

This area is geologically complex, with quartzite outcrops in the western half of the site, and with areas of mica schist. In the east, there are granitoid and migmatite formations, all of Precambrian age (Instituto Nacional de Geológia 1987). This complexity gives rise to a varied topography and soil types, with the areas of quartzite giving rise to coarse sandy soils whilst finer and deeper loamy soils are recorded at Moribane (Timberlake *et al.* 2016b). The foothills of the Chimanimani Mountains are incised by numerous steep-sided river valleys. Climate is variable, with average annual rainfall varying from ca. 1,000 mm to > 1,500 mm per year depending on locality, with most rains falling between late November and late March; average annual temperatures are ca. 19–22°C (Ghiurghi *et al.* 2010).

The lowland forests vary considerably in species composition and in the relative extent of evergreen versus deciduous components, depending on availability of moisture and soil type. The highest proportion of evergreen elements are typically found in valleys, with a gradual to more abrupt transition to deciduous elements on ridges; the latter can transition into miombo woodland. The forest canopy varies from 20–30 m tall with occasional emergents to 40 m. Of particular interest botanically are the mostly evergreen forests and riverine fringes of Makurupini, Maronga and the southwestern-most part of Zomba (Thekeza). Amongst the dominant trees, these areas support potentially the largest population globally of *Maranthes goetzeniana*. Overall, the dominant species is *Newtonia buchananii*, with *Albizia adianthifolia*, *Celtis gomphophylla*, *Erythrophleum suaveolens* and *Millettia stuhlmannii* amongst the other common species. *Terminalia* (formerly *Pteleopsis*) *myrtifolia* is mainly restricted to semi-deciduous forest, for example at Moribane Forest Reserve. Common understorey trees include *Funtumia africana* – particularly abundant at Moribane – and *Aidia micrantha*, *Rawsonia lucida* and *Tabernaemontana* spp. At Maronga, additional common understorey species include *Alchornea hirtella*, *Craterispermum schweinfurthii*, *Drypetes arguta* and *Synsepalum chimanimani*. Lianas are frequent throughout. The species composition of these forests is discussed in detail by Timberlake *et al.* (2016b), with additional observations by Müller *et al.* (2005).

Along rocky riverine fringes at Maronga, *Uapaca lissopyrena* is frequent, easily identified by its stilt roots. This habitat has a characteristic understorey with small trees and shrubs of *Cleistanthus*

polystachyus subsp. *milleri*, *Diospyros natalensis*, *Mascarenhasia arborescens*, *Nuxia oppositifolia*, *Tricalysia coriacea* subsp. *angustifolia* and a dwarf species of *Podocarpus* which has previously been equated with the South African *P. elongatus* or *P. latifolius* but which may prove to be a distinct species restricted to the Chimanimani quartzites (Rokni *et al.* 2019). *Khaya anthotheca* also occurs along riverine fringes.

The forests are interspersed with areas of miombo woodland, dominated by *Brachystegia* and *Uapaca* species, particularly *B. spiciformis* and *U. kirkiana*. Other common miombo trees include *Burkea africana*, *Maprounea africana* and *Pterocarpus angolensis.* Whilst important ecologically, the miombo does not contain high numbers of rare or threatened plant species. Low-elevation outcrops of nutrient-deficient quartzites (Chimanimani sandstones) are most frequent in the Makurupini-Maronga area but are also found further east in Zomba community. These areas are usually associated with light woodland dominated by *Brachystegia tamarindoides* subsp. *microphylla*, and with an interesting herb and succulent flora on the rocks and thin soils which remains understudied.

Small areas of seasonally wet grassland with sandy-peat soils occur within the forest-woodland mosaic. Common grasses include *Hyparrhenia* spp., *Themeda triandra*, *Panicum dregeanum* and *Imperata cylindrica*, with scattered shrubs including *Dissotis princeps* and *Eriosema psoraleoides.* These wetlands can support an interesting wetland herb community, as well as patches of swamp forest with *Garcinia imperialis*, *Uapaca lissopyrena* and

Lowland Chimanimani quartzite outcrops at Maronga, habitat for *Ficus muelleriana* (ID)

Forest destruction at Maronga (ID)

Synsepalum chimanimani (BW)

Vepris drummondii (ID)

Voacanga thouarsii among the common species. At the Zomba community, there are also larger areas of swamp and lowland watercourses that are fringed by large stands of the striking tree *Pandanus livingstonianus* which, whilst fairly widespread, has very isolated and localised populations and is thought to be threatened by habitat loss. The Zomba Centro Swamp is particularly important for this species, and also supports extensive papyrus (*Cyperus papyrus*) stands (Timberlake *et al.* 2016b).

Conservation issues

The entirety of this IPA falls within the Chimanimani National Park (CNP) and Trans-Frontier Conservation Area (TFCA): the northern portion of the site lies within the core CNP/TFCA whilst the Maronga, Moribane and Zomba Forest Reserves and surrounding community lands are within the buffer zone. The CNP has a comprehensive management plan (Ghiurghi *et al.* 2010) but this has not been fully implemented to date. Natural habitats within the core CNP/TFCA are largely intact with only small areas of human encroachment at present, although better demarcation of the core reserve boundary is desirable to prevent further encroachment (Timberlake *et al.* 2016b). However, threats are severe within the buffer zone including within the three Forest Reserves. These reserves are not managed for their biodiversity, having been originally established in 1953 for timber production and possibly for watershed protection (Müller *et al.* 2005). Large areas of forest have either been cleared or degraded for subsistence agriculture, using fire as a means to clear the undergrowth once the large trees have been felled. Excessive and indiscriminate burning prevents forest regrowth and also impacts other key habitats. Regular burning also encourages the continuing spread of the invasive South American shrub *Vernonanthura polyanthes*, which is now dominant over many hectares of disturbed, former forest habitats in the Chimanimani foothills, out-competing native species and preventing regeneration of natural habitats and encroaching into forest margins. A further threat is the impact of gold mining along some of the major rivers that flow from the massif, which pollutes the watercourses and also denudes fringing vegetation. Conservation action is urgently needed in this IPA, particularly in the area that falls outside the core CNP. Work with community leaders to attempt to better balance livelihoods with biodiversity conservation is ongoing, led by the Micaia Foundation. This has led to the establishment of informal community conservation areas in the Maronga, Zomba and Mpunga communities (Timberlake *et al.* 2016b); these areas are included within the core IPA delineation. Recently, plans have been mooted to cultivate coffee as a shade crop at Moribane Forest as part of a habitat restoration plan for the Chimanimani TFCA under the draft "Plano de Restauração paisagem de Chimanimani" (C. de Sousa, pers. comm. 2021). Such a scheme would need to be carefully managed and focused on degraded forest areas in order to prevent damage to the intact forest ecosystem.

The site is included within the Chimanimani Mountains Important Bird Area (IBA) and Parque Nacional de Chimanimani Key Biodiversity Area (KBA), both of which now include the whole of the CNP core and buffer zones. This site is noted to be probably the area of greatest avian diversity within Mozambique, although there are no endemic bird species (BirdLife 2021e). The forests also have an important population of African Elephant (*Loxodonta africana*), these being most frequent at Moribane Forest Reserve. The IPA would qualify as an Alliance for Zero Extinction (AZE) site on the basis of the presence of *Ficus muelleriana* and *Streptocarpus acicularis*.

Key ecosystem services

The lowland forests and woodlands supply a range of key ecosystem services, including the provision and regulation of regional water resources, prevention of severe erosion from the steep slopes of the foothill valleys, and provisioning of a range of important materials for local communities, such as timbers and fibres for construction and local honey production. A number of plant species with potential for sustainable economic use were identified during field surveys in 2015 (Timberlake *et al.* 2016b); these include the plumose seeds of *Funtumia africana* for specialist paper making, the wild coffee *Coffea salvatrix*, the fruits of the miombo tree *Uapaca kirkiana*, the stems of *Cyperus papyrus* and *Phragmites* sp. for mat-making and potentially *Khaya anthotheca* for local carpentry, although the lattermost may not be sustainable.

As part of the wider Chimanimani landscape, this area has high ecotourism potential, particularly for walking tours that combine forest trekking and wildlife watching with mountaineering. The Ndzou Camp, within the Moribane Forest Reserve (Mpunga Community), is an established ecotourist centre, with elephant-watching a particular attraction. The Makurupini Falls near the Zimbabwe border are of particularly high scenic beauty and could be a further tourist attraction, although they are remote.

Crepidorhopalon flavus (BW)

Ecosystem service categories

- Provisioning – Food
- Provisioning – Raw materials
- Provisioning – Fresh water
- Provisioning – Medicinal resources
- Regulating services – Local climate and air quality
- Regulating services – Carbon sequestration and storage
- Regulating services – Erosion prevention and maintenance of soil fertility
- Regulating services – Pollination
- Habitat or supporting services – Habitats for species
- Habitat or supporting services – Maintenance of genetic diversity
- Cultural services – Tourism

IPA assessment rationale

The Chimanimani Lowlands qualify as an IPA under all three criteria. This area supports important populations of 14 criterion A(i) globally threatened species, of which one is Critically Endangered, six are Endangered and seven are Vulnerable. Two of these species – *Streptocarpus acicularis* (CR) and *Ficus muelleriana* (EN) – are known only from within this IPA, and all except *Coffea salvatrix* (EN) are highly range-restricted. The site also contains important populations of one A(iii) species and three A(iv) species that have not yet been assessed on the IUCN Red List. The site also qualifies under Criterion B(ii) for its exceptional richness of range-restricted species, with 20 species with a range of 10,000 km² including two national endemics, significantly above the threshold of 3% of the national list of priority species. Under Criterion C(iii), it qualifies on the basis of containing the largest extent of Low Altitude Moist Forest in Mozambique and important intact areas of Medium Altitude Moist Forest, both highly range-restricted and nationally threatened habitats in the country.

Priority species (IPA Criteria A and B)

FAMILY	TAXON	IPA CRITERION A	IPA CRITERION B	≥ 1% OF GLOBAL POP'N	≥ 5% OF NATIONAL POP'N	IS 1 OF 5 BEST SITES NATIONALLY	ENTIRE GLOBAL POP'N	SPECIES OF SOCIO-ECONOMIC IMPORTANCE	ABUNDANCE AT SITE
Asphodelaceae	*Aloe ballii* var. *makurupiniensis*	A(i)	B(ii)	✓	✓	✓			scarce
Asteraceae	*Gutenbergia westii*	A(i)	B(ii)	✓	✓	✓			frequent
Asteraceae	*Kleinia chimanimaniensis*	A(iv)	B(ii)						unknown
Asteraceae	*Vernonia muelleri* subsp. *muelleri*		B(ii)						occasional
Commelinaceae	*Cyanotis chimanimaniensis* ined.	A(iv)	B(ii)	✓	✓	✓			unknown
Cyperaceae	*Scleria pachyrrhyncha*	A(i)			✓	✓			unknown
Eriocaulaceae	*Mesanthemum africanum*		B(ii)						occasional
Fabaceae	*Tephrosia longipes* var. *swynnertonii*	A(iv)	B(ii)		✓	✓			unknown
Gesneriaceae	*Streptocarpus acicularis*	A(i)	B(ii)	✓	✓	✓	✓		scarce
Lamiaceae	*Syncolostemon flabellifolius*		B(ii)						scarce
Linderniaceae	*Crepidorhopalon flavus*	A(i)	B(ii)	✓	✓	✓			occasional
Loranthaceae	*Englerina swynnertonii*	A(iii)	B(ii)	✓	✓	✓			unknown
Moraceae	*Ficus muelleriana*	A(i)	B(ii)	✓	✓	✓	✓		occasional
Phyllanthaceae	*Phyllanthus bernierianus* var. *glaber*		B(ii)	✓	✓	✓			unknown
Poaceae	*Danthoniopsis chimanimaniensis*	A(i)	B(ii)	✓	✓	✓			occasional
Rubiaceae	*Afrocanthium ngonii*	A(i)	B(ii)		✓	✓			unknown
Rubiaceae	*Coffea salvatrix*	A(i)		✓	✓	✓			occasional
Rubiaceae	*Otiophora lanceolata*	A(i)	B(ii)	✓	✓	✓			frequent
Rubiaceae	*Sericanthe chimanimaniensis*	A(i)	B(ii)	✓	✓	✓			occasional
Rutaceae	*Vepris drummondii*	A(i)	B(ii)	✓	✓	✓			occasional
Sapotaceae	*Synsepalum chimanimani*	A(i)	B(ii)	✓	✓	✓			frequent
Zamiaceae	*Encephalartos chimanimaniensis*	A(i)	B(ii)	✓	✓	✓			scarce
		A(i): 14 ✓ A(iii): 1 ✓ A(iv): 3 ✓	B(ii): 20 ✓						

Threatened habitats (IPA Criterion C)

HABITAT TYPE	IPA CRITERION C	≥ 5% OF NATIONAL RESOURCE	≥ 10% OF NATIONAL RESOURCE	IS 1 OF 5 BEST SITES NATIONALLY	ESTIMATED AREA AT SITE (IF KNOWN)
Medium Altitude Moist Forest [MOZ-02]	C(iii)			✓	
Low Altitude Moist Forest [MOZ-03]	C(iii)		✓	✓	

Protected areas and other conservation designations

CONSERVATION AREA TYPE	CONSERVATION AREA NAME	RELATIONSHIP OF IPA TO CONSERVATION AREA
National Park	Chimanimani	protected/conservation area overlaps with IPA
Trans-Frontier Conservation Area	Chimanimani	protected/conservation area overlaps with IPA
Forest Reserve (production)	Maronga	IPA encompasses protected/conservation area
Forest Reserve (production)	Moribane	IPA encompasses protected/conservation area
Forest Reserve (production)	Zomba	IPA encompasses protected/conservation area
Important Bird Area	Chimanimani Mountains (Mozambique)	protected/conservation area overlaps with IPA
Key Biodiversity Area	Parque Nacional de Chimanimani	protected/conservation area encompasses IPA

Threats

THREAT	SEVERITY	TIMING
Small-holder farming	high	Ongoing – increasing
Logging & wood harvesting	low	Ongoing – stable
Increase in fire frequency/intensity	high	Ongoing – increasing
Invasive non-native/alien species	high	Ongoing – increasing

CHIMANIMANI MOUNTAINS

Assessors: Jo Osborne, Iain Darbyshire

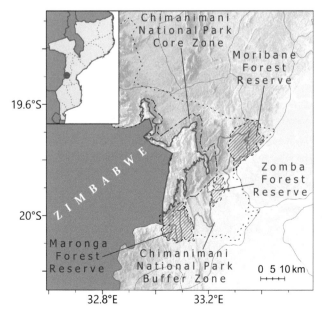
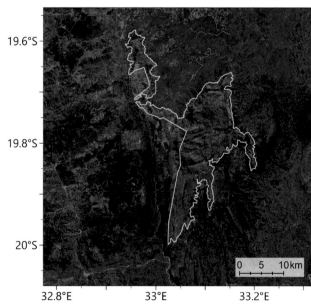

INTERNATIONAL SITE NAME		Chimanimani Mountains	
LOCAL SITE NAME (IF DIFFERENT)		–	
SITE CODE	MOZTIPA003B	PROVINCE	Manica

LATITUDE	-19.80678	LONGITUDE	33.11202
ALTITUDE MINIMUM (m a.s.l.)	1,200	ALTITUDE MAXIMUM (m a.s.l.)	2,436
AREA (km²)	319	IPA CRITERIA MET	A(i), A(iv), B(ii), C(iii)

Site description

The Chimanimani Mountains extend across the Mozambique-Zimbabwe border from Sussendenga District in Manica Province of Mozambique, into Manicaland Province in Zimbabwe. Mozambique's Chimanimani Mountains IPA encompasses the montane area over 1,200 m elevation, up to the border with Zimbabwe, including both the main massif and an area known as 'The Corner' to the north, which is separated from the main massif by the Musapa Gap. The IPA site includes the highest mountain in Mozambique, Monte Binga at 2,436 m elevation. The Chimanimani Mountains are protected on both sides of the border as National Parks, which together form the Chimanimani Trans-Frontier Conservation Area (TFCA). Immediately to the south and east of the Chimanimani Mountains IPA, below 1,200 m, the lower mountain slopes of the Chimanimani National Park, as well as adjacent Forest Reserves and community lands, are included within the Chimanimani Lowlands IPA.

Botanical significance

The Chimanimani Mountains are Mozambique's most valuable site for plant endemism and have high international conservation significance. The IPA includes Mozambique's largest areas of montane grassland and scrubland, here occurring mainly on quartzite rock but also with extensive areas of grassland on schist. In the rugged montane landscape, steep quartzite crags, ledges and boulders form habitats that support species-rich lithophytic plant communities with many endemic species. Of the 74 endemic plants of the cross-border Chimanimani Mountains so far recorded, 61 have been found within this IPA to date, many having only recently been found on the Mozambique side of the border during surveys in the mid-2010s (Timberlake *et al.* 2016; Wursten *et al.* 2017). Many of the endemics are from plant groups typical of

nutrient-poor soils, including three species of *Thesium*, one endemic and two near-endemic *Erica* species, and one of the few tropical African restio species, *Platycaulos quartziticola* (Cheek *et al.* 2018). Among the wide range of other plant families and genera represented in the endemic flora, the genus *Aloe* is particularly notable for having five endemic taxa. Three plant species are known only from within this IPA boundary at present: *Streptocarpus montis-bingae* (DD), a small herb known only from Monte Binga; *Dianthus chimanimaniensis* (VU), a tufted perennial herb; and *Centella obtriangularis* (VU), a small creeping herb, although there are unconfirmed records of the latter from the Zimbabwe side. Many more range-restricted species of the Chimanimani-Nyanga (Sub-)Centre of Plant Endemism also occur; overall, 95 species with a range of less than 10,000 km² are recorded from within this IPA.

Thirty plants that occur within the IPA are assessed as globally threatened, of which eight are Endangered, including two recently described small endemic trees *Empogona jenniferae* and *Olinia chimanimani*, the orchid *Neobolusia ciliata* and the grass *Danthoniopsis chimanimaniensis* (Timberlake *et al.* 2016b; Wursten *et al.* 2017; Cheek *et al.* 2018; Shah *et al.* 2018). Of the Vulnerable species, many are threatened in other parts of their range and the Chimanimani Mountains are considered to be the most secure site for these taxa. A significant proportion of the strict-endemic species are not threatened as their habitats are largely intact and little-disturbed.

A recent checklist of the vascular plants of the massif above 1,200 m elevation on both sides of the Mozambique-Zimbabwe border recorded a total of 977 taxa and noted that, although total species richness is not particularly high in comparison to other mountain sites in the *Flora Zambesiaca* area, this site holds significantly higher numbers of endemics, with 7.7% of the total flora being endemic, compared with 5.4% on Mount Mulanje in Malawi, 1.7% on the Nyika Plateau in Malawi and 1.4% in the Nyanga Highlands of Zimbabwe (Wursten *et al.* 2017).

New species to science are still being discovered and described from the site, such as the recently published *Sericanthe chimanimaniensis* (Wursten *et al.* 2020), and surveys in the mid-2010s uncovered a potential new *Streptocarpus* allied to *S. grandis*, together with unmatched *Erica*, *Coleus* and *Syncolostemon* spp., amongst other potential novelties (I. Darbyshire *et al.*, pers obs.).

Habitat and geology

The Chimanimani Mountains IPA is dominated by montane grassland and scrubland habitats within a rugged landscape of mountain peaks, steep crags and boulders. The surficial geology is predominantly a nutrient-deficient quartzite, with a smaller proportion of more nutrient-rich schists; these formations are primarily of the Frontier (Fronteira) Series of the Umkondo Group, dating from the later Precambrian (Timberlake *et al.* 2016b). Quartzite rock outcrops, crags and boulders give rise to a wide range of microhabitats and support a high species diversity. The habitat and geology of this site are discussed in detail in Timberlake *et al.* (2016b) and are summarised here.

Montane grasslands occupy an area of ca. 200–250 km^2 across the highlands, occurring mainly on areas of level or rolling terrain. Those on quartzite occur on a thin white sandy soil and are often interspersed with scrub and rock outcrops, with the more extensive areas occurring in broad valleys; the dominant grass species is *Loudetia simplex*. Those on schists tend to form rolling hills on a red soil and *Themeda triandra* is the most characteristic species, although *L. simplex* remains common. Scattered bushes occur within the schist grassland, and these can become more frequent in some areas to form a scrubland, 1–2 m tall, with several *Protea* spp. and *Leucospermum saxosum* amongst the most common shrubs, with the near-endemic *Myrica chimanimaniana* also locally frequent. An Ericaceous scrub 0.5–3 m tall is frequent on quartzite and is a species-rich assemblage in which many of the endemics occur; a range of *Erica* species are noted, with *E. hexandra* particularly common on the rock outcrops together with other shrubby species and succulents including the impressive *Aloe munchii*. These Ericaceous

Extensive schist grasslands with quartzite outcrops in the distance (TS)

Montane grassland and quartzite outcrops in the Chimanimani Mountains (TS)

The Chimanimani Mountains with montane forest and bushlande (TS)

scrublands are considered to be fire-sensitive but are somewhat protected from the worst fires by the intervening areas of bare rock. Extensive areas of exposed quartzite support a lithophytic community, in which the clump-forming sedge *Coleochloa setifera* is common together with the endemic *Xerophyta argentea* and *Aloe hazeliana*. Shaded areas amongst the rocks can support interesting herbaceous species such as *Impatiens salpinx*, *Streptocarpus* spp., orchids and ferns.

Across the IPA, boggy areas and streams are frequent, draining into several larger rivers, the largest of which is the Rio Mufomodzi in the central-northern part of the massif. Broad river valleys support level areas of grassland and bog vegetation on nutrient-poor alluvial soils and peat. Areas of seepage and wet depressions can support an interesting herb flora including *Xyris* spp., the endemics *Mesanthemum africanum* and *Platycaulos quartziticola*, and a number of orchids and sedges.

Sheltered stream gullies and river gorges support areas of moist evergreen forest, mostly classed here as medium-altitude moist forest (occurring below 1400 m) with a few smaller patches of montane moist forest (mostly occurring over 1600 m). Whilst mainly very small in area, the largest patch noted by Timberlake *et al.* (2016b) measures ca. 4.2 km². These forests have not been well surveyed across the site, but some of the higher altitude patches have been found to be typical Afromontane forests, with a closed canopy of ca. 10–15 m and with characteristic species including *Ilex mitis*, *Macaranga mellifera*, *Podocarpus milanjianus*,

Schefflera umbellifera and *Syzygium cordatum*. Lianas and epiphytes are frequent, and the ground layer has many ferns and mosses.

Although accurate climate data are lacking for the high mountains, rainfall is estimated at ca. 1,500–2,000 mm per year, but may reach as high as ca. 3,000 mm on the highest peaks. Rainfall occurs all year round but peaks from November to April. Mists are frequent and supply additional moisture during dry periods. Mean average temperatures are below 18°C, and frost is noted to be frequent above 1,500 m elevation (Ghiurghi *et al.* 2010; Timberlake *et al.* 2016b).

Conservation issues

The Chimanimani Mountains IPA lies almost entirely within the core zone of the Chimanimani National Park (CNP) and the Trans-Frontier Conservation Area (TFCA), a protected area that is essentially uninhabited and designated as non-use. This area has a comprehensive management plan (Ghiurghi *et al.* 2010), although this has not all been implemented. The vegetation within the IPA site is mostly intact and considered to be free of any major threat at present. However, illegal small-scale goldmining, fire frequency and invasive species all pose potential threats to the vegetation and both monitoring and management are recommended. Tourism also needs to be considered as a conservation issue for this site.

Illegal gold-mining was first recorded in the CNP in 2004, and increased rapidly with as many as 10,000 miners operating in 2006; by 2016 this had reduced to ca. 1,000 miners, due mainly to the most accessible gold having been exhausted (Dondeyne *et al.* 2009; Timberlake *et al.* 2016b). The mining has been concentrated along water courses and has not directly impacted the populations of most of the endemic and threatened plant species, most of which occur in different habitats. However, the serious negative impact of goldmining on the upland hydrology and ecology cannot be overlooked. A probable indirect impact of the illegal goldmining has been an increased frequency of wild-fires when fire is used for hunting by the miners or set accidentally. Some of the montane habitats, such as the schist grasslands and scrublands, are likely to be adapted to fire to some extent. Nevertheless, increased fire frequency above natural levels is likely to impact scrub vegetation and moist forest edges, preventing recovery between fires and affecting recruitment of young plants. Other issues associated with the mining activity and associated presence of traders in the highlands include the use of caves and sheltered rocky areas as temporary to more permanent shelters which can damage the shaded rock flora, including *Streptocarpus* spp., and the gathering of fuelwood, although this latter threat appears to have been minimal (Timberlake *et al.* 2016b).

The invasive shrub *Vernonanthura polyanthes* was originally introduced into Mozambique from South America as a nectar source for bees and is now becoming widespread in the lower foot-slopes of Chimanimani Mountains. Recently, several individuals have been recorded within the montane area at 1,200–1,400 m elevation. This shrub is a potential future threat to the forest margin and scrubland vegetation in lower elevation areas of this IPA as it can form dense stands on disturbed ground and fire-damaged areas. However, this is a much more serious threat in the Chimanimani Lowlands IPA.

Tourism in the Chimanimani Mountains is considered a potential conservation issue as it may potentially have both a positive and a negative impact. The Chimanimani Mountains have strong potential for eco-tourism, providing a wilderness experience and an opportunity for local communities to benefit from conservation of the CNP. However, tourism must be well-managed to avoid damage to habitats and vegetation through trampling, fire and pollution.

The Chimanimani Mountains IPA lies within the Chimanimani Mountains Important Bird Area which includes both the high altitude massif and surrounding lowlands. The entirety of the CNP core and buffer area is also designated as the Chimanimani Key Biodiversity Area (KBA), based primarily on its rich flora. The Mountains would qualify as an Alliance for Zero Extinction site on the basis of the Endangered endemic plant species noted above.

Habitat damage from artisanal gold mining (TS)

Key ecosystem services

In addition to its high plant biodiversity value, the Chimanimani Mountains have an economic value as a wilderness area for ecotourism, and much of this potential is yet to be exploited. The vegetation contributes to carbon sequestration and storage and provides habitat for montane flora and fauna. The mountains and forests are also an important watershed for the surrounding area.

Ecosystem service categories

- Provisioning – Fresh water
- Regulating services – Carbon sequestration and storage
- Habitat or supporting services – Habitats for species
- Cultural services – Tourism

Myrica chimanimaniana (TS)

IPA assessment rationale

The Chimanimani Mountains qualify as an IPA under all three criteria. Under Criterion A(i) the site supports important populations of 29 globally threatened plant taxa. Half of these are endemic to this mountain range, with the remainder mostly being range-restricted Chimanimani-Nyanga endemics, six of which are only recorded at this site within the Mozambique IPA network. In addition, five potentially threatened range-restricted endemics occur here, qualifying the site under Criterion A(iv). The Chimanimani Mountains support an exceptional number of species of high conservation importance with three site endemics and 93 regional endemics with a restricted range of less than 10,000 km². Many of these are Chimanimani endemics that occur on both sides of the Mozambique-Zimbabwe border. This total of 96 qualifying taxa is just short of 20% of the total list of sub-criterion B(ii) qualifying species for Mozambique, making this the richest site botanically in Mozambique. Under criterion C(iii) the site includes the largest extent of montane grassland in Mozambique, this being one of Mozambique's national priority habitats recognised during the first Mozambique TIPAs workshop in Maputo in January 2018. In addition, the site supports small areas of medium-altitude moist forest and montane moist forest, two further national priority habitats, but it does not qualify as one of the five best sites for these two habitats.

Priority species (IPA Criteria A and B)

FAMILY	TAXON	IPA CRITERION A	IPA CRITERION B	≥ 1% OF GLOBAL POP'N	≥ 5% OF NATIONAL POP'N	IS 1 OF 5 BEST SITES NATIONALLY	ENTIRE GLOBAL POP'N	SPECIES OF SOCIO-ECONOMIC IMPORTANCE	ABUNDANCE AT SITE
Apiaceae	*Centella obtriangularis*	A(i)	B(ii)	✓	✓	✓	✓		common
Apocynaceae	*Asclepias cucullata* subsp. *scabrifolia*		B(ii)						occasional
Apocynaceae	*Asclepias graminifolia*		B(ii)						scarce
Apocynaceae	*Aspidoglossum glabellum*	A(i)	B(ii)	✓	✓	✓			scarce
Apocynaceae	*Ceropegia chimanimaniensis*		B(ii)						scarce
Apocynaceae	*Raphionacme pulchella*	A(i)		✓	✓	✓			scarce

Priority species (IPA Criteria A and B)

FAMILY	TAXON	IPA CRITERION A	IPA CRITERION B	≥ 1% OF GLOBAL POP'N	≥ 5% OF NATIONAL POP'N	IS 1 OF 5 BEST SITES NATIONALLY	ENTIRE GLOBAL POP'N	SPECIES OF SOCIO-ECONOMIC IMPORTANCE	ABUNDANCE AT SITE
Asparagaceae	Asparagus chimanimanensis		B(ii)						frequent
Asparagaceae	Chlorophytum pygmaeum subsp. rhodesianum		B(ii)						occasional
Asparagaceae	Eriospermum mackenii subsp. phippsii		B(ii)						occasional
Asphodelaceae	Aloe hazeliana var. hazeliana		B(ii)						occasional
Asphodelaceae	Aloe hazeliana var. howmanii		B(ii)						occasional
Asphodelaceae	Aloe munchii		B(ii)						frequent
Asphodelaceae	Aloe plowesii	A(i)	B(ii)	✓	✓	✓			occasional
Asphodelaceae	Aloe rhodesiana	A(i)				✓			scarce
Asphodelaceae	Aloe wildii		B(ii)						unknown
Asteraceae	Anisopappus paucidentatus		B(ii)						common
Asteraceae	Aster chimanimaniensis		B(ii)						unknown
Asteraceae	Cineraria pulchra		B(ii)						frequent
Asteraceae	Gutenbergia westii	A(i)	B(ii)	✓	✓	✓			scarce
Asteraceae	Helichrysum africanum		B(ii)						common
Asteraceae	Helichrysum moorei		B(ii)						common
Asteraceae	Helichrysum rhodellum		B(ii)						occasional
Asteraceae	Kleinia chimanimaniensis	A(iv)	B(ii)			✓			scarce
Asteraceae	Lopholaena brickellioides	A(iv)	B(ii)			✓			occasional
Asteraceae	Schistostephium oxylobum	A(i)	B(ii)	✓	✓	✓			occasional
Asteraceae	Senecio aetfatensis		B(ii)						scarce
Asteraceae	Vernonia muelleri subsp. muelleri		B(ii)						frequent
Asteraceae	Vernonia nepetifolia		B(ii)						occasional
Balsaminaceae	Impatiens salpinx		B(ii)						occasional
Campanulaceae	Lobelia cobaltica		B(ii)						frequent
Campanulaceae	Wahlenbergia subaphylla subsp. scoparia	A(iv)	B(ii)	✓	✓	✓			occasional
Caryophyllaceae	Dianthus chimanimaniensis	A(i)	B(ii)	✓	✓	✓	✓		unknown
Commelinaceae	Cyanotis chimanimaniensis ined.	A(iv)	B(ii)	✓	✓	✓			unknown
Crassulaceae	Kalanchoe velutina subsp. chimanimaniensis		B(ii)						occasional
Ericaceae	Erica lanceolifera	A(i)	B(ii)	✓	✓	✓			occasional
Ericaceae	Erica pleiotricha var. blaerioides		B(ii)						frequent

Priority species (IPA Criteria A and B)

FAMILY	TAXON	IPA CRITERION A	IPA CRITERION B	≥ 1% OF GLOBAL POP'N	≥ 5% OF NATIONAL POP'N	IS 1 OF 5 BEST SITES NATIONALLY	ENTIRE GLOBAL POP'N	SPECIES OF SOCIO-ECONOMIC IMPORTANCE	ABUNDANCE AT SITE
Ericaceae	*Erica pleiotricha* var. *pleiotricha*	A(i)	B(ii)	✓	✓	✓			scarce
Ericaceae	*Erica wildii*		B(ii)						unknown
Eriocaulaceae	*Mesanthemum africanum*		B(ii)						common
Euphorbiaceae	*Euphorbia crebrifolia*		B(ii)						occasional
Fabaceae	*Aeschynomene aphylla*	A(i)	B(ii)	✓	✓	✓			unknown
Fabaceae	*Aeschynomene chimanimaniensis*		B(ii)						scarce
Fabaceae	*Aeschynomene grandistipulata*		B(ii)						frequent
Fabaceae	*Aeschynomene inyangensis*		B(ii)						frequent
Fabaceae	*Crotalaria phylicoides*		B(ii)						occasional
Fabaceae	*Otholobium foliosum* subsp. *gazense*		B(ii)						unknown
Fabaceae	*Pearsonia mesopontica*		B(ii)						scarce
Fabaceae	*Rhynchosia chimanimaniensis*	A(i)	B(ii)	✓	✓	✓			unknown
Fabaceae	*Rhynchosia stipata*		B(ii)						unknown
Fabaceae	*Rhynchosia swynnertonii*		B(ii)						unknown
Fabaceae	*Tephrosia chimanimaniana*		B(ii)						scarce
Fabaceae	*Tephrosia longipes* var. *drummondii*	A(iv)	B(ii)			✓			occasional
Gesneriaceae	*Streptocarpus grandis* subsp. *septentrionalis*	A(iv)	B(ii)	✓	✓	✓			occasional
Gesneriaceae	*Streptocarpus hirticapsa*	A(i)	B(ii)	✓	✓	✓			occasional
Gesneriaceae	*Streptocarpus montis-bingae*		B(ii)				✓		scarce
Iridaceae	*Dierama plowesii*	A(i)	B(ii)	✓	✓	✓			scarce
Iridaceae	*Gladiolus zimbabweensis*	A(i)	B(ii)	✓	✓	✓			occasional
Iridaceae	*Hesperantha ballii*		B(ii)						scarce
Lamiaceae	*Aeollanthus viscosus*		B(ii)						frequent
Lamiaceae	*Coleus caudatus*		B(ii)						occasional
Lamiaceae	*Coleus sessilifolius*		B(ii)						frequent
Lamiaceae	*Stachys didymantha*		B(ii)						unknown
Lamiaceae	*Syncolostemon flabellifolius*		B(ii)						frequent
Lamiaceae	*Syncolostemon oritrephes*	A(i)	B(ii)	✓	✓	✓			scarce
Melastomataceae	*Dissotis pulchra*	A(i)	B(ii)	✓	✓	✓			occasional
Melastomataceae	*Dissotis swynnertonii*	A(i)	B(ii)	✓	✓	✓			occasional
Melianthaceae	*Bersama swynnertonii*		B(ii)						scarce

Priority species (IPA Criteria A and B)

FAMILY	TAXON	IPA CRITERION A	IPA CRITERION B	≥ 1% OF GLOBAL POP'N	≥ 5% OF NATIONAL POP'N	IS 1 OF 5 BEST SITES NATIONALLY	ENTIRE GLOBAL POP'N	SPECIES OF SOCIO-ECONOMIC IMPORTANCE	ABUNDANCE AT SITE
Myricaceae	Myrica chimanimaniana	A(i)	B(ii)	✓	✓	✓			frequent
Oleaceae	Olea chimanimani		B(ii)						occasional
Orchidaceae	Disa chimanimaniensis		B(ii)						unknown
Orchidaceae	Neobolusia ciliata	A(i)	B(ii)	✓	✓	✓			occasional
Orchidaceae	Polystachya subumbellata		B(ii)						unknown
Orchidaceae	Schizochilus lepidus	A(i)	B(ii)	✓	✓	✓			unknown
Orobanchaceae	Buchnera chimanimaniensis		B(ii)						common
Orobanchaceae	Buchnera subglabra	A(i)	B(ii)	✓	✓	✓			frequent
Penaeaceae	Olinia chimanimani	A(i)	B(ii)	✓	✓	✓			occasional
Peraceae	Clutia sessilifolia		B(ii)						occasional
Phyllanthaceae	Phyllanthus bernierianus var. glaber		B(ii)						unknown
Poaceae	Danthoniopsis chimanimaniensis	A(i)	B(ii)	✓	✓	✓			unknown
Poaceae	Eragrostis desolata		B(ii)						frequent
Polygalaceae	Polygala zambesiaca	A(i)	B(ii)	✓	✓	✓			occasional
Proteaceae	Faurea rubriflora		B(ii)						occasional
Proteaceae	Protea caffra subsp. gazensis		B(ii)						frequent
Proteaceae	Protea enervis	A(i)	B(ii)	✓	✓	✓			scarce
Restionaceae	Platycaulos quartziticola		B(ii)						frequent
Rubiaceae	Empogona jenniferae	A(i)	B(ii)	✓	✓	✓			scarce
Rubiaceae	Oldenlandia cana		B(ii)						occasional
Rubiaceae	Otiophora inyangana subsp. inyangana		B(ii)						unknown
Rubiaceae	Otiophora inyangana subsp. parvifolia		B(ii)						occasional
Rubiaceae	Pavetta umtalensis		B(ii)						unknown
Rubiaceae	Sericanthe chimanimaniensis	A(i)	B(ii)	✓	✓	✓			occasional
Santalaceae	Thesium chimanimaniense		B(ii)						common
Santalaceae	Thesium dolichomeres		B(ii)						occasional
Santalaceae	Thesium pygmeum		B(ii)						occasional
Scrophulariaceae	Selago anatrichota		B(ii)						occasional
Thymelaeaceae	Struthiola montana		B(ii)						unknown
Velloziaceae	Xerophyta argentea		B(ii)						common
Xyridaceae	Xyris asterotricha	A(i)	B(ii)	✓	✓	✓			scarce
		A(i): 29 ✓ A(iv): 6 ✓	B(ii): 96 ✓						

Lobelia cobaltica (BW)

Impatiens salpinx (TS)

Threatened habitats (IPA Criterion C)

HABITAT TYPE	IPA CRITERION C	≥ 5% OF NATIONAL RESOURCE	≥ 10% OF NATIONAL RESOURCE	IS 1 OF 5 BEST SITES NATIONALLY	ESTIMATED AREA AT SITE (IF KNOWN)
Montane Moist Forest [MOZ-01]	C(iii)				
Medium Altitude Moist Forest [MOZ-02]	C(iii)				
Montane Grassland [MOZ-09]	C(iii)		✓	✓	

Protected areas and other conservation designations

CONSERVATION AREA TYPE	CONSERVATION AREA NAME	RELATIONSHIP OF IPA TO CONSERVATION AREA
National Park	Chimanimani National Park	protected/conservation area encompasses IPA
Trans-Frontier Conservation Area (core zone)	Chimanimani Trans-Frontier Conservation Area	protected/conservation area encompasses IPA
Important Bird Area	Chimanimani Mountains (Mozambique)	protected/conservation area encompasses IPA
Key Biodiversity Area	Parque Nacional de Chimanimani	protected/conservation area encompasses IPA

Threats

THREAT	SEVERITY	TIMING
Mining & quarrying	low	Ongoing – trend unknown
Recreational activities	low	Future – inferred threat
Increase in fire frequency/intensity	unknown	Ongoing – trend unknown
Invasive non-native/alien species/diseases	unknown	Future – inferred threat
Seepage from mining	medium	Ongoing – trend unknown

INHAMBANE
PROVINCE

TEMANE

Assessor(s): Castigo Datizua, Clayton Langa, Iain Darbyshire, Sophie Richards

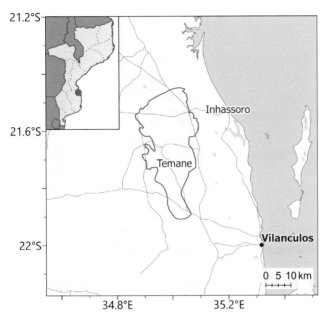

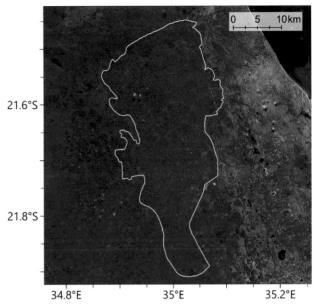

INTERNATIONAL SITE NAME		Temane	
LOCAL SITE NAME (IF DIFFERENT)		–	
SITE CODE	MOZTIPA055	PROVINCE	Inhambane

LATITUDE	-21.67864	LONGITUDE	34.98088
ALTITUDE MINIMUM (m a.s.l.)	20	ALTITUDE MAXIMUM (m a.s.l.)	65
AREA (km²)	678	IPA CRITERIA MET	A(i)

Site description

The Temane IPA lies entirely within Inhassoro District in northern Inhambane Province and covers an area of 678 km² between the latitudes -21.49° to -21.91° and the longitudes 34.89° to 35.04°. The boundaries of this IPA were primarily delineated to encompass important habitats that support both a notable number of plant species endemic to Mozambique, including five threatened species, and a range of ecosystems services that the habitats provide.

Inhassoro District has, in recent years, attracted significant economic interest related to the exploitation of mineral resources. The critical areas for biodiversity within the Temane IPA fall within one of the largest natural gas and heavy sand deposits in Mozambique, centred on

Temane, Maimelane, and Cometela villages (MAE 2005b; Impacto Lda. 2018). These developments, in addition to the activities of local communities, are impacting negatively on the IPA through the transformation and degradation of its ecosystems (MICOA 2012a).

Botanical significance

This IPA is of high botanical importance primarily because of the presence of thicket and dry woodland habitats that support a number of northern Inhambane's endemic and range-restricted species of conservation importance. Five of these species are assessed as globally threatened on the IUCN Red List: *Bauhinia burrowsii* (EN; restricted to Inhassoro, Vilanculos and Mapinhane areas of Inhambane), *Croton aceroides* (EN; encountered in two sites in Inhassoro District, and from between

Mabote and Funhalouro in northern Inhambane, and near Panda and Homoine further south in Inhambane), the woody climber *Triaspis suffulta* (EN: restricted to the Inhassoro and Vilanculos areas), *Croton inhambanensis* (VU; restricted to Inhassoro and Mapinhane) and *Ozoroa gomesiana* (VU; found only in northern Inhambane, mostly concentrated in the Inhassoro, Mapinhane and Vilanculos areas). All of these species are endemics of the proposed Inhambane (sub-)Centre of Plant Endemism (Darbyshire *et al.* 2019a).

This IPA is so far known to support six national endemic plant taxa and one near-endemic plant taxa. The endemic species to Mozambique consist of the five previously mentioned globally threatened species, plus the Least Concern species *Dolichandrone alba*.

Habitat and geology

In the broad sense, the Temane region lies within the Swahilian-Maputaland Regional Transition Zone phytogeographical region according to Clarke (1998), which covers much of the coastal-belt of Mozambique, or according to Schipper and Burgess (2015), the Southern Zanzibar-Inhambane Coastal Forest Mosaic Ecoregion which stretches for around 2,200 km along the eastern coast of the African continent, from southern Tanzania to Xai-Xai (Gaza Province) in Mozambique. In a narrower phytogeographical sense, this IPA falls within the northern extension of the Maputaland Centre of Endemism, recently proposed as the Inhambane (sub-)Centre of Endemism (Darbyshire *et al.* 2019a).

The climate is influenced by the warm current from the Mozambique Channel and is characterised as humid tropical by the coastline and dry tropical inland. The site experiences two seasons; the wet season runs from August to February, whilst the dry and relatively cool season runs from February to July. In the wet season, the average temperatures vary between 28°C and 30°C, while in the dry season, the temperatures vary between 18°C and 27°C. The average annual rainfall ranges from 865mm to 936 mm, with higher rainfall on the coast (Governo do Distrito de Inhassoro 2011; World Resources Institute 2021). The elevation of the IPA ranges from 20–65 m a.s.l. The region is part of the great coastal plains that stretch along a large extent of coastal Mozambique, and is characterized by red clay soils and sodic soils (mananga soils) dominating the inland zones (MICOA 2012a).

Inhassoro District has been subject to several recent botanical surveys, which have helped to build our understanding of plant diversity in the Temane IPA region. Three main types of vegetation can be distinguished at this site. (1) A miombo woodland and grassland mosaic is encountered in the south of the IPA and consists of open woodland of medium sized-trees and shrubs (with ca. 35% canopy cover), with dominant species being *Julbernardia globiflora* and *Brachystegia spiciformis* accompanied by species such as *Afzelia quanzensis*, *Albizia adianthifolia*, *Garcinia livingstonei* and *Pterocarpus angolensis*, and with grassland with ca. 50% ground cover, with for example, *Eragrostis chapelieri*, *Melinis repens*,

Guibourtia conjugata thicket-forest (WM)

Slash-and-burn agriculture within the *Spirostachys* thickets (WM)

Perotis patens, *Schizachyrium sanguineum* and *Sporobolus pyramidalis* dominating the landscape (Deacon 2014). (2) Mixed dry forest-woodland, which is the most extensive type of vegetation found at this site, and also features miombo species, but with the canopy here dominated by tree species such as *Afzelia quanzensis*, *Albizia adianthifolia*, *Balanites maughamii*, *Garcinia livingstonei*, *Guibourtia conjugata*, *Pterocarpus angolensis* and *Suregada zanzibariensis* (MICOA 2012a; Deacon 2014). This habitat is a mosaic of open woodland of medium-sized trees and shrubs (of ca. 30% cover), and an herbaceous ground cover (ca. 60%), with common grasses including *Megathyrsus maximus*, *Schizachyrium sanguineum* and *Sporobolus pyramidalis*. (3) Sand thicket, sometimes interspersed within the miombo woodland and mixed dry forest-woodland vegetation types, this habitat is widespread within the IPA, particularly in the northeast, and comprises dense and short semi-deciduous species dominated by *Hymenocardia ulmoides* and *Spirostachys africana*, with emergent trees of *Adansonia digitata*, *Balanites maughamii* and *Cordyla africana*. Climbers are numerous and include *Ancylobotrys petersiana*, *Apodostigma pallens*, *Artabotrys brachypetalus*, *Artabotrys monteiroae* and *Monodora junodii* var. *junodii* among others (Deacon 2014; Lötter *et al.* in

prep.). These areas of thicket have a canopy cover ranging from 25% to 45% and a sparse herbaceous ground cover of 3–10% (Deacon 2014). This latter vegetation type corresponds to the Pande Sand Thicket of Lötter *et al.* (in prep.), a highly range-restricted vegetation unit the majority of which lies within the Temane IPA, and is of particular importance for most of the range-restricted and threatened species of this site.

Conservation issues

The Temane IPA does not lie within a formal protected area. However, the region is covered by the recently identified Inhassoro-Vilanculos Key Biodiversity Area (WCS *et al.* 2021).

This IPA is currently under high pressure and degradation by local communities because of the harvesting of firewood, charcoal production, livestock grazing, agriculture with an associated increase in fire frequency, and expansion of settlements (A. Massingue, pers. comm. 2020; World Resources Institute 2021; Google Earth 2021). In addition, parts of the IPA are also experiencing habitat degradation due to activities related to natural gas exploration around Temane village and heavy sand exploitation in Maimelane and Cometela villages.

Cultivated areas are concentrated along access roads and paths, and also notably near Sasol's oil flowlines which occur across the vegetation mosaics of this IPA and themselves cause a level of habitat degradation (Deacon 2014; Google Earth 2021). The *Spirostachys*-dominated thickets are particularly targeted for slash-and-burn agriculture by local communities, as the associated soils have a higher nutrient value than those in adjacent habitats. However, these thickets are resilient and soon return to their previous state once the fields are abandoned after ca. 10 years (W. McCleland, pers. comm. 2021).

Key ecosystem services

The vegetation of the Temane IPA is of great importance as a biodiverse area for both plant species and a range of fauna (for example, endemic reptiles such as *Panaspis* and *Atractaspis* spp.) (Deacon 2014). The forests contribute significantly to carbon storage and climate regulation, particularly related to the precipitation cycle. Moreover, the local communities also take advantage of these terrestrial habitats, where they harvest firewood, wild fruits and medicinal plants, construction materials (sand, rock, lime and wood), and wood for fuel or charcoal production (A. Massingue, pers. comm. 2020). At present, however, the extraction of goods provided by this site constitute a threat to its ecosystems due to the high demand and lack of sustainable management.

Ecosystem service categories

- Provisioning – Food
- Provisioning – Medicinal resources
- Regulating services – Local climate and air quality
- Regulating services – Carbon sequestration and storage
- Regulating services – Moderation of extreme events
- Regulating services – Erosion prevention and maintenance of soil fertility
- Habitat or supporting services – Habitats for species
- Habitat or supporting services – Maintenance of genetic diversity

IPA assessment rationale

Temane qualifies as an IPA under Criterion A(i) as this area supports important populations of five globally threatened species: *Bauhinia burrowsii* (EN), *Croton aceroides* (EN), *Triaspis suffulta* (EN), *Croton inhambanensis* (VU) and *Ozoroa gomesiana* (VU). This IPA is so far known to support six national endemic species (Darbyshire *et al.* 2019a); this is below the 3% threshold of Mozambican endemic and range-restricted species needed to qualify this site under Criterion B(ii).

Triaspis suffulta (WM)

Bauhinia burrowsii (WM)

Spirostachys africanus-dominated low thicket (WM)

Croton inhambanensis (WM)

Priority species (IPA Criteria A and B)

FAMILY	TAXON	IPA CRITERION A	IPA CRITERION B	≥ 1% OF GLOBAL POP'N	≥ 5% OF NATIONAL POP'N	IS 1 OF 5 BEST SITES NATIONALLY	ENTIRE GLOBAL POP'N	SPECIES OF SOCIO-ECONOMIC IMPORTANCE	ABUNDANCE AT SITE
Anacardiaceae	Ozoroa gomesiana	A(i)	B(ii)	✓	✓	✓			abundant
Bignoniaceae	Dolichandrone alba		B(ii)						abundant
Euphorbiaceae	Croton aceroides	A(i)	B(ii)	✓	✓	✓			scarce
Euphorbiaceae	Croton inhambanensis	A(i)	B(ii)	✓	✓	✓			frequent
Fabaceae	Bauhinia burrowsii	A(i)	B(ii)	✓	✓	✓			abundant
Malpighiaceae	Triaspis suffulta	A(i)	B(ii)	✓	✓	✓			scarce
		A(i): 5 ✓	B(ii): 6						

Protected areas and other conservation designations

CONSERVATION AREA TYPE	CONSERVATION AREA NAME	RELATIONSHIP OF IPA TO CONSERVATION AREA
No formal protection	N/A	
Key Biodiversity Area	Inhassoro-Vilankulos	protected/conservation area overlaps with IPA

Threats

THREAT	SEVERITY	TIMING
Housing & urban areas	high	Ongoing – trend unknown
Commercial & industrial areas	high	Ongoing – trend unknown
Shifting agriculture	high	Ongoing – trend unknown
Small-holder plantations	low	Ongoing – trend unknown
Small-holder grazing, ranching or farming	low	Ongoing – trend unknown
Oil & gas drilling	high	Ongoing – trend unknown
Mining & quarrying	unknown	Ongoing – trend unknown
Roads & railroads	high	Ongoing – trend unknown
Logging & wood harvesting	high	Ongoing – trend unknown
Increase in fire frequency/intensity	unknown	Ongoing – trend unknown
Invasive non-native/alien species	medium	Ongoing – trend unknown

INHASSORO-VILANCULOS

Assessors: Clayton Langa, Castigo Datizua, Iain Darbyshire, Sophie Richards

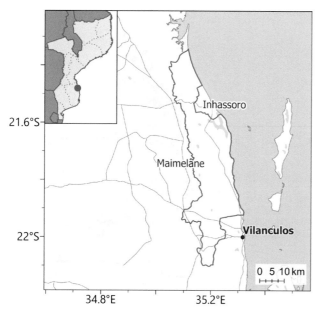
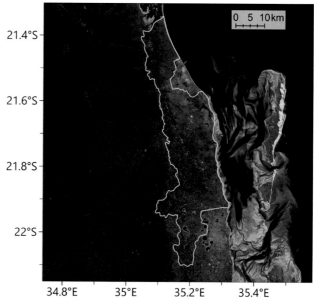

INTERNATIONAL SITE NAME		Inhassoro-Vilanculos	
LOCAL SITE NAME (IF DIFFERENT)		–	
SITE CODE	MOZTIPA053	PROVINCE	Inhambane

Coastal woodland with *Xylia mendoncae* (AM)

Govuro River and floodplain (WM)

LATITUDE	-21.72211	LONGITUDE	35.18269
ALTITUDE MINIMUM (m a.s.l.)	0	ALTITUDE MAXIMUM (m a.s.l.)	55
AREA (km²)	953	IPA CRITERIA MET	A(i), A(iv), B(ii)

Site description

The Inhassoro-Vilanculos IPA is located along the coast of northern Inhambane Province within the Inhassoro and Vilanculos Districts. It covers an area of 953 km² between the latitudes -21.34° to -22.11° and longitudes 35.09° to 35.4°. This IPA falls within the proposed Inhambane (sub-)Centre of Plant Endemism (CoE) (Darbyshire *et al.* 2019a). To the east of this site is the Indian Ocean coast, while in the south is Vilanculos town, in the north the mangroves of Govuro District and to the west the site boundary runs largely parallel to the EN1 road. The town of Inhassoro has been excluded from the site boundary. Despite a range of human pressures, this site contains a rich range of terrestrial habitats that support a notable number of plant species endemic to Mozambique.

Neighbouring this site are Temane IPA [MOZTIPA055] to the west and Mapinhane IPA [MOZTIPA056] to the south-west, while approximately 30–35 km offshore is the Bazaruto Archipelago IPA [MOZTIPA042]. The site also overlaps with the Inhassoro-Vilanculos Key Biodiversity Area (KBA) but is not formally protected at present.

Botanical significance

Inhassoro-Vilanculos hosts seven range-restricted and globally threatened plant species. This includes three Endangered species: *Ecbolium hastatum* and the wild relative of the aubergine, *Solanum litoraneum* (both endemic to southern Mozambique), and the woody climber, *Triaspis suffulta* (a strict endemic of the Inhambane CoE). In addition, there are four Vulnerable species known from this site: *Elaeodendron fruticosum* (endemic to the Inhambane CoE, occurring across coastal areas of Gaza and Inhambane Provinces); *Ozoroa gomesiana* (a strict endemic of northern Inhambane Province, mostly within the Mapinhane, Inhassoro and Vilanculos IPAs); *Psychotria amboniana* subsp. *mosambicensis* (endemic to southern Mozambique); and *Xylia mendoncae* (a strict endemic of northern Inhambane Province, occurring in Vilanculos, Inhassoro and Govuro Districts). The largest population globally of *X. mendoncae* occurs within this IPA, which is therefore the global stronghold for this species, with more than 65% of its population estimated to be within this IPA. From satellite imagery (Google Earth 2021), it is notable that this species prefers coastal deciduous miombo woodland, with *Brachystegia* sp. and *Julbernardia globiflora*, close to freshwater depressions. The habitats around the freshwater depressions are preferred areas for machambas. Despite this, *X. mendoncae* appears to be able to regenerate in secondary regrowth following shifting agriculture.

A further species of conservation interest is the cycad *Encephalartos ferox* subsp. *emersus*. Provisionally assessed as Endangered, this subspecies has now been confirmed to be endemic to the Inhambane CoE, in Inhambane coastal districts of Inhassoro, Vilanculos and Jangamo (Massingue 2019), occurring on ancient termitaria in seasonally flooded coastal plains (Burrows *et al.* 2018). This IPA contains a majority of the global population. This subspecies was previously suggested to be Critically Endangered (Rousseau *et al.* 2015) but is now known to occur in a wider area of the Inhambane coastal districts (Massingue 2019).

This IPA holds within its boundaries 12 national endemic species and six near-endemic species. The Mozambican endemics comprise the seven previously mentioned globally threatened or potentially threatened species, plus four Least Concern species, *Ammannia fernandesiana* (for which this may be the most important site globally), *Chamaecrista paralias*, *Triainolepis sancta* and *Zanthoxylum delagoense*.

Near-endemic plant taxa known from this site include *Commiphora schlechteri* (LC), *Crotalaria dura* subsp. *mozambica*, *Cussonia arenicola* and *Pavetta gracillima*. Additionally of interest is the saltmarsh species *Caroxylon littoralis*, restricted to the coastlines on the Mozambique Channel with an Area of Occupancy of around 48 km^2 (Friis & Holt 2017). This IPA represents one of only three locations for this species in Mozambique.

In addition to this, Vilanculos-Inhassoro hosts significant areas of coastal miombo on primary dunes. Throughout most of Mozambique, coastal miombo occurs towards the rear of the dune system; however, in this IPA, miombo also occurs on primary dunes. Nationally this habitat is highly restricted, limited to a small area around Vilanculos and Inhassoro Districts, and is also threatened by encroaching agriculture and expanding townships (Massingue 2019). This IPA is, therefore, of great conservation importance for this habitat type.

Habitat and geology

In the broad sense, the Inhassoro-Vilanculos IPA region lies within the Swahilian-Maputaland Regional Transition Zone phytogeographical region according to Clarke (1998), covering much of coastal-belt Mozambique, and in the Southern Zanzibar-Inhambane Coastal Forest Mosaic Ecoregion according to Schipper & Burgess (2015), which stretches for ca. 2,200 km along

Coastal dune thicket within the Inhassoro-Vilanculos IPA (WM)

the eastern coast of the African continent, from southern Tanzania to Xai-Xai (Gaza Province) in Mozambique. In a narrower phytogeographical sense, this IPA falls within the northern extension of the Maputaland Centre of Endemism, recently proposed as the Inhambane (sub-)CoE (Darbyshire et al. 2019a), which was proposed due to the high concentration of plant endemism found only within this area (Massingue 2019).

The climate is influenced by the warm current from the Mozambique Channel and is characterized to be a humid to sub-humid tropical climate, with two seasons (Lambrechts 2003; Cumbe 2007; Massingue 2019): a hot rainy season from August to February, and a dry and relatively cool season from February to July. In the rainy season, the average temperatures vary from 28°C to 30°C, while in the dry season the temperatures vary from 18°C to 27°C. Average annual rainfall ranges from 865 mm to 936 mm, with highest rainfall on the coast (Lambrechts 2003, Governo do Distrito de Inhassoro 2011, EOH 2015a, World Resources Institute 2021). The elevation of the IPA ranges from 0 m to 55 m. It is part of the great coastal plains that stretch along a large extent of coastal Mozambique and is characterized by unconsolidated sandy soils of fine texture, with a very low clay content, originated from wind and/ or marine activities, interspersed with areas of hydromorphic soils and soils derived from marine sediments found along the coastline. Red clay soils and sodic soils (mananga soils) dominate the inland zones (MICOA 2012a).

The coastal vegetation of the Inhassoro-Vilanculos IPA is highly fragmented through a combination of threats (see below). Despite this, there are still significant areas of intact coastal habitats that are of high conservation concern. In recent years, this IPA has been subject to several botanical surveys, which have helped to add, update, and confirm our understanding of plant diversity in the Inhassoro IPA region and to improve understanding of the mosaic of habitats within this IPA. It lies within a frequently flooded landscape and so features a range of wetland types notably: riverine floodplains with swamps, seasonal coastal streams, coastal lakes and lagoons, and mangrove forests at the coast. Among these wetlands, the most important are the Nhangonzo stream by the coast in the eastern most corner of this IPA (-21.72°, 35.24°) and the Govuro River floodplain further inland to the south (-21.75°, 35.14°), dominated by emergent and inundated vegetation and characterised by the presence of Phragmites australis and Nymphaea within the water bodies, and by the abundance of Hyphaene coriacea and Phoenix reclinata associated with termite mounds (MICOA 2012a; Deacon 2014).

In the east of the IPA, there is a small extent of mangroves with Rhizophora mucronata, Bruguiera gymnorrhiza, Avicennia marina and Ceriops tagal. Associated with this habitat are coastal saltmarshes, with scattered to dense samphires (Sarcocornia sp.) (MICOA 2012a; Lötter et al. in prep.).

The coastal dunes support a mosaic of semi-deciduous to evergreen thicket and forest, ranging in density and structure. This vegetation type is dominated by Acacia kraussiana, A. robusta var. usambarensis, Acokanthera oblongifolia, and Acridocarpus natalitius, with plants of conservation interest including Commiphora schlechteri and, notably, Ecbolium hastatum (Lötter et al. in prep.).

Littoral miombo woodland covers significant areas of the IPA, dominated by Brachystegia spiciformis, B. torrei and Julbernardia globiflora (Lötter et al. in prep.). Towards the coastline, the canopy is more closed (75% tree and shrub cover) and the understorey is dominated by the grass Halopyrum mucronatum together with shrubby Scaevola plumieri. Further inland, the canopy is more open (35% canopy cover) and herbaceous cover is greater, featuring grasses such as Eragrostis chapelieri, Melinis repens, Perotis patens, Schizachyrium sanguineum and Sporobolus pyramidalis (Deacon 2014; EOH 2015b). Interestingly, this site also hosts miombo on primary dunes, a nationally scarce and threatened habitat known only from Vilanculos to Inhassoro, which is dominated by Julbernardia globiflora and rarely Brachystegia spiciformis (Massingue 2019). This important habitat protects inland terrestrial ecosystems from excessive marine influences and holds important plants for conservation interest including the Endangered Triaspis suffulta and Ecbolium hastatum among others.

Further inland are deciduous forest and woodland mosaics, often with an open woodland of medium-sized trees and shrubs (of 30% cover) and with an herbaceous ground cover (60%) of Megathyrsus maximus, Schizachyrium sanguineum

Ecbolium hastatum (WM)

Xylia mendoncae (WM)

and *Sporobolus pyramidalis*. This habitat holds similar tree species to the miombo woodland, with *Brachystegia spiciformis* and *Julbernardia globiflora*, but is dominated by deciduous species such as *Afzelia quanzensis*, *Balanites maughamii*, *Sideroxylon inerme* subsp. *diospyroides*, and *Suregada zanzibariensis* (MICOA 2012a; Lötter *et al.* in prep.). Moreover, this habitat holds several plants of conservation interest, namely *Chamaecrista paralias*, *Commiphora schlechteri*, *Psychotria amboniana* subsp. *mosambicensis* and *Zanthoxylum delagoense*.

In the west of this site are a number of seasonal wetlands, associated with savanna vegetation. The woody component is dominated by *Hyphaene coriacea* accompanied by *Albizia versicolor*, *Annona senegalensis*, *Dichrostachys cinerea*, *Ozoroa obovata* and *Sclerocarya birrea*, alongside the endemic species *Chamaecrista paralias*; these wetlands are also where the endemic shrubby herb *Ammannia fernandesiana* can be found.

Conservation issues

Despite its importance for plant diversity and endemism, the Inhassoro-Vilanculos IPA does not lie within a formal protected area. However, most of the region is included within the recently identified Inhassoro-Vilankulos KBA (WCS *et al.* 2021). The area is not currently recognised as a RAMSAR site, despite the diversity of wetlands

which are suitable for waterbirds and migratory birds (Lambrechts 2003; Golder Associates 2014; EOH 2015a, 2015b; Google Earth 2021). Impacto Lda. (2018) determined that 63.4 ha of thicket/ coastal dune forest located in a narrow strip along the northern and southern part of the Nhangonzo Estuary qualifies as a "Critical Habitat". Moreover, the restricted primary dune miombo, a habitat type of conservation significance, also occurs within the coastal dune vegetation (Massingue 2019).

Along the coast, the Nhangonzo humid zone has recently been subjected to drilling of pools for prospecting of petroleum and natural gas, and seismic research. These activities have increased vegetation damage by opening new access routes, and intensification of land use by local communities. However, according to the recent report by Impacto Lda. (2018), these areas will be no longer exploited, as it would contravene requirements present in the current exploitation license for Sasol Lda, which prohibits activities related to oil and gas exploitation within a range of 500 m from the coast.

Botanical surveys carried out in some areas of this site, especially the critical area of Nhangonzo, pointed out the existence of invasive plant species in the vicinity of villages and agricultural areas, such as *Agave sisalana*, *Lantana camara*, *Melinis repens* and *Opuntia ficus-indica*. There are also some exotic trees, such as cashew, mango and *Casuarina equisetifolia*, which occur in small numbers in disturbed areas (EOH 2015a, 2015b).

More generally, this IPA is currently under high pressure and degradation by local communities because of the harvesting of firewood, charcoal production, livestock grazing, agriculture with an associated increase in fire frequency, and expansion of settlements (MICOA 2012a; Massingue 2019; Google Earth 2021; World Resources Institute 2021). The cultivated areas are concentrated by wetlands and along access roads and paths, and notably near Sasol's oil flowlines which are widespread along the coastal stretch of this IPA and themselves cause a level of habitat degradation (Deacon 2014; EOH 2015a, 2015b; Google Earth 2021). It is also likely that the expansion of tourism infrastructure observed within this IPA, especially along the coastline, is also negatively affecting the integrity of the dune ecosystems by gradually replacing the primary dune miombo vegetation (Massingue 2019).

Coastal forest and woodland restoration and prevention of further habitat loss or degradation should be considered high priorities to enable the conservation of the threatened plant species of this IPA. Urgent action is required to protect the remaining vegetation given the scale of habitat loss to date.

Key ecosystem services

The ecosystem services of this IPA can be divided into terrestrial and marine/aquatic services. On the terrestrial side (the focus of the IPA), the coastal thickets and forests contribute significantly to carbon storage and climate regulation, particularly related to the precipitation cycle. In addition, they help to maintain the natural integrity of the coast by protecting against coastal erosion from the ocean and winds, as well as being important catchment basins in the protection of groundwater hydrological processes (EOH 2015a, 2015b; Massingue 2019). Further, these habitats are also important for a range of fauna (e.g. endemic reptiles such as *Panaspis* and *Atractaspis* spp.). Moreover, the local human populations also take advantage of these terrestrial habitats, where they harvest firewood, wild fruits, and medicinal plants (EOH 2015a, 2015b; A. Massingue, pers. comm. 2020).

The aquatic/marine ecosystems are represented at Inhassoro-Vilanculos IPA by riverine floodplains, coastal streams, coastal lakes and lagoons, and mangroves. The mangrove forests are mostly associated with estuaries and provide several environmental, economic, and social services. They are important in preventing coastal erosion, in alleviating floods and in the reproduction cycles of various species. Mangroves are also of socio-economic value, not only through their association with fishing activities but also because they are utilised by local households for construction, and as a source of traditional medicines and firewood (MICOA 2012a; EOH 2015a, 2015b). Some estuarine ecosystems play important ecological roles due to their high productivity, providing a source of nutrients and organic matter to other ecosystems, and providing shelter for many species and nurseries for migratory species (MICOA 2012a). The wetland flats also act as corridors for fauna that utilise dense cover to move or migrate (Deacon 2014).

The tourist potential of the IPA region is concentrated on the coast and dunes (Governo do Distrito de Inhassoro 2011; MICOA 2012a). However, there are also notable areas of interest in inland ecosystems, for example the wetlands that support migratory birds and therefore may have some value as a birdwatching spot.

Ecosystem service categories

- Provisioning – Food
- Provisioning – Fresh water
- Provisioning – Medicinal resources
- Regulating services – Local climate and air quality
- Regulating services – Moderation of extreme events
- Regulating services – Erosion prevention and maintenance of soil fertility
- Habitat or supporting services – Habitats for species
- Habitat or supporting services – Maintenance of genetic diversity
- Cultural services – Recreation and mental and physical health
- Cultural services – Tourism

IPA assessment rationale

Inhassoro-Vilanculos qualifies as an IPA under criterion A(i), as this area supports important populations of seven globally threatened species, namely *Ecbolium hastatum* (EN), *Elaeodendron fruticosum* (VU), *Ozoroa gomesiana* (VU), *Psychotria amboniana* subsp. *mosambicensis* (VU), *Solanum litoraneum* (EN), *Triaspis suffulta* (EN) and *Xylia mendoncae* (VU). It also qualifies under Criterion A(iv), due to the occurrence of *Encephalartos ferox* subsp. *emersus*, as this subspecies is range-restricted and potentially threatened, and the majority of its known global population lies within this IPA. This IPA holds 12 national endemic species and falls within the top 15 sites nationally for endemic and range-restricted species; Inhassoro-Vilanculos therefore qualifies under sub-criterion B(ii) of the IPA criteria. The coastal miombo on primary dunes is a range-restricted and threatened habitat covered by this IPA which supports seven endemic species. With further research, this site may qualify under criterion C(iii) of the IPA criteria in future, due to the presence of this habitat type.

Encephalartos ferox subsp. *emersus* (WM)

Priority species (IPA Criteria A and B)

FAMILY	TAXON	IPA CRITERION A	IPA CRITERION B	≥ 1% OF GLOBAL POP'N	≥ 5% OF NATIONAL POP'N	IS 1 OF 5 BEST SITES NATIONALLY	ENTIRE GLOBAL POP'N	SPECIES OF SOCIO-ECONOMIC IMPORTANCE	ABUNDANCE AT SITE
Acanthaceae	*Ecbolium hastatum*	A(i)	B(ii)	✓	✓	✓			unknown
Anacardiaceae	*Ozoroa gomesiana*	A(i)	B(ii)	✓	✓	✓			common
Celastraceae	*Elaeodendron fruticosum*	A(i)	B(ii)	✓	✓	✓			unknown
Fabaceae	*Chamaecrista paralias*		B(ii)						common
Fabaceae	*Xylia mendoncae*	A(i)	B(ii)	✓	✓	✓			occasional
Lythraceae	*Ammannia fernandesiana*		B(ii)						unknown
Malphigiaceae	*Triaspis suffulta*	A(i)	B(ii)	✓	✓	✓			scarce
Rubiaceae	*Psychotria amboniana* subsp. *mosambicensis*	A(i)	B(ii)	✓	✓	✓			occasional
Rubiaceae	*Triainolepis sancta*		B(ii)						common
Rutaceae	*Zanthoxylum delagoense*		B(ii)						scarce
Solanaceae	*Solanum litoraneum*	A(i)	B(ii)	✓	✓	✓			common
Zamiaceae	*Encephalartos ferox* subsp. *emersus*	A(iv)	B(ii)	✓	✓	✓			scarce
		A(i): 7 ✓ A(iv): 1 ✓	B(ii): 12						

Protected areas and other conservation designations

CONSERVATION AREA TYPE	CONSERVATION AREA NAME	RELATIONSHIP OF IPA TO CONSERVATION AREA
No formal protection	N/A	
Key Biodiversity Area	Inhassoro-Vilankulos	protected/conservation area overlaps with IPA

Threats

THREAT	SEVERITY	TIMING
Housing & urban areas	medium	Ongoing – trend unknown
Commercial & industrial areas	medium	Ongoing – trend unknown
Tourism & recreation areas	medium	Ongoing – trend unknown
Shifting agriculture	medium	Ongoing – trend unknown
Small-holder grazing, ranching or farming	low	Ongoing – trend unknown
Oil & gas drilling	medium	Ongoing – trend unknown
Mining & quarrying	unknown	Ongoing – trend unknown
Roads & railroads	high	Ongoing – trend unknown
Logging & wood harvesting	low	Ongoing – trend unknown
Recreational activities	high	Ongoing – trend unknown
Increase in fire frequency/intensity	unknown	Ongoing – trend unknown
Invasive non-native/alien species/diseases	medium	Ongoing – trend unknown

BAZARUTO ARCHIPELAGO

Assessors: Castigo Datizua, Clayton Langa, Iain Darbyshire, Sophie Richards

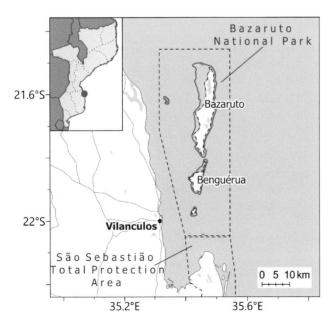

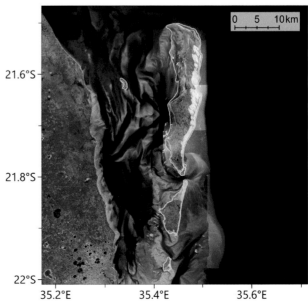

INTERNATIONAL SITE NAME		Bazaruto Archipelago	
LOCAL SITE NAME (IF DIFFERENT)		Arquipélago de Bazaruto	
SITE CODE	MOZTIPA042	PROVINCE	Inhambane

LATITUDE	-21.73061	LONGITUDE	35.44182
ALTITUDE MINIMUM (m a.s.l.)	0	ALTITUDE MAXIMUM (m a.s.l.)	90
AREA (km²)	190	IPA CRITERIA MET	A(i)

Site description

The Bazaruto Archipelago is located between the latitudes 21°30'–22°10' S and longitudes 35°22'–35°30' E, off the Indian Ocean coast of northern Inhambane province, between Vilanculos and Inhassoro districts in southern Mozambique. It consists of five Islands, Bazaruto (120.5 km²), Benguerra (32.86 km²), Magaruque (2.96 km²), Santa Carolina (2.10 km²), and Bangue (0.66 km²) (Everett et al. 2008; Díaz Pelegrín et al. 2016). This Archipelago forms an Important Plant Area (IPA) located within the marine protected area of the Bazaruto Archipelago National Park (BANP), which spans 1,430 km² (African Parks 2021).

These islands contain a range of terrestrial and marine habitats and provide refuge for a great variety of plant and animal species (Everett et al. 2008), including a group of endemic and near-endemic species to Mozambique. Only four of the islands, Bazaruto, Benguerra, Magaruque, and Santa Carolina, are included in the IPA, as according to Everett et al. (2008) Bangue Island comprises only beach and pioneer dune vegetation, and so is unlikely to be of high botanical importance.

Botanical significance

This site is of high botanical importance as it holds a number of range-restricted species of high conservation concern, of which three are globally threatened and endemic species to Mozambique: Memecylon insulare (CR; restricted to Magaruque Island), Jatropha subaequiloba (VU; found on Bazaruto Island and nearby São Sebastião Peninsula), and Ochna beirensis (EN; possibly found on these islands and also nearby Beira and Cheringoma). Ochna beirensis has been documented as present in BANP by Everett et al. (2008), but this record requires confirmation as no herbarium specimen has been seen from this site by the current authors. Although in BANP, the botanical component is currently under-explored, the islands are known so far to support eight national endemic plant taxa (plants that only occur in Mozambique) and 10 near-endemic plant taxa (plants that are restricted to Mozambique and neighbouring countries). The endemics consist of the three previous mentioned globally threatened species, plus four Least Concern species, Chamaecrista paralias, Psydrax moggii, Triainolepis sancta and Zanthoxylum delagoense, and one that has not yet been assessed but is considered likely to be of Least Concern, Spermacoce kirkii. To date, only three of the islands, Bazaruto, Magaruque, and Santa Carolina, are known to hold range-restricted plant species of high conservation importance, but given that Benguerra Island has a similar vegetation composition to Bazaruto Island, future surveys are likely to reveal that Benguerra also holds populations of some of these priority species.

The seagrass communities around the islands are also of importance and contain a population of the globally Vulnerable Zostera capensis.

Habitat and geology

This IPA results from a dynamic process of stacking dunes, originating from coastal deposits from the Indian Ocean. Three dune forms can be recognised, namely ancient dunes, an ocean dune cord, and recent coastal formations (Díaz Pelegrín et al. 2016). The soils are sandy, rocky, and white and are poor for intensive agricultural activity due to the limiting factors of low water retention capacity and low fertility (Díaz Pelegrín et al. 2016). The climate is classified as Humid Tropical Coastal, which is biseasonal, with the peak dry season in June to August and the peak humid season in December to March. The islands have an

Coastal dune and beach flora on Bazaruto Island (OB)

annual rainfall of approximately 1,200 mm, and an annual temperature average ranging from 20° to 26°C according to the season. The geographical elevation ranges from 0 m to 4 m at the sea line and inland wetlands up to 90 m on the highest dunes near the coast. The combination of these different physical elements has significant effects on the biotic composition and diversity of this IPA.

This IPA has been mapped and briefly classified into 11 natural terrestrial vegetation communities (Dutton & Drummond 2008): (1) savanna grassland, maintained principally by a perched water table, with woody species including *Garcinia livingstonei* and *Ozoroa obovata*; (2) marsh or edaphic grassland, with a range of dominant grasses including *Sporobolus virginicus*, *Diplachne fusca* and *Andropogon eucomus*; (3) evergreen dune forest, which is highly degraded and only known from three small remnant patches on Benguerra and Magaruque, with occurrence of *Balanites maughamii* and *Ozoroa obovata*; (4) secondary dune forest, dominated by *Mimusops caffra*, *Olax dissitiflora* and *Bourreria petiolaris*; (5) scrub thicket, dominated by *Eugenia* spp. and *Euclea racemosa*; (6) swamp forest, severely damaged by agriculture; (7) woodland dominated by *Dialium schlechteri* and *Julbernardia globiflora*; (8) thicket associated with the perched water table at the base of west-facing coastal sand dunes, dominated by *Olax dissitiflora*, *Bourreria petiolaris* and *Acacia karroo*; (9) pioneer dune flora, dominated by species such as *Ipomoea pes-caprae*, *Scaevola plumieri* and *Cyperus crassipes*; (10) mangroves, only known from Bazaruto, Benguerra and Santa Carolina, dominated by *Rhizophora mucronata*, *Bruguiera cylindrica* and *Ceriops tagal*; and (11) salt marshes or salinas, also known only from Bazaruto, Benguerra and Santa Carolina, dominated by, *Sesuvium portulacastrum*, *Salicornia perennis* and *Salicornia perrieri*.

Bazaruto Island supports the largest areas of natural habitat, where nine of the 11 vegetation types occur, with only evergreen dune forest and *Dialium-Julbernardia* woodland absent. Four vegetation types can be found across Magaruque Island, namely evergreen dune forest, secondary dune forest, scrub thicket, and pioneer dune vegetation. Santa Carolina Island holds a large mangrove community, secondary dune forest, and salt marsh (Dutton & Drummond 2008), whilst evergreen dune forest and *Dialium-Julbernardia* woodland occur on Benguerra Island (Downs & Wirminghaus 1997; Dutton & Drummond 2008). From these 11 terrestrial vegetation assemblages in the BANP, three are of highest botanical significance, as they hold species of conservation importance: *Memecylon insulare* occurs in evergreen dune forest, whilst *Jatropha subaequiloba* is found in secondary dune forest and edaphic grassland. The secondary dune forests are also suitable for *Ochna beirensis*.

There is also a significant extent of seagrass communities within the IPA, associated with the sandy tidal flats and dominated by *Thalassodendron ciliatum*, *Cymodocea rotundata*, *Halodule uninervis* and *Zostera capensis* (Bandeira *et al.* 2008).

Conservation issues

The Bazaruto Archipelago National Park (BANP) was primarily designated to protect marine mammals (dugongs, dolphins, whales), sharks, sea turtles, corals, Echinoderms (Holoturias), molluscs, and fish species (Vaz *et al.* 2008). However, both marine and terrestrial ecosystems are now benefitting from conservation measures undertaken by African Parks (African Parks 2021).

Since 2017, the BANP has been managed by African Parks in partnership with the Mozambique government under a 25-year agreement. African Parks' stated priorities include strengthening law enforcement in order to reduce threats to the biodiversity of the BANP and building support for the conservation actions through community engagement, training, and local employment. Thirty-four new rangers have been employed and trained as part of this process (African Parks 2021).

With regard to terrestrial biodiversity, birds are the most studied fauna group to date, and they also benefit from international protection status since many are migratory (Díaz Pelegrín *et al.* 2016). Most of the key terrestrial habitats in the BANP, including mangroves forests and some lagoons and swamps, have been given the designation of Total Terrestrial Protection Zones (TTPZ) (Díaz Pelegrín *et al.* 2016). However, some of these areas are currently under pressure owing to unsustainable tourism and human population growth. Therefore, the integrity of the terrestrial biodiversity of this IPA is threatened by settlement expansion and tourism infrastructure and footfall, unsustainable subsistence and consumption activities such as intensive firewood harvesting, agriculture, some livestock grazing (goats), and collection of medicinal plants (Dutton 1990; Downs & Wirminghaus 1997; Everett *et al.* 2008, Díaz Pelegrín *et al.* 2016). There are also many cases of uncontrolled fire events reported in the forests, sometimes being set intentionally for small-scale shifting cultivation and pasture areas (Díaz Pelegrín *et al.* 2016). These issues must be addressed if the botanical importance of these islands is to be maintained.

The seagrass communities are also protected, designated as Marine Total Protection Zones (MTPZ).

Key ecosystem services

Apart from BANP's role in regulating and maintaining marine ecological processes and providing natural resources to local communities, terrestrial ecosystems also provide important ecosystem services. They help to maintain the natural integrity of the islands by protecting against coastal erosion from the ocean and winds. Terrestrial and coastal forests also contribute significantly to carbon storage and climate regulation, particularly related to the precipitation cycle. Seagrass carpets are also important for carbon storage (Fourqurean *et al.* 2012). The mangrove forests, besides storing carbon, constitute an important micro-ecosystem and ecological role (being the reproductive grounds and a refuge for marine fauna) and have socio-economic value. The local communities also take advantage of the terrestrial habitats, where they harvest firewood and practice subsistence agriculture, and collect wild fruits and medicinal plants (Díaz Pelegrín *et al.* 2016). The touristic potential of the Bazaruto Archipelago is also high due to the beauty of the natural resources, beaches, coral reefs, seagrass carpets, and crystal clear waters for diving and marine megafauna; because of these factors, the Islands are regarded as a key area for nature-based tourism in Mozambique (World Bank 2018).

Jatropha subaequiloba (OB)

Ecosystem service categories

- Provisioning – Food
- Provisioning – Raw materials
- Provisioning – Medicinal resources
- Regulating services – Local climate and air quality
- Regulating services – Carbon sequestration and storage
- Regulating services – Moderation of extreme events
- Regulating services – Erosion prevention and maintenance of soil fertility
- Habitat or supporting services – Habitats for species
- Habitat or supporting services – Maintenance of genetic diversity
- Cultural services – Recreation and mental and physical health
- Cultural services – Tourism

Savanna grassland and dune thicket on Bazaruto Island (OB)

IPA assessment rationale

The Bazaruto Archipelago qualifies as an IPA under Criterion A(i), as these islands support populations of two, and potentially three, endemic and globally threatened species. Bazaruto Island is considered likely to be the main global stronghold for the Vulnerable *Jatropha subaequiloba* and Magaruque Island is the only known locality globally for the Critically Endangered *Memecylon insulare*. Everett *et al.* (2008) documented the presence of the globally Endangered *Ochna beirensis* in scrub thickets in secondary dunes in this IPA, but this record requires confirmation. In addition, the seagrass communities surrounding the islands support a population of the globally Vulnerable *Zostera capensis*, which extends around the coast of southern Africa; the Bazaruto population is of national importance. Overall, this IPA supports populations of 18 species that are endemic or near-endemic to Mozambique, according to Darbyshire *et al.* (2019a), although only seven of these qualify under sub-criterion B(ii) and so this site does not currently meet the threshold under that sub-criterion. However, further botanical surveys are required as it is considered likely that there may be further rare and threatened plant species present in under-explored habitats at this IPA.

Priority species (IPA Criteria A and B)

FAMILY	TAXON	IPA CRITERION A	IPA CRITERION B	≥ 1% OF GLOBAL POP'N	≥ 5% OF NATIONAL POP'N	IS 1 OF 5 BEST SITES NATIONALLY	ENTIRE GLOBAL POP'N	SPECIES OF SOCIO-ECONOMIC IMPORTANCE	ABUNDANCE AT SITE
Euphorbiaceae	*Jatropha subaequiloba*	A(i)	B(ii)	✓	✓	✓			unknown
Fabaceae	*Chamaecrista paralias*		B(ii)						common
Melastomataceae	*Memecylon insulare*	A(i)	B(ii)	✓	✓	✓	✓		unknown
Ochnaceae	*Ochna beirensis*	A(i)	B(ii)			✓			unknown
Rubiaceae	*Psydrax moggii*		B(ii)						common
Rubiaceae	*Spermacoce kirkii*		B(ii)						unknown
Rubiaceae	*Triainolepis sancta*		B(ii)						common
Rutaceae	*Zanthoxylum delagoense*		B(ii)						scarce
Zosteraceae	*Zostera capensis*	A(i)			✓	✓			common
		A(i): 4 ✓	B(ii): 8						

Protected areas and other conservation designations

CONSERVATION AREA TYPE	CONSERVATION AREA NAME	RELATIONSHIP OF IPA TO CONSERVATION AREA
National Park	Bazaruto Archipelago National Park	protected/conservation area encompasses IPA
Key Biodiversity Area	Grande Bazaruto	protected/conservation area encompasses IPA
Important Bird Area	Greater Bazaruto	protected/conservation area encompasses IPA

Threats

THREAT	SEVERITY	TIMING
Housing & urban areas	medium	Ongoing – increasing
Tourism & recreation areas	high	Ongoing – increasing
Shifting agriculture	unknown	Ongoing – trend unknown
Small-holder grazing, ranching or farming	low	Ongoing – trend unknown
Roads & railroads	medium	Ongoing – trend unknown
Gathering terrestrial plants	low	Ongoing – trend unknown
Logging & wood harvesting	low	Ongoing – trend unknown
Recreational activities	high	Ongoing – increasing
Increase in fire frequency/intensity	unknown	Ongoing – trend unknown

SÃO SEBASTIÃO PENINSULA

Assessors: Sophie Richards, Iain Darbyshire

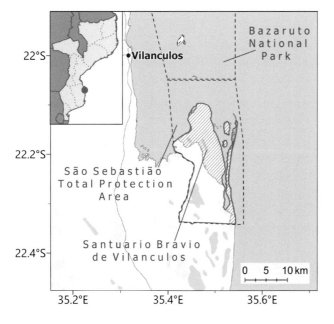
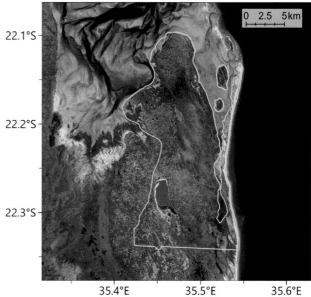

INTERNATIONAL SITE NAME		São Sebastião Pennisula	
LOCAL SITE NAME (IF DIFFERENT)		Península de São Sebastião	
SITE CODE	MOZTIPA045	PROVINCE	Inhambane

LATITUDE	-22.13190	LONGITUDE	35.47420
ALTITUDE MINIMUM (m a.s.l.)	0	ALTITUDE MAXIMUM (m a.s.l.)	92
AREA (km²)	227.0	IPA CRITERIA MET	A(i)

Site description

The São Sebastião Peninsula Important Plant Area is located in coastal Vilanculos District, south-east of Vilanculos (or Vilankulos) town. The IPA covers a total area of 227 km² and consists of the terrestrial zones of the São Sebastião Total Protection Area, including the islands Luene and Chilonzuíne. This site is situated just south of the Bazaruto Archipelago IPA [MOZTIPA042].

The São Sebastião Peninsula contains a number of important coastal habitats, including mangroves, saltmarshes, miombo and dune thicket, which support a range of rare and threatened species. Although the entire IPA falls within a protected area, only the north-west of the peninsula appears to be under conservation management. The Santuario Bravio de Vilanculos ("Vilanculos Sanctuary"), which covers 105 km² of this IPA, was granted private reserve status in 2003 (SBV 2017a). Habitat restoration is being undertaken within Vilanculos Sanctuary and, following this work, there is now a marked difference between vegetation cover either side of the private reserve boundary (Google Earth 2021). Although many of the habitats outside Vilanculos Sanctuary are degraded, they have been included within this IPA as they are also part of the Total Protection Area. Restoration work and the introduction of sustainable development initiatives in this part of the IPA could enable the conservation of species and habitats while providing secure livelihoods for local communities.

Botanical significance

The São Sebastião Peninsula hosts a number of plant species of conservation importance, including four globally threatened species. Of particular note is *Ecbolium hastatum* (EN), known only from coastal areas of southern Mozambique and threatened elsewhere by habitat clearance for tourism and subsistence agriculture. *E. hastatum* is locally common where it occurs in this IPA (*Jacobsen* #6082) but is generally scarce across the site as a whole (Massingue *et al.* 2021), known from only a couple of localities. São Sebastião represents the most secure site for *E. hastatum* and is therefore crucial in preventing the extinction of this species.

In addition to this Endangered species, a further three Vulnerable species occur at this site: *Elaeodendron fruticosum*, *Jatropha subaequiloba* and *Millettia ebenifera*. *Jatropha subaequiloba* is particularly important as it is known only from this site and Bazaruto Island, covering a range of just 75 km². Although more widespread, occurring throughout southern coastal Mozambique, *M. ebenifera* and *E. fruticosum* face threats such as expansion of urban areas, tourism and conversion of land to agriculture throughout their respective ranges. For *M. ebenifera*, São Sebastião represents the only protected area within its range and so is of great importance for the conservation of this species.

An additional Vulnerable taxon, *Psychotria amboniana* subsp. *mosambicensis*, may well occur within this IPA. A specimen that is highly likely to be this sub-species was collected at this site, but further investigation is needed to confirm its presence at São Sebastião (Massingue *et al.* 2021). *P. amboniana* subsp. *mosambicensis* is restricted to southern coastal Mozambique and so its presence would also represent an additional endemic species within this IPA.

Overall, there are nine endemic species known from this IPA, including the three threatened species alongside *Carpolobia suaveolens*, *Chamaecrista paralias*, *Triainolepis sancta*, *Tritonia moggii* and *Zanthoxylum delagoense* (all LC). Although not endemic, or thought to be threatened with extinction, the saltmarsh species *Caryoxylon littoralis* is also of interest. *C. littoralis* has a limited distribution, which is restricted to the coastlines across the Mozambique Channel with an area of occupancy of around 48 km² (Friis & Holt 2017). Previously, material of this species from São Sebastião was thought to be an undescribed species, *Salsola* sp. A. However, Friis and Holt (2017) found this population to be conspecific with *C. littoralis*, a species then thought limited to Madagascar and Île Europe. São Sebastião represents one of only three locations for this species in Mozambique.

The presence of *Pavetta uniflora* may also be of conservation importance. Despite the wide range of this species, from Inhambane Province in Mozambique to Somalia in the north, *P. uniflora* is scarce along the east African coastline and has not yet been assessed for the IUCN Red List but may well be a threatened species.

The Near Threatened species, *Encephalartos ferox*, occurs within this IPA (Read 2020). This is probably the subspecies *ferox* which occurs in sheltered coastal dunes. Another Near Threatened species, *Coffea racemosa*, is known from this site (Read 2020). Also known as Inhambane coffee, this species is a tertiary relative of, and may be a useful gene donor to, commercial coffee species, while seeds of *C. racemosa* itself can also be roasted and used to make coffee (O'Sullivan & Davis 2017).

Although not associated with rare or threatened species, the mangroves at this site are of great importance as an ecological community for the habitats they provide for marine life and for coastal protection, particularly during cyclone season.

Habitat and geology

São Sebastião hosts a range of coastal habitats underlain by sandy soils (Massingue *et al.* 2021). Average temperatures in summer (October to March) are 28–33°C and in winter (April to September) 22–27°C. Rainfall within this IPA averages around 750 mm per annum, most of which falls between December and March, coinciding largely with the cyclone season (January to March), when the spring tides are particularly high (SBV 2017b).

Mangroves dominated by trees of *Rhizophora mucronata* and the rush *Juncus kraussii* (CL)

View of foredunes and coastal thicket (SV)

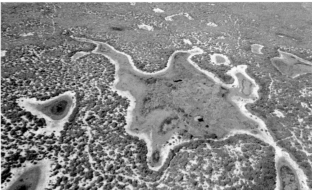

Coastal pans surrounded by open miombo (SV)

Previously, limited botanical collecting was undertaken at this site in 1958 (A.O.D. Mogg) and in 2002 (A.H.G. Jacobsen). However, as part of the conservation work undertaken at Vilanculos Sanctuary, a checklist was compiled by Mark Read (2020), a private resident of Vilanculos Sanctuary, and a botanical survey is now underway by Instituto de Investigação Agrária de Moçambique, Eduardo Mondlane University and Royal Botanic Gardens, Kew to build on the knowledge of the terrestrial habitats and associated plant diversity (Massingue *et al.* 2021).

Owing to the annual tidal surges that occur during cyclone season, the mangroves are of great importance in preventing inundation of the peninsula. The most extensive patch of mangrove forest is found along the eastern coast of the peninsula, while there are smaller patches on Ilha Lunene and in the wetlands on the western coast (Google Earth 2021). Common mangrove species such as *Avicennia marina, Ceriops tagal, Rhizophora mucronata* and *Sonneratia alba* are present in these habitats (Read 2020). Associated with these mangrove habitats and tidal inlets are numerous saltmarshes. These areas have sandy white soils and feature *Caryoxylon littoralis,* which was described in the 1950s as "very prevalent" at this site (*Mogg* #29153). Other saltmarsh species recorded by Read (2020) include *Salicornia perennis* and *S. perrieri,* while *Sesuvium portulacastrum* is likely to occur across both the mangroves and marshes.

Littoral dunes vegetation, towards the coastline, includes small trees and shrubs such as *Barleria delagoensis, Diospyros rotundifolia, Ochna natalitia* and *Tricalysia delagoensis* and larger trees such as *Hyphaene coriacea* and *Mimosops caffra.*

Lötter *et al.* (in prep.) categorises much of the habitat at this site as part of the wider Inhambane Dune Thicket type, a semi-deciduous to evergreen vegetation type found throughout the coastal dunes of this province. Coastal thickets occur inland of the littoral dunes but include many of the same species. The endemic and Vulnerable species *Elaeodendron fruticosum* and *Millettia ebenifera* have been recorded from these thickets, as has *Grewia occidentalis,* which is highly likely to be the endemic variety, *littoralis,* typical of coastal dunes in this part of Mozambique (Read 2020; Lötter *et al.*, in prep.). Patches of *Ecbolium hastatum* (EN) are known to occur in the shade of this thicket (*Jacobsen* #6082), while *Encephalartos ferox* (VU) has also been recorded from this site, probably occurring within the sheltered areas of the dunes.

Miombo occurs interspersed with thicket vegetation, varying from dense patches up to 10 m high to open patches with a canopy around 2 m (Massingue *et al.* 2021). Despite the varying structure of miombo woodland in this IPA, *Julbernadia globiflora* dominates throughout, followed by *Brachystegia spiciformis* and *B. torrei. Coffea racemosa* likely occurs in the understorey, alongside endemic species such as *Chamaecrista paralias, Elaeodendron fruticosum* and *Triainolepis sancta* (Massingue *et al.* 2021). A grassy understorey sparsely populates the ground layer of open miombo, with grass species recorded at this site including *Andropogon schirensis, Eragrostis inamoena, Panicum maximum* and *Tricholaena monachme* (Read 2020).

Of particular conservation interest are the areas of miombo on primary dunes that occur within this IPA. First identified by Massingue (2019), this habitat type has a low canopy and is often associated with wetlands. Restricted to the coastlines of Inhassoro and Vilanculos Districts, miombo on primary dunes is unusual as, nationally, coastal miombo is typically confined to older dunes. This habitat type was identified within this IPA in recent survey work (Massingue *et al.* 2021) and is dominated by *Brachystegia spiciformis*, while the endemic species *Chamaecrista paralias* (LC) occurs in large populations within the understorey (Massingue *et al.* 2021).

Along the south-western boundary of Vilanculos Sanctuary are a number of brackish lagoons with associated Cyperaceae species at the margins.

Outside the boundary of Vilanculos Sanctuary much of the land within this protected area has been fragmented by agriculture, although some intact habitat remains, particularly in the east of the peninsula. Little botanical collecting has been conducted in this area of the IPA and so it is not clear whether species of conservation importance remain. Crops grown in these areas include cassava, maize, wheat, beans and peanuts (SBV 2017b). Previous to the establishment of Vilanculos Sanctuary in 2000, the northern and western parts of this peninsula were also cultivated for subsistence agriculture. Evidence of this remains today with cashew, coconut and mangos tree still growing within the Sanctuary (Massingue *et al.* 2021).

Conservation issues

While the entirety of this IPA falls within São Sebastião Total Protection Area (TPA), some areas of the peninsula are not well protected and have been subject to habitat degradation. The north and east of the peninsula, covering around 105 km² of terrestrial and marine areas, is fenced off as a privately managed by Santuario Bravio de Vilanculos ("Vilanculos Sanctuary"), while the remainder of the TPA is heavily degraded by subsistence agriculture, collecting of firewood and timber. The only high-quality areas of habitat that remain outside Vilanculos Sanctuary are cemeteries, where local customs permit only limited collection of firewood (Massingue *et al.* 2021).

Vilanculos Sanctuary is a concession that was granted by the Mozambican government to a private consortium of developers in 2000 (Ashley & Wolmer 2003). The three stated core functions of the Sanctuary are conservation, community upliftment and eco-tourism development. To limit the population density of the area, there are limits on residential and tourism accommodation capacity and Vilanculos Sanctuary is currently below these thresholds (SBV 2017b). While there is inevitably some habitat disturbance through development, the limited number of visitors and the emphasis on nature and conservation within the tourist experience minimise threats such as clearance and disturbance of habitats which are faced in tourist centres elsewhere in Mozambique.

Jatropha subaequiloba photographed on Bazaruto Island (OB)

Diospyros rotundifolia (CL)

In 2003 the area was granted private reserve status. Previously the area was farmed for subsistence agriculture, and habitats were degraded as a result (SBV 2017b). Alongside restricting agriculture, fishing and the extraction of other resources, conservation work at Vilanculos Sanctuary also includes habitat restoration activities such as control of bush encroachment, regulation of fire regime and reintroduction of herbivores (SBV 2017a). Control of problematic plant species, such as the parasitic *Cassytha filiformis*, is also undertaken within the reserve through selective eradication (Massingue *et al.* 2021).

Chamaecrista paralias on the foredunes (CD)

Brachystegia spiciformis in coastal thicket. (SV)

There is a stark contrast between the vegetation cover within Vilanculos Sanctuary and that in neighbouring areas (Google Earth 2021). Outside the reserve boundary, the remaining area of the São Sebastião TPA continues to be degraded, with a reported 33% decrease in tree cover since 2000 (World Resources Institute 2021). Massingue *et al.* (2021) also observed some signs of continued wood collecting and extraction of fibres within Vilanculos Sanctuary boundary and conclude that there is still some dependence of local people on resources within this area, possibly because resources are scarcer in the degraded habitats elsewhere. Although there have been many conservation and restoration successes within Vilanculos Sanctuary, a strategy across the entire São Sebastião TPA is desirable for balancing nature conservation while meeting the needs of local people, namely the ability to produce sufficient food and access to fuel.

Some progress has already been made in economic development opportunities for local people, with the creation of jobs in tourism and support for healthcare, water security and education. A compensation scheme has also been established for lost crops or fishing opportunities due to conservation (SBV 2017a). While large investments have been made in local communities, with over $3.5 million of investment reported by 2017 (SBV 2017a), and conservation efforts are seeing successes, inevitably some have lost out on livelihood opportunities and object to the restrictions associated with the establishment of the Vilanculos Sanctuary (Ashley & Wolmer 2003; O'Connor 2006). It is of critical importance, therefore, that any further conservation initiatives within the TPA are done in collaboration with local communities.

This IPA falls within the wider Grande Bazaruto Key Biodiversity Area which spans São Sebastião Pennisula, the Bazaruto Archipelago and the coastal waters northwards to the Save estuary. Most of this KBA covers marine areas and is triggered by marine species. However, *Jatropha subaequiloba* (VU) is also a trigger for this KBA site, with the entirety of this species' distribution falling within this KBA (in terms of IPAs, this species is split between this site and Bazaruto Archipelago [MOZTIPA042]). In addition, this KBA contains the entire known global population of two reptile species, *Lygosoma lanceolatum* (LC) and *Scelotes insularis* (LC). Both species have been recorded on São Sebastião within the dune thicket of this IPA (Jacobsen *et al.* 2010), and so protection of these habitats is crucial to the conservation of these reptile taxa.

Bird species of conservation interest include Southern-banded snake eagle (*Circaetus fasciolatus* – NT) and Olive bee-eater (*Merops superciliosus* – LC). For the latter species, the north-west of Vilanculos Sanctuary hosts the second largest breeding occurrence in Africa (SBV 2017b). Inventorying of avian taxa by Vilanculos Sanctuary has so far recorded 300 species (SBV 2017a). This includes new records of avian taxa for Mozambique at this site, including Saunder's Tern (*Sternula saundersi* – LC) and Damara Tern (*Sternula balaenarum* – VU); the latter species breeds mostly in Namibia and was effectively unknown from the east coast of Africa, but over 100 individuals were observed within this IPA between 2019 and 2020 (C. Read pers. comm. 2020).

Key ecosystem services

The extensive mangroves at this site provide a number of ecosystem services, particularly in coastal protection when spring tides are high. As a changing climate may increase the frequency and severity of cyclones (World Bank 2019), these mangroves forests may play an increasingly important role in mitigating high winds and storm surges. Eland are known to visit the mangroves and have been observed using the leaves of these trees as salt licks (SBV 2017b), while there are also a number of bird and marine species that rely on this habitat. The Kewene community, who reside on the north-west coast of the peninsula outside Vilanculos Sanctuary, depend heavily on the mangroves, fishing in the waters and harvesting trees for firewood and construction poles. However, there are indications that the mangroves are becoming heavily degraded in this area (Massingue *et al.* 2021). Timber and fuelwood are also extracted from the miombo woodlands and coastal thicket outside the private sanctuary.

The terrestrial habitats within Vilanculos Sanctuary play an important role in attracting residents and tourists, as well as supporting the conservation of fauna. There are strict rules on capacity and tourist activities at the site to prevent unnecessary disturbances within Vilanculos Sanctuary.

Ecosystem service categories

- Provisioning – Food
- Provisioning – Raw materials
- Regulating services – Moderation of extreme events
- Habitat or supporting services – Habitats for species
- Cultural services – Recreation and mental and physical health
- Cultural services – Tourism

IPA assessment rationale

São Sebastião Pennisula qualifies as an IPA under sub-criterion A(i). One Endangered species, *Ecbolium hastatum*, and three Vulnerable species, *Jatropha subaequiloba, Millettia ebenifera* and *Elaeodendron fruticosum*, have been recorded from this site. Nine species meet sub-criterion B(ii), representing fewer than 3% of the endemic and range-restricted plant species of Mozambique required for this site to qualify under this sub-criterion.

Zanthoxylum delagoense (CL)

Dense miombo dominated by *Julbernardia globiflora* (CL)

Priority species (IPA Criteria A and B)

FAMILY	TAXON	IPA CRITERION A	IPA CRITERION B	≥ 1% OF GLOBAL POP'N	≥ 5% OF NATIONAL POP'N	IS 1 OF 5 BEST SITES NATIONALLY	ENTIRE GLOBAL POP'N	SPECIES OF SOCIO-ECONOMIC IMPORTANCE	ABUNDANCE AT SITE
Acanthaceae	*Ecbolium hastatum*	A(i)	B(ii)	✓	✓	✓			scarce
Euphorbiaceae	*Jatropha subaequiloba*	A(i)	B(ii)	✓	✓	✓			scarce
Iridaceae	*Tritonia moggii*		B(ii)	✓	✓				unknown
Fabaceae	*Millettia ebenifera*	A(i)	B(ii)	✓	✓	✓			frequent
Fabaceae	*Chamaecrista paralias*		B(ii)	✓	✓	✓			abundant
Celastraceae	*Elaeodendron fruticosum*	A(i)	B(ii)	✓	✓	✓			frequent
Polygalaceae	*Carpolobia suaveolens*		B(ii)	✓					occasional
Rubiaceae	*Triainolepis sancta*		B(ii)	✓	✓	✓			occasional
Rutaceae	*Zanthoxylum delagoense*		B(ii)	✓					occasional
		A(i): 4 ✓	B(ii): 9						

Protected areas and other conservation designations

CONSERVATION AREA TYPE	CONSERVATION AREA NAME	RELATIONSHIP OF IPA TO CONSERVATION AREA
Total Protection Area	São Sebastião Total Protection Area	protected/conservation area encompasses IPA
Wildlife Sanctuary	Santuario Bravio de Vilanculos	protected/conservation area overlaps with IPA
Key Biodiversity Area	Grande Bazaruto	protected/conservation area encompasses IPA

Threats

THREAT	SEVERITY	TIMING
Tourism & recreation areas	low	Ongoing – trend unknown
Small-holder farming	medium	Ongoing – trend unknown
Logging & wood harvesting	low	Ongoing – trend unknown
Invasive & other problematic species, genes & diseases	low	Ongoing – declining

MAPINHANE

Assessors: Castigo Datizua, Clayton Langa, Iain Darbyshire, Sophie Richards

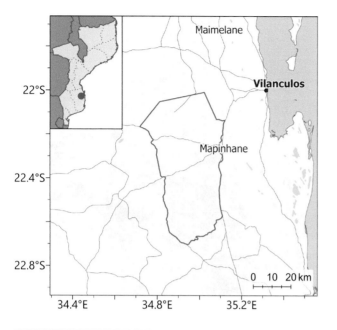

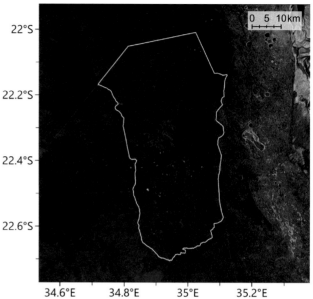

INTERNATIONAL SITE NAME		Mapinhane	
LOCAL SITE NAME (IF DIFFERENT)		–	
SITE CODE	MOZTIPA056	PROVINCE	Inhambane

LATITUDE	-22.44511	LONGITUDE	35.05208
ALTITUDE MINIMUM (m a.s.l.)	20	ALTITUDE MAXIMUM (m a.s.l.)	150
AREA (km²)	2070	IPA CRITERIA MET	A(i)

Site description

The Mapinhane IPA is shared by Vilanculos and Massinga Districts in northern Inhambane Province. It is situated to the west of the EN1 road, extending beyond Mapinhane village in the north and Chicomo village in the south, and covering an area of 2,070 km² between the latitudes -22.01° to -22.71° and longitudes 35.72° to 35.33°. The boundaries of this IPA were delineated to encompass important habitats that support a notable number of plant species endemic to Mozambique, including four threatened species, and a range of ecosystems services that the habitats provide. This site is heavily impacted by deforestation due to timber exploitation, subsistence agriculture and settlement expansion, and is further impacted by fire events associated with local communities, resulting in the transformation and degradation of its ecosystems.

Botanical significance

This IPA is of high botanical importance because of the presence, throughout the miombo woodlands and mixed deciduous forests and woodlands, of several endemic and restricted species of the proposed Inhambane (sub-)Centre of Plant Endemism (Darbyshire *et al.* 2019a). Mapinhane holds some of the most extensive populations of four threatened species endemic to northern Inhambane Province: *Bauhinia burrowsii* (EN), *Croton inhambanensis* (VU), *Ozoroa gomesiana* (VU) and *Xylia mendoncae* (VU), although the latter is rather scarce here.

Overall, this IPA supports eight national endemic plant taxa and six near-endemic plant taxa. The endemic species consist of the four previously mentioned threatened species, plus an additional four Least Concern endemics.

Habitat and geology

In the broad sense, the Mapinhane region lies within the Swahilian-Maputaland Regional Transition Zone phytogeographical region according to Clarke (1998), which covers much of central, coastal-belt of Mozambique, and the Southern Zanzibar-Inhambane Coastal Forest Mosaic Ecoregion according to Schipper & Burgess (2015), which stretches for ca. 2,200 km from southern Tanzania to Xai-Xai (Gaza Province) in Mozambique. In a narrower phytogeographical sense, this area constitutes the northern extension of the Maputaland Centre of Endemism, recently proposed as the Inhambane (sub-)Centre of Endemism (Darbyshire *et al.* 2019a).

The climate in the IPA is influenced by the warm current from the Mozambique Channel, and is characterized as tropical dry, with two seasons. The hot and rainy season runs from October to March, while the cool and dry season runs from April to September. Annual rainfall average ranges from 1,000 mm to 1,200 mm, whilst temperatures peak in January (28.6°C) and reach a minimum in July (19.0°C) (MAE 2005c, 2005d; MICOA 2012b, 2012c). The geographical elevation of the Mapinhane IPA ranges from 20 m to 150 m (Google Earth 2021). A range of soils are present, classified into three groups: (1) sodic soils (mananga soils), (2) sandy soils, and (3) red clay soils (MICOA 2012a, 2012b).

The Environmental Profile Assessment reports by MICOA (2012b, 2012c) for Vilanculos and Massinga Districts, respectively, provide an overview of the habitat mosaics and plant diversity of the Mapinhane IPA. Two main types of vegetation can be distinguished at this site. (1) Miombo woodlands dominated by *Julbernardia globiflora* and *Brachystegia spiciformis* and accompanied by a range of other tree species such as *Afzelia quanzensis*, *Albizia adianthifolia*, *Garcinia livingstonei*, *Pterocarpus angolensis* and the palm *Hyphaene coriacea*. (2) Deciduous forests mixed with woodlands also featuring miombo species noted above but with a number of additional taxa including *Acacia nigrescens*, *Balanites maughamii*, *Cordyla africana*, *Kirkia acuminata*, *Sterculia africana*, and *Suregada zanzibariensis* (MICOA 2012b). The grass communities of the IPA are varied, but particularly dominant species include *Chloris gayana*, *C. virgata*, *Dactyloctenium aegyptium*, *D. giganteum*, *Melinis repens* and *Pogonarthria squarrosa* (A. Massingue, pers. comm. 2021).

In addition, as is noted at Temane IPA (approximately 11 km to the north), the miombo woodland and mixed forest-woodland vegetation types of this site are sometimes interspersed with small patches of sand thicket mosaic (Lötter *et al.* in prep.). More generally, the area covered by Mapinhane IPA encompasses three habitats according to the classification of Lötter *et al.* (in prep.): mainly Urronga Lowland Dry Woodland and Vilanculos Coastal Miombo with small areas of Pande Sand Thicket.

Conservation issues

The Mapinhane IPA does not lie within a formal protected area. However, the northern portion of the IPA is covered by the recently identified Inhassoro-Vilankulos Key Biodiversity Area (WCS *et al.* 2021).

Thicket vegetation within the Mapinhane IPA (AM)

This IPA is heavily subject to habitat loss (deforestation) and fragmentation due to timber exploitation, and subsistence agriculture through slash-and-burn methods. The most widely cultivated crops are maize, peanuts, beans and cassava (MAE 2005c, 2005d). Settlement expansion and increased fire frequency through deliberate burning by local communities are further threats (MICOA 2012b, 2012c; A. Massingue. pers. comm. 2021). The MICOA (2012b) report notes that fire events recorded throughout the Mapinhane IPA are also derived from palm wine extraction from *Hyphaene coriacea*, where fire is used to clear palm leaf thicket and access the sap more easily. Palm wine constitutes one of the main income sources for local households. There is no information available on the threat from invasive plant species on site. However, there are a range of exotic trees planted, such as coconut, citrus fruits, cashew and mango, which occur in small numbers in abandoned areas. All of these above-mentioned activities impact negatively on the IPA through the transformation and degradation of its ecosystems.

Key ecosystem services

This IPA of entirely terrestrial habitats contributes significantly to carbon storage and climate regulation, particularly related to the precipitation cycle. Moreover, the forests and woodlands stretching across the IPA also provide a range of provisioning services for the local community, which could be managed at sustainable levels, notably harvesting of firewood, wild fruits, medicinal plants and palm wine extraction from *Hyphaene coriacea*. These habitats also provide important habitat and supporting services for a range of fauna.

Ecosystem service categories

- Provisioning – Food
- Provisioning – Medicinal resources
- Regulating services – Local climate and air quality
- Regulating services – Carbon sequestration and storage
- Habitat or supporting services – Habitats for species
- Habitat or supporting services – Maintenance of genetic diversity
- Cultural services – Education

IPA assessment rationale

Mapinhane qualifies as an IPA under criterion A(i) in view of the site holding globally important populations of four species of high conservation importance, namely *Bauhinia burrowsii* (EN), *Croton inhambanensis* (VU), *Ozoroa gomesiana* (VU) and *Xylia mendoncae* (VU). In total, this IPA supports 14 species that are endemic or near-endemic to Mozambique according to Darbyshire *et al.* (2019a). However, as only eight of these qualify under sub-criterion B(ii), this site does not meet the threshold (3%) of Mozambican species of high conservation importance within the site, but it is possible that further B(ii) species will be uncovered here following more intensive botanical surveys.

Priority species (IPA Criteria A and B)

FAMILY	TAXON	IPA CRITERION A	IPA CRITERION B	≥ 1% OF GLOBAL POP'N	≥ 5% OF NATIONAL POP'N	IS 1 OF 5 BEST SITES NATIONALLY	ENTIRE GLOBAL POP'N	SPECIES OF SOCIO-ECONOMIC IMPORTANCE	ABUNDANCE AT SITE
Anacardiaceae	*Ozoroa gomesiana*	A(i)	B(ii)	✓	✓	✓			frequent
Bignoniaceae	*Dolichandrone alba*		B(ii)						abundant
Euphorbiaceae	*Croton inhambanensis*	A(i)	B(ii)	✓	✓	✓			common
Fabaceae	*Baphia massaiensis* subsp. *gomesii*		B(ii)						scarce
Fabaceae	*Bauhinia burrowsii*	A(i)	B(ii)	✓	✓	✓			frequent
Fabaceae	*Chamaecrista paralias*		B(ii)						common
Fabaceae	*Xylia mendoncae*	A(i)	B(ii)	✓	✓	✓			scarce
Loranthaceae	*Englerina schlechteri*		B(ii)						unknown
		A(i): 4 ✓	B(ii): 8						

Protected areas and other conservation designations

CONSERVATION AREA TYPE	CONSERVATION AREA NAME	RELATIONSHIP OF IPA TO CONSERVATION AREA
No formal protection	N/A	
Key Biodiversity Area	Inhassoro-Vilankulos	protected/conservation area overlaps with IPA

Threats

THREAT	SEVERITY	TIMING
Housing & urban areas	high	Ongoing – increasing
Shifting agriculture	high	Ongoing – increasing
Small-holder plantations	low	Ongoing – trend unknown
Small-holder grazing, ranching or farming	medium	Ongoing – trend unknown
Roads and railroads	medium	Ongoing – trend unknown
Gathering terrestrial plants	high	Ongoing – trend unknown
Logging & wood harvesting	high	Ongoing – increasing
Increase in fire frequency/intensity	high	Ongoing – trend unknown
Invasive non-native/alien species/diseases	unknown	Ongoing – trend unknown

POMENE

Assessor(s): Sophie Richards, Iain Darbyshire

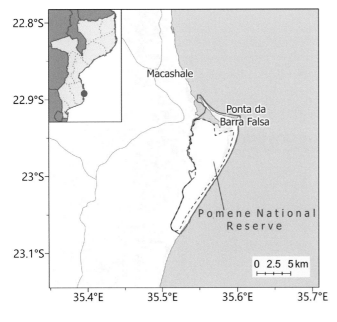
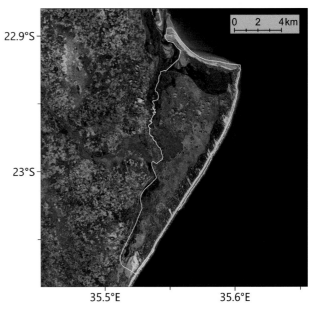

INTERNATIONAL SITE NAME		Pomene	
LOCAL SITE NAME (IF DIFFERENT)		–	
SITE CODE	MOZTIPA041	PROVINCE	Inhambane

Dune scrub featuring the suffrutex *Salacia kraussii* (JEB)

LATITUDE	-22.99790	LONGITUDE	35.55960
ALTITUDE MINIMUM (m a.s.l.)	0	ALTITUDE MAXIMUM (m a.s.l.)	120
AREA (km²)	74	IPA CRITERIA MET	A(i), A(iv), B(ii)

Site description

The Pomene IPA falls within Massinga District of Inhambane Provence. The site is predominantly coastal and lies to the east of the Muducha River, spanning an area of 74 km² from Guma village in the south to the estuary at Pomene Bay in the north-west and Ponta Barra Falsa (False Bar Point) to the north-east. The boundary of this IPA largely follows that of Pomene National Reserve, but the most northerly section, including the mangrove forest and lagoons west of the Muducha estuary, and the eastern boundary, following the shoreline to incorporate the intact dune habitat, are beyond the reserve boundary.

The presence of the National Reserve at Pomene has maintained a large area of intact vegetation, particularly coastal dunes, in contrast to much of the coastline from Maputo to the Save River which has been cleared for subsistence farming (BirdLife International 2020b). The site falls within the proposed Inhambane (sub-)Centre of Plant Endemism (Darbyshire *et al.* 2019a) and there are at least 11 Mozambican endemic species known from the Pomene IPA.

Botanical significance

Pomene falls within the proposed Inhamabane (sub-)Centre of Plant Endemism (Darbyshire *et al.* 2019a). A total of 11 species endemic to Mozambique have been recorded within this IPA. One of these, *Elaeodendron fruticosum* (VU), is only known from this Centre of Endemism (CoE) and has been described as common towards Pomene Bay. *E. fruticosum* is one of three globally threatened species recorded at this site alongside *Euphorbia baylissii* and *Solanum litoraneum*. All three of these Vulnerable species are threatened throughout their ranges by conversion of habitat to subsistence agriculture (Matimele *et al.* 2018a; Richards, in press [c]). Currently, these threatened species are only known from outside the Pomene National Reserve boundary within this IPA, towards Ponta Barra Falsa, where land is moderately threatened by tourism.

Salicornia mossambicensis, one of the eleven endemics at this site, occurs in the salt marshes to the north. Currently assessed as Data Deficient, this species is only known from one other location which, due to its proximity to the city of Inhambane, is highly threatened in this area. The range of *S. mossambicensis*, based on extent of occurrence calculated using an area of habitat approach (Brooks *et al.* 2019), is approximately 200 km², less than the 5,000 km² threshold to qualify as a range-restricted endemic under IPA sub-criterion A(iv).

A number of species recorded at this site are also found across the Maputaland CoE in the broadest sense; for instance, *Trichoneura schlechteri*, has been recorded from this IPA and represents the only collection in Inhambane Province and the most northerly edge of this species' known range. In addition, *Encepharlartos ferox* subsp. *ferox* (assessed as NT at species level) has a distribution from Inhambane to KwaZulu-Natal, South Africa, and occurs within the coastal woodland near Ponta Barra Falsa. This cycad is common throughout its range but is threatened by overcollection and the loss of coastal habitat (Donaldson 2010e).

The intact coastal habitats at this site, including dune vegetation and mangrove salt marshes to the north, are of clear importance to a number of species with limited distributions. The coastal habitat of Pomene is of particular importance as much of this vegetation is under threat or has already been cleared elsewhere to make way for subsistence agriculture, with this IPA representing the longest tract of intact coastal forest between the Save River and Maputo (BirdLife International 2020b).

Habitat and geology

Pomene, as a coastal site, is underlaid by sandy soils with little organic matter or water retention ability (Macandza *et al.* 2015). The area of 1–2 km westward from the coastline is dominated by coastal dune vegetations types, with pioneer communities on the foredunes including species such as *Ipomoea pes-caprae*, *Cyperus crassipes* and *Canavalia rosea* (Macandza *et al.* 2015). Further inland is dense coastal thicket dominated by *Diospyros rotundifolia* and *Mimusops caffra*, hosting the endemic *Elaeodendron fruticosum* (VU) and the near-endemic cycad *Encephalartos ferox* subsp. *ferox* (assessed as NT at species level). On the most inland section of the coastal dunes, the dense thicket transitions to miombo dominated by *Brachystegia spiciformis* and *Afzelia quanzensis* (Macandza *et al.* 2015).

To the west of the dunes, running from north-east to the south-west of the national reserve, is

View of open miombo savanna at Pomene (JEB)

an area of shrubby grassland. Common shrubby species in these areas include *Salacia kraussii*, *Hyphaene coriacea* and *Garcinia livingstonei*, while dominant grasses include *Heteropogon contortus* and *Imperata cylindrica* (Macandza *et al.* 2015). A number of endemic species inhabit the shrubby-grassland mosaic, including *Dracaena subspicata* and *Chamaecrista paralias*.

Towards the north of the IPA, surrounding the lagoon and on the banks of Muducha River, is an area of mangrove forest, most of which lies outside the boundaries of the National Reserve. Five mangrove species are known from this area, in order of dominance, these are: *Rhizophora mucronata*, *Avicennia marina*, *Ceriops tagal*, *Bruguiera gymnorhiza* and *Sonneratia alba* (Louro *et al.* 2017). The mangrove species are used by local communities for construction as the timber is resistant to insect damage (Macandza *et al.* 2015). The salt marshes associated with the mangroves are important habitat for the endemic species *Psydrax moggii* (LC) and *Salicornia mossambicensis* (DD). This IPA has been delineated to include only the mangroves east of the Muchuda River, however, it should be noted that there is also a 3 km stretch of mangrove forest to the north of this site, toward Macashale, which is also likely to be of ecological importance.

Seasonally flooded grasslands are found at the margins of the mangroves on the banks of the Muchada River as it approaches the estuary. These areas have not been extensively studied, but they are known to be dominated by a number of *Cyperus* species and grasses such as *Imperata cylindrica* and *Dichanthium* species (probably *D. annulatum*) (Macandza *et al.* 2015). South of the mangroves, the riverbanks are dominated by reedbeds mostly of the species *Phragmites mauritianus*, while species of the genus *Cyperus* are common. *Coix lacryma-jobi* is described by Macandza *et al.* (2015) as dominant in the reed beds. However, as a non-native species that is not commonly known from this part of Mozambique, it is not clear whether this represents a misidentification or an as yet unrecorded introduction. The herbaceous Rubiaceae *Oldenlandia corymbosa* has also been recorded from these areas and is likely associated with disturbed areas. There are some large trees on the riverbanks including *Ficus* species (Macandza *et al.* 2015). Both the seasonally inundated grasslands

and the riverine vegetation are on clay-rich soils with a high organic matter content and high water retention, in contrast to the sandy soils that cover the rest of the IPA (Macandza *et al.* 2015).

From the centre to the western boundary of the IPA, the vegetation is predominantly miombo, covering nearly 40% of the National Reserve (Macandza *et al.* 2015). The miombo here is dominated by *Julbernardia globiflora*, although, like the dune miombo, both *Brachystegia spiciformis* and *Afzelia quanzensis* feature heavily in the species composition. Most of this woodland is open with seasonal pools and a grassy understorey dominated by *Heteropogon contortus*, *Digitaria eriantha* and, in wetter areas, *Imperata cylindrica* (Macandza *et al.* 2015). It may be of interest to study these pools for the presence of ephemeral wetland species such as *Ammannia*, a genus known to include a number of endemic and near-endemic species in Mozambique (Darbyshire *et al.* 2019a). Denser patches of miombo, predominantly south of the settlements within the reserve (centered on -22.98°, 35.55°), have a thinner grass understorey and show much less disturbance than the open miombo.

Conservation issues

The Pomene IPA expands upon the current boundary of Pomene National Reserve and only the mangroves and dunes to the north near Pomene Bay and the dunes along the eastern shoreline are not currently protected. The national reserve was established in 1964 and 200 km² in area was originally designated; however, at present the reserve covers only 50 km² (Macandza *et al.* 2015). Management structures for the reserve were only created in 2009 and subsequently the first management plan, covering the period 2016–2020, was set out. This plan included a proposed expansion of the reserve which would cover all of the eastern dunes and the dunes and mangroves in the north of this IPA within the core zone, with the mangroves north of this IPA, towards Macashale, in the buffer zone (Impacto Lda. 2016). Separately, the development of a marine reserve, including the marine and coastal landscape of this IPA up to Vilanculos Bay, was also proposed recently by a joint public-private initiative. In any case, the proposed expansions would incorporate the entirety of this IPA, which would be of particular importance for the continued integrity of the dunes and northern mangroves.

The reserve itself is not as heavily populated as surrounding areas. However, there were around 500 residents recorded by the reserve administration, concentrated primarily in the north (Impacto Lda. 2016). Land in the wider Massinga area, like much of the reserve, is underlaid by sandy soils with poor fertility, allowing only 2–3 agricultural cycles before it is abandoned (Macandza et al. 2015). The resulting scarcity of agricultural land outside the reserve has led people to move inside its boundaries. However, some of the homesteads, particularly in the south, have previously been abandoned, with sources suggesting that people left due to the restrictions on permitted activities within the reserve or to find work elsewhere (Impacto Lda. 2016). Machambas in the north of the reserve are sparsely distributed and most households depend on subsistence farming, growing crops such as maize, cowpea and cassava, as well as breeding poultry, goats and pigs. Although poor soils in the region may have driven people to farm land within the reserve, given that soils are similarly poor within the reserve, there is a high threat of shifting agriculture within the reserve as soils exhaust quickly and people are forced to move to gain a sufficient harvest. Agronomic research and support for local people in transitioning to more sustainable agricultural techniques throughout the district could help to alleviate land pressure both inside and outside this IPA.

Linked to agriculture is the threat of uncontrolled fires, most of which stem from the use of fire to open up land for farming (Macandza et al. 2015). With high fuel loads and the highest density of people, the miombo and shrubby grassland to the north are at greatest risk from uncontrolled fires.

Wood for domestic fuel is extracted from miombo while timber is extracted from mangroves, particularly trees of larger diameter, for the construction of homes and the camps of fisherman who visit the site for the wet season (Louro et al. 2017). The mangrove forest is also thought to experience degradation through the fishing of marine invertebrates such as the giant mud crab (*Scylla serrata*). A study by Louro et al. (2017) found that up to 41% of trees near the water's edge were cut; however, some regeneration was also observed following felling. The mangroves at Pomene are of great ecological value, with a number of interesting marine species including sea turtle and dolphin species alongside whale shark and possibly dugong (Louro et al. 2017). Mangroves are also known for providing protection against storm surges. It is therefore of upmost importance that this ecosystem is integrated into the protected area network and that only sustainable usage is permitted.

The establishment of zonation to regulate anthropogenic disturbance was suggested in the 2016–2020 management plan (see Impacto Lda. 2016). Under this proposal, much of the eastern mangroves, including areas currently outside the reserve, would be placed under a "Community and Resource Use" zone, allowing for the continued use of the area for sustainable subsistence activities but preventing the use of resources for commercial purposes. Much of the rest of the reserve, consisting mostly of open miombo and shrubby grasslands, would be placed under "Resource Management", which suggests further limitations on permitted activities compared to the above zone in an effort to restore grazing mammals and promote tourism. "Special Protection" zones cover the dense miombo and the riverine/estuary vegetation including the seasonally flooded grassland, riverine fringes and some of the western patches of mangrove. In these areas, it is proposed that there would be no extraction of resources, with only the collection of medicinal plants and artisanal fishing for local purposes allowed with written permission from the reserve.

Previous to the first management plan, local initiatives were established with a view to conserving local wildlife. The Pomene Co-management Committee (Comité de Co-gestão Pomene) and the Community Fisheries Council (Conselho Comunitário de Pescas) are two community organisations working in partnership with the reserve administration. Both committees promote community engagement with conservation of both coastal and marine ecosystems by educating local people on the threats to local biodiversity and the benefits of protecting this biodiversity for livelihoods and future generations. Alongside education, patrols are undertaken to detect illegal activities such as uncontrolled fires or the cutting of mangrove or reeds for sale (Macandza et al. 2015).

The community engagement prior to the implementation of management plans appears to have had a significant impact on the activities of local communities, with a survey finding that 67% of respondents were aware that there are laws restricting resource use within the reserve (Macandza *et al.* 2015). In addition, the reserve is one of the most intact stretches of coastline in southern Mozambique which further suggests a level of local adherence.

The high-quality coastal habitat is of particular importance for avian taxa, providing habitat for species such as the globally Near Threatened Plain-backed Sunbird (*Cyanomitra verreauxii*) and the range-restricted species (defined by the Important Bird Area criteria) Rudd's Apalis (*Apalis ruddi*; LC). With a number of threatened and range-restricted species, the site was recognised in 2001 as an Important Bird Area.

Key ecosystem services

The mangroves in the north of the IPA provide a range of ecosystem services, including timber for construction, food (primarily seafood), and medicines, as well as environmental regulation through protection of coastal habitats and mitigation against storm surges that may become more frequent and extreme with climate change (Macandza *et al.* 2015). The marine life that inhabits the mangroves provides employment in the rainy season for itinerant fisherman; however, this practice may be restricted in the future if the reserve boundary expands to include the eastern patch of mangrove forest.

The timber from mangroves is preferred for construction due to its insect-resistant properties. However, wood is also extracted from miombo to use as fuel, with around 80% of respondents to a survey of local residents stating that firewood is their main source of domestic energy. Reeds and papyrus obtained from the wetlands are useful sources of fibre to make mats and sieves, and wild-harvested fruits, tubers and roots are important sources of food during times of famine (Macandza *et al.* 2015).

The interesting avifauna and range of habitats provide the site with tourism potential. Although there are already three lodges established to the north of the reserve, and around 20% of local

people received some income from tourism in 2015 (Macandza *et al.* 2015), the site does not receive large numbers of tourists at present, with an average of only 19 visitors per month in 2015 (Impacto Lda. 2016). Currently Pomene National Reserve gains income from the entry fee and so is not receiving substantial income from tourism. However, the 2016–2020 management plan recognised the potential for responsibly planned tourism to create greater financial sustainability for the reserve, and sets out long-term ambitions to this end. Tourism in particular could make an important contribution to the incomes of local people, many of whom live below the poverty line as defined by the World Bank (Macandza *et al.* 2015; World Bank 2020). It is not known how much progress has been made in expanding tourism; however, in 2017 the site became a new stop for MSC Cruises, with excursions including tours of the mangroves and snorkelling (South Africa Travel Online 2021). While tourism would create much-needed income for the local area, activities should be monitored to ensure that there is no undue disturbance of habitats and planning should be done in collaboration with local communities.

Three sacred sites have been recorded from within the IPA, two of which are located in the dunes to the north. From interviews with local leaders, the sites represent founder families within the area with ceremonies including rain request ceremonies, healing request ceremonies and the welcoming of visitors (Macandza *et al.* 2015).

Ecosystem service categories

- Provisioning – Food
- Provisioning – Raw materials
- Provisioning – Medicinal resources
- Regulating services – Moderation of extreme events
- Habitat or supporting services – Habitats for species
- Cultural services – Tourism
- Cultural services – Spiritual experience and sense of place
- Cultural services – Cultural heritage

IPA assessment rationale

Pomene qualifies under IPA sub-criterion A(i) with three Vulnerable species: *Euphorbia baylissii*, *Elaeodendron fruticosum* and *Solanum litoraneum*. This site also qualifies under sub-criterion A(iv) for the range-restricted endemic *Salicornia mossambicensis* (DD). With 11 endemic taxa recorded to date, Pomene also qualifies under sub-criterion B(ii), falling within the top 15 sites nationally for endemic and range-restricted species. While there is currently insufficient information for assessing this site under criterion C(iii), it should also be noted that Pomene hosts an extensive area of intact coastal dune habitat. Much of this habitat in southern Mozambique has been degraded through conversion to agriculture, however, the Pomene Natural Reserve has facilitated the protection of this habitat within the IPA.

Triainolepis sancta on the foredunes (JEB)

Priority species (IPA Criteria A and B)

FAMILY	TAXON	IPA CRITERION A	IPA CRITERION B	≥ 1% OF GLOBAL POP'N	≥ 5% OF NATIONAL POP'N	IS 1 OF 5 BEST SITES NATIONALLY	ENTIRE GLOBAL POP'N	SPECIES OF SOCIO-ECONOMIC IMPORTANCE	ABUNDANCE AT SITE
Amaranthaceae	*Salicornia mossambicensis*	A(iv)	B(ii)	✓	✓	✓			unknown
Asparagaceae	*Dracaena subspicata*		B(ii)	✓	✓	✓			unknown
Celestraceae	*Elaeodendron fruticosum*	A(i)	B(ii)	✓	✓	✓			frequent
Euphorbiaceae	*Euphorbia baylissii*	A(i)	B(ii)	✓	✓	✓			occasional
Fabaceae	*Chamaecrista paralias*		B(ii)	✓	✓	✓			frequent
Malvaceae	*Grewia occidentalis* var. *littoralis*		B(ii)	✓	✓	✓			unknown
Rubiaceae	*Psydrax moggii*		B(ii)	✓	✓				unknown
Rubiaceae	*Spermacoce kirkii*		B(ii)	✓	✓	✓			unknown
Rubiaceae	*Triainolepis sancta*		B(ii)	✓	✓	✓			unknown
Rutaceae	*Zanthoxylum delagoense*		B(ii)	✓					unknown
Solanaceae	*Solanum litoraneum*	A(i)	B(ii)	✓	✓	✓		✓	unknown
		A(i): 3 ✓ A(iv): 1 ✓	B(ii): 11						

Protected areas and other conservation designations

CONSERVATION AREA TYPE	CONSERVATION AREA NAME	RELATIONSHIP OF IPA TO CONSERVATION AREA
National Reserve	Pomene National Reserve	IPA encompasses protected/conservation area
Important Bird Area	Pomene	IPA encompasses protected/conservation area

Threats

THREAT	SEVERITY	TIMING
Tourism & recreation areas	medium	Ongoing – trend unknown
Shifting agriculture	medium	Ongoing – trend unknown
Subsistence/artisinal aquaculture	low	Ongoing – trend unknown
Logging & wood harvesting	low	Ongoing – trend unknown
Increase in fire frequency/intensity	high	Ongoing – trend unknown

PANDA-MANJACAZE

Assessors: Jo Osborne, Iain Darbyshire

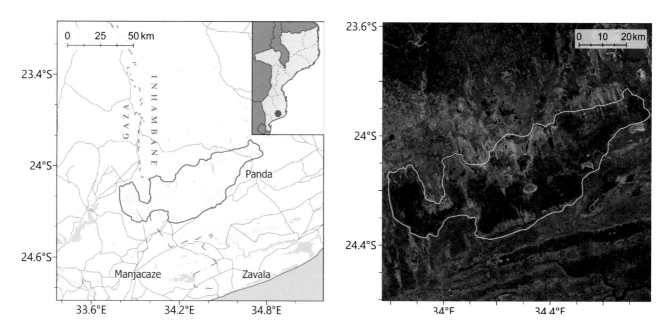

INTERNATIONAL SITE NAME		Panda-Manjacaze	
LOCAL SITE NAME (IF DIFFERENT)		–	
SITE CODE	MOZTIPA016	PROVINCE	Inhambane

LATITUDE	-24.18378	LONGITUDE	34.12489
ALTITUDE MINIMUM (m a.s.l.)	20	ALTITUDE MAXIMUM (m a.s.l.)	150
AREA (km²)	2,599	IPA CRITERIA MET	A(i), A(iv)

Site description

The Panda-Manjacaze IPA covers an area of ca. 2,600 km², in Panda District of southwest Inhambane Province and Manjacaze (Mandlakazi) and Chibuto Districts of southeast Gaza Province. It lies to the northwest of the Panda-Manjacaze road (417), west of Panda town and north of Manjacaze town, ca. 60–120 km inland from the Indian Ocean coastline. This large area supports a rich mosaic of habitats including sizable areas of dry forests on deep sands which are important for a range of rare and threatened plant species, some of which are on the northern edges of their range here. The site overlaps with an Important Bird Area, the Panda *Brachystegia* woodlands IBA, but is not formally protected at present.

Botanical significance

Significant areas of *Androstachys johnsonii*-dominated dry forest occur within the Panda-Manjacaze site. This timber tree, the only species in the genus *Androstachys*, occurs in southern Africa and Madagascar where it forms dense stands, usually in well-drained rocky areas. In Inhambane it forms a distinctive dry forest on raised sands. The timber, known locally as 'mecrusse' or 'cimbirre', is durable and sought after for construction and, as a result, *Androstachys* forest has become fragmented both nationally and within this IPA. The sand forest habitat at Panda-Manjacaze supports several globally threatened plant species. These include the Critically Endangered *Guibourtia sousae*, a tree species apparently endemic to this IPA, where it is known only from the type collection from 1936 (*Gomes e Sousa* #1927; Darbyshire *et al.* 2018c). This species has not been rediscovered despite recent efforts (Osborne *et al.* 2019a; J.E. Burrows, pers. comm.), but it could be easily overlooked as it is vegetatively similar to *G. conjugata* which is a common component of these dry forests. Two Endangered Maputaland endemic species, *Cola dorrii* and *Xylopia torrei*, were first discovered within this IPA during botanical fieldwork in 2019 (Osborne *et al.* 2019a) and this site represents the northernmost limits of their respective ranges. The Vulnerable

taxa *Acridocarpus natalitius* var. *linearifolius* and *Euphorbia baylissii* are also recorded from these forests, and several other notable shrubby species occur, including the localised and uncommon *Ephippiocarpa orientalis*, the Mozambican near-endemic *Microcos* (*Grewia*) *microthyrsa* and the Maputaland endemic *Psydrax fragrantissima* (NT).

The seasonally wet grasslands that occur extensively between the raised sands also support a number of interesting species, including the little-known national endemic *Indigofera mendoncae* (DD) and the scarce *Striga junodii* for which this is, again, one of the northernmost known localities. Other significant species that occur within the IPA include the scarce national endemic *Celosia nervosa* (DD), and important populations of two near-endemic species that are assessed as Near-Threatened, the scarce erect or scrambling shrub *Sclerochiton coeruleus* and the cycad *Encephalartos ferox* subsp. *ferox*. Large areas of the site are not well-studied botanically and other notable plant species are likely to occur here. In addition to the sand forest habitat, the Panda-Manjacaze IPA includes valuable woodlands, wetlands, and seasonally flooded savanna grasslands.

Habitat and geology

The Panda-Manjacaze IPA is in a low-lying area of mostly flat terrain on sandy soils on gently undulating Quaternary dune deposits, with an elevation ranging from 20 m to 150 m a.s.l.. The site consists of a mosaic of habitats depending on the gradual variation in elevation, with large areas of woodland and seasonally wet savanna grassland,

Androstachys-dominated forest in Panda District (ID)

Syzygium cordatum-dominated groundwater forest with the climbing fern *Stenochlaena tenuifolia* (JO)

Sclerochiton coeruleus (ID)

Cola dorrii (ID)

and smaller areas of dry forest, groundwater forest, lakes, and wetlands. Much of the survey work from which the following habitat notes are derived was conducted in the area between Chichococha and Chihuwane to the southwest of Panda (Osborne *et al.* 2019a).

On satellite imagery (Google Earth 2021), *Androstachys*-dominated dry forest fragments are clearly distinguishable by their very dark green colour compared to the surrounding paler green secondary forest and woodland. The *Androstachys* forest tends to support few other tree species, with the exception of *Guibourtia conjugata*, which can be co-dominant and is also the dominant species in surrounding secondary forest. A range of understorey shrubs are frequent, including *Croton pseudopulchellus*, *Drypetes arguta*, *Hyperacanthus microphyllus*, *Salacia leptoclada*, *Suregada zanzibarensis* and *Vepris* sp., with *Boscia foetida*, *Combretum celastroides* and *Margaritaria discoidea* common along margins and clearings. *Warneckea sansibarica* can be frequent in the understorey of *Guibourtia*-dominated forest. Some small patches of a more thicket-like woodland with *Diospyros rotundifolia*, *Mimusops caffra* and *Ochna natalitia* are noted on deep sands, appearing reminiscent of the coastal woodlands and thickets of Maputaland.

The extensive miombo woodlands here are almost exclusively dominated by *Brachystegia spiciformis*, often in pure stands. Much of the woodland is secondary with a canopy height of ca. 8 m, though occasional larger trees remain including to 15 m tall and 50 cm d.b.h.. A range of herbs and shrubs occur in the understorey, and the cycad *Encephalartos ferox* subsp. *ferox* is frequent in some woodland patches. A drier miombo woodland, the Pangue dry miombo of Lötter *et al.* (in prep.), is recorded from the north of the IPA.

Extensive savanna grasslands occur in low-lying, seasonally wet areas and are interspersed with small woodland patches. These areas burn naturally in the dry season, as indicated by an abundance of perennial suffruticose species with signs of previous burning at the stem bases. Common grass species recorded here include *Chrysopogon serrulatus*, *Cymbopogon caesius*, *Diheteropogon amplectens*, *Eragrostis* sp. and *Setaria sphacelata*. Common woody species include the palms *Hyphaene coriacea* and *Phoenix reclinata*, together with *Acacia* sp., *Syzygium cordatum* and *Terminalia sericea*, whilst *Brachystegia spiciformis* occurs in drier areas. The suffruticose shrub *Salacia kraussii* is abundant at the transition between woodland patches and open grassland. The grasslands support a range of herbaceous species, such as *Bergia decumbens*, *Chamaecrista paralias*, *Leucas milanjiana* and *Vahlia capensis* subsp. *vulgaris*.

Areas of groundwater (swamp) forest occur in the lowest-lying areas, usually as ribbons along seasonal rivers and streams. These are dominated by *Syzygium cordatum*, while the forest floor is dominated by the giant climbing fern *Stenochlaena tenuifolia*. Other notable tree species in this habitat are *Ficus trichopoda* and *Voacanga thouarsii*. Some large wetlands occur, with reedbeds and permanent or seasonal lakes, but these have not been well surveyed botanically to date.

Conservation issues

Ongoing and unsustainable logging of *Androstachys johnsonii* has led to fragmentation of *Androstachys* dry forest within the Panda-Manjacaze IPA. As evident from satellite imagery (Google Earth 2021), the fragmentation is particularly apparent in Panda District of Inhambane, while there appears to be larger and more intact forest patches remaining in neighbouring Gaza to the north of Manjacaze, although botanical survey is needed to confirm the quality of this forest. Regeneration of *Androstachys* on cleared land and within secondary forest patches was not seen during fieldwork in 2019 (Osborne *et al.*, pers. obs.). Logging of the miombo woodlands for timber and charcoal production is also resulting in declines in this habitat, although miombo remains extensive at present.

There is currently very little settlement within the site, although agricultural fields surround the site to the east, south and west and may potentially expand into the IPA in the future. Other potential conservation issues include low-intensity grazing of cattle and goats in the savanna grasslands, and the occasional presence of invasive *Opuntia* sp. in some areas of miombo woodland, particularly around the margins (Osborne *et al.* 2019a). However, at present, these do not appear to pose a serious threat to this site. The extension of the IPA across two Provinces, Inhambane (Panda) and Gaza (Manjacaze-Chibuto), may present challenges for management and so this IPA may need to be divided in the future for conservation management purposes.

The IPA overlaps with an Important Bird Area, the Panda *Brachystegia* woodlands IBA (Birdlife International 2021f), reinforcing the biodiversity value and case for formal protection of this site. The miombo woodlands here are of particular importance for holding a disjunct population of Olive-headed Weaver (*Ploceus olivaceiceps*, NT). A population of African Elephant (*Loxodonta africana*, EN) is noteworthy, and evidence for their presence in the groundwater forests southwest of Panda was noted during recent fieldwork (Osborne *et al.* 2019a). This area would also qualify as an Alliance for Zero Extinction (AZE) site based on the presence of *Guibourtia sousae*, although it is not in the current AZE network and was also not included within the recent Key Biodiversity Areas assessment for Mozambique.

Key ecosystem services

The Panda-Manjacaze IPA supports a significant population of the important timber tree species *Androstachys johnsonii*, known as 'mecrusse', 'cimbirre' or 'Lebombo ironwood', which forms a distinctive and valuable dry forest habitat for flora and fauna. The site also contains large areas of woodland and seasonally wet savanna grassland habitats, overall supporting a high plant diversity across a habitat mosaic and providing diverse

Guibourtia conjugata and *Androstachys johnsonii* dry forest (JO)

habitat for wildlife. In addition to *Androstachys johnsonii*, a range of other socio-economically important plants occur within the site including two palm species, *Hyphaene coriacea* and *Phoenix reclinata*; the shrub *Salacia kraussii*; and the tree species *Brachystegia spiciformis*, *Dolichandrone alba*, *Guibourbia conjugata*, *Syzygium cordatum*, and *Terminalia sericea*. The vegetation contributes to carbon sequestration and storage.

Encephalartos ferox subsp. *ferox* within miombo woodland (JO)

Ecosystem service categories

- Provisioning – Raw materials
- Regulating services – Carbon sequestration and storage
- Habitat or supporting services – Habitats for species

IPA assessment rationale

The Panda-Manjacaze area qualifies as an IPA under criteria A and C. Under criterion A(i), the site supports important populations of five globally threatened plants – it is the only known site globally for *Guibourtia sousae* (CR) and is the only site within the Mozambique IPA network for *Cola dorrii* (EN). Under criterion A(iv), the range-restricted Mozambique endemic *Indigofera mendoncae* (DD) occurs at this site; again, this is the only site for this species within the IPA network.

Priority species (IPA Criteria A and B)

FAMILY	TAXON	IPA CRITERION A	IPA CRITERION B	≥ 1% OF GLOBAL POP'N	≥ 5% OF NATIONAL POP'N	IS 1 OF 5 BEST SITES NATIONALLY	ENTIRE GLOBAL POP'N	SPECIES OF SOCIO-ECONOMIC IMPORTANCE	ABUNDANCE AT SITE
Amaranthaceae	*Celosia nervosa*		B(ii)						unknown
Annonaceae	*Xylopia torrei*	A(i)	B(ii)	✓	✓	✓			scarce
Bignoniaceae	*Dolichandrone alba*		B(ii)					✓	unknown
Euphorbiaceae	*Euphorbia baylissii*	A(i)	B(ii)	✓					unknown
Fabaceae	*Chamaecrista paralias*		B(ii)						unknown
Fabaceae	*Guibourtia sousae*	A(i)	B(ii)				✓		scarce
Fabaceae	*Indigofera mendoncae*	A(iv)	B(ii)	✓	✓	✓			unknown
Malpighiaceae	*Acridocarpus natalitius* var. *linearifolius*	A(i)							unknown
Malvaceae	*Cola dorrii*	A(i)		✓	✓	✓			occasional
Rubiaceae	*Psydrax moggii*		B(ii)						occasional
		A(i): 5 ✓ A(iv): 1 ✓	B(ii): 8						

Protected areas and other conservation designations

CONSERVATION AREA TYPE	CONSERVATION AREA NAME	RELATIONSHIP OF IPA TO CONSERVATION AREA
No formal protection	N/A	
Important Bird Area	Panda Brachystegia woodlands	protected/conservation area overlaps with IPA

Savanna grassland with *Hyphaene coriacea* (ID)

Threats

THREAT	SEVERITY	TIMING
Small-holder farming	low	Ongoing – trend unknown
Small-holder grazing, ranching or farming	low	Ongoing – trend unknown
Logging & wood harvesting	high	Ongoing – trend unknown
Invasive non-native/alien species	low	Ongoing – trend unknown

INHARRIME-ZÁVORA

Assessors: Sophie Richards, Iain Darbyshire, Jo Osborne

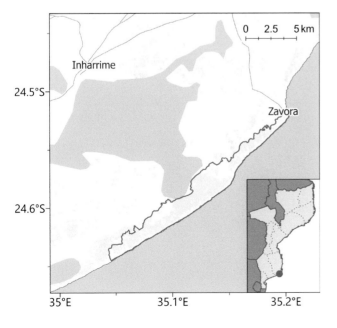

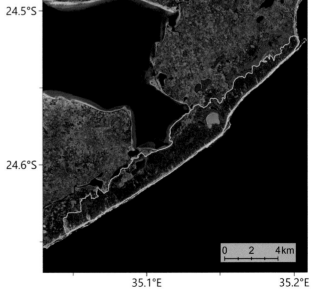

INTERNATIONAL SITE NAME		Inharrime-Závora	
LOCAL SITE NAME (IF DIFFERENT)		–	
SITE CODE	MOZTIPA044	PROVINCE	Inhambane

LATITUDE	-24.58704	LONGITUDE	35.13056
ALTITUDE MINIMUM (m a.s.l.)	0	ALTITUDE MAXIMUM (m a.s.l.)	170
AREA (km²)	31.9	IPA CRITERIA MET	A(i)

Site description

Inharrime-Závora is a coastal IPA spanning either side of the boundary between Inharrime and Zavala Districts of Inhambane Province. The site is 31.9km² in area and falls within the proposed Inhambane (sub-)Centre of Endemism (Darbyshire *et al.* 2019a). Like much of the coastal region of this CoE, the site is under pressure from conversion of habitat to agriculture. However, the coastal dune vegetation here is largely intact. The IPA extends from Ponta Závora in the north-east, 20 km in a south-westerly direction towards Lagoa Maiene, with Lagoa Poelela falling to the south-west of the site.

Botanical significance

The coastal vegetation of southern Inhambane is under high pressure from agriculture and large areas have already been degraded. The stretch of largely intact coastal dunes within this IPA is, therefore, of botanical importance. This habitat hosts three globally Vulnerable species: *Euphorbia baylissii*, an endemic species restricted to the southern coastal areas of Mozambique, alongside *Allophyllus mossambicensis* and *Elaeodendron fruticosum*, which are both endemic to Gaza and Inhambane Provinces. All three species are threatened throughout their range by conversion of coastal habitat to machambas and, to a lesser extent, by the expansion of tourism. While both of these threats are present within this IPA, there is still a significant stretch of intact coastal habitat at this site and so it is a globally significant location for the conservation of these species.

In total, there are five endemic species within this IPA. Most of these species are concentrated on the coastal dunes in the core zone. One species of interest is an as yet undescribed species of *Eugenia*, *Eugenia* sp. A of *Trees and Shrubs Mozambique* (Burrows *et al.* 2018). Recorded at Ponta Závora, this species is only known from the coastlines of Inhambane, Gaza and Maputo Provinces.

Outside the IPA, four endemic species have been recorded towards Inharrime town. Two species,

Coastal dune thicket (JO)

Edge of coastal forest (JO)

Foredune vegetation (JO)

Baphia ovata (NT) and *Psydrax moggii* (LC), were recorded relatively recently, in 2007 (*Burrows* #10109) and 2009 (*Burrows* #11082), respectively, from fragments of thicket vegetation at the lake edges. However, two other endemics recorded around Lagoa Poelela, *Spermacoce kirkii* and *Millettia ebenifera* (VU), were recorded in 1955 (*Exell* #666) and 1944 (*Mendonça* #3372). Owing to the highly degraded habitat beyond the coastal forest, however, these localities around Lagoa Poelela were excluded from the IPA, but some of these species may well be found within the site boundaries with further investigation.

Habitat and geology

The underlying geology of this site is of Quaternary sandstone and the soils are predominantly sandy, with some recent alluvium distributed by the rivers (Impacto Lda. 2012a). Average temperatures range from 19°C in July to 28.6°C in January. The dry season is between May and October, while 74% of annual precipitation falls between November and April (MAE 2005e).

Only very limited botanical surveys have been undertaken within this IPA, although recent botanic collections have been made at Ponta Závora in 2005 (J.E. Burrows and S.M. Burrows), and around Lago Tsene in 2019 (see Osborne *et al.* 2019a). The site is dominated by dense, dune thicket-forest. As with most coastal vegetation, there is a successional gradient from the foredunes to the older dunes further inland. Pioneer communities consist of species such as *Sesuvium portulacastrum*, *Cyperus crassipes*, *Scaevola plumieri* and *Ipomoea pes-caprae*, while shrubs such as *Eugenia capensis* subsp. *capensis* and *Diospyros rotundifolia* occur further back from the coastline at the top of the beach (Impacto Lda. 2012a; Osborne *et al.* 2019a).

Moving inland beyond these pioneer communities, trees such as *Craibia zimmermanii* and *Afzelia quanzensis* occupy the dune slopes, with the occasional *Euphorbia baylissii* in the shaded understorey (Osborne *et al.* 2019a). In the middle of the dune system, thickets dominate the vegetation. The species composition of these thickets includes *Olax dissitiflora* and *Cassia abbreviata*, and on older dunes the thicket canopy is around 4 m tall (*Osborne* #1670). The topology of the dunes provides a variety of micro-habitats, including sheltered dune slacks that host small numbers of *Encephalartos ferox* subsp. *ferox* (NT). As the thicket transitions into coastal dry forest on older dunes,

Elaeodendron fruticosum (JO)

Edge of Lagoa Poelela with coastal forests in the distance (JO)

with a canopy height of around 5 m, *Mimusops caffra* begins to dominate while other species such as *Suregada zanzibariensis* and *Drypetes natalensis* are common (Osborne *et al.* 2019a).

Around the back of the dune system and further inland, there are a number of lagoons and channels present at Inharrime-Závora. The lagoons in this IPA are brackish, although less saline than sea water (Hill *et al.* 1975). The largest body of water, Lagoa Poelela, is connected via 75 km of channels and other lagoons to the Indian Ocean. The edges of these lakes have herbaceous communities consisting mostly of Cyperaceae species, including *Cyperus laevigatus, C. natalensis*, and *Fimbristylis dichotoma*, alongside *Phragmites* (probably *P. mauritianus*) (Impacto Lda. 2012a).

There has been extensive clearance of the vegetation for machambas towards the back of the dune system. Osborne *et al.* (2019a) note that while clearings were more frequent further inland, there were also some smaller clearings observed closer to the foredunes. Crops grown in the Inharrime coastal area include rice, corn, manioc, peanuts, cashews, beans and pineapples (Impacto Lda. 2012a). Small-scale coconut plantations are also frequent, with some occurring near Lagoa Poelela within this IPA (J. Osborne, pers. comm. 2021). Where machambas have been abandoned, secondary vegetation includes cashew trees (*Anacardium occidentale*), *Salacia kraussii* and *Chrysocoma mozambicensis* (Osborne *et al.* 2019a).

Conservation issues

The site does not fall within a protected area, Key Biodiversity Area, Important Bird Area or RAMSAR site.

The primary threat to the vegetation within this IPA is shifting agriculture. Much of the area between Inharrime town and the coastline has already been converted to machambas, leaving only fragments of miombo. The IPA itself is under intense pressure from further expansion of agricultural land. Clearance of coastal forest at the south-eastern edge of Lagoa Poelela commenced in the mid-2000s, with the most significant clearances taking place between 2010 and 2013 (World Resources Institute 2021). In 2019, small patches of burned vegetation, in the process of being cleared, were observed towards the foredunes, further highlighting the continued threat to the coastal dunes (Osborne *et al.* 2019a). It is likely that the use of burning for vegetation clearance also poses the additional threat of uncontrolled fires, clearing wider swathes of vegetation than intended.

In addition to agriculture, the harvesting of trees threatens the integrity of the remnants of natural vegetation in the area. The main source of domestic fuel in Inharrime District is firewood while local timber is used for construction. For some Inharrime residents, depletion of resources has reached such an extent that they must travel over 5 km to find firewood (MAE 2005e). Sustainable resource usage strategies in both Inharrime and Zavala Districts would help relieve pressure on the IPA while also securing essential resources for local people into the future. According to a 2012 report of the Inharrime coastal area, there were plans to introduce a land-use management plan for the district (Impacto Lda. 2012a), however, it is unclear how much progress has been made to this end.

Towards Ponta Závora, further development of tourism may be a threat to habitats (Matimele *et al.* 2018a). Dense coastal forest has already been cleared to make way for tourist accommodation. However, the forest setting contributes to the visitor experience (Nhanombe Lodge 2021) which may, to some extent, limit degradation of habitat.

Euphorbia baylissii growing in the coastal forest understorey (JO)

The vertebrate taxa of this site have not yet been inventoried; however, there is a marine lab based at Závora.

Key ecosystem services

Fruits are harvested from coastal species such *as Diospyros rotundifolia*, *Phoenix reclinata* and *Salacia kraussii* by local people (Osborne *et al.* 2019a). Wood is also used as a source of timber and fuel, although these resources are currently being unsustainably harvested in some areas (MAE 2005e).

Ponta Závora and Lagoa Poelea both host tourists and, while the focus is mostly on water sports and marine wildlife, the dune vegetation in which accommodation is situated likely contributes to the visitor experience (Nhanombe Lodge 2021). There could be scope for expanding the tourist offer, particularly including terrestrial landscapes.

A lighthouse at Ponta Závora, built in 1910, is one of the few remaining, operational, lighthouses in Mozambique (Harrison & Finnegan 2021). There are a number of old buildings at Ponta Závora, built during Portuguese colonial rule, that may be of historic interest.

Ecosystem service categories

- Provisioning – Food
- Provisioning – Raw materials
- Cultural services – Tourism

IPA assessment rationale

Inharrime-Závora qualifies as an IPA under sub-criterion A(i) of the Important Plant Area criteria due to the presence of three globally Vulnerable species. Five endemic species have also been recorded; however, this only represents 1% of B(ii) qualifying species, less than the 3% threshold required. The intact coastal habitat at this site is also of botanical importance as much of this vegetation has been transformed or degraded across Inhambane Province and the Inhambane Centre of Endemism. At present, however, there is insufficient data to assess this site under sub-criterion C(iii).

Brakish edge of Lago Tsene (JO)

Priority species (IPA Criteria A and B)

FAMILY	TAXON	IPA CRITERION A	IPA CRITERION B	≥ 1% OF GLOBAL POP'N	≥ 5% OF NATIONAL POP'N	IS 1 OF 5 BEST SITES NATIONALLY	ENTIRE GLOBAL POP'N	SPECIES OF SOCIO-ECONOMIC IMPORTANCE	ABUNDANCE AT SITE
Celastraceae	*Elaeodendron fruticosum*	A(i)	B(ii)	✓	✓				unknown
Euphorbiaceae	*Euphorbia baylissii*	A(i)	B(ii)	✓	✓	✓			occasional
Malvaceae	*Grewia occidentalis* var. *littoralis*		B(ii)	✓	✓	✓			unknown
Myrtaceae	*Eugenia* sp. A of T.S.M.		B(ii)	✓	✓	✓			unknown
Sapindaceae	*Allophylus mossambicensis*	A(i)	B(ii)	✓	✓	✓			scarce
		A(i): 3 ✓	**B(ii): 5**						

Protected areas and other conservation designations

CONSERVATION AREA TYPE	CONSERVATION AREA NAME	RELATIONSHIP OF IPA TO CONSERVATION AREA
No formal protection	N/A	

Threats

THREAT	SEVERITY	TIMING
Small-holder farming	high	Ongoing – trend unknown
Tourism & recreation areas	medium	Ongoing – trend unknown
Housing & urban areas	low	Ongoing – trend unknown
Logging & wood harvesting	low	Ongoing – trend unknown

GAZA
PROVINCE

CHIDENGUELE

Assessors: Sophie Richards, Iain Darbyshire

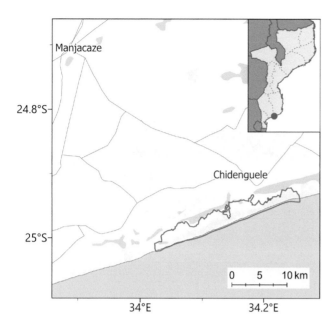

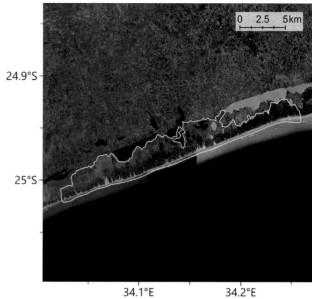

INTERNATIONAL SITE NAME		Chidenguele	
LOCAL SITE NAME (IF DIFFERENT)		–	
SITE CODE	MOZTIPA050	PROVINCE	Gaza

LATITUDE	-24.96651	LONGITUDE	34.15002
ALTITUDE MINIMUM (m a.s.l.)	0	ALTITUDE MAXIMUM (m a.s.l.)	75
AREA (km²)	60.3	IPA CRITERIA MET	A(i)

Intertidal rock platforms at Chidenguele with *Thalassodendron leptocaule* (SB, featuring Vera Bandeira)

Site description

Chidenguele is a coastal IPA, in Manjacaze (Mandlakazi) District of Gaza Province. The site is situated around 2 km south-west of Chidenguele town and covers an area just under 60 km². This IPA encompasses high-quality coastal forest habitat, a stretch of 25 km along the coastline from Lago Matsambe in the east to Praia de Chiziane in the west. Much of this habitat is threatened throughout southern Mozambique by clearance for agriculture while the forests around Chidenguele are, by contrast, largely intact and therefore offer an opportunity to conserve this habitat type.

Beyond the coastal forests, Chidenguele IPA stretches into the intertidal zone, to include the sandstone platforms on which the Near Threatened seagrass, *Thalassodendron leptocaule*, grows. The coastal area, Praia do Chidenguele, is also a small tourist location with a number of lodges in the forests to the west of this IPA.

The site has also been delineated to include the inland floodplain and wetland habitat associated with the globally Vulnerable palm species, *Raphia australis*. Although only a small population of this species is present at this site, Chidenguele is one of the few sites from which this species is known globally and so is of great importance to the conservation of *R. australis*. Only small areas of the IPA support *R. australis*, but the ecological integrity of the landscape as a whole is important to protect this species and wider biodiversity at Chidenguele.

Botanical significance

Chidenguele has been delineated to cover one of the few sites known to host the Vulnerable palm *Raphia australis*. Interestingly, unlike other sites further south where this species inhabits swamp or gallery forests, *Raphia australis* grows within reed beds at Chidenguele (Matimele *et al.* 2016b). This site is also unique as the most northerly location within the species' distribution. To date, only 10 individuals have been observed from this site, but the area is poorly sampled and it is estimated that there are upwards of 20 individuals present (H. Matimele, pers. comm. 2021). Chidenguele represents one of the five best sites in Mozambique for this species, all of which are under heavy pressure from expanding agricultural land (Matimele 2016). The continued presence of

R. australis at this site is of great importance to the overall resilience of the species.

The dune forests of Chidenguele are also of conservation importance, particularly in providing habitat for two globally threatened species, *Ecbolium hastatum* (EN) and *Elaeodendron fruticosum* (VU). *E. hastatum* is a rare species with a patchy distribution, known only from southern Mozambique at between 5 and 7 locations. The coastal dune thicket to which this species is restricted is known to be threatened by the expansion of towns and tourism infrastructure (Darbyshire *et al.* 2018d). Although only three individuals have been observed towards Praia de Chiziane (McCleland & Massingue 2018), there are likely to be more individuals around Chidenguele that have yet to be recorded. This IPA represents one of only three relatively secure sites for this species and, given that the AOO (area of occupancy) is only slightly above the Critically Endangered category threshold, it is of great importance in preventing the extinction of this species.

Another threatened endemic, *Elaeodendron fruticosum* (VU), which is also restricted to the coastal dunes of southern Mozambique, is present within this IPA. This species, like *Ecbolium hastatum* and *Raphia australis*, is threatened by habitat loss through conversion of coastal dune habitat for agriculture.

In total, there are five endemic species recorded from this IPA, including *Elaeodendron fruticosum* and *Ecbolium hastatum*, all of which occur within the coastal dune habitat. For one of these endemics, *Baphia ovata* (NT), Chidenguele may host one of the largest populations throughout this species' range (Langa *et al.* 2019a).

Chidenguele falls within the proposed Inhambane (sub-)Centre of Plant Endemism (CoE) (Darbyshire *et al.* 2019a), and some endemics present at this site, such as *Baphia ovata* and *Elaeodendron fruticosum*, have distributions limited only to this CoE. On a broader scale, Chidenguele also falls within the Coastal Forest of Eastern Africa Biodiversity Hotspot, covering some of the southernmost and westernmost forests within this hotspot. While the coastal forests and thickets of this site are largely intact and host a number of endemic species, possibly more than is currently known, there are insufficient data at this time to assess this habitat type across the proposed Inhambane CoE under the IPA criteria. However, it is possible that this site could also qualify under IPA criterion C in the future.

Delineation of this IPA has also included the intertidal zone, to cover the habitat of the Near Threatened seagrass *Thalassodendron leptocaule*. This species is known to form ecologically dominant stands on sandstone platforms with a number of epiphytic algae and marine invertebrates dependent on the habitat created by *T. leptocaule*. However, through human disturbance and climate change, this species may soon be threatened with extinction (Darbyshire *et al.* 2020b). At Chidenguele the population may be relatively secure, but increased footfall due to tourism at the site may pose a threat to this ecologically important species.

Habitat and geology

There have been few botanical studies in the Chidenguele area, which have focused mainly on a small number of endemic or threatened species. A wide-scale botanical inventory is, however, yet to take place. The following habitat description is based primarily on a report by Impacto Lda. (2012b) on Manjacaze (Mandlakazi) District.

Chidenguele encompasses a range of habitats spanning the intertidal zone and coastal dunes through to miombo woodlands inland, interspersed with a number of lagoons and wetlands associated with these lagoons. Underlying this site is Quaternary sedimentary geology, with mostly sandy soils and some recent alluvial deposits around the wetlands in the core zones and associated with the lagoons.

Furthest out to sea are the seagrass communities in the mid intertidal zone. Dominated by the seagrass *Thalassodendron leptocaule* (NT), these communities occur on sandstone platforms and are either submerged at all times or slightly exposed at low tide (Darbyshire *et al.* 2020b). *T. leptocaule* provides habitat for a number of species. One study of this species in South Africa found 52 taxa of macroalgae and 204 macroinvertebrates living on a population of this seagrass (Browne *et al.* 2013). It is therefore likely that the seagrass around Chidenguele also supports a complex web of ecological interactions.

Raphia australis palm growing in reed bed (HM)

Moving into the coastal zone, the dune vegetation follows a successional gradient, with common pioneer species such as *Cyperus crassipes, Ipomoea pes-caprae, Scaevola thunbergii* and *Sesuvium portulacastrum* inhabiting the foredunes. Moving further inland, shrubby vegetation dominates, including species such as *Diospyros rotundifolia* and *Grewia occidentalis* var. *litoralis*. *Encephalartos ferox* subsp. *ferox*, assessed as Near Threatened at species level, has been recorded from sheltered valleys within the dune system. The dune thicket towards Praia de Chizaine has a low canopy and includes species such as *Croton pseudopulchellus, Zanthoxylum delagoense, Manilkara concolor* and *Mimusops caffra* in the canopy with a sparse understorey that features *Barleria repens*. It is likely that dune thicket of this character is present throughout the dunes of this IPA. Dense coastal forest is present in some areas towards the back of the dune system. The coastal forests in this district are dominated by *Afzelia quanzensis, Mimopsus caffra, Sideroxylon inerme* and *Ficus* species.

Behind the coastal forest-thicket is habitat characterised by Lötter *et al.* (in prep.) as "Inharrime Coastal Palmveld", a poorly drained area of open wooded grassland. Species include palms such as *Phoenix reclinata, Hyphaene coriacea* and *Borassus aethiopum,* alongside various grass species including several of genera *Andropogon, Eragrostis* and *Hyparrhenia* (Lötter *et al.*, in prep.). In drier areas, *Brachystegia spiciformis* dominates with species such as *Albizia adianthifolia* and *Afzelia quanzensis* occurring in regrowth following disturbance (Impacto Lda. 2012b). Patches of reedbed wetlands, some of which are associated with the lagoons, are dominated by *Phragmites australis* and *Typha capensis*. These reedbeds provide important habitat for *Raphia australis* (VU) and this is the only site from which *R. australis* is known to grow in reedbeds (Matimele *et al.* 2016b).

There are a number of pools around the wetland areas of the IPA, some of which appear seasonal from satellite imagery (Google Earth 2021). Some of these pools have been reported to support *Pandanus livingstonianus* on the edges (Impacto Lda. 2012b), although this species may be restricted to the larger bodies of water that retain water year-round. This species may be of note as some sources report it to be endemic to

Mozambique (see Burrows *et al.* 2018), although the delimitation of this species is disputed.

The wetlands and lagoons are important sources of water for agriculture and, as such, a significant proportion of land bordering these areas is used for small-scale farming of crops such as sugar and rice. Much of the miombo in the north and east of this IPA, associated with urban expansion of Chidenguele, has also been converted to farmland. Agriculture in this area is largely for subsistence purposes and common crops include corn, beans, peanuts, cassava and sweet potatoes.

Conservation issues

Chidenguele does not fall within a protected area, Key Biodiversity Area or RAMSAR site. There are no known land management plans or local conservation initiatives within this IPA (Impacto Lda. 2012b). However, between 2009 and 2010, a project funded by the Global Environment Facility was undertaken to promote community awareness about the value of biodiversity, specifically focussing on the shore of Lagoa Inhapavola, to the east of this IPA. Although much of the land surrounding this lake lies north-east of this IPA, there may be some indirect benefits to greater community awareness in the local area.

The main threat to the flora of this site is conversion of habitats for agriculture. Much of the northern and eastern areas of this IPA have been converted to agriculture. Most of this is done on a small-scale for subsistence purposes, with crops including corn, beans, peanuts, cassava and sweet potatoes. However, there are a small number of family-run farm businesses selling crops such as rice, cashews and vegetables. While it is true that clearing of land for agricultural expansion at this site has been limited in the past decade, there continues to be small-scale clearance and degradation of woodland and coastal forest within the IPA (Google Earth 2021; World Resources Institute 2021). Of greatest concern is the farming of land in and around the wetlands of this site, with subsistence crops such as rice and sugarcane grown in these areas (H. Matimele, pers. comm. 2021). While the reliable supply of water in these wetlands supports higher agricultural yields, the water demand of agriculture also changes the hydrology of these wetlands, reducing seed recruitment for *Raphia australis* (VU) and weakening the long-term resilience of the population, particularly as this species flowers only once in its lifetime (Matimele *et al.* 2016b).

Another major factor contributing to habitat loss is the establishment of tourism in the dune forests to the south-east of this IPA. Clearance of this dense vegetation for access roads and new facilities has taken place in recent decades (Google Earth 2021). In addition, increased activity and footfall in the intertidal zone, linked to marine tourism activities, will potentially degrade the habitats for *Thalassodendron leptocaule* (NT), a seagrass on which many marine species depend (Darbyshire *et al.* 2020b). While the site has great tourism potential, and expansion of tourism could support livelihoods for local people, the expansion of this sector near Chidenguele must be done responsibly to prevent unnecessary damage to the natural environment – particularly as the habitats here are a major draw for tourists. A number of hotels within and around this IPA already promote the natural scenery as part of their tourist offer. Tourism could, therefore, incentivise the protection of local habitats, with businesses dependent on these habitats to continue to attract visitors.

The fauna of this IPA has not been inventoried although the coastal areas and wetlands may be important for avifauna.

Key ecosystem services
This IPA currently encompasses several hotels and lodges (Google Earth 2021). Much of the tourism focusses on beach and marine activities; however, a number of these hotels mention the nature-rich setting in promotional materials, with the intact coastal forests of particular interest. Although it is unlikely that there are any large charismatic

mammals in the area (Impacto Lda. 2012b), there could be some potential for birdwatching associated with coastal forest species, such as Rudd's Apalis (*Apalis ruddi* – LC) and Neergaard's Sunbird (*Cinnyris neergaardi* – NT). An inventory of local bird species could help to promote the nature tourism potential of the site.

The lagoons and wetlands are important water sources for both domestic use and for agriculture, although this may be at the expense of the wetland habitats within this IPA (Impacto Lda. 2012b; Matimele *et al.* 2016b).

The seagrass *Thalassodendron leptocaule* provides habitat for numerous marine organisms (Darbyshire *et al.* 2020b). This habitat, along with coral reefs in the subtidal zone, may contribute to resource-rich marine ecosystems on which artisanal fishing activities in the district depend (Impacto Lda. 2012b).

There are no formal forestry operations in the area; however, wood is harvested for domestic timber and fuel needs (Impacto Lda. 2012b).

Ecosystem service categories

- Provisioning – Food
- Provisioning – Raw materials
- Provisioning – Fresh water
- Habitat or supporting services – Habitats for species
- Cultural services – Recreation and mental and physical health
- Cultural services – Tourism

View of lagoon and surrounding "Inharrime Palm Veld" vegetation (HM)

IPA assessment rationale

Chidenguele qualifies as an IPA under sub-criterion A(i), as it hosts important populations of one Endangered species, *Ecbolium hastatum*, and two Vulnerable species, *Raphia australis* and *Elaedendron fruitcosum*. As one of only a handful of sites from which *R. australis* is known, and as the most northerly and only site at which this species occurs in reed beds, Chidenguele is of great importance for the conservation of this Vulnerable palm species. Six endemic species are known from this IPA, representing 1% of species from the national list of species of high conservation importance, below the 3% threshold required to meet sub-criterion B(ii).

Raphia australis regeneration (HM)

Priority species (IPA Criteria A and B)

FAMILY	TAXON	IPA CRITERION A	IPA CRITERION B	≥ 1% OF GLOBAL POP'N	≥ 5% OF NATIONAL POP'N	IS 1 OF 5 BEST SITES NATIONALLY	ENTIRE GLOBAL POP'N	SPECIES OF SOCIO-ECONOMIC IMPORTANCE	ABUNDANCE AT SITE
Acanthaceae	*Ecbolium hastatum*	A(i)	B(ii)	✓	✓	✓			scarce
Arecaceae	*Raphia australis*	A(i)				✓			occasional
Celestraceae	*Elaeodendron fruticosum*	A(i)	B(ii)	✓	✓	✓			unknown
Fabaceae	*Baphia ovata*		B(ii)	✓	✓	✓			frequent
Malvaceae	*Grewia occidentalis* var. *littoralis*		B(ii)	✓					unknown
Myrtaceae	*Eugenia* sp. A of T.S.M.		B(ii)	✓	✓	✓			occasional
Rubiaceae	*Psydrax moggii*		B(ii)	✓					unknown
		A(i): 3 ✓	**B(ii): 6**						

Protected areas and other conservation designations

CONSERVATION AREA TYPE	CONSERVATION AREA NAME	RELATIONSHIP OF IPA TO CONSERVATION AREA
No formal protection	N/A	

Threats

THREAT	SEVERITY	TIMING
Housing & urban areas	medium	Ongoing – trend unknown
Tourism & recreation areas	low	Ongoing – trend unknown
Small-holder farming	high	Ongoing – trend unknown
Recreational activities	low	Ongoing – trend unknown

BILENE-CALANGA

Assessors: Sophie Richards, Iain Darbyshire, Hermenegildo Matimele, Castigo Datizua

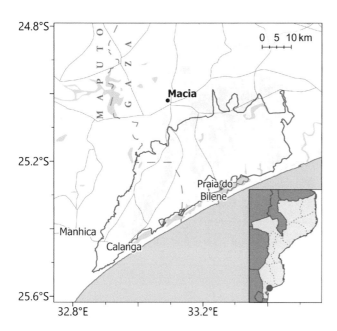

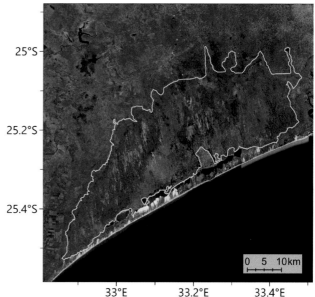

INTERNATIONAL SITE NAME		Bilene-Calanga	
LOCAL SITE NAME (IF DIFFERENT)		–	
SITE CODE	MOZTIPA049	PROVINCE	Gaza

LATITUDE	-25.21076	LONGITUDE	33.16372
ALTITUDE MINIMUM (m a.s.l.)	10	ALTITUDE MAXIMUM (m a.s.l.)	90
AREA (km²)	1366	IPA CRITERIA MET	A(i)

Site description

Bilene-Calanga is an IPA that spans the boundary between two provinces, Gaza, falling within Xai-Xai and Bilene Districts, and Maputo, falling within Manhiça District. Bounded to the north-east by the Limpopo river, this site represents the easterly edge of both the Maputaland Centre of Plant Endemism in the narrow sense (Darbyshire *et al.* 2019a) and the Maputaland-Pondoland-Albany Biodiversity Hotspot (CEPF 2010).

This site is 1,400 km² in area and runs from the Limpopo estuary in the east to Calanga, by the Incomati River, in the west. To the south-east the site is bounded by a number of lagoons, including Lagoa Uembje, that run parallel to the coastline at Praia do Bilene. The town of Praia do Bilene itself has been excluded from the IPA. To the north, the boundary runs south of the EN1 road.

This site has been delineated to include a number of habitat patches of importance for two key species: *Memecylon incisilobum*, a Critically Endangered species known only from this site globally, and *Raphia australis*, a Vulnerable species with the majority of the global population occurring in the wetlands of this site (Matimele *et al.* 2016b, 2018b). The key habitats for these two species have been delineated within the site map for information but should not be treated as core zones as the whole landscape is important for the integrity of this site, particularly for *Raphia australis* which is dependent on hydrology beyond its habitat.

Botanical significance

Bilene-Calanga is of high botanical significance as the only site from which the Critically Endangered species *Memecylon incisilobum* is known. Limited to a dense coastal forest fragment (-25.190°, 33.208°)

within Chihacho sacred forest, this species is known to have a global range of 4 km² in which there are fewer than 250 individuals (Matimele *et al.* 2018b). Searches have been conducted in neighbouring forest patches which appear similarly intact; however, *M. incisilobum* has never been recorded elsewhere (Matimele 2016). While local beliefs surrounding this sacred forest have prevented degradation, such practices are not observed by people from outside the area. In 2010, a cellphone mast was erected in the centre of the forest, with an access road causing additional degradation, while the forest edge continues to be degraded by burning used to clear adjacent agricultural land (Matimele *et al.* 2018b). Conservation action to prevent further degradation of this forest patch is critical to preventing the extinction of *M. incisilobum*.

The IPA as a whole has been delineated to include the best habitat for another globally threatened species, *Raphia australis* (VU). This species, known commonly as Rafia or Kosi Palm, flowers once every 20–30 years and dies soon after (Burrows *et al.* 2018). The best habitat is concentrated within swamp wetlands, with several habitat patches included within the IPA. However, conservation action is required across the site to ensure that the integrity of these wetlands is not indirectly degraded. *R. australis* is dependent on drainage lines in coastal swamp forest (Burrows *et al.* 2018), and the disruption of water availability, through conversion of land to agriculture, would have a strongly detrimental impact on *R. australis* at this

site. Estimates suggest that around 4,000 mature individuals, out of a global population of 5,500–7,000 individuals, are present at this site (Matimele *et al.* 2016b). Bilene-Calanga is, therefore, of prime importance in preventing the extinction of *Raphia australis*.

Another globally Vulnerable taxon, *Psychotria amboniana* subsp. *mosambicensis*, is known from this IPA, occurring in Chihacho sacred forest. This taxon is endemic to southern Mozambique, from Maputo city in the south to the Save River in the north. The final Vulnerable species present at this site is *Millettia ebenifera* which, like *P. amboniana* subsp. *mosambicensis*, is endemic to coastal Mozambique, where habitat is highly threatened by clearance (Richards, in press [d]). There are eight endemic species known from this site in total; these include two as yet undescribed species, *Pachystigma* sp. A of *Flora Zambesiaca* (Bridson 1998), which is known only from this IPA, and *Eugenia* sp. A of *Trees and Shrub Mozambique* (Burrows *et al.* 2018) which is known throughout the coastal region of southern Mozambique.

Habitat and geology

Bilene-Calanga IPA is predominantly underlain by Quaternary interior dunes with sandy soils, and some recent alluvial deposits underlie the swamp wetlands and rivers to the south-west (MAE 2005f; Impacto Lda. 2012c). Average temperatures are in the 24 – 26°C range and average precipitation for the Bilene-Macia District is between 800 and 1,000 mm annually (MAE 2005f).

Raphia australis palms within Bilene wetlands (JEB)

Rachises of *Raphia australis*, harvested for boat-making (CD)

Swamp forests featuring *Pandanus livingstonianus* (CD)

Towards Praia do Bilene town, at the south-east boundary, the vegetation is predominantly coastal grassland dominated by species from the genera *Eragrostis*, *Triraphis* and *Urelytrum*. Inland, there are also patches of scrub, with species including *Albizia adianthifolia*, *Sclerocarya caffra* and *Terminalia sericea* (Impacto Lda. 2012c).

Areas of semi-deciduous, coastal forest occur within the habitat mosaic, including Chicaho sacred forest (-25.183°, 33.178°) and Ngondze forest (-25.086°, 33.172°). These two forests have been sampled as part of a permanent plot by Fernandes *et al.* (2020). These authors found that common tree species include *Afzelia quanzensis*, *Albizia adianthifolia*, *Apodytes dimidiata*, *Dialium schlechteri* and *Strychnos gerrardii*, while shrubs such as *Psydrax locuples*, *Eugenia mossambicensis* and *Artabotrys monteiroae* occur in the understorey. Herbaceous cover varies depending on canopy shade; in areas of deep shade, *Asparagus* species and the fern *Polypodium scolopendria* were recorded to grow in abundance.

Although many of the forest patches in this IPA appear similar, thorough searches have found that only one, a dense fragment of the Chihacho sacred forest, is home to the Critically Endangered species *Memecylon incisilobum*, providing the only known habitat globally for this species. This patch of forest is known to have good leaf litter and numerous lianas (Matimele *et al.* 2018b).

There are a number of wetlands at this site, occurring along drainage lines, with associated swamp forests that are important habitat for *Raphia australis* (VU) (Matimele 2016). *Pandanus livingstonianus* is frequent within these swamp forests, occurring in large colonies across the IPA (C. Datizua, pers. comm. 2021); this may be of conservation interest as some sources recognise *P. livingstonianus* as endemic to Mozambique (Burrows *et al.* 2018), however, other sources dispute this delimitation of the species. Other tree species in the swamp forests include *Syzygium cordatum* and *Voacanga thouarsii*, while the understorey features the fern species *Stenochlaena tenuifolia*, a species characteristic of these swamp forests in Mozambique (Burrows *et al.* 2018; Hyde *et al.* 2021). On the edge of these swamps is open *Syzygium cordatum* woodland, with the herb *Asparagus densiflorus* and the shrub *Vangueria monteiroi* recorded as widespread in the understorey of some of these forest patches (Hyde *et al.* 2021).

Towards both the river Limpopo and Incomati are areas of floodplain. Much of these area have been transformed by agriculture as local people, reliant on rain-fed agriculture, seek land with more reliable moisture levels. The remaining habitat is mostly grassland, and, although the composition of this area has not been fully documented, the endemic species *Tritonia moggii* is known to occur in floodplain slightly north-east of this IPA (*Rulkens* #s.n.) and may well also occur within its boundaries.

Conservation issues

Bilene-Calanga IPA does not fall within a protected area. However, this site represents the eastern edge of the Maputaland-Pondoland-Albany biodiversity hotspot. This hotspot covers the eastern coasts of southern Africa, identified based on the high level of biodiversity and high level of threat to this biodiversity (CEPF 2010). The Critical Ecosystem Fund Partnership (CEPF) went further in highlighting the importance of the Bilene-Calanga area, by designating two, adjacent Key Biodiversity Areas (KBAs) that overlap with this IPA: Manhiça, to the south-west, and the Xai-Xai and Limpopo Floodplain, to the north-east. The sites were identified in 2008 as part of a network across the Maputaland-Pondoland-Albany biodiversity hotspot of priority areas for conservation, with the two KBAs at Bilene-Calanga recognised for their provision of ecosystem services and the high pressure on these areas from conversion of land to agriculture (CEPF 2010). Together the two KBAs at Bilene-Calanga formed part of "the Mozambique Coastal Belt corridor", recognised for its potential to provide ecological

Memecylon incisilobum (JEB)

resilience and connectivity in the face of future perturbations, particularly climate change.

In 2021, a revised assessment of KBA sites was conducted in Mozambique, which resulted in the unification of much of the area covered by the Manhiça and Xai-Xai and Limpopo Floodplain KBAs into a single KBA (Manhiça-Bilene). This site is triggered by three species, including the IPA priority species *Memecylon incisilobum* (CR) and *Raphia australis* (VU), alongside Orange-fringed River Bream (*Chetia* brevis – EN), a highly threatened species limited only to the Incomati river system (Roux and Hoffman 2017).

The range of inundated habitats, including the floodplains at either side of this IPA and the swamp forests, are highly suitable for agriculture such as rice or sugarcane cultivation (Matimele *et al.* 2016b). To the east of this IPA, on the Limpopo River flats, extensive areas of land have already been converted to rice farming, while Xinavane, to the north-west of the IPA in Manhiça District, has several hectares of sugar plantation (Impacto Lda. 2012d). The Ministry of Agriculture and Rural Development is planning further agricultural concessions as part of their development strategy, as well as incentivising sugarcane cultivation as a food crop and for biofuel (CEPF 2010; Matimele 2016). Bilene-Calanga IPA is therefore under great pressure from agricultural expansion, particularly the wetlands on which *Raphia australis* (VU) is dependent.

Associated with this expansion of agriculture, fires used by local people to clear land for arable farming or to renew pasturelands also threaten *Raphia australis,* alongside other swamp forest species such as *Pandanus livingstonianus. Raphia australis* also

Chicaho sacred forest, a semi-decidious coastal forest (JEB)

faces an additional threat of harvesting for market trade. There is rising demand for the rachises of *R. australis* leaves which, due to their buoyancy, are used to construct boats. Rachises from this site are often sent south to the coastal villages around Maputo (C. Datizua, pers. comm. 2021).

The forest of Chicacho, as a sacred forest, has been somewhat protected from degradation. However, these beliefs are not observed by people from outside the local area. Land in this area has been converted to agriculture, with clearance burning observed on a visit in 2015 (Matimele 2016). Continued burning of adjacent land will erode the forests edges which, given the already fragmented nature of this site, may reduce moisture availability and alter the species composition within this forest. In addition, harvesting of *Afzelia quanzensis* and subsequent timber processing were observed during site surveys in 2019 (Fernandes *et al.* 2020). As the population of Bilene has expanded, there has been greater extraction of wood for charcoal and, together with clearance burning, this has caused a ca. 20% loss in forest area between 2011 and 2016 (Matimele *et al.* 2018b).

Although the *Memecylon* forest fragment is strikingly dense compared to the surrounding vegetation, the construction of a telecommunications mast in the centre of this forest in 2010, along with an accompanying access road, has led to degradation within this forest patch (Matimele 2016). With the presence of *M. inicislobum*, a Critically Endangered species known only from this locality, Bilene-Calanga meets Alliance of Zero Extinction site criteria. It has been estimated that the entire global population of *M. incisilobum* could be lost in around 20 years at the current rate of degradation (Matimele *et al.* 2018b). Urgent conservation action is therefore needed here to prevent further losses and restore degraded areas of this unique site if the extinction of this species is to be prevented. A collaboration between Botanic Gardens Conservation International and Instituto de Investigação Agrária de Moçambique, funded by the Franklinia Foundation and the Global Trees Campaign, is investigating how to conserve *M. inicislobum* at this site. These partners plan to survey habitat and evaluate the population size of *M. incisilobum* to devise an appropriate conservation plan for this species (C. de Souza, pers. comm. 2021). *Ex situ* conservation through seed banking, as *M. incisilobum* is predicted to have orthodox seed storage behaviour (Wyse & Dickie 2018), and cultivation in botanic gardens could also complement *in situ* conservation.

Key ecosystem services

The Chihacho sacred forest lies within this IPA and is a place where local people communicate with ancestors. This patch of forest has remained largely intact, despite agricultural conversion in the surrounding area, because of its spiritual importance for local people.

Wood extraction from these coastal forests is for the production of charcoal and timber – *Afzelia quanzensis* in particular has been observed to be selectively harvested for timber (Fernandes *et al.* 2020; Matimele *et al.* 2018b). Although assessed as Least Concern, overharvesting of this species is a threat throughout its range (Hills 2019) and could negatively impact local population numbers if extracted at an unsustainable rate. In addition, this timber species is extracted from Chihacho forest, in which *M. insicilobum* is present, which could have severe consequences for this ecosystem and the continued survival of the only known population of *M. insicilobum*. Support for sustainable forestry practices would greatly benefit both local people and conservation within the IPA.

The wetlands are known to play a role in flood prevention and may also be important in mitigating against future perturbations, particularly linked with climate change (CEPF 2010).

Praia do Bilene is a popular tourist area. Whilst much of this is focused on the coastal area outside this IPA, there may be some scope for nature tourism within the IPA boundary.

Ecosystem service categories

- Provisioning – Raw materials
- Regulating services – Moderation of extreme events
- Regulating services – Erosion prevention and maintenance of soil fertility
- Cultural services – Tourism
- Cultural services – Spiritual experience and sense of place
- Cultural services – Cultural heritage

IPA assessment rationale

Bilene-Calanga qualifies as an IPA under sub-criterion A(i) with four species at this site meeting the required threshold: one Critically Endangered species, *Memecylon incisilobum*, and three Vulnerable taxa, *Raphia australis, Millettia ebenifera* and *Psychotria amboniana* subsp. *mosambicensis*. With the entire known population of *M. incisilobum* known from Bilene-Calanga IPA, conservation of this site is paramount to preventing the extinction of this highly threatened species. Furthermore, as this IPA also hosts the largest known population of *Raphia australis*, protection of wetland habitat here would also make an important contribution to the overall resilience of this species. Conservation of the Bilene-Calanga site is incredibly urgent due to the pressures from expanding agriculture and deforestation faced, particularly while important habitats remain largely intact and may recover from disturbances. At present, only seven species meeting criterion B(ii) are known from this site. This represents less than 2% of Mozambique's endemic and range-restricted species, which is less than the 3% threshold required for this site to qualify under sub-criterion B(ii).

Priority species (IPA Criteria A and B)

FAMILY	TAXON	IPA CRITERION A	IPA CRITERION B	≥ 1% OF GLOBAL POP'N	≥ 5% OF NATIONAL POP'N	IS 1 OF 5 BEST SITES NATIONALLY	ENTIRE GLOBAL POP'N	SPECIES OF SOCIO-ECONOMIC IMPORTANCE	ABUNDANCE AT SITE
Arecaceae	*Raphia australis*	A(i)		✓	✓	✓			frequent
Fabaceae	*Millettia ebenifera*	A(i)	B(ii)	✓	✓	✓			unknown
Iridaceae	*Tritonia moggii*		B(ii)	✓					frequent
Melastomataceae	*Memecylon incisilobum*	A(i)	B(ii)	✓	✓	✓	✓		occasional
Myrtaceae	*Eugenia* sp. A of T.S.M.		B(ii)	✓	✓	✓			unknown
Rubiaceae	*Pachystigma* sp. A of F.Z.		B(ii)	✓	✓	✓	✓		scarce
Rubiaceae	*Psychotria amboniana* subsp. *mosambicensis*	A(i)	B(ii)	✓	✓	✓			unknown
Rubiaceae	*Psydrax moggii*		B(ii)	✓					unknown
		A(i): 4 ✓	B(ii): 7						

Protected areas and other conservation designations

CONSERVATION AREA TYPE	CONSERVATION AREA NAME	RELATIONSHIP OF IPA TO CONSERVATION AREA
Key Biodiversity Area	Manhiça-Bilene	IPA encompasses protected/conservation area

Threats

THREAT	SEVERITY	TIMING
Housing & urban areas	medium	Ongoing – trend unknown
Tourism & recreation areas	low	Ongoing – increasing
Shifting agriculture	high	Ongoing – increasing
Roads & railroads	medium	Past, likely to return
Increase in fire frequency/intensity	high	Ongoing – trend unknown
Logging & wood harvesting	high	Ongoing – trend unknown
Agro-industry farming	unknown	Future – inferred threat

MAPUTO
PROVINCE

BOBOLE

Assessors: Sophie Richards, Iain Darbyshire, Camila de Sousa

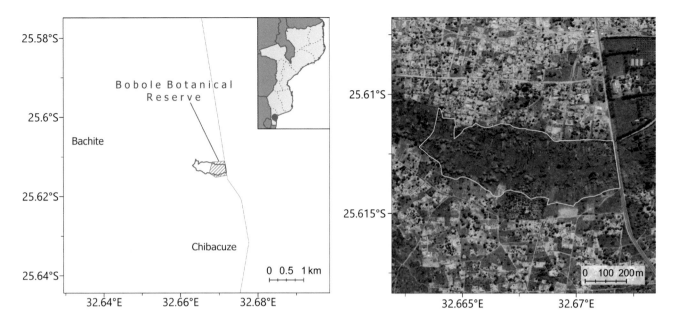

INTERNATIONAL SITE NAME		Bobole	
LOCAL SITE NAME (IF DIFFERENT)		–	
SITE CODE	MOZTIPA054	PROVINCE	Maputo

LATITUDE	-25.61386	LONGITUDE	32.67010
ALTITUDE MINIMUM (m a.s.l.)	10	ALTITUDE MAXIMUM (m a.s.l.)	20
AREA (km²)	0.23	IPA CRITERIA MET	A(i)

Site description

Bobole IPA of Marracuene District, Maputo Province is based upon the Bobole Botanical Reserve (Reserva Botânica de Bobole), the only botanical reserve designated in Mozambique. Located 30 km north of Maputo city on Bobole stream, a tributary of the Incomati River, this IPA is historically associated with floodplain vegetation. However, much of this has been cleared to make way for agriculture.

The IPA is around double the size of Bobole Botanical Reserve. While this IPA extends beyond the reserve in the west, continuing along Bobole stream for a further 450 km beyond the reserve boundary, it also excludes the northerly and southerly edges of the botanical reserve to avoid residential areas.

Botanical significance

The primary aspect of botanical importance within this IPA is the population of the globally Vulnerable species *Raphia australis*. There are 85 individuals of this near-endemic species present at 19 localities within Bobole Botanical Reserve (Pais 2011). There is no information for the population size of *R. australis* outside the reserve and there appear to be few, or possibly no, large individuals visible from satellite imagery (Google Earth 2021). However, the continued integrity of the known populations in the botanical reserve are dependent on wider landscape integrity and hydrology, therefore conservation work upstream will help to support the resilience of these *R. australis* individuals.

Although this site does not host the largest population of this species, there is great opportunity for conserving it here given its status as a botanical reserve. This species is also of particular importance socio-economically, as the local population have a long history of using *R. australis* in construction and for cultural purposes.

The *Raphia australis* population at Bobole is under great pressure from agriculture, as the reserve is already heavily degraded and the persistent disturbances caused by surrounding agriculture is having a detrimental impact on this population, particularly inhibiting the regeneration of this species (Pais 2011). As a result, the density of this species within Bobole Botanical Reserve has decreased from 160 individuals per hectare in 1999 to 39 individuals per hectare in 2010 (Pais 2011).

Habitat and geology

Bobole is a wetland site, centred around Bobole stream, and is underlain by peat alluvium soils with some sandy soils on inland dunes. Mean average temperatures range from 18 to 25°C in June and July to 26–32°C in December and January, while average annual precipitation is 654 mm, with the majority of this falling between November and March (Lötter *et al.*, in prep.). Moisture at this site is also gained through lateral infiltration of aquifers in the sand dunes. This helps to maintain a high water table throughout the year, with sub-surface irrigation of the peat soils also contributing to an increased rate of decay and nutrient cycling, resulting in high-fertility soils (Pais 2011).

Raphia australis occurs on these peat soils, alongside other tree species such as *Afzelia quanzensis*, *Bridelia cathartica*, *Myrica serrata*, *Strychnos spinosa*, *Syzigium cordatum*, *Trichilia emetica* and *Voacanga thouarsii* (Pais 2011). The *Raphia* palms themselves create micro-habitats as water that collects in the leaf axils soon allows the development of humus. A number of epiphytes, particularly ferns, grow in this humus (João 2011). In the understorey, shrubs include *Barringtonia racemosa*, *Phyllanthus reticulatus* and *Sesbania sesban* alongside a number of non-native shrubs such as *Cajanus cajan*, *Lantana camara* and *Ricinus communis*, while the herbaceous layer is dominated by *Typha capensis* (Pais 2011).

Due to the high productivity of the soils within this IPA, however, much of this habitat has been heavily transformed by agriculture. A number of vegetable crops are grown such as onion, carrot, tomato, green beans, lettuce and cabbage. Established *Raphia australis* palms occur within the machambas while the edges of these fields are delineated by drainage channels, where sugarcane and banana are grown (João 2011).

Raphia australis at Bobole (DN)

Conservation issues

Bobole IPA includes much of Bobole Botanical Reserve. First designated in 1945, under Portuguese governance, Bobole Botanical Reserve was established in recognition of the need to protect swamp forest and to provide opportunities to study this ecologically interesting habitat. The swamp forests were formerly extensive around the Incomati River but had been cleared for farming, including commercial rice and banana farms (Pais 2011). A 200 ha area was initially designated, to protect habitats and to provide an opportunity for ecological research, particularly studying the ecologically interesting *Raphia australis*. However, this was reduced to 12 ha by 1967 due to the continued degradation of habitat. Although the site is referred to as a botanical reserve, it is not part of the protected area network. With this lack of formal protection, the site is heavily degraded, with the entire area transformed for agriculture.

There has previously been limited conservation activities in the area, including guards (although there are few and they were previously reported to be unpaid roles) and a netting fence installed in recent decades which was quickly damaged and removed by local people (Manhice 2010). The guards and some residents have raised awareness of the importance of *Raphia australis* within the local community (Manhice 2010). In a survey of local residents, with 47 respondents, over half stated that they actively take care to avoid damage or destruction of this species when undertaking agricultural activities within the reserve (João 2011). These findings suggest there is a good level of awareness surrounding the importance of *R. australis* within this IPA.

Despite this awareness, the species is still exposed to a number of threats because of the economic reliance of local residents on farming this land (Manhice 2010). Burning to clear land for a new agricultural cycle often damages leaves and trunks of established *R. australis*, while seedlings are actively removed during the weeding process and left by the roadside (Manhice 2010; João 2011). Seedlings are also negatively impacted by the creation of drainage channels which increase water runoff and lower the recruitment rate (Matimele 2016). Low seedling recruitment was found to be a particular issue within this IPA, with over 902 seeds recorded per hectare of habitat and only 147 seedlings per hectare, constituting the biggest drop in abundance between stages in the *R. australis* lifecycle (Pais 2011). This is of particular concern for a monocarpic species that only reaches maturity after 20–40 years; the loss of seedlings in the present day could have a profoundly detrimental impact on the population viability, the full effects of which may not be revealed for many decades given its lengthy lifecycle.

As the only botanical reserve in Mozambique, there is a lot of interest in conserving this site and a restoration project, led by Instituto de Investigação Agrária de Moçambique and funded by Biofund, began in 2021. While this project is still in its early stages, with surveys being undertaken to support future conservation action, the end goal is to establish a syntropic agricultural system at this site (C. de Souza, pers. comm. 2021). As local people are heavily dependent on the highly productive land of Bobole, it is not feasible to prevent them from farming the area. However, the development of ecologically sensitive farming practices, particularly with regard to *Raphia australis* regeneration, could allow the integration of livelihoods and conservation within this IPA.

The faunal taxa of this IPA have not been fully documented, although the Palm-nut vulture (*Gypohierax angolensis*), which disperses seeds of *Raphia australis*, is known to occur within Bobole Botanical Reserve. Despite being assessed as globally Least Concern, the presence of this ecologically important bird species is thought to be at risk in this reserve, which may further exacerbate regeneration of the palm species (Pais 2011).

Entrance to Bobole Botanical Reserve (DN)

Machambas surrounding *Raphie* palms (DN)

Raphia australis provides a number of services itself, providing materials that are required for construction and ornaments with cultural value. Leaves are also woven into animal cages, which a small number of local residents sell at market. Otherwise, *R. australis* products are produced primarily for household use (João 2011). This palm species is also known to play a number of roles in the ecosystems in which it resides, including habitat for epiphytes and food for birds, particularly the Palm-nut vulture (Pais 2011).

Ecosystem service categories

- Provisioning – Raw materials
- Provisioning – Fresh water
- Provisioning – Medicinal resources
- Habitat or supporting services – Habitats for species
- Cultural services – Cultural heritage

Key ecosystem services

Given the proximity of Bobole to Maputo there are great opportunities to bring tourism to the site, particularly if restoration also brought the return of wetland bird species. There could also be recreational usage of the site for local people if the area was restored. However, this restoration of habitats would likely require the reallocation of land for those who currently cultivate the area.

At present, the stream is of high importance to local communities, not only contributing to the agricultural productivity of the area through irrigation but also being used for fishing, personal hygiene, and washing clothes and household items (João 2011).

IPA assessment rationale

Bobole qualifies as an IPA under sub-criterion A(i), as one of the five best sites nationally for the Vulnerable species *Raphia australis*. As this site has been recognised as a botanical reserve for several decades, there is a great opportunity to restore habitats and to protect *R. australis* at this site.

Priority species (IPA Criteria A and B)

FAMILY	TAXON	IPA CRITERION A	IPA CRITERION B	≥ 1% OF GLOBAL POP'N	≥ 5% OF NATIONAL POP'N	IS 1 OF 5 BEST SITES NATIONALLY	ENTIRE GLOBAL POP'N	SPECIES OF SOCIO-ECONOMIC IMPORTANCE	ABUNDANCE AT SITE
Arecaceae	*Raphia australis*	A(i)	B(iii)			✓		✓	frequent

Protected areas and other conservation designations

CONSERVATION AREA TYPE	CONSERVATION AREA NAME	RELATIONSHIP OF IPA TO CONSERVATION AREA
Botanical Reserve	Reserva Botânica de Bobole	protected/conservation area matches IPA

Threats

THREAT	SEVERITY	TIMING
Small-holder farming	high	Ongoing – stable
Persecution/control of terrestrial plants	medium	Ongoing – stable

INHACA ISLAND

Assessors: Hermenegildo Matimele, Sophie Richards, Salomão Bandeira, Iain Darbyshire

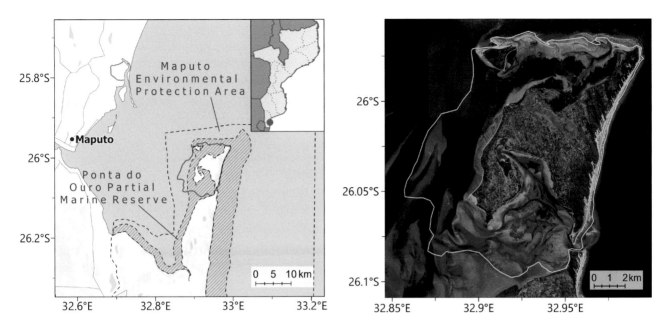

INTERNATIONAL SITE NAME		Inhaca Island	
LOCAL SITE NAME (IF DIFFERENT)		KaNyaka Island	
SITE CODE	MOZTIPA051	PROVINCE	Maputo

LATITUDE	-26.02860	LONGITUDE	32.92730
ALTITUDE MINIMUM (m a.s.l.)	-25	ALTITUDE MAXIMUM (m a.s.l.)	100
AREA (km²)	132.0	IPA CRITERIA MET	A(i), B(ii)

Site description

KaNyaka (also known as Inhaca Island), is situated in Maputo Bay (-26.02°, 32.94°) off the coast of southeast Mozambique. It forms a barrier separating Maputo Bay from the Indian Ocean (Mogg 1967). To the south is the Machangulo Peninsula separated from the Island by an inlet about 800 m wide and up to 15 m deep (Hobday 1977). To the north-west, is Portuguese Island of about 3.7 km² which is included within the boundary of this IPA. The intertidal zone of KaNyaka within Maputo Bay is also included within the boundary of this IPA as it is home to important seagrass communities. In administrative terms, KaNyaka Island, along with Portuguese Island, are under the Maputo municipality and is recognised as a separate municipal district.

KaNyaka Island has a high-energy wind- and wave-dominated oceanward shore with a steep sandy beach and high vegetated coastal dunes. In terms of elevation, the highest point, Mount Inhaca, is over 100 m high (Hobday 1977). The total area of the Island, which resembles a distorted "N" (Muacanhia 2004), is just over 40 km², with a maximum length of about 12 km from Ponta Mazondue to Ponta Torres in the south. The total area of the IPA, which also includes the Portuguese Island and the foreshore of the KaNyaka Island, is about 132 km². It is home to 6,100 inhabitants distributed in three main areas including Nhakene (over 1,500 residents), Ribjêne (2,100) and Inguane (2,500) (Sörbom & Gasim 2018). Alongside the local inhabitants, the island and the adjacent Machangulo Peninsula is a prime tourist destination in southern Mozambique, a particular highlight for cruise liner tourism, hence various tourist facilities have been established, particularly along the coastline.

During the colonial rule, the Islands (KaNyaka and Dos Portugueses) and Maputo Bay had a European presence, especially British and Dutch-Portuguese and, as a result, Ilha dos Portugueses is known by several different names (Adam *et al.* 2014). KaNyaka Island was occupied by the British for over half a century, from 1823 to 1875 (Portugal & Matos 2018). The Island was used to patrol and control activities related to the trade in ivory and enslaved people in the region.

Botanical significance

There are 455 species recorded from KaNyaka (Matimele & Timberlake 2020). Darbyshire *et al.* (2019a) places KaNyaka among sites with more than 20 endemic or range-restricted taxa in Mozambique, therefore recognising it as one of the most important sites nationally for botanical richness.

KaNyaka Island falls entirely within the Maputaland Centre of Endemism (CoE) and several Maputaland endemic or near-endemic plant species have been recorded. A particularly significant species is *Helichrysum moggii* (LC), only known from KaNyaka Island and Santa Maria Cape of the Machangulo Peninsula. Other Maputaland endemic or near-endemic species include Coastal Jackal-berry (*Diospyros inhacaensis*), Coastal Bitter-tea (*Distephanus inhacensis*), Tritonia (*Tritonia moggii*), and Dune Knobwood (*Zanthoxylum delagoense*).

Although *T. moggii* has been recorded as far northwards as Inhambane, the species is highly concentrated on the KaNyaka Island. *Zanthoxylum delagoense*, a Maputaland near-endemic and a Mozambican endemic (Matimele 2016), is also present within the open forests of this IPA.

There are eight globally threatened species present on KaNyaka Island. One of these species, *Ecbolium hastatum* (EN), is only known from about five localities, all restricted to southern Mozambique, including Ponta Ponduine on this island. A second Endangered species, *Solanum litoraneum*, is also endemic to coastal southern Mozambique and is threatened by the development of its coastal dune habitat (Knapp 2021), although the population within this IPA is relatively secure and therefore may represent an important opportunity to conserve *S. litoraneum*.

Alongside these Endangered species are six Vulnerable taxa, including two subspecies of *Tephrosia forbesii*, subsp. *forbesii* and subsp. *inhacensis*. The latter taxon, *Tephrosia forbesii* subsp. *inhacensis*, is known only from the western dunes of KaNyaka Island and is therefore endemic to this IPA. Much of this subspecies' habitat appears intact, but encroaching agriculture and housing (both residential and tourist) threaten this taxon with extinction (Langa *et al.* 2019b). Two other Vulnerable taxa, *Psychotria amboniana* subsp. *mosambicensis* and *Adenopodia schlechteri*, also have restricted ranges, with both species endemic to southern coastal Mozambique (Burrows *et al.* 2018).

View across southeast Inhaca Island (ID)

Thalassodendron leptocaule in intertidal zone off Inhaca Island (JP)

Mangrove communities in southeast Inhaca Island (ID)

Dioscorea sylvatica (VU), contrastingly, has a wide distribution, occurring from South Africa to Zambia, and has been collected in woodland and around abandoned machambas within this IPA. Despite its broad distribution, this species has a history of overharvesting. Tubers of this species have medicinal properties and, in the 1950s, they were harvested on an industrial scale to manufacture cortisone and other steroid hormones (Williams *et al.* 2008). Today this species is still harvested and sold locally, particularly in South Africa, although there is no record of this species being harvested at this site.

The coastal dune communities of KaNyaka are a major botanical importance, hosting most of this site's threatened and endemic species, but the seagrass communities in the intertidal zone are also significant. The Vulnerable species *Zostera capensis* occurs in both the south (Saco and Banco) and north bays of KaNyaka, as well as in the northern intertidal areas of Maputo city, in areas with fine, muddy sediments exposed at low tide. Although found across coastal areas of southern and eastern Africa, this species colonises areas slowly and is sensitive to pollution and sedimentation, while in Mozambique there is

a specific threat of disturbance associated with shellfish harvesting (Short 2010; Bandeira & Gell 2003). The *Z. capensis* meadow in the southern bay of KaNyaka is the largest in the world, and falls within Ponta do Ouro Marine Partial Reserve, and so this IPA offers a great opportunity to conserve this species (Bandeira *et al.* 2014).

Elsewhere in the intertidal zone, on the west coast of the island, is a population of the near-endemic seagrass *Thalassodendron leptocaule*. This species has been assessed as Near Threatened and, although threatened elsewhere by tourism and other coastal activities, the KyNaka populations are relatively secure as all the waters off this island fall within Ponta do Ouro Marine Partial Reserve (Duarte *et al.* 2014). Nevertheless, this species occurs at only one site at Inhaca, at the northernmost point over-looked by the lighthouse (Farol de Inhaca). Both *Z. capensis* and *T. leptocaule* are of great ecological importance, providing habitat, shelter, nurseries and foraging grounds for marine invertebrates and fish (Adams 2016; Browne *et al.* 2013). In total, nine different species of seagrass have been recorded around KaNyaka, representing over three-quarters of Mozambican and over 16% of global seagrass species (Bandeira 2002).

There are a number of useful species present within this IPA. Although none of these species are of conservation concern or restricted to KaNyaka, they are highly important for the local communities. The mangroves, found on the Maputo Bay coastlines of KaNyaka, consist of species common throughout these habitats in Mozambique but are locally important as a source of timber and fuelwood while also providing coastal stabilisation by preventing erosion, regulating sedimentation and protecting against tidal surges (Paula *et al.* 2014). In addition, a number of species are voluntarily protected on the island, including *Strychnos spinosa*, *Syzygium cordatum* and *Sclerocarya birrea* subsp. *caffra* (Marrula), as fruits of these trees are consumed locally. Of these species, particular emphasis is given to the Marrula tree, which has a long history of being encouraged by local people. The fruits of this species are used to make a traditional beverage associated with celebrations in the community, as such it is regarded one of the most important indigenous trees (Mogg 1967).

Although the island has been well-studied over the years, botanical surveys still yield new records for KaNyaka; for instance, the widespread African tree species *Cassipourea malosana* was only recently recorded on the island (Massingue 2019). This suggests that further new records, potentially including further species of conservation concern, may be uncovered in the future.

Habitat and geology

The emergence of KaNyaka is linked to the formation of Maputo Bay that resulted from a recent Holocene transgression (Achimo *et al.* 2014). This island consists of a calcareous sandstone base which has been overlain with notably high dune ridges (Muacanhia 2004). The geomorphological dynamics of the coastal dune systems of KaNyaka together with a prevailing sand deposition and erosion dynamics in shallow waters, continues to shape these islands and the entire Maputo Bay.

The landscape of the island is made up of two long ridges (Mount Inhaca in the north-east and the Barreira Vermelha in the west) trending north-south with an undulating plain between them presenting smaller ridges separated by low sandflats or swampy terrain (Macnae & Kalk 1962; Hobday 1977). The smaller ridges can reach 40 m high, and they are about 5 to 6 km apart west to east. The

soils are mainly sandy, and vary from brown in non-disturbed forest patches, to light yellowish brown in other vegetation types at different phases of development (Campbell *et al.* 1988).

The climate of KaNyaka Island is tropical (Macnae & Kalk 1962) with two main seasons over the year, including a rainy and warm season from October to March, followed by a dry and cooler season from April to September (Muacanhia 2004). The island is usually humid but has a surprisingly low rainfall of around 600 mm per annum. The mean annual temperature is 22 to 23°C, although temperature varies considerably throughout the year, with a maximum of 37°C and a minimum around 12°C (Muacanhia 2004).

Dune vegetation is most dominant on the east coast of KaNyaka, although it is present to a lesser degree on the westerly coastlines (Bandeira *et al.* 2014). On the upper beach, at the edge of the dune thicket, species such as *Canavalia rosea*, *Cissus quadrangularis* and *Cynanchum gerrardii* have been recorded (Hyde *et al.* 2021). This vegetation then transitions to coastal scrub further inland, featuring *Diospyros rotundifolia* and *Euclea natalensis*. Within this IPA, this vegetation type is most defined around Ponta Torres, the most south-easterly point of the island, however, dune scrub is usually continuous with the adjacent coastal thicket (Bandeira *et al.* 2014). Coastal thicket features the Near Threatened species *Encephalartos ferox* alongside species such as *Brexia madagascariensis* and *Brachylaena discolor*. This latter species also occurs within the dune forest further inland, where trees such as *Afzelia quanzensis*, *Eugenia capensis*, *Mimusops caffra* and *Sideroxylon inerme* dominate (Bandeira *et al.* 2014).

According to Paula *et al.* (2014), Maputo Bay is home to six mangrove species: *Avicennia marina*, *Rhizophora mucronata*, *Ceriops tagal*, *Bruguiera gymnorhiza*, *Xylocarpus granatum*, and *Lumnitzera racemosa*. The eastern coastlines of Maputo Bay at KaNyaka (and Machangulo Peninsula to the south) hold extensive mangrove communities. A dwarf form of the mangrove species *A. marina* is the dominant species covering the outer edges of the island particularly in the less inundated areas. The muddy areas with rather less variable salinity have been colonised by *Rhizophora mucronata*. There are also thicket formations, within the mangrove mosaic, which are dominated by *C. tagal* and *B. gymnorhiza*.

The mangroves are bordered inland by saltmarshes (Lötter *et al.*, in prep.), which include sedges such as *Cyperus papyrus* and grasses such as *Phragmites australis*, alongside other herbs including *Hibiscus cannabinus* and *Persicaria decipiens* and the succulent *Sesuvium portulacastrum* (Hyde *et al.* 2021).

Much of the rest of KaNyaka consists of open woodland and savanna (Bandeira *et al.* 2014). Trees such as *Acacia*, likely *A. karroo* as there are numerous mentions of this species in habitat descriptions within this site (*Groenendijk* #1353, #1532, #1942), *Afzelia quanzensis, A. adianthifolia, A. versicolor* and *Dichrostachys cinerea* dominate this habitat around Maputo Bay, with grasses in the understorey including *Hiperthelia dissoluta* and *Cymbopogon* sp. (Bandeira *et al.* 2014). Mogg (1967) noted the conspicuous absence of *Brachystegia* species from the woodland on KaNyaka, with only one individual of *B. tamaridoides* present on Portuguese Island, suggesting that the absence of this genus, which is ubiquitous throughout much of Mozambique, was due to the relatively recent emergence of this island. The woodland at this site is the most impacted by conversion of land to agriculture and includes scrub areas that have been previously used for subsistence agriculture but were later abandoned (Campbell *et al.* 1988). In such areas, dominant species include low shrubs and herbs such as *Helichrysum kraussii, Cassytha filiformis, Digitaria eriantha, Tephrosia purpurea, Dicerocaryum zanguebarium,* and *Imperata cylindrica* (Campbell *et al.* 1988).

In the intertidal zone surrounding KaNyaka are extensive seagrass meadows, covering around half of these areas of the coast of this island (Bandeira *et al.* 2014). In total there are nine seagrass species documented from these waters, largely occurring within Maputo Bay (see Bandeira *et al.* 2014 for full species zonation patterns). Most significant is the large area of *Zostera capensis* (VU) in the bay and sand bank between KaNyaka and the Machangulo Peninsula.

Conservation issues

KaNyaka is an area of significant conservation importance which has long been recognised with the first form of formal conservation of this area established in 1965. The importance of this IPA is goes beyond the national level, falling within the Maputaland CoE (van Wyk 1996), which is part of a global biodiversity hotspot Maputaland-Pondoland-Albany (CEPF 2010). In recognition of the island's tropical biodiversity, a Marine Biological Research Station was established in 1951 (Muacanhia 2004).

Beach and dune thicket in northern Inhaca Island (ID)

Despite being a sanctuary for biodiversity, this island has been experiencing ongoing pressure for many years. Over half a century ago, Mogg (1967) found that forests together with freshwater swamps were under threat due to human encroachment, mainly for subsistence farming. In addition, the island has a dynamic environment that exhibits varying rates of erosion and sedimentation. For example, ongoing erosion resulting from strong wind is progressively depleting the southern east point known as Ponta Torres. The western section at Barreira Vermelha is experiencing degradation due to both tidal and freshwater erosion which causes landslides, particularly in the rainy season (Muacanhia 2004).

After the establishment of the research base at KaNyaka Island in 1951 and, given the continuing increase of local population within this island, Portuguese authorities during the colonial era established the Forest and Marine Reserves in 1965 to protect the ecosystems and biological richness of the island (Muacanhia 2004). However, an increase in the local population, together with the extreme poverty experienced by local communities, poor soils and limited land due to the establishment of forest reserves, have increased pressure on land and resources in the IPA (Muacanhia 2004). To address these issues, in 2009 Inhaca Reserves were incorporated into the newly created POMPR (Ponta do Ouro Marine Partial Reserve). Subsequent to this, in late 2021, approval was given for the merger of POMPR with Maputo Special Reserve to form a new national park, Maputo National Park. This change in status should afford greater protection to the dune and intertidal habitats in the IPA.

POMPR starts at the border with South Africa and extends north for 86 km following the coast into Maputo Bay, including KaNyaka Island, covering the base of the dunes to three nautical miles offshore throughout (Lucrezi *et al.* 2016). In 2019, the Maputo Environmental Protection Area (APA) was designated, covering the area from the POPMR northwards through the Maputo Special Reserve to as far north as KaNyaka Island. An APA is a conservation category under what is regarded as an "Area of Conservation for Sustainable Development" in accordance with the new Conservation Law 5/2017. This conservation category covers a broad landscape within which there may be included some existing protected areas and communities. Therefore, it allows an integrated management of landscapes (including managing existing protected areas or establishing new ones within it) to facilitate implementation of conservation, industrial development, among other development initiatives. An application to UNESCO has been prepared proposing that the area from Ponta do Ouro to KaNyaka should be recognised as a World Heritage Site (Matimele & Timberlake 2020). The full application covers various habitats (terrestrial and marine) and would naturally link with iSimangaliso World Heritage Site across the border in South Africa.

Key ecosystem services

As with other islands in Mozambique, communities on KaNyaka Island rely on artisanal fishing and tourism as their primary livelihood (Book 2012). Because the island is part of the Ponta do Ouro Partial Marine Reserve (POPMR), there are "multiple use zones" where communities' fishing ground are located.

In ecological terms, mangroves provide habitat for a wide range of fauna species including coastal and offshore fish and shellfish which have the mangroves as their main sanctuary for breeding, spawning, and hatching. The mangrove communities, along with primary dune vegetation, provide a buffer between the marine and terrestrial areas as well as protecting shorelines from destructive winds and waves. Mangrove and seagrass communities also contribute to climate regulation due to their role in carbon sequestration. For dugong that still exist around KaNyaka, their diet primarily consists of seagrass species found in the meadows around the island. The mangrove forests enhance water quality by filtering pollutants and terrestrial sediments. In addition, the mangroves and dune vegetation on the island serve as the main barrier protecting coastal erosion. The presence of forest patches contributes to carbon sequestration providing clean air.

Fruits of the trees *Strychnos spinosa*, *Syzygium cordatum*, *Sclerocarya birrea* subsp. *caffra* are consumed locally, while naturalised species such as guava (*Psidium guajava*) are also grown for their fruits or can be used as shade trees or for wind protection. As mentioned in the "Botanical Significance" section, *Sclerocarya birrea* subsp. *caffra* (Marrula) alongside with *Vangueria infausta*,

Strychnos spinosa and *Garcinia livingstonei*, are of cultural significance for local communities. Marrula has a history of being deliberately encouraged by local people, it is used to make a beverage associated with celebrations and as a shade tree, and its soft timber is used to make utensils (Mogg 1967).

Ecosystem service categories

- Provisioning – Food
- Provisioning – Raw materials
- Provisioning – Medicinal resources
- Regulating services – Carbon sequestration and storage
- Regulating services – Moderation of extreme events
- Regulating services – Erosion prevention and maintenance of soil fertility
- Cultural services – Tourism
- Cultural services – Cultural heritage

IPA assessment rationale

KaNyaka Island qualifies as an IPA under criterion A. With its recognised botanical importance at both national and international levels, the island is home to eight threatened taxa that trigger criterion A(i), including two Endangered species, *Ecbolium hastatum* and *Solanum litoraneum*, and six Vulnerable taxa, *Adenopodia schlechteri, Dioscorea sylvatica, Psychotria amboniana* subsp. *mosambicensis, Tephrosia forbesii* subsp. *forbesii, Tephrosia forbesii* subsp. *inhacensis* and *Zostera capensis*. Although not an IPA trigger, it is important to highlight the presence of a near-endemic cycad, *Encephalartos ferox* subsp. *ferox*, assessed as Near Threatened at species level. Overall, there are 12 endemic species within this IPA, falling within the top 15 sites for Mozambique's endemic and range-restricted species and therefore triggering sub-criterion B(ii) for this IPA.

Priority species (IPA Criteria A and B)

FAMILY	TAXON	IPA CRITERION A	IPA CRITERION B	≥ 1% OF GLOBAL POP'N	≥ 5% OF NATIONAL POP'N	IS 1 OF 5 BEST SITES NATIONALLY	ENTIRE GLOBAL POP'N	SPECIES OF SOCIO-ECONOMIC IMPORTANCE	ABUNDANCE AT SITE
Acanthaceae	*Ecbolium hastatum*	A(i)	B(ii)	✓	✓	✓			unknown
Asparagaceae	*Dracaena subspicata*		B(ii)	✓					unknown
Asteraceae	*Helichrysum moggii*		B(ii)	✓	✓	✓			scarce
Dioscoreaceae	*Dioscorea sylvatica*	A(i)			✓	✓		✓	unknown
Euphorbiaceae	*Tragia glabrata* var. *hispida*		B(ii)	✓	✓				unknown
Fabaceae	*Adenopodia schlechteri*	A(i)	B(ii)	✓	✓	✓			unknown
Fabaceae	*Tephrosia forbesii* subsp. *forbesii*	A(i)			✓	✓			unknown
Fabaceae	*Tephrosia forbesii* subsp. *inhacensis*	A(i)	B(ii)	✓	✓	✓	✓		occasional
Hydrocharitaceae	*Halophila ovalis* subsp. *linearis*		B(ii)	✓	✓	✓			unknown
Iridaceae	*Tritonia moggii*		B(ii)	✓	✓	✓			frequent
Rubiaceae	*Psychotria amboniana* subsp. *mosambicensis*	A(i)	B(ii)	✓	✓				common
Rubiaceae	*Psydrax moggii*		B(ii)	✓					unknown
Rutaceae	*Zanthoxylum delagoense*		B(ii)	✓					unknown
Solanaceae	*Solanum litoraneum*	A(i)	B(ii)	✓	✓	✓			occasional
Zosteraceae	*Zostera capensis*	A(i)		✓					abundant
		A(i): 8 ✓	B(ii): 12 ✓						

Protected areas and other conservation designations

CONSERVATION AREA TYPE	CONSERVATION AREA NAME	RELATIONSHIP OF IPA TO CONSERVATION AREA
National Reserve	Ponta do Ouro Partial Marine Reserve	protected/conservation area overlaps with IPA
Key Biodiversity Area	Ponta do Ouro	protected/conservation area overlaps with IPA

Threats

THREAT	SEVERITY	TIMING
Housing & urban areas	medium	Ongoing – trend unknown
Commercial & industrial areas	medium	Ongoing – trend unknown
Small-holder farming	high	Ongoing – increasing
Subsistence/artisinal aquaculture	low	Ongoing – trend unknown
Logging & wood harvesting	medium	Ongoing – trend unknown

NAMAACHA

Assessors: Hermenegildo Matimele, Jo Osborne, Clayton Langa

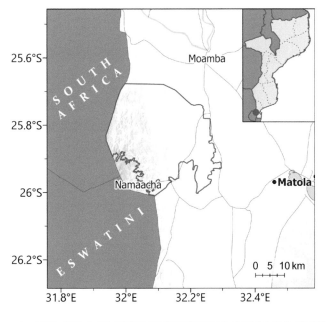
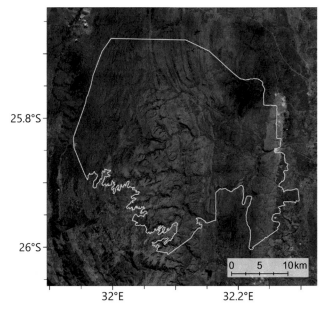

INTERNATIONAL SITE NAME		Namaacha	
LOCAL SITE NAME (IF DIFFERENT)		–	
SITE CODE	MOZTIPA006	PROVINCE	Maputo

LATITUDE	-25.84061	LONGITUDE	32.11057
ALTITUDE MINIMUM (m a.s.l.)	30	ALTITUDE MAXIMUM (m a.s.l.)	630
AREA (km²)	854.0	IPA CRITERIA MET	A(i), A(iv)

Rocky outcrops with succulents at Namaacha (JO)

Site description

The Namaacha Important Plant Area lies in Namaacha District, Maputo Province, Mozambique, next to the border with Eswatini to the south-west and South Africa to the west. It forms the eastern part of the Lebombo Mountains which fall within the Maputaland Centre of Endemism where a high number of endemic plants are known to occur (van Wyk 1996). Maputaland is a centre of endemism (CoE) within the Maputaland-Pondoland-Albany global biodiversity hotspot (CEPF 2010). An analysis by Darbyshire *et al.* (2019a) has treated the Lebombo Mountains as a potential sub-centre of the Maputaland CoE with 17 Mozambican endemic and near-endemic species restricted to this cross border sub-CoE. The boundaries of this IPA were primarily delineated to encompass the majority of known records of IPA trigger species within this region and were subsequently refined using Google Earth Engine (Gorelick *et al.* 2017) to identify and exclude degraded areas.

This IPA covers 854 km², encompassing a montane landscape ranging from 40 to 550 m in elevation, supporting a mosaic of forest on rocky slopes and cliffs together with arid woodland and rock outcrops. There are several springs including Bobo,

Chambadejovo, Maxibobo, Movene, Gumbe and Impaputo that cascade down the rocks, particularly during the rainy season between October and March. The Bobo River, situated in the northern part of the site, flows into the Major River which, in turn, is a tributary of the important Incomati River. In the central part of the proposed area, three rivers – Chambadejovo, Maxibobo and Gumbe – flow into the Movene River, a tributary of the Umbelúzi, another major river in this region. The Impamputo River runs through the southernmost section of the proposed site into the Pequenos Lebombo dam which provides the main water supply for Maputo city. The Namaacha District is famous for its waterfalls, attracting tourists to this part of Maputo Province.

Botanical significance

This section of the Lebombos has been been overlooked botanically, but the main botanical significance of the Namaacha IPA are the undisturbed forest patches along the rocky slopes and rivers, together with the succulent species that occur in the rock outcrops, including *Aloe* and *Euphorbia* species. This IPA is home to species of conservation concern including the cycad *Encephalartos umbeluziensis* (EN), a species restricted to the Lebombo Mountains (particularly along the Umbeluzi River) that is threatened by ongoing illegal removal of plants and habitat loss. *Ceropegia aloicola* (EN) is also threatened as a result of habitat degradation, this species is only known from this IPA. *Barleria oxyphylla* (VU globally but assessed as nationally EN for South Africa), a range-restricted species, is threatened due to habitat loss and degradation (von Staden & Lötter 2018).

Adenium swazicum, assessed as Vulnerable for the Red List of South African Plants, is also threatened due to habitat loss and collecting for medicinal and ornamental uses (Lötter & von Staden 2018). The IPA supports the largest subpopulation of *Adenium swazicum* throughout the species' known distribution in southern Mozambique (H. Matimele, pers. obs). Additional species of very narrow distribution include *Jatropha latifolia* var. *subeglandulosa,* an endemic to Mozambique, and *Tragia glabrata* var. *hispida* also an endemic species known only from Maputo Province, southern Mozambique. *Cyphostemma barbosae*, a Lebombo endemic, also occurs within the IPA.

Namaacha also contains a number of species that are endemic to Maputaland CoE, in the broad sense, of high conservation value including *Asparagus radiatus*, *Australluma ubomboensis* and *Blepharis swaziensis*. Woodlands in the area contain *Acacia swazica, Caesalpinia rostrata* and

Erythroxylum delagoense which, although not endemic, are only known from the southern region within Mozambique (Burrows *et al.* 2018). Several plant species are valued by people as sources of income, nutrition and medicines, and for aesthetic uses. Useful species include: *Warburgia salutaris* (EN, harvested for medicinal uses) (Senkoro *et al.* 2019, 2020), *Androstachys johnsonii* (widely used in construction and fencing of large areas for livestock), *Acacia swazica* (used for charcoal), *Sclerocarya birrea* (used to make a traditional beverage and also provides edible nuts), and *Adenium swazicum* (medicinal and ornamental uses).

Habitat and geology

Vegetation in the Namaacha IPA is variable depending on the proximity of a water course and elevation. The eastern area of the IPA falls within the foothill section of the Lebombo Mountains with elevations as low as 30 m along the Movene River. Under the landcover classification system of Smith *et al.* (2008), the dominant vegetation of the IPA is Lebombo woodland. The tree height ranges between 4 and 8 m, with *Acacia* and *Combretum* species being dominant in some sections. Species found include *Acacia swazica, A. exuvialis, A. burkei, A. caffra, A. davyi, A. nigrescens* and *A. senegal* var. *rostrata*, together with *Combretum apiculatum, C. molle, C. zeyheri, Lannea discolor, Pterocarpus rotundifolius, Sclerocarya birrea* and *Terminalia phanerophlebia*. There are also rock outcrops

Jatropha latifolia var. *subeglandulosa* (TR)

Ceropegia aloicola (JO)

Androstachys johnsonii forest (JO)

dominated by succulent species such as *Euphorbia cooperi, Cussonia natalensis* and *Aloe* spp. The river margins and the cliffs are dominated by forests and thickets with various species including cycads, *Asparagus* spp., and economically important species such as *Androstachys johnsonii,* which is valued for its timber. Along the Movene River to the east at lower altitudes, there are alluvial zones with riverine forests or woodland, typically with tree species such as *Acacia xanthophloea* and *Ficus sycomorus.*

Geological studies of the IPA site are limited, but the Lebombo Mountains are composed of a sequence of volcanic rocks – basaltic lavas and rhyolitic flows – from the Jurassic period about 180 to 179 million years ago (du Randt 2018). These rocks lie on horizontal Karoo supergroup sedimentary rocks to the west and are overlain by Cretaceous to recent sediments to the east. Rhyolite, a resistant rock, is arranged in an alternating manner with basalt, a more readily eroded rock, resulting in a series of parallel sharp ridges with a gentle slope on one side separated by plains or water courses. The whole of the Lebombo Mountains area is relatively low, with the highest peak being no more than 800 m elevation (du Randt 2018). Within the IPA the average elevation is about 270 m, with a highest elevation of 630 m.

The soils in the Namaacha IPA site are derived from rhyolite and basalt and are relatively fertile with high clay contents (du Randt 2018). Red soils dominate the site, but black clays with alluvium are also present to the east on the plains. Subsistence farming is common in the vicinity of human settlement areas, particularly in the southwest near Namaacha village and in the areas near Namaacha waterfall. The climate is tropical humid with two main seasons: a dry and cold season ranging from April to September, followed by a wet, hot and rainy season from October to March.

Conservation issues

The IPA is not under any sort of formal conservation as it falls entirely outside of the existing network of conservation areas in the country. However, it forms part of the proposed Goba conservancy, which is part of a wider regional initiative, the Lubombo Conservancy–Goba, which is a "Trans-frontier Conservation Area" from Eswatini to Mozambique and South Africa (Üllenberg *et al.* 2014, 2015).

Threats to biodiversity within this IPA are well-understood. In the past, the area was heavily impacted by charcoal production and, although at present none of the IPA trigger species are targeted for charcoal production, the impact of

Warburgia salutaris (JEB)

associated habitat destruction on the vegetation and wider biodiversity is expected to cause significant declines in species numbers. Charcoal production here consists of cutting thick woody stems and clearing areas for piling and burning of the stems. Areas that have been cleared for charcoal kilns then become the entry point for invasive plants including *Agave sisalana, Lantana camara, Opuntia ficus-indica,* and *Zinnia peruviana*. This is particularly prominent in the western Macuacua area and in the northern Livevene area (H. Matimele, pers. obs.). However, regulations imposed by Government, coupled with scarcity of accessible suitable species for charcoal-making, have reduced much of this production.

A further threat of particular concern is cattle grazing. Field observation suggests there has been a considerable increase over the past 15 years in the number of areas grazed by livestock. Moreover, the hunting of animals such as Bushpig has been reported, together with the harvesting of medicinal plants to fulfil basic livelihoods, but

also as a source of income for communities in the area. In the Matsequenha area in the east, one of the largest military bases for RENAMO soldiers was located. With the peace agreement achieved in 1992, some communities turned to charcoal production and subsistence farming. Other members of communities have been employed in the livestock industry which has expanded considerably since the peace agreement.

In the Bemassango area, in the northern section of the IPA, Mike Persson has a concession for cattle production. He has employed members of local communities, thus providing them with some income to cover the cost of food, health and education for children. This generation of income has in turn relieved some level of pressure on the natural vegetation. Persson has also shown his willingness to turn the cattle farm into a biodiversity conservation-oriented business with emphasis on ecotourism. In addition, there is a private Namaacha Zoo located in the south-west of the IPA, which has become a tourist attraction for people in the nearby cities and towns.

The inclusion of this IPA in Mozambique's network of conservation areas would not only be greatly beneficial for biodiversity but could also be an opportunity to promote sustainable livelihoods in local communities. Most of the remnants of native forest and woodland are confined to cliffs, gorges and other sites with limited access. The IPA has a relatively high number of rivers together with cliffs, hence, the extent of natural areas in good condition is large. Some of the larger rivers in the region such as the Incomati and Umbeluzi have their tributaries within this IPA. Apart from the plant species triggering IPA, this site is home to *Platysaurus lebomboensis* (Lebombo Flat Lizard), an endemic Lizard only known from the Lebombo Mountains. This section of the Lebombo Mountains, particularly in the Matsequenha area to the northeast, forms one of the best sites for flora and fauna species such as *Asparagus radiatus, Pyrenacantha kaurabassana, Adenium swazicum, Warburgia salutaris,* and the Lebombo Flat Lizard.

Based on the site's biodiversity features, this IPA would have the greatest benefit if it were conserved under the Protection, Conservation and Sustainable use of Biological Diversity Act (Decree No. 16/2014 of the 20th of June). Because it is located near the capital city Maputo (75 km away), there is high potential for ecotourism. Also, because there are communities residing in the area, this IPA could potentially be protected under one of the Conservation Areas of Sustainable Use categories, which permit integrated management allowing some level of harvest in accordance with the limits to be set by the management authority. Those categories include, for example, Sanctuary, Area of Environmental Protection (APA) or Community Conservation Area.

Key ecosystem services

Being a rural area, where infrastructure for provision of services is limited, entire communities within the Namaacha IPA site and its vicinity depend on the streams coming off the mountains as their only source of clean water. Water from the streams is also essential in supporting subsistence agriculture through bucket watering for vegetables, particularly in the dry season from April to September.

The streams or small rivers arising from within the IPA site are also highly important for commercial agriculture and water supply to the nearby cities. For example, streams from this IPA including

Pools within rocky areas of the Namaacha hills (LL)

Chambadejovo, Maxibobo and Gumbe, drain water into the Movene River, which is one of the tributary rivers for the Umbeluzi River. Another stream, the Bobo River, drains water into the Major River, which is one of the tributary rivers to the Incomati River. In addition, the Impaputo River runs through the IPA site before it drains water into the Pequenos Lebombos dam, which is the main water source supplying Maputo city, Matola, and Boane. The Umbeluzi River is the main water source supporting irrigation for larger-scale agriculture in the areas of Boane and Goba. The Incomati River is highly important as it sustains small and larger-scale agriculture around Moamba and Marracuene Districts.

The Namaacha waterfalls fall within the IPA and are a tourist attraction. Tourism has the potential to create an economic cascade effect through which local communities may find opportunities to provide services including selling beverages and food.

The presence of forest patches contributes to carbon sequestration providing clean air. In addition, given the limited access to hospitals, the communities rely on harvesting medicinal plants to combat diseases. Timber is rarely harvested from within the IPA boundary, and collection of plants for medicinal properties occurs mostly in the surrounding woodland areas. Bushmeat hunting, however, does take place on higher slopes, although this practice is limited to a small group of local residents. Bushbuck (*Tragelaphus scriptus*), Bushpig (*Potamochoerus larvatus*), and Duiker (*Cephalophus* spp.) are caught using gin traps. Honey is also reported to be collected in the forest.

Ecosystem service categories

- Provisioning – Food
- Provisioning – Raw materials
- Provisioning – Fresh water
- Provisioning – Medicinal resources
- Regulating services – Local climate and air quality
- Regulating services – Carbon sequestration and storage
- Habitat or supporting services – Habitats for species
- Cultural services – Tourism

IPA assessment rationale

Namaacha qualifies as an IPA under criteria A. The undulating landscape with sharp cuesta ridges, gorges, cliffs and plains is home to four species of conservation concern which trigger criterion A(i): *Encephalartos umbeluziensis* (EN), *Ceropegia aloicola* (EN), *Warburgia salutaris* (EN) and *Barleria oxyphylla* (VU). Overall, there are six endemic and near-endemic species that qualify under criterion B(ii), with two of these species, *Asparagus radiatus* and *Jatropha latifolia* var. *subeglandulosa*, highly range-restricted and triggering criterion A(iv).

View of Namaacha hills showing both forest and wooded grasslands (CL)

Priority species (IPA Criteria A and B)

FAMILY	TAXON	IPA CRITERION A	IPA CRITERION B	≥ 1% OF GLOBAL POP'N	≥ 5% OF NATIONAL POP'N	IS 1 OF 5 BEST SITES NATIONALLY	ENTIRE GLOBAL POP'N	SPECIES OF SOCIO-ECONOMIC IMPORTANCE	ABUNDANCE AT SITE
Acanthaceae	*Barleria oxyphylla*	A(i)	B(ii)	✓	✓	✓			unknown
Apocynaceae	*Ceropegia aloicola*	A(i)	B(ii)	✓	✓	✓	✓		unknown
Asparagaceae	*Asparagus radiatus*	A(iv)	B(ii)	✓					unknown
Canellaceae	*Warburgia salutaris*	A(i)		✓				✓	occasional
Euphorbiaceae	*Jatropha latifolia* var. *subeglandulosa*	A(iv)	B(ii)	✓	✓	✓	✓		unknown
Euphorbiaceae	*Tragia glabrata* var. *hispida*		B(ii)	✓					unknown
Zamiaceae	*Encephalartos umbeluziensis*	A(i)	B(ii)	✓	✓	✓			unknown
		A(i): 4 ✓ A(iv): 2 ✓	B(ii): 6						

Protected areas and other conservation designations

CONSERVATION AREA TYPE	CONSERVATION AREA NAME	RELATIONSHIP OF IPA TO CONSERVATION AREA
No formal protection	N/A	

Threats

THREAT	SEVERITY	TIMING
Logging & wood harvesting – subsistence/small scale	medium	Ongoing – declining
Gathering terrestrial plants	medium	Ongoing – trend unknown
Small-holder grazing, ranching or farming	medium	Ongoing – increasing
Agro-industry grazing, ranching or farming	low	Ongoing – stable
Tourism & recreation areas	low	Ongoing – trend unknown

GOBA

Assessors: Hermenegildo Matimele, Linda Loffler

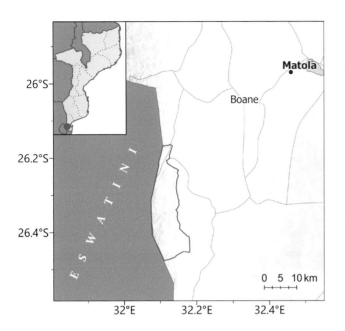

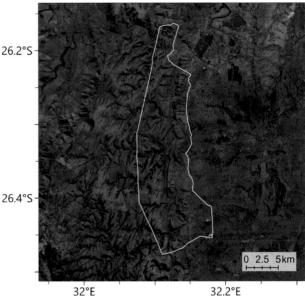

INTERNATIONAL SITE NAME		Goba	
LOCAL SITE NAME (IF DIFFERENT)		–	
SITE CODE	MOZTIPA043	PROVINCE	Maputo

LATITUDE	-26.35252	LONGITUDE	32.10925
ALTITUDE MINIMUM (m a.s.l.)	70	ALTITUDE MAXIMUM (m a.s.l.)	520
AREA (km²)	217	IPA CRITERIA MET	A(i), B(ii), C(iii)

Site description

The Goba Important Plant Area lies in Namaacha District, Maputo Province, Mozambique, on the border with Eswatini to the west at (-26.35°, 32.10°). It forms the eastern part of the Lebombo Mountains, a core area within the Maputaland Centre of Endemism where several endemic plants occur (van Wyk 1996). The boundaries of this IPA were mainly delineated to encompass the majority of known records of the IPA trigger species within this region and were subsequently refined using Google Earth Engine (Gorelick *et al.* 2017) to identify and exclude degraded or transformed and urbanised areas.

This IPA covers 217 km², and extends for approximately 35 km from north to south, encompassing a montane landscape from 70 to 520 m elevation, with most areas lying at around 250 m. It supports a mosaic of forest on rocky slopes and cliffs together with woodland, wooded grassland and rock outcrops. There is one large and regionally important river running through the northern section of the IPA, the Umbeluzi River, in addition to two streams, namely the Changalane and Mazeminhane, in the southern section. The two streams drain their water into Tembe River, another important water source running towards the northeast into Maputo Bay.

Botanical significance

The Lebombo Mountains as a whole are of recognised botanical significance; they fall within the Maputaland Centre of Plant Endemism which is thought to be home to 203 endemic plant species or infraspecific taxa (van Wyk 1996; van Wyk &

Smith 2001). A thorough analysis by Darbyshire *et al.* (2019a) has proposed the Lebombo Mountains as a separate (sub-)Centre of plant endemism within Maputaland, and it is believed to contain 17 Mozambican endemics and near-endemic species restricted only to this sub-centre. Of the 17 taxa, three are known only from Mozambique.

Species of conservation significance in the Goba IPA include: *Indigofera gobensis* (CR), only known from this locality worldwide; and the cycads *Encephalartos lebomboensis* (EN), *Encephalartos senticosus* (EN), *Encephalartos umbeluziensis* (EN) and *Encephalartos aplanatus* (VU), all of which are endemic to the Lebombos. These cycads are threatened due to habitat loss and over-collecting as a result of poaching for different purposes. *Encephalartos umbeluziensis* is highly concentrated at the Goba IPA, which has the second largest population of this species after Mlawula Game Reserve in Eswatini. *Euphorbia baylissii* (VU), threatened due to habitat destruction, also occurs within this IPA and it is not known to occur in any protected area. *Asparagus radiatus* and *Tephrosia gobensis* are among the other Lebombos endemic species confined to the forests of the Lebombo Mountains and included within this IPA. This site is also home to *Warburgia salutaris* (EN) which is threatened due to habitat loss and over-exploitation of parts of the plant, such as bark, stems, and roots, for medicinal usage (Senkoro *et al.* 2019; 2020). *Thesium jeaniae*, which has been assessed regionally as Rare (Raimondo & Scott-Shaw 2007), also occurs

here and is a highly range-restricted species in the southern Lebombo Mountains. Additional species to highlight include *Stapelia unicornis*, *Euphorbia keithii* (known from a range of less than 1,500 km^2), *Gladiolus brachyphyllus* (with a range smaller than 10,000 km^2) and *Cyphostemma barbosae*. These species are endemic or near-endemic to the Lebombo Mountains.

Another species of interest found in Goba IPA, although not endemic or range-restricted nor of conservation concern as per the IUCN Red List, is *Excoecaria madagascariensis* (LC), previously known from Madagascar and Tanzania and so here representing rather a disjunct occurrence.

There are also several species that are important for socio-economic reasons, including *Acacia swazica* (used for charcoal), *Androstachys johnsonii* (widely used in construction and fencing of large areas for livestock), and *Sclerocarya birrea* (source of a traditional beverage and nuts), among others.

Habitat and geology

Vegetation patterns in Goba IPA are in accordance with topography, varying depending on whether an area is in the vicinity of a water course or is on free-draining slopes, with elevation also being an important factor. Forests are confined to river margins and cliffs or slopes, holding species of conservation concern such as *Asparagus radiatus*, *Encephalartos umbeluziensis* and *Erythrophleum lasianthum*, together with economically important

View of Goba hills (HM)

Open wooded grassland on plateau (LL)

Rocky area with cycad, thought to be *Encephalartos senticosus*, in the foreground (LL)

species such as *Androstachys johnsonii*. Away from water courses, the landscape is comprised of woodland dominated by *Acacia* and *Combretum* species, including *Acacia swazica, A. exuvialis, A. burkei, A. caffra, A. davyi, A. nigrescens* and *A. senegal* var. *rostrata*, together with *Combretum apiculatum, C. molle* and *C. zeyheri*. Other important woodland species include *Lannea discolor, Pterocarpus rotundifolius, Sclerocarya birrea* and *Terminalia phanerophlebia*.

In geological terms, the Goba IPA is part of the Lebombo Mountains which consists of a sequence of volcanic rocks – basaltic lavas and rhyolitic flows – from the Jurassic period about 180 to 179 million years ago (du Randt 2018). Rhyolite, a resistant rock, is arranged in an alternating manner with basalt, a more readily eroded rock, resulting in a series of parallel sharp ridges with a gentle slope on one side separated by plains or water courses. The whole of the Lebombo Mountains area is relatively low with the highest peak no more than 800 m in elevation (du Randt 2018). Based on Google Earth imagery, Goba IPA peaks at about 500 m elevation.

The soils in the Goba IPA site are derived from rhyolite and basalt and are relatively fertile with high clay contents (du Randt 2018). Red soils are dominant throughout the area, but black alluvial clays are associated with drainage lines. The area has a tropical humid climate with two main seasons: a dry and cold season from April to September, followed by a wet, hot and rainy season ranging from October to March.

Conservation issues

The Goba IPA is not part of the current network of conservation areas of Mozambique. However, this site encompasses the Goba Ntava Yedzu which is an area of about 9,000 ha managed by the community, though with no legal conservation status. Moreover, it falls entirely within a proposed Goba conservancy which is part of a wider regional initiative, the Lubombo Conservancy–Goba, which is a Trans-Frontier Conservation Area from Eswatini to Mozambique and South Africa (Üllenberg *et al.* 2014, 2015).

Goba IPA experiences habitat destruction resulting primarily from charcoal production. At present, none of the IPA trigger species are targeted for charcoal, however, the impact of habitat clearing for piling and burning of woody stems in the production process is expected to cause significant declines in species of conservation importance at this site. Areas that have been cleared for charcoal kilns then become the entry point for invasive plants including *Agave sisalana*, *Lantana camara*, *Opuntia ficus-indica* and *Zinnia peruviana*.

A further threat of particular concern is the illegal harvesting of plant species for trade in markets in the cities including Maputo, Matola and Boane. With rapid urban expansion over the past 15 years, demand for these plants for ornamental reasons has increased steadily which is likely to cause severe declines in some species, particularly the slow-growing cycad species such as *Encephalartos umbeluziensis* (EN) and *E. lebomboensis* (EN). Some plants in this IPA are also harvested for their medicinal properties, for example, *E. lebomboensis* (Donaldson 2010f) and *Warburgia salutaris* (Senkoro *et al.* 2019, 2020).

In addition to charcoal production and plant poaching, there has been ongoing increase over the past 15 years of concessions granted for livestock grazing. Grazing areas have been fenced, causing an increased demand for poles from species with hard, resistant wood such as *Androstachys johnsonii* which, in turn, is causing significant habitat destruction. Being in the vicinity of protected areas in Eswatini, notably the Mlawula Nature Reserve, there is occasional movement of animals into the unprotected Mozambique lands, and in some instances, these animals are hunted illegally. In addition, there have been reports that artisanal fishing takes place in the bigger rivers such as the Umbeluzi.

By taking advantage of the existing community initiative, the Goba Ntava Yedzu, this site could potentially be protected under one of the Conservation Areas of Sustainable Use categories, aligned with the "Protection, Conservation and Sustainable use of Biological Diversity" Act (Decree No. 16/2014), which permit integrated management, allowing some level of harvest of natural resources in accordance with the limits to be set by the management authority.

Key ecosystem services

Provision of clean water to fulfil the needs of local communities, and those of the surrounds of the Goba IPA, is among the key services delivered by the ecosystems found in this site. Local populations get fresh water from the streams, Changalane and Mazeminhane, for drinking, cooking, and washing. In addition, the water is also used for subsistence agriculture, mainly in the dry season, through bucket watering of vegetables near river margins.

Encephalartos aplanatus (LL)

Asparagus radiatus (LL)

Moreover, the Umbeluzi supplies water to support irrigation for larger-scale agriculture of bananas, rice and vegetables in Mozambique. In Eswatini, the Umbeluzi River is the main water source for irrigation of large-scale sugar plantations (H. Matimele, pers. obs.). Changalane and Mazeminhane streams provide water to the rural communities in the areas of Changalane in the southern part of the Goba IPA. The two streams are tributaries of the Tembe River which drains into Maputo Bay, and provides water for communities living around this river in some sections of Matutuíne and Boane Districts.

Species such as *Warburgia salutaris* are widely known and harvested for their medicinal properties. Forests are the primary source of building materials locally for housing as well as fences for livestock. However, timber is rarely harvested from within the IPA boundary because suitable timber tree species are not easily accessible because they are confined to rocky or gulley areas.

Ecosystem service categories

- Provisioning – Raw materials
- Provisioning – Fresh water
- Provisioning – Medicinal resources

IPA assessment rationale

Goba qualifies as an IPA under criterion A as it supports species of global conservation concern. A total of 7 species trigger criterion A(i): *Indigofera gobensis* (CR), *Warburgia salutaris* (EN), *Encephalartos lebomboensis* (EN), *Encephalartos senticosus* (EN), *Encephalartos umbeluziensis* (EN), *Encephalartos aplanatus* (VU) and *Euphorbia baylissii* (VU). Although not yet formally IUCN assessed, *Asparagus radiatus, Euphorbia keithii, Tephrosia gobensis* and *Thesium jeaniae* are range-restricted endemic species that trigger criterion A(iv).

Priority species (IPA Criteria A and B)

FAMILY	TAXON	IPA CRITERION A	IPA CRITERION B	≥ 1% OF GLOBAL POP'N	≥ 5% OF NATIONAL POP'N	IS 1 OF 5 BEST SITES NATIONALLY	ENTIRE GLOBAL POP'N	SPECIES OF SOCIO-ECONOMIC IMPORTANCE	ABUNDANCE AT SITE
Asparagaceae	*Asparagus radiatus*	A(iv)	B(ii)	✓	✓	✓			unknown
Canellaceae	*Warburgia salutaris*	A(i)			✓	✓		✓	unknown
Euphorbiaceae	*Euphorbia baylissii*	A(i)	B(ii)	✓		✓			unknown
Euphorbiaceae	*Euphorbia keithii*	A(iv)	B(ii)	✓	✓	✓			unknown
Fabaceae	*Indigofera gobensis*	A(i)	B(ii)	✓	✓	✓	✓		unknown
Fabaceae	*Tephrosia gobensis*	A(iv)	B(ii)	✓	✓	✓			unknown
Iridaceae	*Gladiolus brachyphyllus*		B(ii)	✓	✓	✓			unknown
Santalaceae	*Thesium jeaniae*	A(iv)	B(ii)	✓	✓	✓			unknown
Zamiaceae	*Encephalartos aplanatus*	A(i)	B(ii)	✓	✓	✓			unknown
Zamiaceae	*Encephalartos lebomboensis*	A(i)		✓	✓	✓			unknown
Zamiaceae	*Encephalartos senticosus*	A(i)	B(ii)	✓	✓	✓			unknown
Zamiaceae	*Encephalartos umbeluziensis*	A(i)	B(ii)	✓	✓	✓			frequent
		A(i): 7 ✓ A(iv): 4 ✓	B(ii): 10						

Overlooking *Androstachys johnsonii* forest on slopes of the hills (LL)   Rocky areas and pools on plateau (CL)

Protected areas and other conservation designations

CONSERVATION AREA TYPE	CONSERVATION AREA NAME	RELATIONSHIP OF IPA TO CONSERVATION AREA
No formal protection	N/A	
Non-legislative community conservation area	Goba Ntava Yedzu	protected/conservation area encompasses IPA

Threats

THREAT	SEVERITY	TIMING
Small-holder grazing, ranching or farming	medium	Ongoing – increasing
Logging & wood harvesting: subsistence/small scale	medium	Ongoing – trend unknown
Gathering terrestrial plants	medium	Ongoing – increasing
Invasive non-native/alien species/diseases	medium	Ongoing – increasing

LICUÁTI FOREST

Assessors: Hermenegildo Matimele, Jonathan Timberlake

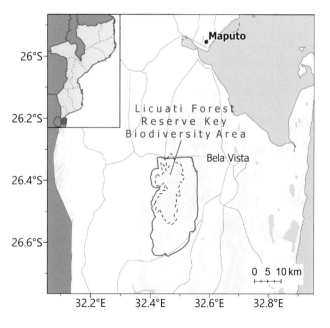 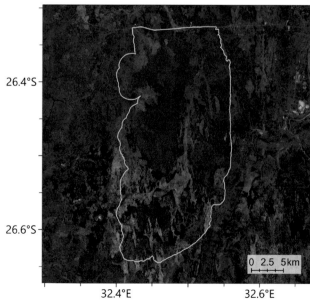

INTERNATIONAL SITE NAME		Licuáti Forest	
LOCAL SITE NAME (IF DIFFERENT)		Floresta de Licuáti	
SITE CODE	MOZTIPA009	PROVINCE	Maputo

LATITUDE	-26.46680	LONGITUDE	32.46030
ALTITUDE MINIMUM (m a.s.l.)	30	ALTITUDE MAXIMUM (m a.s.l.)	75
AREA (km²)	470	IPA CRITERIA MET	A(i), C(iii)

Site description

Licuáti Forest, which contains part of the Licuáti Forest Reserve, is situated in Matutuine District of Maputo Province in southern Mozambique. Located at approximately -26.47°, 32.46° with altitude ranging between 30 and 75 m, this IPA has a total extent of around 470 km². The Lebombo Mountains and the Eswatini border lie about 40 km to the west, the Maputo Special Reserve on the Indian Ocean coast is 30 km to the east, and Maputo city is about 50 km to the north. The northern boundary is formed by the Porto Henrique-Bela Vista Road. This IPA contains one seasonal spring, Puchene Esculo, that flows into the Tembe River, one of the most important rivers in the south of Maputo Province.

"Licuáti" denotes an extremely thick and impenetrable bush in Ronga, the local language (Izidine *et al.* 2009). In biological terms, the Licuáti Forest lies within the core zone of the Maputaland Centre of Plant Endemism (CoE) which is home to a high number of endemic and near-endemic plant species (van Wyk 1996).

Botanical significance

The Licuáti IPA forms a part of the core zone of the Maputaland CoE. While the CoE as a whole is home to a large number of endemic and near-endemic plants, this IPA is of particular significance, with 32 Maputaland (excluding the Lebombos sub-centre) endemics and 2,500 plant taxa in total (van Wyk 1996). Darbyshire *et al.* (2019a) indicate that 13 of these are restricted to the Mozambique section of this CoE.

While the Licuáti IPA is of particular botanical significance owing to its endemic taxa (Matimele 2016), it is also the best and largest remaining extent of a unique vegetation type – Licuáti Thicket. Examples of endemic and threatened taxa within this habitat include *Xylopia torrei* (EN), *Empogona maputensis* (EN), *Warneckea parvifolia* (EN), *Sclerochiton apiculatus* (VU), *Acridocarpus natalitius* var. *linearifolius* (VU), *Polygala francisci* (VU) and *Psychotria amboniana* subsp. *mosambicensis* (VU). Overall, these species are threatened due to habitat loss. In addition, *Acridocarpus natalitius* var.

Grassland surrounding Licuáti Thicket (HM)

Licuáti Thicket (HM)

linearifolius is harvested and traded for medicinal use in Maputo in Mozambique, and in Durban and Johannesburg in South Africa.

Additional examples of Maputaland (in the broad sense) endemic plant species recorded in the Licuáti IPA include *Psydrax fragrantissima* (NT), *Encephalartos ferox* (NT), *Dicerocaryum forbesii* (LC), *Diospyros inhacaensis* (LC), *Pavetta vanwykiana* (LC), *Vangueria monteiroi* (LC) and *Zanthoxylum delagoense* (LC). Alongside the threatened taxa listed above, these species are mostly confined to the Licuáti thicket vegetation type throughout their known range in both Mozambique and just over the border in KwaZulu-Natal Province, South Africa.

Recent burning in thicket (HM)

Species of economic importance found in the Licuáti IPA include *Afzelia quanzensis* for high class timber; *Dialium schlechteri* (LC), *Garcinia livingstonei* and *Vangueria monteiroi* (LC) for edible fruits; and *Acridocarpus natalitius* var. *linearifolius* (VU), *Warburgia salutaris* (EN), *Dicerocaryum forbesii* (LC), *Bridelia cathartica* (LC), *Securidaca longipedunculata*, *Erythrophleum lasianthum* (NT), *Brachylaena huillensis* (NT) and *Xylotheca kraussiana* (LC) for medicines. Although it is also threatened by habitat loss, the biggest threat to *Warburgia salutaris* is human exploitation as it is harvested for its popular medicinal uses (Dludlu *et al.* 2017; Senkoro *et al.* 2019, 2020).

Habitat and geology

The Licuáti IPA area is a mixture of tree savanna and woodland, sand forests, and patches of grassland (Myre 1971; Matimele & Timberlake 2020). The IPA is dominated by Licuáti thicket, also known as short sand forest in South Africa (du Randt 2018), which is mainly dense woody vegetation from 4 to 6 m tall. The characteristic species of the Licuáti thicket include *Warneckea parvifolia*, *Psydrax fragrantissima*, *Sclerochiton apiculatus*, *Croton pseudopulchellus*, *Brachylaena huillense*, *Hymenocardia ulmoides*, *Terminalia* (formerly *Pteleopsis*) *myrtifolia* and *Monodora junodii*. These mostly evergreen thickets are associated with a characteristic emergent tree layer of *Afzelia quanzensis*, *Balanites maughamii*, *Dialium schlechteri* and *Newtonia hildebrandtii*.

Pavetta vanwykiana (JEB)

Empogona maputensis (JEB)

This thicket is interspersed with "tall sand forest" as it is known in South Africa (du Randt 2018), also known as Licuáti Forest (Tokura *et al.* 2020), which has a more open structure, with more species in the tree layer and a canopy above 8 m tall. Characteristic species include *Ptaeroxylon obliquum*, *Erythrophleum lasianthum*, *Cleistanthus schlechteri*, and *Uvaria lucida*. Both Licuáti Thicket (short sand forest) and Licuáti Forest (tall sand forest) are found within a matrix of open woodland dominated by *Albizia adianthifolia* var. *adianthifolia*, *Albizia versicolor* and *Terminalia sericea* (Myre 1971; Siebert *et al.* 2002; Izidine 2003). Other common species dispersed throughout the woodland areas include *Strychnos spinosa*, *Strychnos madagascariensis* and *Vangueria infausta*. Grassland is not extensive within the Licuáti IPA, but where it occurs it contains scattered trees of *Syzygium cordatum* and the palm *Hyphaene coriacea*. Grasslands are also favoured places for *Dicerocaryum forbesii*, a prostrate herb with perennial taproot.

In geological terms, the Licuáti area lies on ancient dunes resulting from geomorphological processes that were operating for millennia over the Pliocene to the Pleistocene periods about 3–5 million years ago (du Randt 2018). The climate is humid tropical with two main seasons, a dry and cold season from April to September, followed by a wet, hot, humid and rainy season from October to March. The Licuáti IPA shows an altitudinal gradient increasing from east to west with annual precipitation of nearly 600 mm (Izidine *et al.* 2003; van Wyk 1996). Because the rainfall is low, the vegetation is maintained by moisture obtained from the south-easterly winds that carry moist coastal air from the sea (Matimele 2016). Species found in the Licuáti IPA are adapted to well-drained nutrient-poor sandy soils.

Conservation issues

Parts of the Licuáti Forest Reserve (LFR) and most of the associated Key Biodiversity Area fall within the Licuáti IPA. In general terms, forest reserves in Mozambique are not regarded as areas for conservation and, therefore, they are not managed by the National Administration of Conservation (Administração Nacional das Áreas de Conservação, ANAC), the government body overseeing nature conservation in the country. Instead, they fall under the National Forestry Directorate with very limited focus on biodiversity issues. Licuáti Thicket, the most unique vegetation of this IPA, occurs nowhere else except for small pockets in Tembe National Park in KwaZulu-Natal, and it forms the core of the Maputaland Centre of Endemism (van Wyk & Smith 2001). The Licuáti Thicket represents the only large area of this thicket type, and this site is therefore irreplaceable.

The Licuáti Forest Reserve was proclaimed as a Forest Reserve in 1943, with focus on protecting the stands of Chamfuta (*Afzelia quanzensis*) and ensuring its sustainable harvesting (Gomes e Sousa 1968). Despite being a forest reserve, historical timber concessions and agricultural and livestock rangeland concessions were issued due to unclear boundary definitions of where the actual reserve was situated. Over the past 10 years, the Licuáti Thicket has become one of Maputo's nearest sources of trees and shrubs for charcoal, with the extraction of larger species such as *Newtonia hildebrandtii*, *Erythrophleum lasianthum*, *Balanites maughamii* and *Manilkara discolor*. Charcoal production involves cutting of thick woody stems, pilling them and covering with sand and grass and then igniting these traditional charcoal kilns (Tokura *et al.* 2020). In the process of cutting and cleaning large stems, many small branches and twigs are left in the forest. These dry out and create a source of fuel for fire. The combination of a seasonal drought, increased fuel-load from branches left during the charcoal production and ignition sources from lighting of charcoal kilns is resulting in more frequent fires within the Licuáti Thicket. This vegetation type is not fire tolerant, and the slow-growing nature of

species within this system suggests that fire and cutting for charcoal will result in severe habitat degradation and possibly an ecological shift to grassy savannas, such as those found in previously disturbed farmed areas around the Licuáti Thicket (Matimele 2016).

With 80% of Mozambique's population depending on charcoal as a source of energy, and all of the wood to make charcoal coming from indigenous vegetation and with little indication of this trend changing (Chavana 2014), further loss and severe degradation of at least 80% of the Licuáti Thicket is expected to occur within the next 25 years (Matimele 2016).

Important bird species found in the IPA include Cape Vulture (*Gyps coprotheres*, EN) (BirdLife International 2017b). Neergaard's Sunbird (*Cinnyris neergaardi*, NT), only known from Mozambique and South Africa, is an uncommon species found particularly in Maputaland sand forest (du Randt 2018).

Key ecosystem services

Ecosystems of the Licuáti IPA provide several essential services such as provision of clean air, climate regulation, carbon sequestration and shelter for a wide range of fauna and flora species (Tokura *et al.* 2020). Because the main infrastructure for provision of basic services is not readily available, communities rely on harvesting plants as their primary source of medicine. Being a sacred place, Licuáti supports traditional values and the inspiration of local people, enhancing the pride of the local communities. In addition, it supports livelihoods through the provision of food, fuelwood and building materials (Matimele 2016).

Sclerochiton apiculatus (JEB)

Xylopia torrei (JEB)

Given the geographical location of the Licuáti on dunes between the Maputo and Tembe Rivers, the IPA plays a major role in water filtration before reaching the two rivers. The Maputo River is the main water source for the local population and for large-scale agriculture, particularly the production of rice in Zitundo and lychee southwards to Tinonganine. The Puchene Esculo stream, with its watershed in this IPA, is a major water source for local communities and livestock. Along the streambanks some local communities grow vegetables and use watering cans to water the plants by hand.

Ecosystem service categories

- Provisioning – Food
- Provisioning – Raw materials
- Provisioning – Medicinal resources
- Regulating services – Local climate and air quality
- Regulating services – Carbon sequestration and storage
- Regulating services – Erosion prevention and maintenance of soil fertility
- Regulating services – Pollination
- Habitat or supporting services – Habitats for species
- Habitat or supporting services – Maintenance of genetic diversity
- Cultural services – Recreation and mental and physical health
- Cultural services – Aesthetic appreciation and inspiration for culture, art and design
- Cultural services – Spiritual experience and sense of place
- Cultural services – Cultural heritage

IPA assessment rationale

Licuáti qualifies as an IPA under criteria A and C. The Licuáti Thicket is unique and is also home to plant species of conservation concern including threatened, rare and range-restricted species that trigger criterion A(i): *Empogona maputensis* (EN), *Warneckea parvifolia* (EN), *Xylopia torrei* (EN), *Sclerochiton apiculatus* (VU), *Acridocarpus natalitius* var. *linearifolius* (VU), *Polygala francisci* (VU), *Psychotria amboniana* subsp. *mosambicensis* (VU), *Rytigynia celastroides* var. *australis* (VU), *Tephrosia forbesii* subsp. *forbesii* (VU) and *Warburgia salutaris* (EN). Licuáti also qualifies under criterion C(iii) as Licuáti Thickets are a range-restricted and nationally threatened habitat that does not occur elsewhere in the country.

Woodpile made in preperation for charcoal production (HM)

Priority species (IPA Criteria A and B)

FAMILY	TAXON	IPA CRITERION A	IPA CRITERION B	≥ 1% OF GLOBAL POP'N	≥ 5% OF NATIONAL POP'N	IS 1 OF 5 BEST SITES NATIONALLY	ENTIRE GLOBAL POP'N	SPECIES OF SOCIO-ECONOMIC IMPORTANCE	ABUNDANCE AT SITE
Acanthaceae	*Sclerochiton apiculatus*	A(i)	B(ii)	✓	✓	✓			frequent
Annonaceae	*Xylopia torrei*	A(i)	B(ii)	✓	✓	✓			frequent
Canellaceae	*Warburgia salutaris*	A(i)			✓	✓			occasional
Fabaceae	*Tephrosia forbesii* subsp. *forbesii*	A(i)		✓					unknown

Priority species (IPA Criteria A and B)

FAMILY	TAXON	IPA CRITERION A	IPA CRITERION B	≥ 1% OF GLOBAL POP'N	≥ 5% OF NATIONAL POP'N	IS 1 OF 5 BEST SITES NATIONALLY	ENTIRE GLOBAL POP'N	SPECIES OF SOCIO-ECONOMIC IMPORTANCE	ABUNDANCE AT SITE
Malpighiaceae	*Acridocarpus natalitius* var. *linearifolius*	A(i)		✓	✓	✓			common
Melastomataceae	*Warneckea parvifolia*	A(i)	B(ii)	✓	✓	✓			frequent
Pedaliaceae	*Dicerocaryum forbesii*		B(ii)	✓	✓	✓			common
Polygalaceae	*Polygala franciscii*	A(i)	B(ii)	✓	✓	✓			unknown
Rubiaceae	*Empogona maputensis*	A(i)	B(ii)	✓	✓	✓			scarce
Rubiaceae	*Pavetta vanwykiana*		B(ii)	✓	✓				scarce
Rubiaceae	*Psychotria amboniana* subsp. *mosambicensis*	A(i)	B(ii)	✓	✓	✓			frequent
Rubiaceae	*Rytigynia celastroides* var. *australis*	A(i)		✓	✓	✓			unknown
Rutaceae	*Zanthoxylum delagoense*		B(ii)	✓					unknown
		A(i): 10 ✓	B(ii): 9						

Threatened habitats (IPA Criterion C)

HABITAT TYPE	IPA CRITERION C	≥ 5% OF NATIONAL RESOURCE	≥ 10% OF NATIONAL RESOURCE	IS 1 OF 5 BEST SITES NATIONALLY	ESTIMATED AREA AT SITE (IF KNOWN)
MOZ: Mozambique Licuáti Thicket [MOZ-07]	C(iii)		✓	✓	

Protected areas and other conservation designations

CONSERVATION AREA TYPE	CONSERVATION AREA NAME	RELATIONSHIP OF IPA TO CONSERVATION AREA
Forest Reserve	Licuáti Forest Reserve	protected/conservation area overlaps with IPA
Key Biodiversity Area	Licuáti Forest Reserve	protected/conservation area overlaps with IPA

Threats

THREAT	SEVERITY	TIMING
Logging & wood harvesting – large scale	high	Ongoing – stable
Logging & wood harvesting – subsistence/small scale	medium	Ongoing – stable
Increase in fire frequency/intensity	high	Ongoing – increasing
Shifting agriculture	high	Ongoing – increasing
Small-holder grazing, ranching or farming	medium	Ongoing – increasing
Gathering terrestrial plants	high	Ongoing – stable

ADDITIONAL SITES OF BOTANICAL INTEREST

The following sites were considered in the development of the IPA network in Mozambique, and may be given IPA status in a future iteration of this work, but they require further investigation.

Cabo Delgado Peninsula (-10.68, 40.62)

This peninsula, situated northeast of Palma in Cabo Delgado Province, supports one of the most intact and extensive examples of coral rag thicket on the Mozambique mainland. To date, only one threatened species has been recorded, *Erianthemum lindense* (VU), but further studies may reveal other rare coral rag specialists. It is also noted as the type locality for *Clerodendrum cephalanthum* var. *torrei*. The nearby Tecomaji and Rongui Islands to the south also have good intact coral rag thicket, and the former island has a record of *Paracephalis trichantha* (VU). These three sites could potentially be combined with the Vamizi Island IPA as they are united by their shared coral rag habitats.

Sakaje Plateau (-11.91, 40.24)

This site to the southwest of the Quiterajo IPA in northern Cabo Delgado Province supports extensive areas of intact dense woody vegetation which are likely to include areas of coastal dry forest and support some of the rare and threatened species recorded at Quiterajo (Timberlake *et al.* 2010). As the two sites are ± contiguous, it might be desirable to extend the Quiterajo IPA in the future. However, almost no botanical exploration of the Sakaje Plateau has been conducted to date.

Mount Mareja (-12.86, 40.14)

This small inselberg is situated within the southern Quirimbas National Park in Cabo Delgado Province and has a nearby tourist camp. On a two-day visit in 2016, accompanying an entomological survey by the California Academy of Sciences, botanical enthusiast Tracey Parker recorded populations of both *Allophylus torrei* (EN) and *Pavetta mocambicensis* (EN) at this site. The Faculty of Natural Sciences at the University of Lúrio (Pemba) intend to conduct further botanical surveys at this site, which may reveal other species of interest (M.I. Caravela, pers. comm. 2021).

Serra Mesa (-14.75, 40.65)

This flat-topped inselberg to the west of the Nacala to Matibane road in Nampula Province has a well preserved dry forest, with a large stand of *Icuria dunensis* (EN) having been recorded there for the first time in early 2022 (Massingue *et al.*, in. prep.), where it occurs together with *Androstachys johnsonii* and *Brachystegia oblonga* (EN). Other areas of the inselberg have a more humid forest assemblage. This site is sure to qualify as an IPA following further exploration.

Mount Ile, Errego (-16.02, 37.21)

This is a small series of inselbergs close to the town of Errego in Zambézia Province. It is the only known locality for the herb *Oldenlandia verrucitesta* (DD), and a number of other scarce species have been recorded including *Euphorbia ramulosa* (provisionally DD), *Faroa involucrata* (DD), and *Searsia acuminatissima* (NT). It is also one of only two known sites for the forest shrub *Polysphaeria harrisii* (EN). However, the area around these inselbergs is heavily populated and evidence from satellite imagery suggests that much natural vegetation has been cleared, particularly the forests and woodland. The records from this site are all from the 1940s–1960s, and further information is required on the current status of this site and its flora.

Mocuba (-16.90, 36.83)

The small inselbergs to the west of Mocuba town in Zambézia Province are notable for being the only known locality for the recently described succulent *Euphorbia stenocaulis* (EN). Other succulent species to be recorded here include the Mozambique endemic *Huernia erectiloba* (LC) and it is likely that other species of interest will be discovered following further surveys.

Gobene (-17.41, 37.70)

This site, situated in the coastal lowlands of

Zambézia Province between the towns of Maganja (Olinga) to the northwest and Bajone and Pebane to the east, formerly supported an extensive semi-deciduous forest and thicket on a series of linear sand dunes (chenier dunes). This forest was the only known site globally for *Huberantha mossambicensis* (CR), and also supported important populations of *Brachystegia oblonga* (EN), *Pavetta dianeae* (EN) and *Scorodophloeos torrei* (EN). However, the vast majority of the natural habitat has been replaced by plantations of coconut, cashew, mango and other crops. The *Brachystegia* and *Scorodophloeos* are still present in the small fragments of natural vegetation that remains, but are heavily cut for firewood (Alves *et al.* 2014b; Darbyshire & Rokni 2020b). The *Huberantha* has not been relocated during recent surveys and may be extinct. In light of these issues, the site is not recognised as an IPA here, but there would still be merit in conserving the small patches of remaining thicket and woodland.

Lupata Gorge and Plateau (-16.69, 34.07)

The Zambezi River cuts through the Lupata Plateau (Serra de Lupata) as a narrow gorge between Capesse and Tambara on the border of Tete and Manica Provinces. This site was visited by John Kirk during his travels up the Zambezi with Dr. Livingstone, where they first found *Bussea xylocarpa* (VU) which is endemic to this area; it has not been collected recently but this is more a result of inaccessibility. The plateau supports extensive areas of a thicket-like vegetation of potential high interest and so is well worthy of future botanical exploration (J. Burrows, pers. comm., 2021).

Levasflor (-18.72, 34.95)

Levasflor is a sustainable forestry concession on the Cheringoma Plateau neighbouring Gorongosa National Park (GNP). The forestry operation is a good example of balancing conservation initiatives with the livelihoods and well-being of local people. Levasflor is Forest Stewardship Council certified and also supports local communities through employment, education and healthcare (Hyde *et al.* 2021). Habitat is largely moist miombo, dominated by *Brachystegia spiciformis,* with records of three endemic species: *Grewia transzambesica*, *Ochna angustata* (NT) and *Pavetta* sp. I of *Flora Zambesiaca* (Bridson & Verdcourt 2003). In addition, the widespread and threatened species *Khaya anthotheca* (VU) occurs in riverine habitat at this site. Levasflor is of particular conservation interest due to the suspected presence of *Cola cheringoma* (EN), an endemic to the Cheringoma

Montane habitats at Serra Choa (JO)

area. This species was recorded in 1957 close to a sawmill at Condué (*Gomes e Sousa* #4441), possibly referring to this site, although the exact locality remains unclear. In any case, the gallery forests at Levasflor are likely to provide suitable habitat for *C. cheringoma* as the underlying limestone substrate here, on which the species depends, becomes exposed by the river cutting (Cheek *et al.* 2019). Further research is required to confirm the presence of *C. cheringoma* at Levasflor.

Serra Choa (-17.99, 33.04)

This extensive montane area to the northwest of Catandica in Manica Province comprises the Mozambique portion of the Nyanga massif, an area that is much better known on the Zimbabwe side of the border (Clark *et al.* 2017). The site is topographically complex and supports a mosaic of moist forests, woodland, rock outcrops and undulating expanses of montane grassland (Osborne & Matimele 2018). As such, it contains some of the most extensive montane habitats in Mozambique, although some areas are disturbed by cattle grazing and cultivation. The Nyanga massif supports 21 strict endemic taxa, but most are known only from Zimbabwe at present (Clark *et al.* 2017). No threatened plant species have so far been recorded from Serra Choa, but there are a number of interesting species present that are restricted to the Chimanimani-Nyanga Centre of Endemism, including *Maytenus chasei* (NT), *Tulbaghia friesii* and *Justicia subcordatifolia*, the lattermost having been found in Mozambique for the first time during fieldwork in 2018. It is likely that many more range-restricted species will be found here following further field research.

Tandara Plateau (-19.62, 32.88)

This montane plateau, rising to over 2,000 m a.s.l., spans the Mozambique-Zimbabwe border to the west of the northern Chimanimani Mountains. It falls within the buffer zone of the Chimanimani Mountains National Park. The site is underlain by dolerite/schist deposits and comprises steep, gulleyed escarpments with some areas of intact forest, and an extensive gently undulating grassy plateau. Several species restricted to the Chimanimani-Nyanga Centre of Endemism were recorded here during botanical exploration in the 1940s and 1960s, including *Rhynchosia chimanimaniensis* (EN) and the only known site in Mozambique for *Afrosciadium rhodesicum*

(VU). However, the grasslands that support these species have since been impacted by both potato cultivation and cattle grazing, and it is not known whether these rare species persist.

Serra Vumba (-18.99, 32.88)

The Bvumba massif, a part of the Chimanimani-Nyanga Centre of Plant Endemism, lies primarily within eastern Zimbabwe but extends into Manica Province as Serra Vumba. It is situated immediately to the south of Manica town, but despite the proximity to extensive settlement, the upper slopes appear to support extensive areas of moist forest and rocky montane grasslands. Very few botanical collections are known from this site, but the much better studied Zimbabwean portion of the massif is known to contain several range-restricted species including the endemic epiphytic orchid *Aeranthes africana* (CR) and the rare herb *Barleria fissimuroides* (EN), otherwise known only from private farmland to the northwest of Manica town in Mozambique (Timberlake *et al.* 2020).

Maputo Special Reserve and Ponto do Ouro (-26.19, 32.72, south to -26.86, 32.89)

This large area of coastal lowlands between Maputo Bay and the border with KwaZulu Natal encompasses a rich mosaic of habitats including intact dune thicket, sand forest, wooded and open grasslands, permanent and seasonal wetlands and lagoons, riverine vegetation and mangroves (WCS *et al.* 2021). The habitats are particularly well preserved within the Maputo Special Reserve, a site of international renown for its important population of migratory elephants (*Loxodonta africana*) amongst other species. There is also some good intact habitat to the south of the reserve, notably around Zitundo and at Ponta do Ouro. The Reserve is assessed as a KBA, primarily on the basis of its fauna but with the shrub *Sclerochiton apiculatus* (VU) included within the assessment. A number of other interesting plant species occur in this broad region; it is the key site in Mozambique for the herb *Dicliptera quintasii* (EN) and supports other rare species such as *Ceropegia vahrmeijeri* and *Thesium vahrmeijeri*, whilst Ponta do Ouro is the only known site in Mozambique for the cycad *Stangeria eriopus* (VU). However, botanical coverage of this vast area is currently low and more information is needed on which specific sites are of highest importance. It may be that one to several IPAs can be identified following further research.

BIBLIOGRAPHY

Achar, J. (2012). ZAMBÉZIA – Queimadas e agricultura rudimentar podem provocar aluimento de terras em Gúruè. In: J. Hanlon (ed.), *MOZAMBIQUE: News reports & clippings.* Available at: https://www.open.ac.uk/technology/mozambique/sites/www.open.ac.uk.technology.mozambique/files/pics/d136953.pdf

Achimo, M., Mugabe, J.A, Momade, F. & Haldorsen, S. (2014). Geomorphology and evolution of Maputo Bay. In: Bandeira, S. & Paula, J. (eds.). *The Maputo Bay Ecosystem.* WIOMSA, Zanzibar Town.

Achten, W.M.J., Dondeyne, S., Mugogo, S., Kafiriti, E., Poesen, J., Deckers, J. & Muys, B. (2008). Gully erosion in South Eastern Tanzania: spatial distribution and topographic thresholds. *Zeitschrift für Geomorphologie* 52: 225 – 235. https://doi.org/10.1127/0372-8854/2008/0052-0225

Adam, I., Machele J. & Saranga, O. (2014). Human setting in Maputo Bay. pp. 67 – 86 in: Bandeira, S. & Paula, J. (eds.). The Maputo Bay Ecosystem. WIOMSA, Zanzibar Town.

Adams, J.B. (2016). Distribution and status of *Zostera capensis* in South African estuaries – A Review. *South African Journal of Botany* 107: 63 – 73.

African Parks (2021). Bazaruto. Available at: https://www.africanparks.org/the-parks/bazaruto

Agencia de Informacao de Mocambique (2020). Mozambique: Forestry Company to Abandon Rights to 54,000 Hectares. AllAfrica. Available at: https://allafrica.com/stories/202011060844.html

Alves, T. & Sousa, C. (2007). Preliminary assessment of the coastal vegetation and mangrove forests of the proposed conservation area of the Primeiras (1as) and Segundas (2as) Island Archipelago. IIAM, Mozambican Institute for Agrarian Research, Maputo. Available at: http://www.biofund.org.mz/biblioteca_virtual/preliminary-assessment-of-the-coastal-vegetation-and-mangrove-forests-of-the-proposed-conservation-area-of-the-primeiras-1as-and-segundas-2as-islands-archipelago/

Alves, M.T., Burrows, J.E., Coates Palgrave, F.M., Hyde, M.A., Luke, W.R.Q., Massingue, A.O., Matimele, H.A., Raimondo, D., Timberlake, J. & von Staden, L. (2014a). *Viscum littorum.* The IUCN Red List of Threatened Species 2014: e.T62497268A62497282. https://dx.doi.org/10.2305/IUCN.UK.2014-3.RLTS.T62497268A62497282.en

Alves, M.T., Burrows, J.E., Timberlake, J., Coates Palgrave, F.M., Hyde, M.A., Massingue, A.O., Matimele, H.A., Raimondo, D., Hadj-Hammou, J. & Osborne, J. (2014b). *Brachystegia oblonga.* The IUCN Red List of Threatened Species 2014: e.T62494198A62494201. https://dx.doi.org/10.2305/IUCN.UK.2014-3.RLTS.T62494198A62494201.en

ANAC (2018). Nature-based Tourism: Mozambique Conservation Areas. Available at: http://pubdocs.worldbank.org/en/881051531337811300/Fichário-ENG-LOW.pdf

Anderson, S. (2002). Identifying Important Plant Areas: a site selection manual for Europe. Plantlife International, Salisbury. Available at: www.plantlife.org.uk/publications/identifying_important_plant_areas_a_site_selection_manual_for_europe.

Araújo, J.R., Afonso, R.S., & Pinto, M.S. (1973). Contribuição para o conhecimento da geologia da área de Morrumbala-Mutarara (Folha SUL-E-36/L, Grau Quadrado 1735). *Boletim Dos Serviços de Geologia e Minas 37:* 1 – 76.

Aremu, A.O., Cheesman, L., Finnie, J.F., Van Staden, J. (2011). *Mondia whitei* (Apocynaceae): a review of its biological activities, conservation strategies and economic potential. *South African Journal of Botany* 77: 960 – 971.

Ashley, C. & Wolmer, W. (2003). Transforming or Tinkering? New Forms of Engagement between Communities and the Private Sector in Tourism and Forestry in Southern Africa. *Sustainable Livelihoods in Southern Africa* 18. Institute of Development Studies, Brighton.

AZE (2018). Alliance for Zero Extinction 2018 Global AZE map. Available at: https://zeroextinction.org/site-identification/2018-global-aze-map/

Bachman, S., Moat, J., Hill, A.W., de la Torre, J. & Scott, B. (2011). Supporting red list threat assessments with GeoCAT: geospatial conservation assessment tool. *ZooKeys* 150: 117 – 126. https://doi.org/10.3897/zookeys.150.2109

Bandeira, S.O. (2002). Diversity and distribution of seagrasses around Inhaca Island, Southern Mozambique. *South African Journal of Botany* 68: 191 – 198.

Bandeira, S.O. & Gell, F. (2003). The seagrasses of Mozambique and southeastern Africa. pp. 93 – 100 in: E.P. Green & F.T. Short (eds.), World Atlas of Seagrasses. Prepared by the UNEP World Conservation Monitoring Centre, University of California Press, Berkeley, USA.

Bandeira, S., Muiocha, D. & Schleyer, M. (2008). Seagrass beds. In: Everett, B.I., van der Elst, R.P. & Schleyer, M.H. (eds.), A Natural History of the Bazaruto Archipelago, Mozambique. Oceanographic Research Institute, Special publication No. 8: 65 – 69, Durban. Available at: https://biofund.org.mz/wp-content/uploads/2015/03/Je02-275.pdf

Bandeira, S., Gullström, M., Balidy, H., Davide, S. & Cossa, D. (2014). Seagrass meadows in Maputo Bay. pp. 147 – 186 in: Bandiera, S. & Paula, J. (eds.). The Maputo Bay Ecosystem. WIOMSA, Zanzibar Town.

Barker, N.P., Faden, R.B., Brink, E. & Dold, A.P. (2001). Rediscovery of *Triceratella drummondii,* and comments on its relationships and position within the family. *Bothalia* 31: 37 – 39.

Barnes, D.K.A. (2001). Hermit crabs, humans and Mozambique mangroves. *African Journal of Ecology* 39: 241 – 248. https://doi.org/10.1046/j.1365-2028.2001.00304.x

Bayliss, J., Monteiro, J., Fishpool, L., Congdon, T.C., Bampton, I., Bruessow, C., Matimele, H., Banze, A. & Timberlake, J. (2010). Biodiversity and conservation of Mount Inago, Mozambique. Available at: https://www.researchgate.net/publication/302973476

Bayliss, J., Timberlake, J., Branch, W., Bruessow, C., Collins, S., Congdon, C., Curran, M., de Sousa, C., Dowsett, R., Dowsett-Lemaire, F., Fishpool, L., Harris, T., Herrmann, E., Georgiadis, S., Kopp, M., Liggitt, B., Monadjem, A., Patel, H., Ribeiro, D., Spottiswoode, C., Taylor, P., Willcock, S. & Smith, P. (2014). The discovery, biodiversity and conservation of Mabu forest – the largest medium-altitude rainforest in southern Africa. *Oryx* 48: 177 – 185. https://doi.org/10.1017/S0030605313000720

Bayliss, J., Brattström, O., Bampton, I. & Collins, S.C. (2019). A new species of *Leptomyrina* Butler, 1898 (Lepidoptera: Lycaenidae) from Mts Mecula, Namuli, Inago, Nallume and Mabu in Northern Mozambique. *Metamorphosis* 30: 19 – 24.

Available at: https://metamorphosis.org.za/articlesPDF/1498/Bayliss%20et%20al.%20Leptomyrina%20congdoni.pdf

Beilfuss, R. (2007). Adaptive management of the invasive shrub *Mimosa pigra* at Gorongosa National Park. Parque Nacional da Gorongosa. Available at: https://www.biofund.org.mz/biblioteca_virtual/adaptive-management-of-the-invasive-shrub-mimosa-pigra-at-gorongosa-national-park/

BGCI (2021). State of the World's Trees. BGCI, Richmond, UK. Available at: https://www.bgci.org/our-work/projects-and-case-studies/global-tree-assessment/

Bingen, B., Bjerkgård, T., Boyd, R., Grenne, T., Henderson, I., Lutro, O., Melezhik, V., Motuza, G., Nordgulen, Ø., Often, M., Sandstad, J.S., Smelror, M., Solli, A., Stein, H., Sæther, O.M., Thorsnes, T., Tveten, E., Bauer, W., Dunkley, P., Gonzalez, E., Hollick, L., Jacobs, J., Key, R., Smith, R., Thomas, R.J., Jamal, D., Catuane, F., de Azavedo, S., Feitio, P., Manhica, V., Manuel, S., Moniz, A., Njange, F., Rossi, de S., Soares, H., Tembe, D., Uachave, B., Viola, G. & Zandamela, E. (2007). The geology of Niassa and Cabo Delgado Provinces with parts of Zambesia and Nampula Provinces, Mozambique. Ministry of Mineral Resources and Energy, National Directorate of Geology. Maputo. Available at: http://nora.nerc.ac.uk/id/eprint/6673/

Biofund (2013). Gorongosa. Platform of the Conservation Areas. Available at: http://www.biofund.org.mz

Biofund (2021). "PROMOVE Biodiversidade" project presents implementation partner to the Zambezia provincial government and to Lugela district. Available at: https://www.biofund.org.mz/en/promove-biodiversidade-project-presents-implementation-partner-to-the-zambezia-provincial-government-and-to-lugela-district/

BirdLife International (2017a). *Sheppardia gunningi* (amended version of 2016 assessment). The IUCN Red List of Threatened Species 2017: e.T22709650A111057443. https://dx.doi.org/10.2305/IUCN.UK.2017-1.RLTS.T22709650A111057443.en.

BirdLife International (2017b). *Gyps coprotheres* (amended version of 2016 assessment). The IUCN Red List of Threatened Species 2017: e.T22695225A118592987. https://dx.doi.org/10.2305/IUCN.UK.2017-3.RLTS.T22695225A118592987.en

BirdLife International (2018). *Chamaetylas choloensis.* The IUCN Red List of Threatened Species 2018: e.T22709004A131333396. https://dx.doi.org/10.2305/IUCN.UK.2018-2.RLTS.T22709004A131333396.en.

BirdLife International (2019). Important Bird Areas factsheet: Njesi plateau. Available at: http://www.birdlife.org

BirdLife International (2020a). Important Bird Areas factsheet: Mount Chiperone. Available at: http://www.birdlife.org/

BirdLife International (2020b). Important Bird Areas factsheet: Pomene. Available at: http://www.birdlife.org/

BirdLife International (2021a). Important Bird Areas factsheet: Mount Namuli. Available at: http://datazone.birdlife.org/site/factsheet/mount-namuli-iba-mozambique/details

BirdLife International (2021b). Important Bird Areas factsheet: Mount Mabu. Available at: http://datazone.birdlife.org/site/factsheet/mount-mabu-iba-mozambique

BirdLife International (2021c). Important Bird Areas factsheet: Primeiras and Segundas Environmental Protection Area (APAIPS). Available at: http://datazone.birdlife.org/site/factsheet/primeiras-and-segundas-environmental-protection-area-(apaips)-iba-mozambique

BirdLife International (2021d). Important Bird Areas factsheet: Gorongosa Mountain and National Park. Available at: http://datazone.birdlife.org/site/factsheet/gorongosa-mountain-and-national-park-iba-mozambique

BirdLife International (2021e). Important Bird Areas factsheet: Chimanimani Mountains (Mozambique). Available at: http://datazone.birdlife.org/site/factsheet/chimanimani-mountains-(mozambique)-iba-mozambique

BirdLife International (2021f). Important Bird Areas factsheet: Panda *Brachystegia* woodlands. Available at: http://datazone.birdlife.org/site/factsheet/panda-brachystegia-woodlands-iba-mozambique

Book, F. (2012). Possible impacts of a marine protected area on the artisanal fisheries on Inhaca Island, with a focus on fishing grounds and transportation. Miljövetenskapligt Program, Göteborgs Universitet. Available at: https://biofund.org.mz/wp-content/uploads/2019/01/1548938303-F0844.Possible%20impacts%20of%20%20a%20marine%20protected%20area%20on%20the%20artisanal%20fisheries%20on%20Inhaca%20Island,%20Mozambique.pdf

Borghesio, L. & Gagliardi, A. (2015). A waterbird survey on the coast of Quirimbas National Park, northern Mozambique. *Bulletin of the African Bird Club* 18: 61 – 67.

Bösenberg, J.D. (2010). *Encephalartos turneri.* The IUCN Red List of Threatened Species 2010: e.T41946A10608314. https://dx.doi.org/10.2305/IUCN.UK.2010-3.RLTS.T41946A10608314.en.

Boyd, R., Nordgulen, Ø., Thomas, R.J., Bingen, B., Bjerkgard, T., Grenne, T., Henderson, I., Melezhik, V.A., Often, M., Sandstad, J.S., Solli, A., Tveten, E., Viola, G., Key, R.M., Smith, R.A., Gonzalez, E., Hollick, L.J., Jacobs, J., Jamal, D., Motuza, G., Bauer, W., Daudi, E., Feitio, P., Manhica, V., Moniz A. & Rosse, D. (2010). The Geology and Geochemistry of The East African Orogen In Northeastern Mozambique. *African Journal of Geology* 113: 1 – 87. https://doi.org/10.2113/gssajg.113.1.87

Bridson, D. (1998). Rubiaceae (part 2). In: G.V. Pope (ed.), *Flora Zambesiaca,* Vol. 5(2). Royal Botanic Gardens, Kew.

Bridson, D. & Verdcourt, B. (2003). Rubiaceae (part 3). In: G.V. Pope (ed.), *Flora Zambesiaca,* Vol. 5(3). Royal Botanic Gardens, Kew.

Brooks, T.M., Pimm, S.L., Akçakaya, H.R., Buchanan, G.M., Butchart, S.H.M., Foden, W., Hilton-Taylor, C., Hoffmann, M., Jenkins, C.N., Joppa, L., Li, B.V., Menon, V., Ocampo-Peñuela, N. & Rondinini, C. (2019). Measuring Terrestrial Area of Habitat (AOH) and Its Utility for the IUCN Red List. *Trends in Ecology and Evolution* 34: 977 – 986. https://doi.org/10.1016/j.tree.2019.06.009

Browne, C.M., Milne, R., Griffiths, C., Bolton, J.J. & Anderson, R.J. (2013). Epiphytic seaweeds and invertebrates associated with South African populations of the rocky shore seagrass *Thalassodendron leptocaule* — a hidden wealth of biodiversity. *African Journal of Marine Sciences* 35: 523 – 531.

Burgess, N., D'Amico Hales, J., Underwood, E., Dinerstein, E., Olson, D., Itoua, I., Schipper, J., Ricketts, T. & Newman, K. (2004a). Terrestrial ecoregions of Africa and Madagascar: a conservation assessment. WWF/Island Press, Washington, USA.

Burgess, N., Salehe, J., Doggart, N., Clarke, G.P., Gordon, I., Sumbi, P. & Rodgers, A. (2004b). Coastal Forests of Eastern Africa. In: Mittermeier, R.A., Robles Gil, P. Hoffman, M., Pilgrim, J., Brooks, T., Goettsch Mittermeier, C., Lamoreux, J. & da Fonseca, G.A.B. (eds.) Hotspots revisited: Earth's biologically richest and most endangered eco-systems. Conservation International.

Burrows, J.E. & Timberlake, J.R. (2011). Mozambique's centres of endemism, with special reference to the Rovuma Centre of Endemism of NE Mozambique and SE Tanzania. *South African Journal of Botany* 77: 518. https://doi.org/10.1016/j.sajb.2011.03.003

Burrows, J.E. & Burrows, S.M. (2012). A preliminary report on the vegetation of Vamizi Island. Unpubl. report. Buffelskloof Herbarium, Lydenburg.

Burrows, J.E., McCleland, W., Bester, P. & Schmidt, E. (2012). Check-list of the plants recorded at the limestone gorges, Cheringoma Plateau. Unpubl. report. Gorongosa National Park.

Burrows, J.E., Timberlake, J., Alves, M.T., Coates Palgrave, F.M., Hyde, M.A., Luke, W.R.Q., Massingue, A.O., Matimele, H.A., Raimondo, D., Osborne, J. & Hadj-Hammou, J. (2014a). *Micklethwaitia carvalhoi.* The IUCN Red List of Threatened Species 2014: e.T62494244A62494265. https://dx.doi.org/10.2305/IUCN.UK.2014-3.RLTS.T62494244A62494265.en

Burrows, J.E., Timberlake, J., Alves, M.T., Coates Palgrave, F.M., Hyde, M.A., Luke, W.R.Q., Massingue, A.O., Matimele, H.A., Raimondo, D., Osborne, J. & Hadj-Hammou, J. (2014b). *Acacia latispina*. The IUCN Red List of Threatened Species 2014: e.T62494299A62494308. https://dx.doi.org/10.2305/IUCN.UK.2014-3.RLTS.T62494299A62494308.en

Burrows, J. E., Burrows, S., Lötter, M. & Schmidt, E. (2018). Trees and Shrubs Mozambique. Print Matters Heritage, Cape Town.

Byfield, A., Atay, S. & Özhatay, N. (2010). Important Plant Areas in Turkey: 122 key Turkish botanical sites. WWF Turkey, Istanbul (first published in Turkish in 2005).

Byrne, J. (2013). An Expedition Back in Time in Mozambique. National Geographic. Available at: https://blog.nationalgeographic.org/2013/05/15/an-expedition-back-in-time-in-mozambique/

Cabo, F. (2020). Mozambique: Gilé and Chimanimani become National Parks, Niassa becomes Special Reserve. WCS Mozambique: In the News. Available at: https://mozambique.wcs.org/About-Us/News/ID/14236.aspx

Campbell B.M., Attwell C.A.M., Hatton J.C., de Jager P., Gambiza J., Lynam T., Mizutani F. & Wynter P. (1988). Secondary Dune Succession on Inhaca Island, Mozambique. *Vegatatio* 78: 3 – 11. https://doi.org/10.1007/BF00045633

Capela, P. (2006). Speculations on *Encephalartos* Species of Mozambique. Ndjira.

Carvalho, A.M. & Bandeira, S.O. (2003). Seaweed flora of Quirimbas Archipelago, northern Mozambique. pp: 319 – 324 in: Chapman, A.R.O., Anderson, R.J., Vreeland, I.R. & Davison, V.J. (eds.). *Proceedings of the XVIIth International Seaweed Symposium*, Cape Town, South Africa. 28 Jan. – 2 Feb. 2001. Oxford University Press, Oxford.

catapu.net (2020). Catapú. Available at: https://www.catapu.net/index.php?id=415&lang=en.

CEPF (2010). Ecosystem Profile: Maputaland-Pondoland-Albany Biodiversity Hotspot. Available at: https://www.cepf.net/sites/default/files/apo_mpah_2011.pdf

CEPF (2020). Critical Ecosystems Partnership Fund. Coastal Forests of Eastern Africa. Available at: https://www.cepf.net/our-work/biodiversity-hotspots/coastal-forests-eastern-africa

CEPF (2021). Grantee projects: Mount Mabu Conservation Project. Available at: https://www.cepf.net/grants/grantee-projects/mount-mabu-conservation-project

Chavana, R. (2014). Estudo da cadeia de valor de carvão vegetal no sul de Moçambique. Relatório preliminar de pesquisa No. 10P. IIAM, Maputo.

Cheek, M. & Lawrence, P. (2019) *Cola clavata*. The IUCN Red List of Threatened Species 2019: e.T34975A111448906. https://dx.doi.org/10.2305/IUCN.UK.2019-1.RLTS.T34975A111448906.en.

Cheek, M., Chipanga, H. & Darbyshire, I. (2018). Notes on the plant endemics of the quartzitic slopes of Mt Chimanimani (Mozambique & Zimbabwe), and a new, Critically Endangered species, *Empogona jenniferae* (Rubiaceae-Coffeeae). *Blumea* 63: 87 – 92. https://doi.org/10.3767/blumea.2018.63.01.08

Cheek, M., Luke, Q., Matimele, H., Banze, A. & Lawrence, P. (2019). *Cola* species of the limestone forests of Africa, with a new, endangered species, *Cola cheringoma* (Sterculiaceae), from Cheringoma, Mozambique. *Kew Bulletin* 74: 1 – 14. https://doi.org/10.1007/s12225-019-9840-3

Chevallier, R. (2018). Livelihood interventions and biodiversity conservation in Quirimbas National Park. South African Institute for International Affairs (SAIIA), *Policy Insights* 57. Available at: https://www.africaportal.org/publications/livelihood-interventions-and-biodiversity-conservation-quirimbas-national-park/#:~:text=Livelihood%20Interventions%20and%20Biodiversity%20Conservation%20in%20Quirimbas%20National%20Park,-Romy%20Chevallier&text=Coastal%20livelihood%20interventions%20can%20help,maintaining%20resources%20and%20the%20environment.

Clarke, G.P. (1998). A new regional centre of endemism in Africa. pp. 53 – 65 in: D.F. Cutler, C.R. Huxley, J.M. Lock. (eds.) Aspects of the ecology, taxonomy and chorology of the floras of Africa and Madagascar. *Kew Bulletin Additional Series*. Royal Botanic Gardens, Kew.

Clarke, G.P. (2001). The Lindi local centre of endemism in SE Tanzania. *Systematics and Geography of Plants* 71: 1063 – 1072. https://doi.org/10.2307/3668738

Clarke, G.P. (2010). Report on a reconnaissance visit to Lupangua Hill, Quissanga District, Cabo Delgado Province, Mozambique, with notes about *Micklethwaitia carvalhoi*. ProNatura International & Instituto de Investigação Agrária de Moçambique Cabo Delgado Expedition 2009. Available at: http://www.coastalforests.org/LupanguaReconnaissanceReport2010medium.pdf

Clarke, G.P. (2011). Observations on the Vegetation and Ecology of Palma and Nangade Districts, Cabo Delgado Province, Mozambique. Available at: http://coastalforests.org/PalmaNangadeVegetationEcologyReport2011medium.pdf

Clark, V.R., Timberlake, J.R., Hyde, M.A., Mapaura, A., Coates Palgrave, M., Wursten, B.T., Ballings, P., Burrows, J.E., Linder, H.P., McGregor, G.K., Chapano, C., Plowes, D.C.H., Childes, S.L., Dondeyne, S., Müller, T. & Barker, N.P. (2017). A first comprehensive account of floristic diversity and endemism on the Nyanga Massif, Manica Highlands (Zimbabwe–Mozambique). *Kirkia* 19: 1 – 53.

Coates Palgrave, M., Van Wyk, A.E., Jordaan, M., White, J.A. & Sweet, P. (2007). A reconnaissance survey of the woody flora and vegetation of the Catapú logging concession, Cheringoma District, Mozambique. *Bothalia*, 37: 57 – 73. https://doi.org/10.4102/abc.v37i1.303

Coates Palgrave, F.M., Hyde, M.A., Alves, M.T., Burrows, J.E., Massingue, A.O., Matimele, H.A., Raimondo, D. & Timberlake, J. (2014a). *Dorstenia zambesiaca*. The IUCN Red List of Threatened Species 2014: e.T63707797A63707800. https://dx.doi.org/10.2305/IUCN.UK.2014-3.RLTS.T63707797A63707800.en.

Coates Palgrave, F.M., Burrows, J.E., Timberlake, J., Alves, M.T., Contu, S., Hyde, M.A., Luke, W.R.Q., Massingue, A.O., Matimele, H.A., Raimondo, D., Osborne, J. & Hadj-Hammou, J. (2014b). *Acacia torrei*. The IUCN Red List of Threatened Species 2014: e.T19891788A63707954. https://dx.doi.org/10.2305/IUCN.UK.2014-3.RLTS.T19891788A63707954.en.

Coelho, A.V.P. (1959). Reconhecimentos petrográficos sumários dos maciços da Lupata, Morrumbala, Chiperone-Derre e Milange. *Boletim Dos Serviços de Geologia e Minas 26:* 1 – 47.

Congdon, T.C.E. & Bayliss, J. (2013). Butterflies of Mt Mecula and Mt Yao, Niassa Province, Northern Mozambique. *Metamorphosis* 23: 26 – 34.

Conneely, B. (2013). Uncharted Territory: Scientists Discover New and Incredible Species. National Geographic. Available at: https://blog.nationalgeographic.org/2013/06/05/uncharted-territory-scientists-discover-new-and-incredible-species/

Couch, C., Cheek, M., Haba, P., Molmou, D., Williams, J., Magassouba, S., Doumbouya, S. & Diallo, M.Y. (2019). Threatened habitats and Tropical Important Plant Areas (TIPAs) of Guinea, West Africa. Royal Botanic Gardens, Kew.

Cumbe, A.N.F. (2007). O Património Geológico de Moçambique: Proposta de Metodologia de Inventariação, Caracterização e Avaliação. Tese de Mestrado em Património Geológico e Geoconservação. Departamento de Ciências da Terra. Universidade do Minho. Braga.

Crawford, F.M. & Darbyshire, I. (2015). *Ochna dolicharthros* (Ochnaceae): a new species from northern Mozambique. *Kew Bulletin* 70: 1-7.

Dani Sanchez, M., Clubbe, C. & Hamilton, M.A. (eds.) (2019). Identifying and Conserving Tropical Important Plant Areas in the British Virgin Islands (2016-2019): Final technical report. Royal Botanic Gardens, Kew.

Daniels, S.R., Phiri, E.E. & Bayliss, J. (2014). Renewed sampling of inland aquatic habitats in southern Africa yields two novel freshwater crab species (Decapoda: Potamonautidae: Potamonautes). *Zoological Journal of the Linnean Society* 171: 356 – 369. https://doi.org/10.1111/zoj.12139

Darbyshire, I. (2009). The *Barleria fulvostellata* (Acanthaceae) complex in east Africa. *Kew Bulletin* 64: 673 – 679.

Darbyshire, I. (2018). *Barleria setosa*. The IUCN Red List of Threatened Species 2018: e.T120940735A120980053. http://dx.doi.org/10.2305/IUCN.UK.2018-2.RLTS. T120940735A120980053.en

Darbyshire, I. & Rokni, S. (2019). *Vepris macedoi*. The IUCN Red List of Threatened Species 2019: e.T136536037A136538318. https://dx.doi.org/10.2305/IUCN.UK.2019-2.RLTS. T136536037A136538318.en

Darbyshire, I. & Rokni, S. (2020a). *Streptocarpus erubescens*. The IUCN Red List of Threatened Species 2020: e.T149256393A153685869. https://dx.doi.org/10.2305/IUCN. UK.2020-2.RLTS.T149256393A153685869.en.

Darbyshire, I. & Rokni, S. (2020b). *Scorodophloeus torrei*. The IUCN Red List of Threatened Species 2020: e.T149257100A153685894. https://dx.doi.org/10.2305/IUCN. UK.2020-2.RLTS.T149257100A153685894.en

Darbyshire, I., Vollesen, K. & Kelbessa, E. (2015). Acanthaceae (part 2). In: J.R. Timberlake & E.S. Martins (eds.) *Flora Zambesiaca*, Vol. 8(6). Royal Botanic Gardens, Kew.

Darbyshire, I., Anderson, S., Asatryan, A., Byfield, A., Cheek, M., Clubbe, C., Ghrabi, Z., Harris, T., Heatubun, C.D., Kalema, J., Magassouba, S., McCarthy, B., Milliken, W., Montmollin, B. de, Nic Lughadha, E., Onana, J.M., Saidou, D., Sarbu, A., Shrestha, K. & Radford, E.A. (2017). Important Plant Areas: revised selection criteria for a global approach to plant conservation. *Biodiversity & Conservation* 26: 1767 – 1800. https://doi.org/10.1007/s10531-017-1336-6

Darbyshire, I., Matimele, H.A., Alves, M.T., Chelene, I., Cumbula, S., Datizua, C., De Sousa, C., Langa, C., Massingue, A.O., Mucaleque, P.A., Odorico, D., Osborne, J., Rokni, S., Timberlake, J., Viegas, A. & Vilanculos, A. (2018a). *Gladiolus zambesiacus*. The IUCN Red List of Threatened Species 2018: e.T108615648A108620157. https://dx.doi.org/10.2305/IUCN. UK.2018-2.RLTS.T108615648A108620157.en.

Darbyshire, I., Matimele, H.A., Alves, M.T., Chelene, I., Cumbula, S., Datizua, C., De Sousa, C., Langa, C., Massingue, A.O., Mucaleque, P.A., Odorico, D., Osborne, J., Rokni, S., Timberlake, J., Viegas, A. & Vilanculos, A. (2018b). *Faurea racemosa*. The IUCN Red List of Threatened Species 2018: e.T108615447A108620152. https://dx.doi.org/10.2305/IUCN. UK.2018-2.RLTS.T108615447A108620152.en

Darbyshire, I., Matimele, H.A., Alves, M.T., Baptista, O.J., Bezeng, S., Datizua, C., De Sousa, C., Langa, C., Massingue, A.O., Mtshali, H., Mucaleque, P.A., Odorico, D., Osborne, J., Raimondo, D., Rokni, S., Sitoe, P., Viegas, A. & Vilanculos, A. (2018c). *Guibourtia sousae*. The IUCN Red List of Threatened Species 2018: e.T34500A120980003. https://dx.doi. org/10.2305/IUCN.UK.2018-2.RLTS.T34500A120980003.en

Darbyshire, I., Matimele, H.A., Alves, M.T., Baptista, O.J., Bezeng, S., Datizua, C., De Sousa, C., Langa, C., Massingue, A.O., Mtshali, H., Mucaleque, P.A., Odorico, D., Osborne, J., Raimondo, D., Rokni, S., Sitoe, P., Timberlake, J., Viegas, A. & Vilanculos, A. (2018d). *Ecbolium hastatum*. The IUCN Red List of Threatened Species 2018: e.T120941569A120980108. https://dx.doi.org/10.2305/IUCN.UK.2018-2.RLTS. T120941569A120980108.en.

Darbyshire, I., Timberlake, J., Osborne, J., Rokni, S., Matimele, H., Langa, C., Datizua, C., de Sousa, C., Alves, T., Massingue, A., Hadj-Hammou, J., Dhanda, S., Shah, T. & Wursten, B. (2019a). The endemic plants of Mozambique: diversity and conservation status. *PhytoKeys* 136: 45 – 96. https://doi. org/10.3897/phytokeys.136.39020

Darbyshire, I., Martínez Richart, A.I., Rulkens, T. & Rokni, S. (2019b). *Aloe mossurilensis*. The IUCN Red List of Threatened Species 2019: e.T110760328A110760337. https://dx.doi.org/10.2305/ IUCN.UK.2019-3.RLTS.T110760328A110760337.en.

Darbyshire, I., Rulkens. T. & Rokni, S. (2019c). *Eriolaena rulkensii*. The IUCN Red List of Threatened Species 2019: e.T134844673A134844770. https://dx.doi.org/10.2305/IUCN. UK.2019-2.RLTS.T134844673A134844770.en

Darbyshire, I., Burrows, J.E., Alves, M.T., Chelene, I., Datizua, C., De Sousa, C., Fijamo, V., Langa, C., Massingue, A.O., Massunde, J., Matimele, H.A., Mucaleque, P.A., Osborne, J., Rokni, S. & Sitoe, P. (2019d). *Allophylus torrei*. The IUCN Red List of Threatened Species 2019: e.T136536604A136538323. https://dx.doi.org/10.2305/ IUCN.UK.2019-3.RLTS.T136536604A136538323.en

Darbyshire, I., Massingue, A.O., Osborne, J., De Sousa, C., Matimele, H.A., Alves, M.T., Burrows, J.E., Chelene, I., Datizua, C., Fijamo, V., Langa, C., Massunde, J., Mucaleque, P.A., Rokni, S. & Sitoe, P. (2019e). *Icuria dunensis*. The IUCN Red List of Threatened Species 2019: e.T136532836A136538183. https://dx.doi.org/10.2305/ IUCN.UK.2019-2.RLTS.T136532836A136538183.en

Darbyshire, I., Alves, M.T., Burrows, J.E., Chelene, I., Datizua, C., De Sousa, C., Fijamo, V., Langa, C., Massingue, A.O., Massunde, J., Matimele, H.A., Mucaleque, P.A., Osborne, J., Rokni, S. & Sitoe, P. (2019f). *Blepharis dunensis*. The IUCN Red List of Threatened Species 2019: e.T120941013A120980068. https://dx.doi.org/10.2305/ IUCN.UK.2019-3.RLTS.T120941013A120980068.en

Darbyshire, I., Burrows, J.E., Alves, M.T., Chelene, I., Datizua, C., De Sousa, C., Fijamo, V., Langa, C., Massingue, A.O., Massunde, J., Matimele, H.A., Mucaleque, P.A., Osborne, J., Rokni, S. & Sitoe, P. (2019g). *Rytigynia torrei*. The IUCN Red List of Threatened Species 2019: e.T136535757A136538308. https://dx.doi.org/10.2305/ IUCN.UK.2019-2.RLTS.T136535757A136538308.en.

Darbyshire, I., Langa, C. & Romeiras, M.M. (2019h). A synopsis of *Polysphaeria* (Rubiaceae) in Mozambique, including two new species. *Phytotaxa* 414: 1 – 18. https://doi. org/10.11646/phytotaxa.414.1.1

Darbyshire, I., Wursten, B., Luke, Q. & Fischer, E. (2019i). A revision of the *Crepidorhopalon whytei* complex (Linderniaceae) in eastern Africa. *Blumea* 64: 165 – 176. https://doi.org/10.3767/blumea.2019.64.02.07

Darbyshire, I., Burrows, J.E., Alves, M.T., Chelene, I., Datizua, C., De Sousa, C., Fijamo, V., Langa, C., Massingue, A.O., Massunde, J., Matimele, H.A., Mucaleque, P.A., Osborne, J., Rokni, S. & Sitoe, P. (2019j). *Tarenna longipedicellata*. The IUCN Red List of Threatened Species 2019: e.T136535802A136538313. https://dx.doi.org/10.2305/ IUCN.UK.2019-2.RLTS.T136535802A136538313.en.

Darbyshire, I., Wursten, B. & Rokni, S. (2019k). *Justicia* sp. nov. "B = Bester 11112". The IUCN Red List of Threatened Species 2019: e.T120941681A120980133. https://dx.doi.org/10.2305/IUCN.UK.2019-3.RLTS. T120941681A120980133.en

Darbyshire, I., Rokni, S., Alves, M.T., Burrows, J.E., Chelene, I., Datizua, C., De Sousa, C., Fijamo, V., Langa, C., Massingue, A.O., Massunde, J., Matimele, H.A., Mucaleque, P.A., Osborne, J. & Sitoe, P. (2019l). *Euphorbia graniticola*. The IUCN Red List of Threatened Species 2019: e.T136532350A136538178. https://dx.doi.org/10.2305/ IUCN.UK.2019-2.RLTS.T136532350A136538178.en.

Darbyshire, I., Goyder, D.J., Wood, J.R.I., Banza, A. & Burrows, J.E. (2020a). Further new species and records from the coastal dry forests and woodlands of the Rovuma Centre of Endemism. *Plant Ecology and Evolution* 153: 427 – 445. https://doi.org/10.5091/plecevo.2020.1727

Darbyshire, I., Bandeira, S. & Rokni, S. (2020b). *Thalassodendron leptocaule*. The IUCN Red List of Threatened Species 2020: e.T149255832A149275898. https://dx.doi.org/10.2305/ IUCN.UK.2020-3.RLTS.T149255832A149275898.en.

Darbyshire, I., Polhill, R.M., Magombo, Z. & Timberlake, J.R. (2021). Two new species from the mountains of southern Malawi and northern Mozambique. *Kew Bulletin* 76: 63 – 70. https://doi.org/10.1007/s12225-021-09926-7

Datizua, C. (2020). *Moraea niassensis*. The IUCN Red List of Threatened Species 2020: e.T149256396A153685874. https://dx.doi.org/10.2305/IUCN.UK.2020-3.RLTS. T149256396A153685874.en.

Deacon, A.R. (2014). Environmental Impact Assessment for Sasol PSA and LPG Project: Terrestrial Fauna Impact Assessment. Specialist Report 10. SASOL Petroleum Mozambique Limitada & Sasol Petroleum Temane Lda. Available at: https://www.biofund.org.mz/biblioteca_virtual/environmental-impact-assessment-for-sasol-psa-and-lpg-project-terrestrial-fauna-impact-assessment/

Degreef, J. (2006). Revision of continental African *Tarenna* (Rubiaceae-Pavetteae). *Opera Botanica Belgica* 14: 1 – 150.

Deroin, T. & Lotter, M. (2013). A new *Uvaria* L. species (Annonaceae) from northern Mozambique. *Adansonia* 35: 227 – 234. https://doi.org/10.5252/a2013n2a4

Díaz Pelegrín, I., Luís, L.D., Mafambissa, M., Uetimane, A.E., Madeira, P.D., Chambal, E.M., Gubudo, F.S., Zibane, J.B. & Liberato, N.V.F. (2016). Parque Nacional do Arquipélago de Bazaruto (PNAB). Plano de Maneio 2016 – 2025 de uma Área de Conservação Marinha, Província de Inhambane, Moçambique. EIA & Services, Lda. (Projectos, Consultoria e Auditoria Ambiental). ANAC, MITADER, Maputo.

Dludlu, M., Dlamini, P., Sibandze, G., Vilane, V. & Dlamini, C. (2017). Distribution and conservation status of the Endangered pepperbark tree *Warburgia salutaris* (Canellaceae) in Swaziland. Oryx 51: 451 – 454. https://doi.org/10.1017/S0030605316000302

Donaldson, J.S. (2010a). *Encephalartos gratus*. The IUCN Red List of Threatened Species 2010: e.T41916A10594309. https://dx.doi.org/10.2305/IUCN.UK.2010-3.RLTS.T41916A10594309.en

Donaldson, J.S. (2010b). *Encephalartos pterogonus*. The IUCN Red List of Threatened Species 2010: e.T41897A10574244. https://dx.doi.org/10.2305/IUCN.UK.2010-3.RLTS.T41897A10574244.en

Donaldson, J.S. (2010c). *Encephalartos manikensis*. The IUCN Red List of Threatened Species 2010: e.T41919A10596129. https://dx.doi.org/10.2305/IUCN.UK.2010-3.RLTS.T41919A10596129.en

Donaldson, J.S. (2010d). *Encephalartos munchii*. The IUCN Red List of Threatened Species 2010: e.T41895A10573291. https://dx.doi.org/10.2305/IUCN.UK.2010-3.RLTS.T41895A10573291.en

Donaldson, J.S. (2010e). *Encephalartos ferox*. The IUCN Red List of Threatened Species 2010: e.T41943A10607271. https://dx.doi.org/10.2305/IUCN.UK.2010-3.RLTS.T41943A10607271.en

Donaldson, J.S. (2010f). Encephalartos lebomboensis. The IUCN Red List of Threatened Species 2010: e.T41907A10589133. https://dx.doi.org/10.2305/IUCN.UK.2010-3.RLTS.T41907A10589133.en

Dondeyne, S., Ndunguru, E., Rafael, P. & Bannerman, J. (2009). Artisanal mining in central Mozambique: policy and environmental issues of concern. *Resources Policy* 34: 45 – 50. https://doi.org/10.1016/j.resourpol.2008.11.001

Dorr, L.J. & Wurdack, K.J. (2018). A new disjunct species of *Eriolaena* (Malvaceae, Dombeyoideae) from Continental Africa. *PhytoKeys* 111: 11 – 16. https://doi.org/10.3897/phytokeys.111.29303

Downes, E. & Darbyshire, I. (2018). *Coleus namuliensis* and *Coleus caudatus* (Lamiaceae): a new species and a new combination in the Afromontane flora of Mozambique and Zimbabwe. *Blumea* 62: 168 – 173. https://doi.org/10.3767/blumea.2017.62.03.02

Downs, C. T. & Wirminghaus, J. O. (1997). The terrestrial vertebrates of the Bazaruto Archipelago, Mozambique: a biogeographical perspective. *Journal of Biogeography* 24: 591 – 602. https://doi.org/10.1111/j.1365-2699.1997.tb00071.x

Dowsett-Lemaire, F. (1988). The forest vegetation of Mt Mulanje (Malawi): a floristic and chorological study along an altitudinal gradient (650 – 1950 m). *Bulletin du Jardin Botanique National de Belgique 58*: 77 – 107. https://doi.org/10.2307/3668402

Dowsett-Lemaire, F. (2008). Survey of birds on Namuli Mountain (Mozambique), November 2007, with notes on vegetation and mammals. Misc. Report 60, prepared for the Darwin Initiative. Royal Botanic Gardens, Kew, BirdLife International, Instituto de Investigação Agrária de Moçambique and Mount Mulanje Conservation Trust. Available at: http://citeseerx.ist.psu.edu/viewdoc/download?doi=10.1.1.552.7403&rep=rep1&type=pdf

Dowsett-Lemaire, F. & Dowsett, R.J. (2009). The avifauna and forest vegetation of Mt. Mabu, northern Mozambique, with notes on mammals. Available at: https://biofund.org.mz/wp-content/uploads/2019/01/1548769382-F0876.Dowsett-Lemaire_Mabu%20report,%20Oct%202009.doc.pdf

du Randt, F. (2018) The Sand Forest of Maputaland. South African National Biodiversity Institute. Pretoria, South Africa.

Duarte M.C., Bandeira S. & Romeiras M. (2014). *Thalassodendron leptocaule* – a new species of seagrass from rocky habitats. pp. 175 – 180, In: Bandiera, S. & Paula, J. (eds.). The Maputo Bay Ecosystem. WIOMSA, Zanzibar Town.

Dudley, N. (ed.) (2013). Guidelines for applying protected area management categories including IUCN WCPA best practice guidance on recognising protected areas and assigning management categories and governance types. IUCN, Gland, Switzerland. Available at: https://portals.iucn.org/library/node/30018

Dutton, T.P. (1990). Report to the Honourable Minister of Agriculture on a conservation master plan for sustainable development of the Bazaruto Archipelago. People's Republic of Mozambique. Oceanographic Research Institute, Durban.

Dutton, P. & Drummond, B. (2008). Terrestrial habitats and vegetation. In: Everett, B.I., van der Elst, R.P. & Schleyer, M.H. (eds.), A Natural History of the Bazaruto Archipelago, Mozambique. Oceanographic Research Institute, Special publication No. 8: 37 – 40, Durban. Available at: https://biofund.org.mz/wp-content/uploads/2015/03/Je02-275.pdf

EOH – Coastal and Environmental Services (CES) (2015a). Nhangonzo Coastal Stream Critical Habitat Biodiversity Assessment: Integrated Summary Report. Authors: Avis, T., Martin, T., Massingue, A. & Buque, L. Report Number: 1521646-13552-26. Volume 3, Report 1 of Golder, 2015c.

EOH – Coastal and Environmental Services (CES) (2015b). Nhangonzo Coastal Stream Critical Habitat Biodiversity Assessment: Vegetation and Floristic Baseline Survey. Avis, T., Martin, T., Massingue, A. & Buque, L. Report Number: 1521646-13550-24. Volume 3, Report 2 of Golder, 2015c.

ERDAS (2018). ERDAS Imagine 2018. Hexagon Geospatial, Peachtree Corners Circle Norcross.

ESRI (2019). ArcGIS Pro: Release 2.8. Redlands, CA: Environmental Systems Research Institute.

Everett, B.I., van der Elst, R.P. & Schleyer, M.H. (eds.) (2008). A Natural History of the Bazaruto Archipelago, Mozambique. Oceanographic Research Institute, Special publication No. 8, Durban. Available at: https://biofund.org.mz/wp-content/uploads/2015/03/Je02-275.pdf

Exell, M.A. (1937). Leguminosae From Mozambique, Collected by Gomes e Sousa. *Boletim Da Sociedade Broteriana* 12: 6 – 92.

Fernandes, A., de Sousa, C., Mafalacusser, J., Soares, M. & Alves, T. (2020). Relatório preliminar da Instalação e 1a Medição das Parcelas de Amostragem Permanentes: GB01 e GB02. Fundo Nacional de Desenvolvimento Sustentável. Ministério da Agricultura e Desenvolvimento Rural.

Filimão, E., Mansur, E. & Namanha, L. (1999). Tchuma Tchato: an evolving experience of community-based natural resource management in Mozambique. pp. 145 – 152 in: Proceedings of the International Workshop on Community Forestry in Africa. Participatory forest management: a strategy for sustainable forest management in Africa. 26 – 30 April 1999, Banjul, The Gambia. FAO.

Forbes, K. & Broadhead, J. (2013). Forests and landslides: the role of trees and forests in the prevention of landslides and rehabilitation of landslide-affected areas in Asia. Food and Agriculture Organization of the United Nations. Available at: http://www.fao.org/3/ba0126e/ba0126e.pdf.

Fourqurean, J.W., Duarte, C.M., Kennedy, H., Marba, N., Holmers, M., Mateo, M.A., Apostoloaki, E.T., Kendrick, G.A., Krause-Jensen, D., McGlathery, K.J. & Serrano, O. (2012). Seagrass ecosystems as a globally significant carbon stock. *Nature Goescience* 5: 505 – 509. https://doi.org/10.1038/ngeo1477

Friends of Vamizi (2020). Friends of Vamizi. Conservation and Community. Available at: http://www.vamizi.com/

Friis, I. & Holt, S. (2017). *Salsola* sp. A of Flora Zambesiaca from the coast of Mozambique is *Caroxylon littoralis* (Amaranthaceae subfam. Salsoloideae), hitherto only known from Madagascar. *Webbia* 72: 63 – 69. https://doi.org/10.1080/00837792.2016.1258788

Gaston, K.J. & Fuller, R.A. (2009). The sizes of species' geographic ranges. *Journal of Applied Ecology* 46: 1 – 9. https://doi.org/10.1111/j.1365-2664.2008.01596.x

GBIF.org (2021a). GBIF Occurrence Download: Mount Massangulo. https://doi.org/10.15468/dl.mdnyzj

GBIF.org (2021b). GBIF Occurrence Download. Mount Morrumbala. https://doi.org/10.15468/dl.ur2ssn

GBIF.org (2021c). GBIF Occurrence Download: Mount Muruwere. https://doi.org/10.15468/dl.z38bhw

Ghiurghi, A., Dondeyne, S. & Bannerman, J.H. (2010). Chimanimani National Reserve Management Plan (3 volumes). Report prepared by AgriConsulting for Ministry of Tourism, Mozambique. Available at: https://www.biofund.org.mz/wp-content/uploads/2019/01/1548244323-CHIMANIMANI%20MANAGEMENT%20PLAN%20VOLUME%201%20-%20Jan%208%202010_AF.pdf

Goldblatt, P. (1993). Iridaceae (Part 4). In: Pope, G.V. (ed.) *Flora Zambesiaca,* Vol. 12(4). Royal Botanic Gardens, Kew.

Goldblatt, P., Manning, J.C., Von Blittersdorff, R. & Weber, O. (2014). New species of *Gladiolus* L. and *Moraea* Mill. (Iridaceae) from Tanzania and Mozambique. *Kew Bulletin* 69: 1 – 8. https://doi.org/10.1007/S12225-014-9496-Y

Golder Associates (2014). PSA Development and LPG Project, Final Impact Assessment Report. Sasol Petroleum Mozambique Lda and Sasol Petroleum Temane Lda.

Gomes e Sousa, A. (1968). Reserva Florestal de Licuati. Instituto de Investigação Agrária de Moçambique Comunicações. Vol 18. IIAM, Maputo.

Google Earth (2021). Google Earth Pro. Available at: https://www.google.com/earth/

Gorelick, N., Hancher, M., Dixon, M., Ilyushchenko, S., Thau, D. & Moore, R. (2017). Google Earth Engine: Planetary-scale geospatial analysis for everyone. Remote Sensing of Environment 202: 18 – 27. https://doi.org/10.1016/j.rse.2017.06.031

Governo do Distrito de Inhassoro (2011). Plano Estratégico de Desenvolvimento Distrital – Pedd (2011 – 2015). Inhassoro.

Goyder, D.J., Gilbert, M.C. & Venter, H.J.T. (2020). Apocynaceae (Part 2). In: García, M.A. (ed.) *Flora Zambesiaca*, Vol. 7(3). Royal Botanic Gardens, Kew.

Guyton, J.A., Pansu, J., Hutchinson, M.C., Kartzinel, T.R., Potter, A.B., Coverdale, T.C., Daskin, J.H., da Conceição, A.G., Peel, M.J.S., Stalmans, M.E. & Pringle, R.M. (2020). Trophic rewilding revives biotic resistance to shrub invasion. *Nature Ecology and Evolution* 4: 712 – 724. https://doi.org/10.1038/s41559-019-1068-y

Hancox, J.P., Brandt, D. & Edwards, H. (2002). Sequence stratigraphic analysis of the Early Cretaceous Maconde Formation (Rovuma basin), northern Mozambique. *Journal of African Earth Sciences* 34: 291 – 297. https://doi.org/10.1016/S0899-5362(02)00028-3

Hansen, M.C., Potapov, P.V., Moore, R., Hancher, M, Turubanova, S.A., Tyukavina, A., Thau, D., Stehman, S.V., Goetz, S.J., Loveland, T.R., Kommareddy, A. Egorov, L. Chini, C.O. Justice & Townshend J.R.G. (2013). High-resolution global maps of 21st-century forest cover change. *Science* 342: 850 – 853. https://doi.org/10.1126/science.1244693

Harari, N. (2005). Literature Review on the Quirimbas National Park, Northern Mozambique. Report prepared for the Centre for Development and Environment, Department of Geography, University of Bern. https://doi.org/10.7892/boris.71799

Hardaker, T. & Sinclair, I. (2001). Sasol Birding Map of Southern Africa. Struik.

Harris, T., Darbyshire, I. & Polhill, R. (2011). New species and range extensions from Mt Namuli, Mt Mabu and Mt Chiperone in northern Mozambique. *Kew Bulletin* 66: 241 – 251. https://doi.org/10.1007/s12225-011-9277-9

Harrison, T. & Finnegan, K. (2021). Lighthouse Explorer Database: Ponta Zavora Light. *Lighthouse Digest*. Available at: http://www.lighthousedigest.com/digest/database/unique lighthouse.cfm?value=5256

Hawthorne, W. (1998). *Khaya anthotheca*. The IUCN Red List of Threatened Species 1998: e.T32235A9690061. https://dx.doi.org/10.2305/IUCN.UK.1998.RLTS.T32235A9690061.en.

Hill, B.J., Blaber, S.J.M. & Boltt, R.E. (1975). The Limnology of Lagoa Poelela. *Transactions of the Royal Society of South Africa*, 41: 263 – 271. https://doi.org/10.1080/00359197509519442

Hills, R. (2019). *Afzelia quanzensis*. The IUCN Red List of Threatened Species 2019: e.T60757666A60757681. https://dx.doi.org/10.2305/IUCN.UK.2019-3.RLTS.T60757666A60757681.en

Hobday, D.K. (1977). Late Quaternary sedimentary history of Inhaca Island, Mozambique. *Transactions of the Geological Society of South Africa* 80: 183 – 191.

Howard G., Kamau P., Kindeketa W., Luke W.R.Q., Lyaruu H.V.M., Malombe I., Maunder M., Mwachala G., Njau E.-F., Peres Q., Schatz G.E., Siro Masinde P., Ssegawa P., Wabuyele E. & Wilkins V.L. (2020). *Celosia patentiloba*. The IUCN Red List of Threatened Species 2020: e.T157997A756253.

Hyde, M.A., Wursten, B.T., Ballings, P. & Coates Palgrave, M. (2021). Flora of Mozambique. Available at: https://www.mozambiqueflora.com

Impacto Lda. (2012a). Perfil Ambiental e Mapeamento do Uso Actual da Terra Nos Distritos da Zona Costeira de Moçambique: Distrito de Inharrime Província (Versão Preliminar). Available at: http://www.biofund.org.mz/wp-content/uploads/2019/01/1547461631-Perfil_Inharrime.pdf

Impacto Lda. (2012b). Perfil Ambiental e Mapeamento do Uso Actual da Terra nos Distritos da Zona Costeira De Moçambique: Distrito de Mandlakazi. Available at: https://www.biofund.org.mz/biblioteca_virtual/perfil-ambiental-e-mapeamento-do-uso-actual-da-terra-nos-distritos-da-zona-costeira-de-mocambique-distrito-de-mandlakazi/

Impacto Lda. (2012c). Perfil Ambiental e Mapeamento do Uso Actual da Terra nos Distritos da Zona Costeira de Moçambique: Distrito de Bilene. Available at: https://www.biofund.org.mz/biblioteca_virtual/perfil-ambiental-e-mapeamento-do-uso-actual-da-terra-nos-distritos-da-zona-costeira-de-mocambique-distrito-de-bilene/

Impacto Lda. (2012d). Perfil Ambiental e Mapeamento do Uso Actual da Terra nos Distritos da Zona Costeira de Moçambique (Versão Preliminar): Distrito de Manhiça Província de Maputo. Available at: https://www.biofund.org.mz/biblioteca_virtual/perfil-ambiental-e-mapeamento-do-uso-actual-da-terra-nos-distritos-da-zona-costeira-de-mocambique-distrito-de-manhica/

Impacto Lda. (2016). Plano de Maneio da Reserva Nacional de Pomene. Available at: http://www.biofund.org.mz/biblioteca_virtual/plano-de-maneio-da-reserva-nacional-de-pomene-volume-i-plano-de-maneio/

Impacto Lda. (2018). Sasol Petroleum Mozambique. Categorização da Área de Nhangonzo, Inhambane, Moçambique. Referência do Documento: MSSP1701-IMP180407 – Rev 01.

Inguaggiato, C., Navarra, C., Vailati, A. (2009) The Role of Rural Producers' Organizations within Development Processes: a Case Study on Morrumbala District. *Dynamics of Poverty and Patterns of Economic Accumulation in Mozambique Conference.* Maputo, 22 – 23 of April, 2009, IESE.

Instituto Nacional de Estatistica Moçambique (2021). Instituto Nacional de Estatistica - Moçambique. Available at: http://www.ine.gov.mz/

Instituto Nacional de Geológia (1987). Carta Geológica, scale 1: 1 million. Instituto Nacional de Geológia, Maputo.

Israel, P. (2006). Kummwangalela Guebuza. The Mozambican general elections of 2004 in Muidumbe and the roots of the loyalty of Makonde People to Frelimo. *Lusotopie* 13: 103 – 125. https://doi.org/10.1163/176830806778698150

IUCN (2012). IUCN Red List Categories and Criteria. Version 3.1, 2nd edition. IUCN Species Survival Commission, Gland, Switzerland. Available at: http://www.iucnredlist. org/technical-documents/categories-and-criteria/2001-categories-criteria

IUCN (2016). A global standard for the identification of Key Biodiversity Areas, Version 1.0. First edition. IUCN, Gland, Switzerland. Available at: https://portals.iucn.org/ union/sites/union/files/doc/a_global_standard_for_the_ identification_of_key_biodiversity_areas_final_web.pdf

IUCN SSC Amphibian Specialist Group (2019). *Nothophryne inagoensis.* The IUCN Red List of Threatened Species 2019: e.T149286395A149288435. https://dx.doi.org/10.2305/IUCN. UK.2019-3.RLTS.T149286395A149288435.en.

IUCN (2021). The IUCN Red List of Threatened Species. Version 2021-2. Available at: https://www.iucnredlist.org

Izidine, S.A. (2003). Licuáti forest reserve, Mozambique: Flora, utilization and conservation. MSc thesis. University of Pretoria. Available at: http://hdl.handle.net/2263/56038

Izidine, S. & Bandeira, S.O. (2002). Mozambique. pp. 43 – 60 in: Golding, J.S. (ed.) Southern African Plant Red Data Lists. Southern African Botanical Diversity Network Report No. 14. SABONET, Pretoria.

Izidine, S. & Cándido, A. (2004). Botanical Diversity & Endemism Areas in Mozambique. Proceedings of the Mozambique IPA Workshop. Maputo, Mozambique.

Izidine, S. & Siebert, S., Wyk, A.E. & Zobolo, A.M. (2009). Threats to Ronga custodianship of a sacred grove In Southern Mozambique. *Indilinga: African Journal of Indigenous Knowledge Systems.* 7: 182 – 197. https://doi.org/10.4314/ indilinga.v7i2.26435

Jacobsen, N.H.G., Pietersen, E.W. & Pietersen, D.W. (2010). A preliminary herpetological survey of the Vilanculos Coastal Wildlife Sanctuary on the San Sebastian Peninsula, Vilankulo, Mozambique. *Herpetology Notes* 3: 181 – 193.

Jimu, L. (2011). Threats and conservation strategies for the African cherry (*Prunus africana*) in its natural range – a review. *Journal of Ecology and The Natural Environment*, 3: 118 – 130. https://doi.org/10.5897/jene.9000002

João, F.E. (2011). Análise da influência da prática de agricultura na regeneração e manutenção da espécie *Raphia australis* na Reserva Botânica de Bobole em Marracuene, província de Maputo. Universidade Eduardo Mondlane.

Joaquim, G.B. & Caravela, M.I. (2019). Caracterização de habitats na concessão de Taratibu, Parque Nacional das Quirimbas-PNQ, distrito de Ancuabe. Unpubl. report. Departamento de Botânica, Universidade Lúrio, Pemba.

Jones, S.E., Jamie, G.A., Sumbane, E., & Jocque, M. (2020). The avifauna, conservation and biogeography of the Njesi Highlands in northern Mozambique, with a review of the country's Afromontane birdlife. *Ostrich*: 45 – 56. https://doi. org/10.2989/00306525.2019.16757

Kabanza, A.K., Dondeyne, S., Kimaro, D.N., Kafiriti, E., Poesen, J. & Deckers, J.A. (2013). Effectiveness of soil conservation measures in two contrasting landscape units of South Eastern Tanzania. *Zeitschrift für Geomorphologie* 57: 269 – 288. https://doi.org/10.1127/0372-8854/2013/0102

Kassam, A.H., Van Velthuizen, H. T., Higgins, G.M., Christoforides, A., Voortman, R.L. & Spiers, B. (1981). Assessment of land resources for rainfed crops production in Mozambique. Climate data bank and length of growing period analysis. Project Moz/75/011. FAO

Kenmare Resources (2018). Kenmare Resources plc Annual Report and Accounts 2018. Available at: https://www. kenmareresources.com/application/files/8215/5420/0299/ Kenmare_Resources_plc_Annual_Report__Accounts_2018.pdf

Key Biodiversity Areas Partnership (2020). *Key Biodiversity Areas factsheet: Pomene.* World Database of Key Biodiversity Areas. Available at: http://www.keybiodiversityareas.org

Knapp, S. (2021). *Solanum litoraneum.* The IUCN Red List of Threatened Species 2021: e.T101527720A101527747. https://dx.doi.org/10.2305/IUCN.UK.2021-3.RLTS. T101527720A101527747.en.

Kill, J. (2013). Carbon Discredited: Why the EU Should Steer Clear of Forest Carbon Offsets. Available at: https://www.fern.org/ fileadmin/uploads/fern/Documents/Nhambita_internet.pdf

Lambrechts, A. (2003). Biodiversity Management Plan for Vilanculos Coastal Wildlife Sanctuary. Vol. 1, Condensed Plan, Mozambique. Vilanculos Coastal Wildlife Sanctuary (Pty) Ltd. & Global Environment Facility.

Langa, C., Datizua, C., Matimele, H.A., Rokni, S., Alves, M.T., Burrows, J.E., Chelene, I., Darbyshire, I., De Sousa, C., Fijamo, V., Massingue, A.O., Massunde, J., Mucaleque, P.A., Osborne, J. & Sitoe, P. (2019a). *Baphia ovata*. The IUCN Red List of Threatened Species 2019: e.T120960184A120980303. https://dx.doi.org/10.2305/IUCN.UK.2019-3.RLTS. T120960184A120980303.en.

Langa, C., Datizua, C., Rokni, S., Alves, M.T., Burrows, J.E., Chelene, I., Darbyshire, I., De Sousa, C., Fijamo, V., Massingue, A.O., Massunde, J., Matimele, H.A., Mucaleque, P.A., Osborne, J. & Sitoe, P. (2019b). *Tephrosia forbesii* subsp. *inhacensis.* The IUCN Red List of Threatened Species 2019: e.T120979692A120980463. https://dx.doi.org/10.2305/IUCN. UK.2019-3.RLTS.T120979692A120980463.en

Lawrence, P. & Cheek, M. (2019). *Cola discoglypremnophylla.* The IUCN Red List of Threatened Species 2019: e.T111391854A111449262. https://dx.doi.org/10.2305/ IUCN.UK.2019-1.RLTS.T111391854A111449262.en

Leão, T.C.C., Fonseca, C.R., Peres, C.A. & Tabarelli, M. (2014). Predicting extinction risk of Brazilian Atlantic Forest angiosperms. *Conservation Biology* 28: 1349 – 1359. https:// doi.org/10.1111/cobi.12286

Legado (2021). Legado: Namuli. https://www.legadoinitiative. org/legado-namuli/

Loffler, L. & Loffler, P. (2005). Swaziland Tree Atlas — including selected shrubs and climbers. Southern African Botanical Diversity Network Report No. 35. SABONET, Pretoria.

Lötter, M. & von Staden, L. (2018). *Adenium swazicum* Stapf. National Assessment: Red List of South African Plants version 2020.1. Available at: http://redlist.sanbi.org/species. php?species=997-7

Lötter, M., Burrows, J., McCleland, W., Stalmans, M., Schmidt, E., Soares, M., Grantham, H., Jones, K., Duarte, E., Matimele, H. & Costa, H.M. (In prep.). Historical vegetation map and red list of ecosystems assessment for Mozambique – Version 1.0 – Final report. USAID / SPEED+. Maputo.

Louro, C.M.M., Litulo, C., Pereira, M.A.M. & Pereira, T.I.F.C. (2017). Investigação e Monitoria de Espécies e Ecossistemas nas Áreas de Conservação Marinhas em Moçambique: Reserva Nacional do Pomene. https://doi.org/10.13140/ RG.2.2.33152.56326

Lubke, R.A., Dold, A.D., Brink, E., Avis, A.M. & Wieringa, J.J. (2018). A new species of tree, *Icuria dunensis* (Icurri), of undescribed coastal forests in north-eastern Mozambique. *South African Journal of Botany* 115: 292 – 293. https://doi. org/10.1016/j.sajb.2018.02.063

Lucrezi, S., Milanese M., Markantonatou, V., Cerrano, C., Sara, A., Palma, M. & Saayman, M. (2017). Scuba diving tourism systems and sustainability: Perceptions by the scuba diving industry in two Marine Protected Areas. *Tourism Management* 59: 385 – 403. https://doi.org/10.1016/j. tourman.2016.09.004

Luke, Q., Bangirinama, F., Beentje, H.J., Darbyshire, I., Gereau, R., Kabuye, C., Kalema, J., Kelbessa, E., Kindeketa, W., Minani, V., Mwangoka, M. & Ndangalasi, H. (2015a). *Justicia attenuifolia.* The IUCN Red List of Threatened Species 2015: e.T48153888A48154789. https://dx.doi.org/10.2305/IUCN. UK.2015-2.RLTS.T48153888A48154789.en

Luke, Q., Bangirinama, F., Beentje, H.J., Darbyshire, I., Gereau, R., Kabuye, C., Kalema, J., Kelbessa, E., Minani, V., Mwangoka, M. & Ndangalasi, H. (2015b). *Barleria laceratiflora*. The IUCN Red List of Threatened Species 2015: e.T48153936A48154273. https://dx.doi.org/10.2305/IUCN.UK.2015-2.RLTS.T48153936A48154273.en

Luwire Wildlife Conservancy (2019). Saving the Luwire Wildlife Conservancy. Available at: https://luwire.org/wp-content/uploads/2019/03/Saving-the-Luwire-Conservancy.pdf

Macandza, V., Mamugy, F., Manjate, A.M. & Nacamo, E. (2015). Estudo das Condições Ecológicas e Socioeconómicas da Reserva Nacional de Pomene. Available at: http://www.biofund.org.mz/wp-content/uploads/2018/11/1543394499-F0887.ESTUDO DAS CONDIÇÕES ECOLÓGICAS E SOCIOECONÓMICAS DA RESERVA NACIONAL DE POMENE.pdf

Macauhub (2014). Mozambican government plans to build hydroelectric plant on Lúrio River. Available at: https://macauhub.com.mo/2014/04/01/mozambican-government-plans-to-build-hydroelectric-plant-on-lurio-river/

Macey, P.H., Thomas, R.J., Grantham, G.H., Ingram, B.A., Jacobs, J., Armstrong, R.A., Roberts, M.P., Bingen, B., Hollick, L., de Kock, G.S., Viola, G., Bauer, W., Gonzales, E., Bjerkgård, T., Henderson, I.H.C., Sandstad, J.S., Cronwright, M.S., Harley, S., Solli, A., Nordgulen, Ø., Motuza G., Daudi, E. & Manhiça, V. (2010). Mesoproterozoic geology of the Nampula Block, northern Mozambique: tracing fragments of Mesoproterozoic crust in the heart of Gondwana. *Precambrian Research* 182: 124 – 148. https://doi.org/10.1016/j.precamres.2010.07.005

Macnae, W. & Kalk, M. (1962). The Fauna and Flora of Sand Flats at Inhaca Island, Moçambique. *Journal of Animal Ecology* 31: 93 – 128. https://doi.org/10.2307/2334

Manhica, A.D.S.T. (2012). The geology of the Mozambique Belt and the Zimbabwe Craton around Manica, western Mozambique. Doctoral Thesis, University of Pretoria. Available at: https://repository.up.ac.za/handle/2263/28883

Manhice, A. (2010). Planta rara em risco de extinção em Moçambique. *Notícias*. Available at: https://arseniomanhice.wordpress.com/2014/07/05/planta-rara-em-risco-de-extincao-em-mocambique/

Manuel, I.R.V. (2007). Reduction and management of geo-hazards in Mozambique. *International Journal for Disaster Management & Risk Reduction* 1: 18 – 23. Available at: https://d1wqtxts1xzle7.cloudfront.net/44546832/International_Journal_of_Disaster_Management_and_Risk_Reduction_Vol.1_No.1_2007_1.pdf?1460141273=&response-content-disposition=inline%3B+filename%3DInternational_Journal_for_Disaster_Manag.pdf&Expires=1601551

Manzitto-Tripp, E.A., Darbyshire, I., Daniel, T.F., Kiel, C.A. & McDade, L.A. (2021). Revised classification of Acanthaceae and worldwide dichotomous keys. *Taxon* 71: 103 – 153. https://doi.org/10.1002/tax.12600

Martínez Richart, A.I., Darbyshire, I. & Rulkens, T. (2019). *Aloe argentifolia*. The IUCN Red List of Threatened Species 2019: e.T142844664A142844686. https://dx.doi.org/10.2305/IUCN.UK.2019-3.RLTS.T142844664A142844686.en

Massingue, A.O. (2019). Ecological Assessment and Biogeography of Coastal Vegetation and Flora in Southern Mozambique. Doctor of Philosophy Thesis. Department of Botany, Faculty of Science, Nelson Mandela University.

Massingue, A.O., Datizua, C., Alves, M.T., Burrows, J.E., Chelene, I., Darbyshire, I., De Sousa, C., Fijamo, V., Langa, C., Massunde, J., Matimele, H.A., Mucaleque, P.A., Osborne, J., Rokni, S. & Sitoe, P. (2019). *Ozoroa gomesiana*. The IUCN Red List of Threatened Species 2019: e.T120942095A120980153. https://dx.doi.org/10.2305/IUCN.UK.2019-3.RLTS.T120942095A120980153.en

Massingue, A.O., Datizua, C. Langa, C. & Bruno, C. (2021). A Preliminary Botanical Survey to Provide a Base Knowledge for Biodiversity Conservation in the Vilanculos Coastal Wildlife Sanctuary, Mozambique. Unpubl. report. Royal

Botanic Gardens, Kew and Instituto de Investigação Agrária de Moçambique.

Matimele, H. (2016). An Assessment of the Distribution and Conservation Status of Endemic and Near Endemic Plant Species in Maputaland. University of Cape Town. Available at: https://open.uct.ac.za/handle/11427/20995

Matimele, H. (2021). Mozambique Endemic and Near-Endemic Red Listed Plant Species. Version 1.8. Herbarium LMA: Agricultural Research Institute of Mozambique. Occurrence dataset. https://doi.org/10.15468/8enzjm

Matimele, H. & Timberlake, J. (2020). Maputaland World Heritage Application: Terrestrial Plants and Vegetation. Unpubl. report.

Matimele, H.A., Raimondo, D., Bandeira, S., Burrows, J.E., Darbyshire, I., Massingue, A.O. & Timberlake, J. (2016a). *Emicocarpus fissifolius*. The IUCN Red List of Threatened Species 2016: e.T85955108A85955412. https://dx.doi.org/10.2305/IUCN.UK.2016-3.RLTS.T85955108A85955412.en.

Matimele, H.A., Massingue, A.O., Raimondo, D., Bandeira, S., Burrows, J.E., Darbyshire, I. & Timberlake, J. (2016b). *Raphia australis*. The IUCN Red List of Threatened Species 2016: e.T30359A85955288. https://dx.doi.org/10.2305/IUCN.UK.2016-3.RLTS.T30359A85955288.en.

Matimele, H.A., Alves, M.T., Baptista, O.J., Bezeng, S., Darbyshire, I., Datizua, C., De Sousa, C., Langa, C., Massingue, A.O., Mtshali, H., Mucaleque, P.A. Odorico, D., Osborne, J., Raimondo, D., Rokni, S., Sitoe, P., Timberlake, J., Viegas, A. & Vilanculos, A. (2018a). *Euphorbia baylissii*. The IUCN Red List of Threatened Species 2018: e.T120955807A120980243.

Matimele, H.A., Raimondo, D., Bandeira, S., Burrows, J.E., Darbyshire, I., Massingue, A.O. & Timberlake, J. (2018b). *Memecylon incisilobum* (amended version of 2016 assessment). The IUCN Red List of Threatened Species 2018: e.T85955255A125331050. https://dx.doi.org/10.2305/IUCN.UK.2018-1.RLTS.T85955255A125331050.en.

Mbalaka, J.Y. (2016). Exploring the Migration Experiences of Muslim Yao Women in KwaZulu-Natal, 1994 – 2015. University of KwaZulu-Natal, Durban. Available at: https://researchspace.ukzn.ac.za/handle/10413/16351

McCleland, W. & Massingue, A. (2018). New populations and a conservation assessment of *Ecbolium hastatum* Vollesen. *Bothalia* 48: 1 – 3. https://doi.org/10.4102/abc.v48i1.2282

McCoy, T.A. & Baptista, O.J. (2016). A new species of cremnophytic *Aloe* from Mozambique. *Cactus and Succulent Journal* 88: 172 – 176. https://doi.org/10.2985/015.088.0402

McCoy, T.A., Rulkens, A.J.H. & Baptista, O.J. (2014). An extraordinary new species of *Aloe* from the Republic of Mozambique. *Cactus and Succulent Journal* 89: 214 – 218. https://doi.org/10.2985/015.089.0502

McCoy, T.A., Rulkens, A.J. & Baptista, O.J. (2017). A new species of *Aloe* from the Lúrio waterfalls in Mozambique. *Cactus and Succulent Journal* 89: 214 – 218. https://doi.org/10.2985/015.089.0502

Melezhik, V.A., Kuznetsov, A.B., Fallick, A.F., Smith, R.A., Gorokhov, I.M., Jamal, D. & Catuane, F. (2006). Depositional environments and an apparent age for the Geci meta-limestones: constraints on the geological history of northern Mozambique. *Precambrian Research* 148: 19 – 31. https://doi.org/10.1016/j.precamres.2006.03.003

MAE (Ministério da Administração Estatal) (2005a). Perfil do Distrito de Morrumbala: Província da Zambézia. Available at: www.portaldogoverno.gov.mz

MAE (Ministério da Administração Estatal) (2005b). Perfil do Distrito de Inhassoro, Província de Inhambane. Available at: www.portaldogoverno.gov.mz

MAE (Ministério da Administração Estatal) (2005c). Perfil do Distrito de Vilanculos, Província de Inhambane. Available at: www.portaldogoverno.gov.mz

MAE (Ministério da Administração Estatal) (2005d). Perfil do Distrito de Massinga, Província de Inhambane. Available at: www.portaldogoverno.gov.mz

MAE (Ministério da Administração Estatal) (2005e). Perfil do Distrito de Inharrime, Província de Inhambane. Available at: www.portaldogoverno.gov.mz

MAE (Ministério da Administração Estatal) (2005f). Perfil Do Distrito Do Bilene Macia Província De Gaza. Available at: www.portaldogoverno.gov.mz

MICOA (Ministério para a Coordenação da Acção Ambiental) (2012a). Perfil Ambiental e Mapeamento do uso Actual da Terra nos Distritos da Zona Costeira de Moçambique: Distrito de Inhassoro. Direcção Nacional de Gestão Ambiental, Maputo.

MICOA (Ministério para a Coordenação da Acção Ambiental) (2012b). Perfil Ambiental e Mapeamento do Uso Actual da Terra nos Distritos da Zona Costeira de Moçambique: Distrito de Vilankulos. Versão Preliminar.

MICOA (Ministério para a Coordenação da Acção Ambiental) (2012c). Perfil Ambiental e Mapeamento do Uso Actual da Terra nos Distritos da Zona Costeira de Moçambique: Distrito de Massinga. Versão Preliminar.

MITADER (Ministério da Terra, Ambiente d Desenvolvimento Rural) (2015). Estratégia e Plano de Acção Para a Conservação da Diversidade Biológica em Moçambique. MITADER, Maputo.

Mizuno, M., Wang, C., Gonda, Y., Marui, H., Nishikawa, D., Hirata, I., Sango, D. & Morita, Y. (2018). Landslide Survey and Scal Estimate by DInSAR, GNSS, and Airborne Laser Before Landslide Failure – Landslide Survey of Mt. Inago. Available at: http://www.interpraevent.at/palm-cms/upload_files/Publikationen/Tagungsbeitraege/2018_EA_176.pdf

Moat, J. & Bachman, S. (2020). rCAT: conservation assessment tools. R package version 0.1.6. Available at: https://CRAN.R-project.org/package=rCAT

Mogg, A.O.D. (1967). Comments on the flora of Inhaca Island, Moçambique. *South African Journal of Science* 63: 440.

Montfort, F. (2019). Land use and land cover map of Ribaue Mountains (Mount Ribaue and Mount M'paluwe). Nitidae. Available at: https://www.nitidae.org/files/0afa4c85/land_use_and_land_cover_map_of_ribaue_mountains_mount_ribaue_and_mount_m_paluwe_.pdf

Montfort, F. (2020). Historical and future deforestation analysis of Ribaue Mountains (Mount Ribaue and Mount M'paluwe). Nitidae. Available at: https://www.nitidae.org/files/4d43bf48/historical_and_future_deforestation_analysis_of_ribaue_mountains_mount_ribaue_and_mount_m_paluwe_.pdf

Mozambique News Agency (2016). Gorongosa National Park to expand. *AIM Reports.* Available at: http://www.poptel.org.uk/mozambique-news/newsletter/aim538.pdf

Muacanhia, T. (2004). Environmental changes on Inhaca Island, Mozambique: development versus degradation. In: Momade, F. Achimo, M. Haldorsen, S. (eds.), The Impact of Sea-level Change, Past, Present, Future. *Boletim Geológica* 43: 28 – 33.

Mucaleque, P.A. (2020a). Mozambique TIPAs Fieldwork Report: Goa and Sena Islands, Mozambique Island District, Nampula Province, September 2020. Unpubl. report. Instituto de Investigação Agrária de Moçambique (IIAM).

Mucaleque, P.A. (2020b). *Ammannia moggii*. The IUCN Red List of Threatened Species 2020: e.T149257990A153685939. https://dx.doi.org/10.2305/IUCN.UK.2020-3.RLTS.T149257990A153685939.en

Müller, T., Sitoe, A. & Mabunda, R. (2005). Assessment of the Forest Reserve Network in Mozambique. WWF Mozambique, Maputo. Available at: http://cgcmc.gov.mz/attachments/article/100/548946e10cf2ef344790ae27.pdf

Müller, T., Mapaura, A., Wursten, B., Chapano, C., Ballings, P. & Wild, R. (2012). Vegetation Survey of Mount Gorongosa. Occasional Publications in Biodiversity No. 23, Biodiversity Foundation for Africa, Bulawayo. Available at: https://www.gorongosa.org/sites/default/files/research/041-bfa_no.23_gorongosa_vegetation_survey.pdf

Mynard, P. & Rokni, S. (2019). *Cordia megiae*. The IUCN Red List of Threatened Species 2019: e.T141800272A141800288. https://dx.doi.org/10.2305/IUCN.UK.2019-2.RLTS.T141800272A141800288.en. Accessed 15 June 2021.

Myre, M. (1971). As pastagens da regiao do Maputo. Memorias: 3. IIAM, Maputo.

Nagy, B. & Watters, B. (2019). *Nothobranchius niassa*. The IUCN Red List of Threatened Species 2019: e.T131471671A131471686. https://dx.doi.org/10.2305/IUCN.UK.2019-3.RLTS.T131471671A131471686.en. Accessed 24 March 2021.

NASA Shuttle Radar Topography Mission (SRTM) (2013). Shuttle Radar Topography Mission (SRTM) Global. Distributed by OpenTopography. https://doi.org/10.5069/G9445JDF

Nhanombe Lodge (2021). Activities at Nhanombe. Available at: www.nhanombelodge.com.

Nitidæ (2021). Namuli Sky Island – Creation of a new protected area around Mount Namuli. Available at: https://www.nitidae.org/en/actions/namuli-creation-d-une-nouvelle-aire-protegee-autour-du-mont-namuli

Njagi, D. (2019). 'A crisis situation': Extinctions loom as forests are erased in Mozambique. Mongabay Series: Forest Trackers. Available at: https://news.mongabay.com/2019/12/a-crisis-situation-extinctions-loom-as-forests-are-erased-in-mozambique/

O'Connor, M. (2006). After the war, eco construction. Financial Times. Available at: https://www.ft.com/content/3511f9ee-0206-11db-a141-0000779e2340

O'Sullivan, R. J. & Davis, A. (2017). *Coffea salvatrix*. The IUCN Red List of Threatened Species 2017: e.T18290408A18539335. https://doi.org/https://dx.doi.org/10.2305/IUCN.UK.2017-3.RLTS.T18290408A18539335.en.

Osborne, J. & Matimele, H. (2018). Mozambique TIPAs Fieldwork Summary Report. Manica Highlands: Garuzo Forest, Tsetserra and Serra Choa, June 2018. Unpubl. report. Royal Botanic Gardens, Kew and Instituto de Investigação Agrária de Moçambique (IIAM).

Osborne, J. & Rokni, S. (2020). *Barleria torrei*. The IUCN Red List of Threatened Species 2020: e.T120940515A120980028. https://dx.doi.org/10.2305/IUCN.UK.2020-2.RLTS.T120940515A120980028.en

Osborne, J., Rokni, S., Matimele, H., Langa, C., Zandamela, J., Macanzi, B., Tembe, E., Machuama, B., Zakueu Munwane, C. & Cumbane, I. (2018a). Lebombo Mountains reconnaissance expedition, 8–15 March 2018. Unpubl. report. Royal Botanic Gardens, Kew and Instituto de Investigação Agrária de Moçambique.

Osborne, J., Matimele, H. & Timberlake, J. (2018b) Zambezia Province: Mount Lico and Pico Muli, May 2018. Unpubl. botanical report prepared for the project "Scientific expedition to Mt. Lico and adjacent mountains".

Osborne, J., Langa, C., Datizua, C. & Darbyshire, I. (2019a). Mozambique TIPAs Fieldwork Report: Inhambane Province – Panda, Mabote and Lagoa Poelela, Jan–Feb 2019. Unpubl. report. Royal Botanic Gardens, Kew and Instituto de Investigação Agrária de Moçambique (IIAM).

Osborne, J., Datizua, C., Banze, A., Mamba, A., Mucaleque, P., & Rachide., T. (2019b). Mozambique TIPAs Fieldwork Summary Report. Niassa Province: Lago District mountains and Njesi. Royal Botanic Gardens, Kew and Instituto de Investigação Agrária de Moçambique (IIAM). https://doi.org/10.13140/RG.2.2.30330.72648

Osborne, J., Darbyshire, I., Matimele, H.A., Alves, M.T., Chelene, I., Datizua, C., De Sousa, C., Langa, C., Massingue, A.O., Mucaleque, P.A., Odorico, D., Rokni, S., Rulkens, A.J.H., Timberlake, J. & Viegas, A. (2019c). *Aloe ribauensis*. The IUCN Red List of Threatened Species 2019: e.T110780332A110780364. https://dx.doi.org/10.2305/IUCN.UK.2019-1.RLTS.T110780332A110780364.en

Osborne, J., Banze, A., Mtshali, H., Mucaleque, P. A., Rokni, S. & Vilanculos, A. (2019d). *Euphorbia decliviticola*. The IUCN Red List of Threatened Species 2019: e.T120955505A120980228. https://dx.doi.org/10.2305/IUCN.UK.2019-1.RLTS.T120955505A120980228.en

Osborne, J., Matimele, H.A., Alves, M.T., Chelene, I., Darbyshire, I., Datizua, C., De Sousa, C., Langa, C., Massingue, A.O., Mucaleque, P.A., Odorico, D., Rokni, S., Rulkens, A.J.H., Timberlake, J. & Viegas, A. (2019e). *Streptocarpus myoporoides*. The IUCN Red List of Threatened Species 2019: e.T120956335A120980268. https://dx.doi.org/10.2305/IUCN.UK.2019-1.RLTS.T120956335A120980268.en

Osborne, J., Matimele, H.A., Alves, M.T., Chelene, I., Darbyshire, I., Datizua, C., De Sousa, C., Langa, C., Massingue, A.O., Mucaleque, P.A., Odorico, D., Rokni, S., Rulkens, A.J.H., Timberlake, J. & Viegas, A. (2019f) *Euphorbia grandicornis* subsp. *sejuncta*. The IUCN Red List of Threatened Species 2019: e.T120955804A120980238. https://dx.doi.org/10.2305/IUCN.UK.2019-1.RLTS.T120955804A120980238.en

Osborne, J., Rulkens, T., Alves, M.T., Burrows, J.E., Chelene, I., Darbyshire, I., Datizua, C., De Sousa, C., Fijamo, V., Langa, C., Massingue, A.O., Massunde, J., Matimele, H.A., Mucaleque, P.A., Rokni, S. & Sitoe, P. (2019g). *Aloe decurva*. The IUCN Red List of Threatened Species 2019: e.T110713829A110713841. https://dx.doi.org/10.2305/IUCN.UK.2019-3.RLTS.T110713829A110713841.en

Osborne, J., Rokni, S., Alves, M.T., Burrows, J.E., Chelene, I., Darbyshire, I., Datizua, C., De Sousa, C., Fijamo, V., Langa, C., Massingue, A.O., Massunde, J., Matimele, H.A., Mucaleque, P.A. & Sitoe, P. (2019h). *Raphionacme pulchella*. The IUCN Red List of Threatened Species 2019: e.T136528489A136538103. https://dx.doi.org/10.2305/IUCN.UK.2019-3.RLTS.T136528489A136538103.en

Osborne, J., Rulkens, T., Alves, M.T., Burrows, J.E., Chelene, I., Darbyshire, I., Datizua, C., De Sousa, C., Fijamo, V., Langa, C., Massingue, A.O., Massunde, J., Matimele, H.A., Mucaleque, P.A., Rokni, S. & Sitoe, P. (2019i). *Aloe cannellii*. The IUCN Red List of Threatened Species 2019: e.T110697369A110697395. https://dx.doi.org/10.2305/IUCN.UK.2019-3.RLTS.T110697369A110697395.en

Osborne, J., Datizua, C., Mucaleque, P., Fischer, E. (2022). *Hartliella txitongensis* (Linderniaceae), a new species from Mozambique. *Kew Bulletin*. https://doi.org/10.1007/s12225-022-10034-3

Pais, A. de J.R. (2011). Estudo da ocorrência e estado de conservação da *Raphia australis* Oberm. Strey na Reserva Botânica de Bobole. Universidade Eduardo Mondlane.

Parker, V. (2001). Mozambique. pp. 627 – 638 in: Fishpool, L.D.C. & Evans, M.I. (eds.), Important Bird Areas in Africa and associated islands : priority sites for conservation. BirdLife Conservation Series, No. 11, Pisces Publications and BirdLife International, Newbury and Cambridge. Available at: https://www.biofund.org.mz/wp-content/uploads/2017/03/BirdLife-Intl-Important-Bird-Areas-in-Mozambique.pdf

Parque Nacional da Gorongosa (2016). Gorongosa Map of Life. Available at: https://gorongosa.org/map-of-life/

Parque Nacional da Gorongosa (2019). Our Gorongosa – A Park for the People. Available at: https://www.gorongosa.org/sites/default/files/research/2019_highlights_corrected.pdf

Parque Nacional da Gorongosa (2020). Our Gorongosa – Together we create real impact 2020. Available at: https://gorongosa.org/wp-content/uploads/2020/12/12-10-2020-Eng-Highlights-document-reduced-size.pdf

Paula J., Macamo C. & Bandeira S. (2014). Mangroves of Maputo Bay. pp. 109 – 146. in: Bandeira, S. & Paula, J. (eds). The Maputo Bay Ecosystem. WIOMSA, Zanzibar Town.

Paula, A., Litulo, C., Costa, H. et al. (2015). As maravilhas de Taratibu / The wonders of Taratibu. Biodinamica / Universidade Lúrio / Parque Nacional das Quirimbas / WWF Mozambique. Available at: https://biodinamica.co.mz/wp-content/uploads/2015/08/brochura_taratibu_20150605_v1.pdf

Plantlife International (2004). Identifying and protecting the world's most Important Plant Areas. Plantlife International, Salisbury. Available at: https://www.plantlife.org.uk/uk/our-work/publications/identifying-and-protecting-worlds-most-important-plant-areas

Plantlife International (2018). Identifying and conserving Important Plant Areas (IPAs) around the world: A guide for botanists, conservationists, site managers, community groups and policy makers. Plantlife, Salisbury, UK. Available at: https://www.plantlife.org.uk/uk/our-work/publications/identifying-and-conserving-important-plant-areas-ipas-around-the-world

Polhill, R.M. & Wiens, D. (1998). Mistletoes of Africa. Royal Botanic Gardens, Kew.

Portugal, S. & Matos, A. (eds.) (2018). Escalas e espaços: IX edição do congresso Ibérico de estudos Africanos. Vol. III. Centro de Estudos Sociais, Universidade de Coimbra, Portugal. Available at: https://estudogeral.sib.uc.pt/bitstream/10316/80924/1/cescontexto_debates_xx.pdf

POWO (2021). Plants of the World Online. Facilitated by the Royal Botanic Gardens, Kew. Available at: http://www.plantsoftheworldonline.org/

Premier African Minerals (2020). Catapu Limestone Project. Available at: https://www.premierafricanminerals.com/mozambique/catapu-limestone-project

Radford, E.A. & Odé, B. (eds) (2009). Conserving Important Plant Areas: investing in the green gold of South East Europe. Plantlife International, Salisbury. Available at: http://www.plantlife.org.uk/uploads/documents/IPAa_SEE_report_web.pdf

Raimondo, D. & Scott-Shaw, C.R. (2007). *Thesium jeanae* Brenan. National Assessment: Red List of South African Plants version 2020.1. Available at: http://redlist.sanbi.org/species.php?species=699-94

Rainforest Trust (2021). Safeguard the Highest Peaks of Mount Namuli. Available at: https://www.rainforesttrust.org/projects/safeguard-the-highest-peaks-of-mount-namuli/

Ramsar (2011). Lake Niassa Ramsar Site. Available at: https://rsis.ramsar.org/ris/1964

Read, M. (2020). São Sebastião Plant Species Checklist. Unpubl. report. Santuario Bravio de Vilanculos.

Remane, I.A.D. & Therrell, M.D. (2019). Tree-ring analysis for sustainable harvest of *Millettia stuhlmannii* in Mozambique. *South African Journal of Botany* 125: 120 – 125. https://doi.org/10.1016/j.sajb.2019.07.012

Richards, S.L. (2021a). *Streptocarpus leptopus*. The IUCN Red List of Threatened Species 2021: e.T184921539A184921757. https://dx.doi.org/10.2305/IUCN.UK.2021-2.RLTS.T184921539A184921757.en.

Richards, S.L. (2021b). *Gyrodoma hispida*. The IUCN Red List of Threatened Species 2021: e.T172305210A172352491. https://dx.doi.org/10.2305/IUCN.UK.2021-2.RLTS.T172305210A172352491.en.

Richards, S.L. (In press [a]). *Celosia pandurata*. IUCN Red List of Threatened Species.

Richards, S.L. (In press [b]). *Triceratella drummondii*. The IUCN Red List of Threatened Species.

Richards, S.L. (In press [c]). *Elaeodendron fruticosum*. The IUCN Red List of Threatened Species.

Richards, S.L. (In press [d]). *Millettia ebenifera*. IUCN Red List of Threatened Species.

Riddell, I., Lockwood, G., Marais, E., Davis, G., Parker, V., The Mutare Bird Club & BirdLife Zimbabwe. (n.d.). The Birds of Catapú. The Birds and Trees of Catapú and environs. Available at: https://static1.squarespace.com/static/54004981e4b0cd9fe3d19b85/t/55af7dbbe4b04fd6bca7d7c4/1437564347400/cataputreeandbirdsforweb.pdf

Rodrigues, C.J., Bettencourt, A.J. & Rijo, L. (1975). Races of the pathogen and resistance to coffee rust. *Annual Review of Phytopathology* 13: 49 – 70.

Røhnebæk Bjergene, L. (2015). Promised jobs that never materialised: Forestry investments in Niassa Province, Mozambique – benefits and challenges. Masters Thesis, Norwegian University of Life Sciences. Available at: http://hdl.handle.net/11250/2368469

Rokni, S., Wursten, B. & Darbyshire, I. (2019) *Synsepalum chimanimani* (Sapotaceae), a new species from the

Chimanimani Mountains of Mozambique and Zimbabwe, with notes on the botanical importance of this area. *PhytoKeys* 133: 115 – 132. https://doi.org/10.3897/phytokeys.133.38694

Rousseau, P., Vorster, P.J., Afonso, A.V. & Van Wyk, A. (2015). Taxonomic notes on *Encephalartos ferox* (Cycadales: Zamiaceae), with the description of a new subspecies from Mozambique. *Phytotaxa* 204: 99 – 115. http://dx.doi.org/10.11646/phytotaxa.204.2.1

Roux, F. & Hoffman, A. (2017). *Chetia brevis*. The IUCN Red List of Threatened Species 2017: e.T4626A99450207. https://dx.doi.org/10.2305/IUCN.UK.2017-3.RLTS.T4626A99450207.en.

Rulkens, A.J.H. & Baptista, O.J. (2009). Field observations and local uses of the poorly known *Sansevieria pedicellata* from Manica province in Mozambique. *Sansevieria* 20: 2 – 7.

Ryan, P.G., Bento, C., Cohen, C., Graham, J., Parker, V. & Spottiswoode, C. (1999). The avifauna and conservation status of the Namuli Massif, northern Mozambique. *Bird Conservation International* 9: 315 – 331. https://doi.org/10.1017/S0959270900003518

SBV (Santuario Bravio de Vilanculos Lda.) (2017a). A Review of Co-Management Models for Conservation Areas in Mozambique. Available at: https://biofund.org.mz/wp-content/uploads/2017/08/Cabo-Sao-Sebastiao--Santuario-Bravio-25-July.pdf

SBV (Santuario Bravio de Vilanculos Lda.) (2017b). The Sanctuary Brochure. Available at: https://mozsanctuary.com/wp-content/uploads/2017/07/The-Sanctuary-Brochure.pdf

Schaefer, H. (2009). *Momordica mossambica* sp. nov. (Cucurbitaceae) from miombo woodland in northern Mozambique. *Nordic Journal of Botany* 27: 359 – 361. https://doi.org/10.1111/j.1756-1051.2009.00515.x

Schipper, J. & Burgess N. (2015) Ecoregions: Southern-east Africa: Mozambique, Tanzania, Malawi and Zimbabwe. Available at: https://www.worldwildlife.org/ecoregions/at0128

Senkoro, A., Shackleton, C., Voeks, R. & Ribeiro, A. (2019). Uses, knowledge, and management of the threatened Pepper-bark tree (*Warburgia salutaris*) in southern Mozambique. *Economic Botany* 73: 304 – 324.

Senkoro, A., Talhinhas, P., Simões, F., BatistaSantos, P., Shackleton, C., Voeks, R., Marques, I. & RibeiroBarros, A. (2020). The genetic legacy of fragmentation and overexploitation in the threatened medicinal African pepperbark tree, *Warburgia salutaris. Scientific Reports* 10: 19725. https://doi.org/10.1038/s41598-020-76654-6 . Downloaded on 25 June 2021.

Shah, T., Darbyshire, I. & Matimele, H. (2018). *Olinia chimanimani* (Penaeaceae), a new species endemic to the Chimanimani Mountains of Mozambique and Zimbabwe. *Kew Bulletin* 73: 36.. https://doi.org/10.1007/s12225-018-9757-2

Shapiro, A., Poursanidis, D., Traganos, D., Teixeira, L., Muaves, L. (2020). Mapping and Monitoring the Quirimbas National Park Seascape. WWF-Germany, Berlin. Available at: https://www.researchgate.net/publication/342626184

Short, F.T., Coles, R., Waycott, M., Bujang, J.S., Fortes, M., Prathep, A., Kamal, A.H.M., Jagtap, T.G., Bandeira, S., Freeman, A., Erftemeijer, P., La Nafie, Y.A., Vergara, S., Calumpong, H.P. & Makm, I. (2010). *Zostera capensis*. The IUCN Red List of Threatened Species 2010: e.T173370A7001305. https://dx.doi.org/10.2305/IUCN.UK.2010-3.RLTS.T173370A7001305.en

Siebert, S.J., Bandiera, S.O., Burrows, J.E. & Winter, P.J. (2002). SABONET southern Mozambique expedition 2001. *SABONET News* 7: 6 – 18.

Silveira, P. & Paiva, J. (2009) Second report on the floristic survey conducted at Vamizi and Rongui Islands, Cabo Delgado, Mozambique. Unpubl. report. Universidade de Aveiro.

Smith, T. J. (2005). Important plant areas (IPAs) in southern Africa. Combined proceedings of workshops held in Mozambique, Namibia and South Africa. Southern African Botanical Diversity Network Report No. 39. SABONET, Pretoria.

Smith, R. J., Easton, J., Nhancale, B. A., Armstrong, A. J., Culverwell, J., Dlamini, S.D., Goodman, P.S., Loffler, L., Matthews, W. S., Monadjem, A., Mulqueeny, C.M., Ngwenya, P., Ntumi, C.P., Soto, B. & Leader-Williams, N. (2008). Designing a transfrontier conservation landscape for the Maputaland centre of endemism using biodiversity, economic and threat data. *Biological Conservation* 141: 2127 – 2138.

Sörbom, J. & Gasim, A. (2018). Solid Waste Management at Inhaca Island. School of Architecture and the Built Environment. KTH Royal Institute of Technology, Stockholm.

South Africa Travel Online (2021). Cruises to Pomene from Durban 2021/2022. Available at: https://www.southafrica.to/transport/cruises/Durban/pomene/pomene.php

Spottiswoode, C.N., Patel, I.H., Herrmann, E., Timberlake, J. & Bayliss, J. (2008). Threatened bird species on two little-known mountains (Chiperone and Mabu) in northern Mozambique. *Ostrich* 79: 1 – 7. https://doi.org/10.2989/OSTRICH.2008.79.1.1.359

Spottiswoode, C.N., Fishpool, L.D.C. & Bayliss, J.L. (2016). Birds and biogeography of Mount Mecula in Mozambique's Niassa National Reserve. *Ostrich* 87: 281 – 284. https://doi.org/10.2989/00306525.2016.1206041

Stahl, M. (2020). Pyric Herbivory: Understanding Fire-Herbivore Interactions in Gorongosa National Park. Department of Ecology and Evolutionary Biology. Princeton University.

Stalmans, M. & Beilfuss, R. (2008). Landscapes of the Gorongosa National Park. Gorongosa Research Center. Available at: https://www.researchgate.net/publication/314878798

Stalmans, M., Davies, G.B.P., Trollip, J. & Poole, G. (2014). A major waterbird breeding colony at Lake Urema, Gorongosa National Park, Moçambique. *Durban Natural Science Museum Novitates* 37: 54 – 57.

Stalmans, M.E., Massad, T.J., Peel, M.J.S., Tarnita, C.E. & Pringle, R.M. (2019). War-induced collapse and asymmetric recovery of large-mammal populations in Gorongosa National Park, Mozambique. *PLoS ONE* 14: e0212864. https://doi.org/10.1371/journal.pone.0212864

Steinbruch, F. (2010). Geology and geomorphology of the Urema Graben with emphasis on the evolution of Lake Urema. *Journal of African Earth Sciences* 58: 272 – 284.

Strugnell, A.M. (2002). Endemics of Mt. Mulanje. The Endemic Spermatophytes of Mt. Mulanje, Malawi. *Systematics and Geography of Plants 72:* 11 – 26. Available at: http://www.jstor.org/stable/3668760

Symes, C. (2012). Mangrove Kingfishers (*Halcyon senegaloides;* Aves: Alcedinidae) nesting in arboreal Nasutitermes (Isoptera: Termitidae Nasutitermitinae) termitaria in central Mozambique. *Annals of the Ditsong National Museum of Natural History* 2: 146 – 152.

TCT Dalmann (2020). *Catapu.* Available at: https://www.dalmann.com/

TEEB (2010). The Economics of Ecosystems and Biodiversity: Mainstreaming the Economics of Nature: A Synthesis of the Approach, Conclusions and Recommendations of TEEB. Available at: http://teebweb.org/publications/teeb-for/synthesis/

Thiers, B. [continuously updated]. Index Herbariorum: A global Directory of Public Herbaria and Associated Staff. New York Botanical Garden's Virtual Herbarium. Available at: http://sweetgum.nybg.org/science/ih/

Timberlake, J. (2017). Mt Namuli – a conservation update. Unpubl. report for Legado, Mozambique. Sussex, UK.

Timberlake, J. (2019). *Erythrococca zambesiaca*. The IUCN Red List of Threatened Species 2019: e.T146427908A146819180. https://dx.doi.org/10.2305/IUCN.UK.2019-3.RLTS.T146427908A146819180.en.

Timberlake, J. (2020). *Pavetta chapmanii*. The IUCN Red List of Threatened Species 2020: e.T146652565A146819426. https://doi.org/https://dx.doi.org/10.2305/IUCN.UK.2020-2.RLTS.T146652565A146819426.en

Timberlake, J. (2021a). A first plant checklist for Mt. Namuli, northern Mozambique. *Kirkia* 19: 191 – 225.

Timberlake, J.R. (2021b). *Vepris myrei*. The IUCN Red List of Threatened Species 2021: e.T146722255A146819491. https://dx.doi.org/10.2305/IUCN.UK.2021-3.RLTS.T146722255A146819491.en.

Timberlake, J. & Chidumayo, E. (2011). Miombo Ecoregion Vision Report. *Occasional Publications in Biodiversity* 20. https://doi.org/10.1109/mcs.1983.1104758

Timberlake, J., Golding, J. & Clarke, P. (2004). Niassa Botanical Expedition. *Occasional Publications in Biodiversity* 12. Available at: http://www.biodiversityfoundation.org/documents/BFA No.12_Niassa Botany.pdf

Timberlake, J., Bayliss, J., Alves, T., Baena, S., Harris, T. & Sousa, C. da. (2007). Biodiversity and Conservation of Mount Chiperone, Mozambique. Report for Darwin Initiative Award 15/036: Monitoring and Managing Biodiversity Loss in South-east Africa's Montane Ecosystems. Available at: https://biofund.org.mz/wp-content/uploads/2018/12/1544778472-F2339.Darwin%20Initiative%20Award%2015%20036%20Monitoring%20and%20Managing%20Biodiversity%20Loss%20in%20Sout_2007_Timberlake_Et_Al_Chiperone.Pdf

Timberlake, J., Dowsett-Lemaire, F., Bayliss, J., Alves, T., Baena, S., Bento, C., Cook, K., Francisco, J., Harris, T., Smith, P. & de Sousa, C. (2009). Mt Namuli, Mozambique: Biodiversity and Conservation. Report produced under Darwin Initiative Award 15/036. Royal Botanic Gardens, Kew. Available at: http://www.biofund.org.mz/wp-content/uploads/2019/09/1568639660-F1232.2009-Timberlake-Et-Al-Namuli.Pdf

Timberlake, J., Goyder, D., Crawford, F. & Pascal, O. (2010). Coastal dry forests in Cabo Delgado Province, northern Mozambique: Botany and vegetation. Report for ProNatura International. Royal Botanic Gardens, Kew.

Timberlake, J., Goyder, D., Crawford, F., Burrows, J.E., Clarke, G.P., Luke, Q., Matimele, H., Müller, T., Pascal, O., de Sousa, C. & Alves T. (2011). Coastal dry forests in northern Mozambique. *Plant Ecology and Evolution* 144: 126 – 137. https://doi.org/10.5091/plecevo.2011.549

Timberlake, J., Bayliss, J., Dowsett-Lemaire, F., Congdon, C., Branch, B., Collins, S., Curran, M., Dowsett, R.J., Fishpool, L., Francisco, J., Harris, T., Kopp, M. & Sousa, C. de (2012). Mt Mabu, Mozambique: Biodiversity and Conservation. Report for Darwin Initiative Award 15/036: Monitoring and Managing Biodiversity Loss in South-East Africa's Montane Ecosystems. Royal Botanic Gardens, Kew. https://www.kew.org/sites/default/files/Mabu%20report_Final%202012_0.pdf

Timberlake, J., Matimele, H. & Massingue, A. (2014). Environmental Assessment of Proposed Road Alignment – Pemba to Mocimboa da Praia, Northern Mozambique: Plants and Vegetation. Unpubl. report prepared for ERM (Southern Africa).

Timberlake, J.R., Darbyshire, I., Wursten, B., Hadj-Hammou, J., Ballings, P., Mapaura, A., Matimele, H., Banze, A., Chipanga, H., Muassinar, D., Massunde, M., Chelene, I., Osborne, J. & Shah, T. (2016a). Chimanimani Mountains: Botany and Conservation. Report produced under CEPF Grant 63512. Royal Botanic Gardens, Kew. Available at: https://www.birdlife.org/sites/default/files/attachments/kew_chimanimani_cepf_report_revised-lr.pdf

Timberlake, J.R., Darbyshire, I., Cheek, M., Banze, A., Fijamo, V., Massunde, J., Chipanga, H. & Muassinar, D. (2016b). Plant conservation in communities on the Chimanimani footslopes, Mozambique. Report for Darwin Initiative Award 2380: Balancing Conservation and Livelihoods in the Chimanimani Forest Belt, Mozambique. Royal Botanic Gardens, Kew. Available at: https://www.kew.org/sites/default/files/Chimanimani%20Darwin%20report%2C%20FINAL.pdf

Timberlake, J., Matimele, H.A., Alves, M.T., Banze, A., Chelene, I., Darbyshire, I., Datizua, C., De Sousa, C., Langa, C., Mtshali, H., Mucaleque, P.A., Odorico, D., Osborne, J., Rokni, S., Viegas, A. & Vilanculos, A. (2018). *Maranthes goetzeniana*. The IUCN Red List of Threatened Species 2018:

e.T120955453A120980208. https://dx.doi.org/10.2305/IUCN.UK.2018-2.RLTS.T120955453A120980208.en

Timberlake, J., Ballings, P., de Deus Vidal Jr, J., Wursten, B., Hyde, M., Mapaura, A., Childes, S., Palgrave, M.C. & Clark, V.R. (2020). Mountains of the Mist: a first plant checklist for the Bvumba Mountains, Manica Highlands (Zimbabwe-Mozambique). *PhytoKeys* 145: 93 – 129. https://doi.org/10.3897/phytokeys.145.49257

Tokura, W., Matimele, H., Smit, J. & Hoffman, M. (2020). Long-term changes in forest cover in a global biodiversity hotspot in southern Mozambique. *Bothalia* 50: 95 – 97. https://doi.org/10.38201/btha.abc.v50.i1.1.

Tolley, K. (2017). Hidden Under the Clouds: Species Discovery in the Unexplored Montane Forests of Mozambique to Support New Key Biodiversity Areas. Report for CEPF Small Grant Project. Available at: https://www.cepf.net/grants/grantee-projects/identify-new-eastern-afromontane-key-biodiversity-areas-ribaue-and-inago-and

Tolley, K. (2018). Into the Clouds: Surveying the Sky Islands of Mozambique (Part 1). South African Biodiversity Institute. Available at: https://www.sanbi.org/news/into-the-clouds-surveying-the-sky-islands-of-mozambique-part-1/

Tolley, K.A., Farooq, H., Verburgt, L., Alexander, G.J., Conradie, W., Raimundo, A., & Sardinha, C.I.V. (2019a). *Cordylus meculae*. The IUCN Red List of Threatened Species 2019: e.T177561A120594696. https://dx.doi.org/10.2305/IUCN.UK.2019-2.RLTS.T141800272A141800288.en

Tolley, K.A., Verburgt, L., Alexander, G.J., Conradie, W., Farooq, H., Raimundo, A., Sardinha, C.I.V. & Bayliss, J. (2019b). *Rhampholeon bruessoworum*. The IUCN Red List of Threatened Species 2019: e.T61366030A149766721. https://dx.doi.org/10.2305/IUCN.UK.2019-3.RLTS.T61366030A149766721.en.

Tolley, K., Farooq, H., Verburgt, L., Alexander, G.J., Conradie, W., Raimundo, A., Sardinha, C.I.V. & Bayliss, J. (2019c). *Rhampholeon nebulauctor*. The IUCN Red List of Threatened Species 2019: e.T61365784A149767278.

Üllenberg, A., Buchberger, C., Meindl, K., Rupp, L., Springsguth, M. & Straube, B. (2014). Evaluating Cross-borders Natural Resource Management Projects: Mhlumeni Goba Community Tourism and Conservation initiative Lubombo Conservancy – Goba TFCA. Berlin, Germany. Unpubl. report. https://tfcaportal.org/system/files/resources/Evaluationreport_part_LCG.pdf

Üllenberg, A., Buchberger, C., Meindl, K., Rupp, L., Springsguth, M. & Straube, B. (2015). Evaluating Cross-Borders Natural Resource Management Projects: Community-based Tourism Development and Fire Management in Conservation Areas of the SADC Region. Berlin, Germany. Unpubl. report.

UNEP (2021). The Species+ Website. Nairobi, Kenya. UNEP-WCMC, Cambridge, UK. Available at: www.speciesplus.net

UNEP-WCMC (2021). Protected Area Profile for Mozambique from the World Database of Protected Areas, November 2021. Available at: www.protectedplanet.net

UNESCO (2020). UNESCO World Heritage Sites: Island of Mozambique. Available at: https://whc.unesco.org/en/list/599/multiple=1&unique_number=709

URS/Scott Wilson (2011). Scoping Study Report on the Moebase and Naburi Mineral Sands Deposits, Mozambique. Report prepared for Pathfinder Minerals plc. URS/Scott Wilson, Chesterfield. Available at: http://www.pathfinderminerals.com/~/media/Files/P/Pathfinders-ECW/Attachments/pdf/pathfinder-Minerals-Scoping-Study-Report.pdf

van Berkel, T., Sumbane, E., Jones, S.E. & Jocque, M. (2019). A mammal survey of the Serra Jeci Mountain Range, Mozambique, with a review of records from northern Mozambique's inselbergs. *African Zoology* 54: 31 – 42. https://doi.org/10.1080/15627020.2019.1583081

van der Weijden, W., Leewis, R. & Bol, P. (2004). 100 of the World's Worst Invasive Alien Species: a selection from the Global Invasive Species Database. *Aliens* 12. https://doi.org/10.1163/9789004278110_019

van Velzen, R., Collins, S.C., Brattstrom, O. & Congdon, C.E. (2016). Description of a new *Cymothoe* Hübner, 1819 from northern Mozambique (Lepidoptera: Nymphalidae: Limenitidinae). *Metamorphosis* 27: 34 – 41. Available at: https://www.researchgate.net/publication/304526494

van Wyk, A. (1996). Biodiversity of Maputaland Centre. pp. 198 – 207. in: van der Maesen, L., van der Burgt, X., van Medenbach, R. (eds). The Biodiversity of African Plants: Proceedings XIVth AETFAT congress 22 – 27 August 1994, Wageningen, The Netherlands. Springer Netherlands, Dordrecht.

van Wyk, A.E. & Smith, G.F. (2001). Regions of Floristic Endemism in Southern Africa. A Review With Emphasis on Succulents. Umdaus Press, Hatfield, South Africa.

Vaz, K., Norton, P., Avaloi, R., Chambal, H., Afonso, P.S., Falcão, M.P., Pereira, M. & Videira, E. (2008). Plano de Maneio do Parque Nacional do Arquipélago do Bazaruto 2008 – 2012. Ministério do Turismo Direcção Nacional das Áreas de Conservação, Maputo. Available at: https://www.biofund. org.mz/wp-content/uploads/2019/01/1548237006-VOL%20 1%20-%20PNAB%20PM.pdf

Verdcourt, B. (2000). Leguminosae (part 6). In: Pope, G.V. (ed.), *Flora Zambesiaca*, Vol. 3(6). Royal Botanic Gardens, Kew.

Virtanen, P. (2002). The role of customary institutions in the conservation of biodiversity: sacred forests in Mozambique. *Environmental Values* 11: 227 – 241.

von Staden, L. & Lötter, M. (2018). *Barleria oxyphylla* Lindau. National Assessment: Red List of South African Plants version 2020.1. Available at: http://redlist.sanbi.org/species. php?species=3909-58

Wabuyele E., Sitoni D., Njau E.-F., Mboya E.I., Lyaruu H.V.M., Kindeketa W., Kalema J., Kabuye C., Kamau P., Luke W.R.Q., Malombe I., Mollel N., Schatz G.E. & Ssegawa P. (2020). *Hugonia grandiflora*. The IUCN Red List of Threatened Species 2020: e.T158188A765731.

Warren, M. (2019). Why Cyclone Idai is one of the Southern Hemisphere's most devastating storms. *Nature News*. https:// doi.org/10.1038/d41586-019-00981-6

WCS, Government of Mozambique & USAID (2021). Key Biodiversity Areas (KBAs) identified in Mozambique: Factsheets VOL. II. USAID / SPEED+, Maputo. Available at: https://www.biofund.org.mz/wp-content/uploads/2021/ 05/1622195386-2021_KBAs_Moz_vol_ii_Factsheets_EN.pdf

White, F. (1983a). Vegetation of Africa. A Descriptive Memoir to Accompany the UNESCO/AETFAT/UNSO Vegetation Map of Africa. Natural Resources Research 20. UNESCO, Paris.

White, F. (1983b). Ebenaceae. pp. 248 – 300. In: Launert, E. (ed.), *Flora Zambesiaca*, Vol. 7(1). Flora Zambesiaca Managing Committee, London.

Wieringa, J.J. (1999). Monopetalanthus exit. A systematic study of *Aphanocalyx, Bikinia, Isuria, Michelsonia* and *Tetraberlinia* (Leguminosae, Caesalpinioideae). *Wageningen Agricultural University Papers* 99: 1 – 320. Available at: https://edepot. wur.nl/162697

Wild, H. & Barbosa, L.A.G. (1968). Vegetation map of the Flora Zambesiaca area (1: 250,000 scale). Supplement to *Flora Zambesiaca*. M.O. Collins, Salisbury [Harare], Zimbabwe.

Wildlife Conservation Society Mozambique (2021). Niassa Special Reserve. Available at: https://mozambique.wcs.org/ Wild-Places/Niassa-National-Reserve

Williams, V.L., Raimondo, D., Crouch, N.R., Cunningham, A.B., Scott-Shaw, C.R., Lötter, M. & Ngwenya, A.M. (2008). *Dioscorea sylvatica* Eckl. National Assessment: Red List of South African Plants version 2020.1. Available at: http:// redlist.sanbi.org/species.php?species=1777-4002

Willis, K.J. (ed.) (2017). State of the World's Plants 2017. Report. Royal Botanic Gardens, Kew.

Wisborg, P. & Jumbe, C.B.L. (2010). Mulanje Mountain Biodiversity Conservation Project: Mid-Term Review for the Norwegian Government. Noragric Report No. 57. Norwegian University of Life Science. Available at: https://hdl.handle. net/11250/2646182

Woodcock, C.E., Allen, A.A., Anderson, M., Belward, A.S., Bindschadler, R., Cohen, W.B., Gao, F., Goward, S.N., Helder, D., Helmer, E., Nemani, R., Oreapoulos, L., Schott, J., Thenkabail, P.S., Vermote, E.F., Vogelmann, J., Wulder, M.A. & Wynne, R. (2008). Free Access to Landsat Imagery. *Science* 320: 1011. https://doi.org/10.1126/science.320.5879.1011a

Woolley, A.R. (1987). Alkaline Rocks and Carbonatites of the World: Africa. Geological Society of London.

World Bank (2018). Mozambique Country Forest Note. Report No: AUS0000336. Available at: http://documents.worldbank. org/curated/en/693491530168545091/Mozambique-Country-forest-note

World Bank (2019). Disaster Risk Profile: Mozambique. Available at: https://www.gfdrr.org/en/publication/disaster-risk-profile-mozambique

World Bank (2020). Poverty and Shared Prosperity 2020: Reversals of Fortune. https://doi.org/10.1596/978-1-4648-1602-4.

World Bank (2021). World Bank Open Data. Available at: https:// data.worldbank.org/

World Resources Institute (2020 – 2021). Global Forest Watch. Available at: www.globalforestwatch.org

World Weather Online (2021). https://www.worldweatheronline. com

Wursten, B., Timberlake, J. & Darbyshire, I. (2017). The Chimanimani Mountains: An updated checklist. *Kirkia* 19: 70 – 100.

Wursten, B., Bridson, D., Janssens, S.B. & De Block, P. (2020). A new species of *Sericanthe* (Coffeeae, Rubiaceae) from Chimanimani Mountains, Mozambique-Zimbabwe border. *Phytotaxa* 430: 109 – 118. https://doi.org/10.11646/ phytotaxa.430.2.3

WWF (2011). Lake Niassa Declared a Reserve. Available at: https://www.worldwildlife.org/stories/lake-niassa-declared-a-reserve

WWF Mozambique (2016). A triste história do massacre de elefantes na Reserva de Taratibu. Available at: https://www. wwf.org.mz/noticias/?2500/A-triste-histria-do-massacre-de-elefantes-na-Reserva-de-Taratibu

Wyse, S.V. & Dickie, J.B. (2018). Taxonomic affinity, habitat and seed mass strongly predict seed desiccation response: a boosted regression trees analysis based on 17,539 species. *Annals of Botany* 121: 71 – 83. https://doi.org/10.1093/aob/ mcx128

MAP REFERENCES

IMAGERY MAPS

Base map: World Imagery. (2021). Esri, Maxar, GeoEye, Earthstar Geographics, CNES/Airbus DS, USDA, USGS, AeroGRID, IGN and the GIS User Community.

REFERENCE MAPS

Country and Province boundaries: Global Administrative Areas. (2021). GADM database of Global Administrative Areas, version 2.8. Available at: www.gadm.org.

Hillshade: Williams, J.J. (Unpublished). 90m hillshade DEM. Royal Botanic Gardens Kew. Made with NASA Shuttle Radar Topography Mission (SRTM) (2013). Shuttle Radar Topography Mission (SRTM) Global. Distributed by OpenTopography.

Localities: National Geospatial-Intelligence Agency. (2021). Complete Files of Geographic Names for Geopolitical Areas: Mozambique. Available at: https://geonames.nga.mil/gns/html/namefiles.html.

Major localities: Natural Earth. (2021). Populated Places: Large scale data 1:10m, Version 4.1.0. Available at: naturalearth.com.

Protected areas: UNEP-WCMC and IUCN (2021). Protected Planet: The World Database on Protected Areas (WDPA), September 2021. Cambridge, UK. Available at: www.protectedplanet.net.

Roads: Roads and Defense Mapping Agency (DMA). (1992). Digital Chart of the World. Defense Mapping Agency, Fairfax, Virginia.

Natural Earth. (2021). Roads: Large scale data 1:10m. Available at: naturalearth.com

Tree cover: Williams, J.J. (Unpublished). Mozambique Tree Cover WTL1. Royal Botanic Gardens, Kew. Made with Landsat archive, courtesy of the U.S. Geological Survey in Google Earth Engine.

Water areas-Lakes: Natural Earth. (2021). Lakes and Reservoirs: Large scale data 1:50 m. Available at: naturalearth.com

Water areas- Rivers: Natural Earth. (2021). Rivers and Lake Centrelines: Large scale data 1:50 m. Available at: naturalearth.com

APPENDIX: LIST OF A(I) AND B(II) PLANT TAXA OF MOZAMBIQUE

The table below lists all Mozambican taxa that could trigger sub-criterion A(i) or contribute to triggering B(ii) of the IPA criteria (see chapter "Identifying Important Plant Areas in Mozambique: methods and resources" for full details on each sub-criterion). A(i) qualifying taxa are those assessed as threatened on the IUCN Red List of Threatened Species, although certain thresholds must be met at each site for it to qualify as an IPA under A(i). Assessments that have been submitted via the IUCN Species Information Service (SIS) but have not yet been published are included, as are assessments that require updating as they were assessed under previous iterations of the IUCN Red List criteria and/or are more than 10 years old. The B(ii) taxa listed are either national endemics or have a range

of less than 10,000 km². In total, there are 507 taxa that can contribute to triggering B(ii) at a site.

The following species have been removed from the below table:

- *Cola discoglypremnophylla* (EN), while this species likely occurs in Mozambique, there is no fertile material to confirm its presence;

- *Grevea eggelingii* var. *echinocarpa* (EN), possibly a synonym of Near Threatened species *Grevea eggelingii* var. *eggelingii*; and

- *Ixora scheffleri* subsp. *scheffleri* (VU), as this subspecies is much more widespread than is considered in the Red List assessment.

Family	Taxon	A(i) species IUCN Red List Assessment	B(ii) species endemism status	IPAs
Acanthaceae	*Asystasia malawiana* Brummitt & Chisumpa	VU B2ab(iii)		Mount Mabu, Mount Namuli
Acanthaceae	*Barleria fissimuroides* I.Darbysh.	EN B2ab(iii)	<10k km²	
Acanthaceae	*Barleria fulvostellata* C.B.Clarke subsp. *mangochiensis* I.Darbysh.	EN B1ab(iii)	<10k km²	
Acanthaceae	*Barleria laceratiflora* Lindau	EN B2ab(iii)	<10k km²	Goa and Sena Islands
Acanthaceae	*Barleria oxyphylla* Lindau	VU B1ab(ii,iii) +2ab(ii,iii)	<10k km²	Namaacha
Acanthaceae	*Barleria rhynchocarpa* Klotzsch	VU B2ab(ii,iii,iv)		Quirimbas Archipelago
Acanthaceae	*Barleria setosa* (Klotzsch) I.Darbysh.	EN B1ab(i,ii,iii,iv) +2ab(i,ii,iii,iv)	Endemic	Goa and Sena Islands
Acanthaceae	*Barleria torrei* I.Darbysh.	EN B1ab(iii)+2ab(iii)	Endemic	Njesi Plateau
Acanthaceae	*Barleria vollesenii* I.Darbysh.	EN B2ab(iii)	<10k km²	
Acanthaceae	*Barleria whytei* S.Moore	EN B2ab(iii)		Vamizi Island
Acanthaceae	*Blepharis dunensis* Vollesen	EN B1ab(iii)+2ab(iii)	Endemic	Quinga
Acanthaceae	*Blepharis gazensis* Vollesen		Endemic	
Acanthaceae	*Blepharis swaziensis* Vollesen		<10k km²	
Acanthaceae	*Blepharis torrei* Vollesen		<10k km²	
Acanthaceae	*Cephalophis lukei* Vollesen	EN B2ab(iii)		Inhamitanga Forest

Family	Taxon	A(i) species IUCN Red List Assessment	B(ii) species endemism status	IPAs
Acanthaceae	*Dicliptera quintasii* Lindau	VU B1ab(ii,iii,iv,v) +2ab(ii,iii,iv,v)	<10k km²	
Acanthaceae	*Dicliptera* sp. B of F.Z.		Endemic	Inhamitanga Forest
Acanthaceae	*Duosperma dichotomum* Vollesen	VU D2	Endemic	Muàgámula River
Acanthaceae	*Ecbolium hastatum* Vollesen	EN B2ab(iii)	Endemic	Chidenguele, Inhassoro-Vilanculos, Inhaca (KaNyaka) Island, São Sebastião Peninsula
Acanthaceae	*Isoglossa namuliensis* I.Darbysh. & T.Harris	CR B1ab(iii)+2ab(iii)	Endemic	Mount Namuli
Acanthaceae	*Justicia attenuifolia* Vollesen	VU D2		Serra Mecula and Mbatamila
Acanthaceae	*Justicia gorongozana* Vollesen		Endemic	Catapú, Cheringoma Limestone Gorges
Acanthaceae	*Justicia niassensis* Vollesen	EN B1ab(ii,iii) +2ab(ii,iii)	Endemic	Pemba?
Acanthaceae	*Justicia* sp. A of F.Z.	EN B2ab(ii,iii,v)	Endemic	Mount Gorongosa
Acanthaceae	*Justicia* sp. B of F.Z.		Endemic	Cheringoma Limestone Gorges
Acanthaceae	*Justicia subcordatifolia* Vollesen & I.Darbysh.		<10k km²	
Acanthaceae	*Lepidagathis plantaginea* Mildbr.	EN B2ab(iii)		
Acanthaceae	*Rhinacanthus submontanus* T.Harris & I.Darbysh.	VU B2ab(iii)		Mount Gorongosa
Acanthaceae	*Sclerochiton apiculatus* Vollesen	VU B1ab(i,ii,iii,v) + 2ab(i,ii,iii,v)	<10k km²	Licuáti Forest
Acanthaceae	*Sclerochiton hirsutus* Vollesen	VU D2	Endemic	Mount Mabu, Mount Namuli
Aizoaceae	*Trianthema mozambiquensis* H.E.K.Hartmann & Liede		Endemic	
Amaranthaceae	*Caroxylon littoralis* (Moq.) Akhani & Roalson		<10k km²	
Amaranthaceae	*Celosia nervosa* C.C.Towns.		Endemic	Panda-Manjacaze
Amaranthaceae	*Celosia pandurata* Baker	VU B1ab(iii)+2ab(iii)	Endemic	Inhamitanga Forest, Mount Morrumbala, Urema Valley and Sangarassa Forest
Amaranthaceae	*Celosia patentiloba* C.C.Towns.	CR B2ab(iii), D	<10k km²	Mueda Plateau and Escarpments
Amaranthaceae	*Salicornia mossambicensis* (Brenan) Piirainen & G.Kadereit		Endemic	Pomene
Amaryllidaceae	*Tulbaghia friesii* Suess.		<10k km²	

Family	Taxon	A(i) species IUCN Red List Assessment	B(ii) species endemism status	IPAs
Anacardiaceae	*Lannea welwitschii* (Hiern) Engl. var. *ciliolata* Engl.	EN B2ab(iii)		Mueda Plateau and Escarpments
Anacardiaceae	*Ozoroa gomesiana* R.Fern. & A.Fern.	VU B1ab(iii)+2ab(iii)	Endemic	Inhassoro-Vilanculos, Mapinhane, Temane
Annonaceae	*Hexalobus mossambicensis* N.Robson	VU B2ab(iii)	Endemic	Lower Rovuma Escarpment, Matibane Forest, Quiterajo
Annonaceae	*Huberantha mossambicensis* (Vollesen) Chaowasku	CR B2ab(iii)	Endemic	
Annonaceae	*Monanthotaxis suffruticosa* P.H.Hoekstra	VU B2ab(iii)		Lower Rovuma Escarpment
Annonaceae	*Monanthotaxis trichantha* (Diels) Verdc.	VU B2ab(ii,iii,v)		Lower Rovuma Escarpment, Matibane Forest, Quiterajo
Annonaceae	*Monodora carolinae* Couvreur	EN B1ab(iii)+2ab(iii)	<10k km²	Mueda Plateau and Escarpments
Annonaceae	*Monodora stenopetala* Oliv.	VU B2ab(i,ii,iii,iv)		Catapú, Inhamitanga Forest
Annonaceae	*Uvaria rovumae* Deroin & Lötter	CR B1ab(iii)+2ab(iii); D	Endemic	Mueda Plateau and Escarpments
Annonaceae	*Xylopia lukei* D.M.Johnson & Goyder	EN B1ab(ii,iii) +2ab(ii,iii)	<10k km²	Lower Rovuma Escarpment
Annonaceae	*Xylopia tenuipetala* D.M.Johnson & Goyder	EN B1ab(ii,iii) +2ab(ii,iii)	Endemic	Quiterajo
Annonaceae	*Xylopia torrei* N.Robson	EN B2ab(ii,iii,iv,v)	Endemic	Licuáti Forest, Panda-Manjacaze
Apiaceae	*Afrosciadium rhodesicum* (Cannon) P.J.D.Winter	VU B1ab(iii)+2ab(iii)	<10k km²	
Apiaceae	*Centella obtriangularis* Cannon	VU D2	Endemic	Chimanimani Mountains
Apiaceae	*Pimpinella mulanjensis* C.C.Towns.		<10k km²	Mount Namuli
Apocynaceae	*Asclepias cucullata* (Schltr.) Schltr. subsp. *scabrifolia* (S.Moore) Goyder		<10k km²	Chimanimani Mountains, Tsetserra
Apocynaceae	*Asclepias graminifolia* (Wild) Goyder		<10k km²	Chimanimani Mountains
Apocynaceae	*Aspidoglossum glabellum* Kupicha	EN B1ab(iii)+2ab(iii)	<10k km²	Chimanimani Mountains
Apocynaceae	*Aspidoglossum hirundo* Kupicha	VU B2ab(iii)		
Apocynaceae	*Ceropegia aloicola* M.G.Gilbert	EN B1ab(iii)	Endemic	Namaacha
Apocynaceae	*Ceropegia chimanimaniensis* M.G.Gilbert		<10k km²	Chimanimani Mountains
Apocynaceae	*Ceropegia cyperifolia* Bruyns		Endemic	Mount Massangulo
Apocynaceae	*Ceropegia gracilidens* Bruyns		Endemic	

Family	Taxon	A(i) species IUCN Red List Assessment	B(ii) species endemism status	IPAs
Apocynaceae	*Ceropegia muchevensis* M.G.Gilbert	CR B1ab(i,ii,iii)	Endemic	
Apocynaceae	*Ceropegia nutans* (Bruyns) Bruyns	VU D1+D2	Endemic	Mount Namuli
Apocynaceae	*Ceropegia vahrmeijeri* (R.A.Dyer) Bruyns		<10k km²	
Apocynaceae	*Cynanchum oresbium* (Bruyns) Goyder	VU D2	Endemic	Mount Inago and Serra Merripa, Ribáuè-M'paluwe
Apocynaceae	*Emicocarpus fissifolius* K.Schum.& Schltr.	CR D	Endemic	
Apocynaceae	*Huernia erectiloba* L.C.Leach & Lavranos		Endemic	Mount Inago and Serra Merripa, Ribáuè-M'paluwe
Apocynaceae	*Huernia leachii* Lavranos		<10k km²	
Apocynaceae	*Huernia verekeri* Stent subsp. *pauciflora* (L.C.Leach) Bruyns		Endemic	
Apocynaceae	*Landolphia watsoniana* Rombouts	VU B2ab(iii)		Lower Rovuma Escarpment
Apocynaceae	*Marsdenia gazensis* S.Moore		<10k km²	
Apocynaceae	*Orbea halipedicola* L.C.Leach		Endemic	Urema Valley and Sangarassa Forest
Apocynaceae	*Pachycarpus concolor* E.Mey. subsp. *arenicola* Goyder		<10k km²	
Apocynaceae	*Pleioceras orientale* Vollesen	VU D2		Inhamitanga Forest
Apocynaceae	*Raphionacme pulchella* Venter & R.L.Verh.	EN B2ab(iii)		Chimanimani Mountains, Serra Mocuta
Apocynaceae	*Stapelia unicornis* C.A.Luckh.		<10k km²	
Apocynaceae	*Stomatostemma pendulina* Venter & D.V.Field	VU D2	Endemic	Ribáuè-M'paluwe
Apocynaceae	*Strophanthus hypoleucos* Stapf.	VU B2ab(iii)		Quirimbas Inselbergs, Ribáuè-M'paluwe
Apocynaceae	*Vincetoxicum monticola* Goyder		<10k km²	Mount Gorongosa, Tsetserra
Araceae	*Gonatopus petiolulatus* (Peter) Bogner	VU B2ab(iii)		Lower Rovuma Escarpment
Araceae	*Stylochaeton euryphyllus* Mildbr.	VU B2ab(iii)		Quiterajo
Araceae	*Stylochaeton tortispathus* Bogner & Haigh	VU D2	Endemic	Quiterajo
Araliaceae	*Polyscias albersiana* Harms	EN B1ab(iii)		
Arecaceae	*Raphia australis* Oberm. & Strey	VU A3c+4c; B1ab(iii,v) +2ab(iii,v); C1		Bilene-Calanga, Bobole, Chidenguele
Asparagaceae	*Asparagus chimanimanensis* Sebsebe		<10k km²	Chimanimani Mountains

Family	Taxon	A(i) species IUCN Red List Assessment	B(ii) species endemism status	IPAs
Asparagaceae	*Asparagus humilis* Engl.	EN B2ab(iii)		
Asparagaceae	*Asparagus radiatus* Sebsebe		<10k km²	Goba, Namaacha
Asparagaceae	*Chlorophytum pygmaeum* (Weim.) Kativu subsp. *rhodesianum* (Rendle) Kativu		<10k km²	Chimanimani Mountains
Asparagaceae	*Dracaena subspicata* (Baker) Byng & Christenh. (*Sansevieria subspicata* Baker)		Endemic	Cheringoma Limestone Gorges, Inhaca (KaNyaka) Island, Pomene, Urema Valley and Sangarassa Forest
Asparagaceae	*Eriospermum mackenii* (Hook.f.) Baker subsp. *phippsii* (Wild) P.L.Perry		<10k km²	Chimanimani Mountains
Asphodelaceae	*Aloe argentifolia* T.A.McCoy, Rulkens & O.J.Baptista	VU D1	Endemic	Lúrio Waterfalls, Chiure
Asphodelaceae	*Aloe cannellii* L.C.Leach		Endemic	Serra Mocuta
Asphodelaceae	*Aloe decurva* Reynolds	CR B1ab(iii,v) +2ab(iii,v)	Endemic	Mount Zembe
Asphodelaceae	*Aloe menyharthii* Baker subsp. *ensifolia* S.Carter		Endemic	Mount Inago and Serra Merripa
Asphodelaceae	*Aloe mossurilensis* Elert	CR B2ab(iii)	Endemic	
Asphodelaceae	*Aloe ribauensis* T.A.McCoy, Rulkens & O.J.Baptista	EN B1ab(iii,v) +2ab(iii,v)	Endemic	Mueda Plateau and Escarpments, Ribáuè-M'paluwe
Asphodelaceae	*Aloe rulkensii* T.A.McCoy & O.J.Baptista	CR B1ab(iii)+2ab(iii); D	Endemic	Ribáuè-M'paluwe
Asphodelaceae	*Aloe torrei* I.Verd. & Christian		Endemic	Mount Namuli
Asphodelaceae	*Aloe ballii* Reynolds var. *makurupiniensis* Ellert	VU D2	<10k km²	Chimanimani Lowlands
Asphodelaceae	*Aloe excelsa* A.Berger var. *breviflora* L.C.Leach		<10k km²	
Asphodelaceae	*Aloe hazeliana* Reynolds var. *hazeliana*		<10k km²	Chimanimani Mountains
Asphodelaceae	*Aloe hazeliana* Reynolds var. *howmanii* (Reynolds) S.Carter		<10k km²	Chimanimani Mountains
Asphodelaceae	*Aloe inyangensis* Christian var. *kimberleyana* S.Carter		<10k km²	Tsetserra
Asphodelaceae	*Aloe munchii* Christian		<10k km²	Chimanimani Mountains
Asphodelaceae	*Aloe plowesii* Reynolds	VU D2	<10k km²	Chimanimani Mountains
Asphodelaceae	*Aloe rhodesiana* Rendle	VU B1ab(iii)+2ab(iii)		Chimanimani Mountains, Mount Gorongosa
Asphodelaceae	*Aloe wildii* (Reynolds) Reynolds		<10k km²	Chimanimani Mountains
Asteraceae	*Adelostigma athrixioides* Steetz [uncertain species]		Endemic	

Family	Taxon	A(i) species IUCN Red List Assessment	B(ii) species endemism status	IPAs
Asteraceae	*Anisopappus paucidentatus* Wild		<10k km²	Chimanimani Mountains
Asteraceae	*Aster chimanimaniensis* Lippert		<10k km²	Chimanimani Mountains
Asteraceae	*Blepharispermum brachycarpum* T.Erikss.	EN B1ab(iii)+2ab(iii)		
Asteraceae	*Bothriocline glomerata* (O.Hoffm. & Muschl.) C.Jeffrey	EN B2ab(i,ii,iii,iv,v)		
Asteraceae	*Bothriocline moramballae* (Oliv.& Hiern) O.Hoffm.		Endemic	Mount Morrumbala, Mount Nállume, Mount Namuli, Ribáuè-M'paluwe
Asteraceae	*Bothriocline steetziana* Wild & G.V.Pope		Endemic	Mount Inago and Serra Merripa
Asteraceae	*Cineraria pulchra* Cron		<10k km²	Chimanimani Mountains, Mount Gorongosa, Tsetserra
Asteraceae	*Gutenbergia westii* (Wild) Wild & G.V.Pope	VU B1ab(iii)+2ab(iii)	<10k km²	Chimanimani Lowlands, Chimanimani Mountains, Serra Mocuta
Asteraceae	*Gyrodoma hispida* (Vatke) Wild		Endemic	Urema Valley and Sangarassa Forest
Asteraceae	*Helichrysum acervatum* S.Moore		<10k km²	Tsetserra
Asteraceae	*Helichrysum africanum* (S.Moore) Wild		<10k km²	Chimanimani Mountains
Asteraceae	*Helichrysum chasei* Wild		<10k km²	Tsetserra
Asteraceae	*Helichrysum lastii* Engl.		<10k km²	Mount Namuli
Asteraceae	*Helichrysum moggii* Wild		Endemic	Inhaca (KaNyaka) Island
Asteraceae	*Helichrysum moorei* Staner		<10k km²	Chimanimani Mountains
Asteraceae	*Helichrysum rhodellum* Wild		<10k km²	Chimanimani Mountains
Asteraceae	*Kleinia chimanimaniensis* van Jaarsv.		<10k km²	Chimanimani Lowlands, Chimanimani Mountains
Asteraceae	*Lopholaena brickellioides* S.Moore		<10k km²	Chimanimani Mountains, Tsetserra
Asteraceae	*Schistostephium oxylobum* S.Moore	VU B1ab(iii)+2ab(iii)	<10k km²	Chimanimani Mountains, Tsetserra
Asteraceae	*Senecio aetfatensis* B.Nord.		<10k km²	Chimanimani Mountains
Asteraceae	*Senecio forbesii* Oliv. & Hiern [uncertain species]		Endemic	
Asteraceae	*Senecio peltophorus* Brenan		<10k km²	Mount Mabu, Mount Namuli
Asteraceae	*Vernonia calvoana* (Hook.f.) Hook.f. subsp. *meridionalis* (Wild) C.Jeffrey (*Baccharoides calvoana* (Hook.f.) Isawumi subsp. meridionalis (Wild) Isuwami, El-Ghazaly & B.Nord.)		<10k km²	Mount Gorongosa

Family	Taxon	A(i) species IUCN Red List Assessment	B(ii) species endemism status	IPAs
Asteraceae	*Vernonia muelleri* Wild subsp. *muelleri*		<10k km²	Chimanimani Lowlands, Chimanimani Mountains
Asteraceae	*Vernonia nepetifolia* Wild		<10k km²	Chimanimani Mountains
Balsaminaceae	*Impatiens psychadelphoides* Launert	VU B2ab(iii)		Mount Namuli
Balsaminaceae	*Impatiens salpinx* Schulze & Launert		<10k km²	Chimanimani Mountains
Balsaminaceae	*Impatiens wuerstenii* S.B.Janssens & Dessein	VU D2	Endemic	Mount Gorongosa
Bignoniaceae	*Dolichandrone alba* (Sim) Sprague		Endemic	Mapinhane, Panda-Manjacaze, Temane
Boraginaceae	*Cordia mandimbana* E.S.Martins		Endemic	
Boraginaceae	*Cordia megiae* J.E.Burrows	VU D2	Endemic	Catapú, Inhamitanga Forest
Boraginaceae	*Cordia stuhlmannii* Gürke	VU B2ab(iii)	Endemic	Catapú, Inhamitanga Forest
Boraginaceae	*Cordia torrei* E.S.Martins	EN B2ab(iii)		Catapú
Burseraceae	*Canarium madagascariense* Engl.	EN B1ab(i,ii,iii,iv) +2ab(i,ii,iii,iv)		
Campanulaceae	*Lobelia blantyrensis* E. Wimm.		<10k km²	Mount Namuli
Campanulaceae	*Lobelia cobaltica* S.Moore		<10k km²	Chimanimani Mountains
Campanulaceae	*Wahlenbergia subaphylla* (Baker) Thulin subsp. *scoparia* (Wild) Thulin		<10k km²	Chimanimani Mountains
Canellaceae	*Warburgia salutaris* (G.Bertol.) Chiov.	EN A1acd		Goba, Licuáti Forest, Namaacha
Capparaceae	*Capparis viminea* Hook.f. & Thomson ex Oliv. var. *orthacantha* (Gilg & Gilg-Ben.) DeWolf		<10k km²	
Capparaceae	*Maerua andradae* Wild		Endemic	Muàgámula River, Mueda Plateau and Escarpments
Capparaceae	*Maerua brunnescens* Wild		Endemic	Catapú, Inhamitanga Forest, Urema Valley and Sangarassa Forest
Capparaceae	*Maerua scandens* (Klotzsch) Gilg		Endemic	
Caprifoliaceae	*Pterocephalus centennii* M.J.Cannon	CR B1ab(iii)+B2ab(iii)	Endemic	Tsetserra
Caryophyllaceae	*Dianthus chimanimaniensis* S.S.Hooper	VU D2	Endemic	Chimanimani Mountains
Celastraceae	*Crossopetalum mossambicense* I.Darbysh.	EN B1ab(iii)+2ab(iii)	<10k km²	Lower Rovuma Escarpment

Family	Taxon	A(i) species IUCN Red List Assessment	B(ii) species endemism status	IPAs
Celastraceae	*Elaeodendron fruticosum* N.Robson	VU B2ab(ii,iii)	Endemic	Chidenguele, Inharrime-Závora, Inhassoro-Vilanculos, Pomene, São Sebastião Peninsula
Celastraceae	*Gymnosporia gurueensis* (N.Robson) Jordaan	EN B1ab(iii)+2ab(iii)	Endemic	Mount Namuli
Celastraceae	*Gymnosporia oxycarpa* (N.Robson) Jordaan		<10k km²	
Celastraceae	*Salacia orientalis* N.Robson	VU B2ab(iii)		Lower Rovuma Escarpment, Mueda Plateau and Escarpments
Clusiaceae	*Garcinia acutifolia* N.Robson	VU B1+2c		Lower Rovuma Escarpment
Combretaceae	*Combretum caudatisepalum* Exell & J.G.García	VU D2	Endemic	Muàgámula River, Pemba
Combretaceae	*Combretum lasiocarpum* Engl.& Diels		Endemic	
Combretaceae	*Combretum lindense* Exell & Mildbr.	CR B2ab(iii)	<10k km²	Lower Rovuma Escarpment
Combretaceae	*Combretum stocksii* Sprague		Endemic	Lower Rovuma Escarpment, Mueda Plateau and Escarpments, Quiterajo
Combretaceae	*Terminalia barbosae* (Exell) Gere & Boatwr. (*Pteleopsis barbosae* Exell)	VU B1ab(iii)	Endemic	Muàgámula River
Commelinaceae	*Aneilema arenicola* Faden		<10k km²	
Commelinaceae	*Aneilema mossambicense* (Faden) Faden		Endemic	Matibane Forest
Commelinaceae	*Cyanotis chimanimaniensis* Faden ined.		<10k km²	Chimanimani Lowlands, Chimanimani Mountains
Commelinaceae	*Cyanotis namuliensis* Faden ined.		Endemic	Mount Namuli
Commelinaceae	*Triceratella drummondii* Brenan	CR B1ab(iii)+2ab(iii)	Endemic	Moebase
Connaraceae	*Vismianthus punctatus* Mildbr.	VU B1ab(iii)+2ab(iii)	<10k km²	Lower Rovuma Escarpment, Mueda Plateau and Escarpments, Quiterajo
Convolvulaceae	*Convolvulus goyderi* J.R.I.Wood	EN B1ab(iii)	Endemic	Lower Rovuma Escarpment
Convolvulaceae	*Ipomoea ephemera* Verdc.		Endemic	
Convolvulaceae	*Ipomoea venosa* (Desr.) Roem. & Schult. subsp. *stellaris* (Baker) Verdc. var. *obtusifolia* Verdc.		Endemic	
Convolvulaceae	*Turbina longiflora* Verdc.		Endemic	
Crassulaceae	*Crassula leachii* R.Fern.		Endemic	
Crassulaceae	*Crassula morrumbalensis* R.Fern.	CR B1ab(iii)+2ab(iii)	Endemic	Mount Morrumbala

Family	Taxon	A(i) species IUCN Red List Assessment	B(ii) species endemism status	IPAs
Crassulaceae	*Crassula zombensis* Baker f.		<10k km²	Mount Namuli
Crassulaceae	*Kalanchoe fernandesii* Raym.-Hamet		Endemic	
Crassulaceae	*Kalanchoe hametiorum* Raym.-Hamet		Endemic	Mount Inago and Serra Merripa, Ribáuè-M'paluwe
Crassulaceae	*Kalanchoe velutina* Welw. ex Britten subsp. *chimanimaniensis* (R.Fern.) R.Fern.		<10k km²	Chimanimani Mountains
Cucurbitaceae	*Diplocyclos tenuis* (Klotzsch) C.Jeffrey	VU B2ab(ii,iii,v)		
Cucurbitaceae	*Momordica henriquesii* Cogn.	EN B2ab(iii)		Mueda Plateau and Escarpments
Cucurbitaceae	*Momordica mosambica* H.Schaef.		Endemic	Eráti
Cucurbitaceae	*Peponium leucanthum* (Gilg) Cogn.	VU B1ab(iii)+2ab(iii)		Lower Rovuma Escarpment
Cyperaceae	*Scleria pachyrrhyncha* Nelmes	EN B2ab(iii)		Chimanimani Lowlands
Dioscoreaceae	*Dioscorea sylvatica* Eckl.	VU A2d		Inhaca (KaNyaka) Island, Mount Gorongosa
Droseraceae	*Aldrovanda vesiculosa* L.	EN B2ab(iii,iv,v)		
Ebenaceae	*Diospyros magogoana* F.White	CR D		Lower Rovuma Escarpment
Ebenaceae	*Diospyros shimbaensis* F.White	VU B2ab(iii)		Lower Rovuma Escarpment
Ericaceae	*Erica lanceolifera* S.Moore	VU B1ab(iii)+2ab(iii)	<10k km²	Chimanimani Mountains
Ericaceae	*Erica pleiotricha* S.Moore var. *blaerioides* (Wild) R.Ross		<10k km²	Chimanimani Mountains
Ericaceae	*Erica pleiotricha* S.Moore var. *pleiotricha*	VU D2	<10k km²	Chimanimani Mountains
Ericaceae	*Erica wildii* Brenan		<10k km²	Chimanimani Mountains
Eriocaulaceae	*Eriocaulon infaustum* N.E.Br.		Endemic	
Eriocaulaceae	*Mesanthemum africanum* Moldenke		<10k km²	Chimanimani Lowlands, Chimanimani Mountains
Erythroxylaceae	*Nectaropetalum carvalhoi* Engl.	VU B1ab(iii)+2ab(iii)	<10k km²	Quirimbas Archipelago, Quiterajo
Euphorbiaceae	*Croton aceroides* Radcl.-Sm.	EN B2ab(iii)	Endemic	Temane
Euphorbiaceae	*Croton inhambanensis* Radcl.-Sm.	VU B1ab(ii,iii,iv) +2ab(ii,iii,iv)	Endemic	Mapinhane, Temane
Euphorbiaceae	*Croton kilwae* Radcl.-Sm.	EN B2ab(iii)		Eráti, Quiterajo
Euphorbiaceae	*Croton leuconeurus* Pax subsp. *mossambicensis* Radcl.-Sm.		Endemic	
Euphorbiaceae	*Croton megalocarpoides* Friis & M.G.Gilbert	VU B2ab(iii)		

Family	Taxon	A(i) species IUCN Red List Assessment	B(ii) species endemism status	IPAs
Euphorbiaceae	*Crotonogynopsis australis* Kenfack & Gereau		<10k km²	Mount Mabu
Euphorbiaceae	*Erythrococca zambesiaca* Prain	VU D2	<10k km²	Urema Valley and Sangarassa Forest
Euphorbiaceae	*Euphorbia ambroseae* L.C.Leach var. *ambrosae*		Endemic	Cheringoma Limestone Gorges, Urema Valley and Sangarassa Forest
Euphorbiaceae	*Euphorbia angularis* Klotzsch	VU D2	Endemic	Goa and Sena Islands
Euphorbiaceae	*Euphorbia baylissii* L.C.Leach	VU B2ab(iii)	Endemic	Goba, Inharrime-Závora, Panda-Manjacaze, Pomene
Euphorbiaceae	*Euphorbia bougheyi* L.C.Leach		Endemic	Cheringoma Limestone Gorges
Euphorbiaceae	*Euphorbia citrina* S.Carter		<10k km²	Mount Gorongosa, Tsetserra
Euphorbiaceae	*Euphorbia contorta* L.C.Leach		Endemic	
Euphorbiaceae	*Euphorbia corniculata* R.A.Dyer		Endemic	Mount Inago and Serra Merripa, Quirimbas Inselbergs
Euphorbiaceae	*Euphorbia crebrifolia* S.Carter		<10k km²	Chimanimani Mountains
Euphorbiaceae	*Euphorbia crenata* (N.E.Br.) Bruyns		Endemic	
Euphorbiaceae	*Euphorbia decliviticola* L.C.Leach		<10k km²	Mount Inago and Serra Merripa, Ribáuè-M'paluwe
Euphorbiaceae	*Euphorbia depauperata* A.Rich. var. *tsetserrensis* S.Carter		<10k km²	Tsetserra
Euphorbiaceae	*Euphorbia grandicornis* N.E.Br. subsp. *sejuncta* L.C.Leach	EN B1ab(iii,v) +2ab(iii,v)	Endemic	Mount Nállume
Euphorbiaceae	*Euphorbia graniticola* L.C.Leach		Endemic	Mount Muruwere-Bossa, Mount Zembe
Euphorbiaceae	*Euphorbia keithii* R.A.Dyer		<10k km²	Goba
Euphorbiaceae	*Euphorbia knuthii* Pax subsp. *johnsonii* (N.E.Br.) L.C.Leach		Endemic	
Euphorbiaceae	*Euphorbia marrupana* Bruyns	EN B1ab(iii)+2ab(iii)	Endemic	
Euphorbiaceae	*Euphorbia namuliensis* Bruyns		Endemic	Mount Namuli
Euphorbiaceae	*Euphorbia neohalipedicola* Bruyns		Endemic	
Euphorbiaceae	*Euphorbia neorugosa* Bruyns		<10k km²	
Euphorbiaceae	*Euphorbia plenispina* S.Carter		Endemic	
Euphorbiaceae	*Euphorbia ramulosa* L.C.Leach		Endemic	
Euphorbiaceae	*Euphorbia schlechteri* Pax		Endemic	
Euphorbiaceae	*Euphorbia stenocaulis* Bruyns	EN B2ab(iii)	Endemic	

Family	Taxon	A(i) species IUCN Red List Assessment	B(ii) species endemism status	IPAs
Euphorbiaceae	*Euphorbia unicornis* R.A.Dyer	EN B1ab(iii)+2ab(iii)	Endemic	Quirimbas Inselbergs
Euphorbiaceae	*Jatropha latifolia* Pax var. *subeglandulosa* Radcl.-Sm.		Endemic	Namaacha
Euphorbiaceae	*Jatropha scaposa* Radcl.-Sm.		Endemic	Urema Valley and Sangarassa Forest
Euphorbiaceae	*Jatropha subaequiloba* Radcl.-Sm.	VU D2	Endemic	Bazaruto Archipelago, São Sebastião Peninsula
Euphorbiaceae	*Mallotus oppositifolius* (Geiseler) Müll.Arg. var. *lindicus* (Radcl.-Sm.) Radcl.-Sm.	VU B1+2b		
Euphorbiaceae	*Micrococca scariosa* Prain	VU B2ab(iii)		
Euphorbiaceae	*Omphalea mansfeldiana* Mildbr.	EN B2ab(iii)		Quiterajo
Euphorbiaceae	*Paranecepsia alchorneifolia* Radcl.-Sm.	VU B1+2b		
Euphorbiaceae	*Mildbraedia carpinifolia* (Pax) Hutch. (*Plesiatropha carpinifolia* (Pax) Breteler)	VU B1+2b		Inhamitanga Forest, Lower Rovuma Escarpment, Quiterajo
Euphorbiaceae	*Ricinodendron heudelotii* (Baill.) Pierre ex Heckel var. *tomentellum* (Hutch. & E.A. Bruce) Radcl.-Sm.	VU B1+2b		
Euphorbiaceae	*Tannodia swynnertonii* (S.Moore) Prain	VU B1+2bc, D2		Mount Gorongosa, Serra Garuzo
Euphorbiaceae	*Tragia glabrata* (Müll.Arg.) Pax & K.Hoffm. var. *hispida* Radcl.-Sm.		Endemic	Inhaca (KaNyaka) Island, Namaacha
Euphorbiaceae	*Tragia shirensis* Prain var. *glabriuscula* Radcl.-Sm.		Endemic	
Fabaceae	*Acacia latispina* J.E.Burrows & S.M.Burrows (*Vachellia latispina* (J.E.Burrows & S.M.Burrows) Kyal. & Boatwr.)	VU B1ab(ii,iii)	Endemic	Muàgámula River, Pemba, Quiterajo
Fabaceae	*Acacia latistipulata* Harms (*Senegalia latistipulata* (Harms) Kyal. & Boatwr.)	VU B2ab(iii)		Lower Rovuma Escarpment, Muàgámula River, Mueda Plateau and Escarpments, Quiterajo
Fabaceae	*Acacia quiterajoensis* Timberlake & Lötter		Endemic	Muàgámula River, Mueda Plateau and Escarpments, Quiterajo, Vamizi Island
Fabaceae	*Acacia torrei* Brenan (*Vachellia torrei* (Brenan) Kyal. & Boatwr.)		Endemic	Urema Valley and Sangarassa Forest
Fabaceae	*Adenopodia schlechteri* (Harms) Brenan	VU B1ab(ii,iii,iv) +2ab(ii,iii,iv)	Endemic	Inhaca (KaNyaka) Island
Fabaceae	*Aeschynomene aphylla* Wild	VU D2	<10k km²	Chimanimani Mountains

Family	Taxon	A(i) species IUCN Red List Assessment	B(ii) species endemism status	IPAs
Fabaceae	*Aeschynomene chimanimaniensis* Verdc.		<10k km²	Chimanimani Mountains
Fabaceae	*Aeschynomene grandistipulata* Harms		<10k km²	Chimanimani Mountains
Fabaceae	*Aeschynomene inyangensis* Wild		<10k km²	Chimanimani Mountains
Fabaceae	*Aeschynomene minutiflora* Taub. subsp. *grandiflora* Verdc.		Endemic	
Fabaceae	*Aeschynomene mossambicensis* Verdc. subsp. *mossambicensis*		Endemic	
Fabaceae	*Aeschynomene pawekiae* Verdc.		<10k km²	
Fabaceae	*Baphia macrocalyx* Harms	VU B1ab(i,ii,iii,iv,v) +2ab(i,ii,iii,iv,v)		Lower Rovuma Escarpment, Mueda Plateau and Escarpments
Fabaceae	*Baphia massaiensis* Taub. subsp. *gomesii* (Baker f.) Brummitt		Endemic	Mapinhane, Ribáuè-M'paluwe, Serra Mecula and Mbatamila
Fabaceae	*Baphia ovata* Sim		Endemic	Chidenguele
Fabaceae	*Baphia punctulata* Harms subsp. *palmensis* Soladoye		Endemic	
Fabaceae	*Bauhinia burrowsii* E.J.D.Schmidt	EN B1ab(iii)+2ab(iii)	Endemic	Mapinhane, Temane
Fabaceae	*Berlinia orientalis* Brenan	VU B1ab(iii)+2ab(iii)		Lower Rovuma Escarpment, Quiterajo
Fabaceae	*Brachystegia oblonga* Sim	CR A2acd; B1ab(ii,iii,v) +2ab(ii,iii,v)	Endemic	Mulimone
Fabaceae	*Bussea xylocarpa* (Sprague) Sprague & Craib	VU D2	Endemic	
Fabaceae	*Chamaecrista paralias* (Brenan) Lock		Endemic	Bazaruto Archipelago, Inhassoro-Vilanculos, Mapinhane, Panda-Manjacaze, Pomene, São Sebastião Peninsula
Fabaceae	*Craibia brevicaudata* (Vatke) Dunn subsp. *schliebenii* (Harms) J.B.Gillett	VU B1+2b		
Fabaceae	*Crotalaria insignis* Polhill	VU B1ab(iii)+2ab(iii)	<10k km²	Tsetserra
Fabaceae	*Crotalaria misella* Polhill		Endemic	
Fabaceae	*Crotalaria mocubensis* Polhill		Endemic	
Fabaceae	*Crotalaria namuliensis* Polhill & T.Harris		Endemic	Mount Namuli
Fabaceae	*Crotalaria paraspartea* Polhill	EN B2ab(ii,iii)	Endemic	
Fabaceae	*Crotalaria phylicoides* Wild		<10k km²	Chimanimani Mountains

Family	Taxon	A(i) species IUCN Red List Assessment	B(ii) species endemism status	IPAs
Fabaceae	*Crotalaria schliebenii* Polhill	VU D2	<10k km²	
Fabaceae	*Crotalaria torrei* Polhill		Endemic	Mount Namuli
Fabaceae	*Entada mossambicensis* Torre	VU D2	Endemic	
Fabaceae	*Erythrina haerdii* Verdc.	VU B1ab(iii)+2ab(iii)		
Fabaceae	*Gelrebia rostrata* (N.E.Br.) Gagnon & G.P.Lewis (*Caesalpinia rostrata* N.E.Br.)		<10k km²	
Fabaceae	*Guibourtia schliebenii* (Harms) J.Leonard	VU B2ab(iii)		Lower Rovuma Escarpment, Quiterajo
Fabaceae	*Guibourtia sousae* J.Leonard	CR B2ab(iii)	Endemic	Panda-Manjacaze
Fabaceae	*Icuria dunensis* Wieringa	EN B2ab(i,ii,iii,iv,v)	Endemic	Matibane Forest, Moebase, Mogincual, Mulimone, Quinga
Fabaceae	*Indigofera emarginella* A.Rich. var. *marrupaënsis* Schrire		Endemic	
Fabaceae	*Indigofera gobensis* Schrire	CR B2ab(iii)	Endemic	Goba
Fabaceae	*Indigofera graniticola* J.B.Gillett		<10k km²	
Fabaceae	*Indigofera mendoncae* J.B.Gillett		Endemic	Panda-Manjacaze
Fabaceae	*Indigofera namuliensis* Schrire		Endemic	Mount Namuli
Fabaceae	*Indigofera pseudomoniliformis* Schrire	VU B2ab(ii,iii,v)	Endemic	Eráti
Fabaceae	*Indigofera torrei* J.B.Gillett	VU B1ab(ii,iii,iv) +2ab(ii,iii,iv)	Endemic	
Fabaceae	*Indigofera vicioides* Jaub. & Spach subsp. *excelsa* Schrire		<10k km²	Tsetserra
Fabaceae	*Lotus wildii* J.B.Gillett		<10k km²	Mount Gorongosa
Fabaceae	*Macrotyloma decipiens* Verdc.		Endemic	
Fabaceae	*Micklethwaitia carvalhoi* (Harms) G.P.Lewis & Schrire	VU B1ab(i,iii,iv,v)	Endemic	Lupangua Peninsula, Matibane Forest, Pemba, Quiterajo
Fabaceae	*Millettia ebenifera* (Bertol.) J.E.Burrows & Lötter	VU B2ab(ii,iii)	Endemic	Bilene-Calanga, São Sebastião Peninsula
Fabaceae	*Millettia impressa* Harms subsp. *goetzeana* (Harms) J.B.Gillett	VU B2ab(iii)		Lower Rovuma Escarpment, Quiterajo
Fabaceae	*Millettia makondensis* Harms	VU B2ab(iii)		Lower Rovuma Escarpment, Muàgámula River
Fabaceae	*Millettia mossambicensis* J.B.Gillett		Endemic	Catapú, Inhamitanga Forest, Matibane Forest, Urema Valley and Sangarassa Forest

Family	Taxon	A(i) species IUCN Red List Assessment	B(ii) species endemism status	IPAs
Fabaceae	*Ormocarpum sennoides* subsp. *zanzibaricum* Brenan & J.B.Gillett	VU B1+2b		Lower Rovuma Escarpment
Fabaceae	*Otholobium foliosum* (Oliv.) C.H.Stirt. subsp. *gazense* (Baker f.) Verdc.		<10k km²	Chimanimani Mountains
Fabaceae	*Pearsonia mesopontica* Polhill		<10k km²	Chimanimani Mountains
Fabaceae	*Platysepalum inopinatum* Harms	VU B2ab(iii)		Lower Rovuma Escarpment
Fabaceae	*Rhynchosia chimanimaniensis* Verdc.	EN B1ab(iii)+2ab(iii)	<10k km²	Chimanimani Mountains
Fabaceae	*Rhynchosia clivorum* S.Moore subsp. *gurueensis* Verdc.		Endemic	Mount Namuli
Fabaceae	*Rhynchosia genistoides* Burtt Davy		<10k km²	
Fabaceae	*Rhynchosia stipata* Meikle		<10k km²	Chimanimani Mountains
Fabaceae	*Rhynchosia swynnertonii* Baker f.		<10k km²	Chimanimani Mountains
Fabaceae	*Rhynchosia torrei* Verdc.		Endemic	Mount Namuli
Fabaceae	*Rhynchosia velutina* Wight & Arn. var. *discolor* (Baker) Verdc.	VU B2ab(iii)		
Fabaceae	*Scorodophloeus torrei* Lock	EN B2ab(iii,v)	Endemic	Mogincual, Mulimone
Fabaceae	*Sesbania speciosa* Taub.	VU B2ab(iii)		
Fabaceae	*Tephrosia chimanimaniana* Brummitt		<10k km²	Chimanimani Mountains, Serra Mocuta
Fabaceae	*Tephrosia faulknerae* Brummitt	EN B2ab(ii,iii,v)	Endemic	
Fabaceae	*Tephrosia forbesii* Baker subsp. *forbesii*	VU B1ab(ii,iii) +2ab(ii,iii)		Inhaca (KaNyaka) Island, Licuáti Forest
Fabaceae	*Tephrosia forbesii* Baker subsp. *inhacensis* Brummitt	VU D2	Endemic	Inhaca (KaNyaka) Island
Fabaceae	*Tephrosia gobensis* Brummitt		<10k km²	Goba
Fabaceae	*Tephrosia longipes* Meisn. var. *drummondii* (Brummitt) Brummitt		<10k km²	Chimanimani Mountains
Fabaceae	*Tephrosia longipes* Meisn. var. *swynnertonii* (Baker f.) Brummitt		<10k km²	Chimanimani Lowlands
Fabaceae	*Tephrosia miranda* Brummitt		Endemic	
Fabaceae	*Tephrosia montana* Brummitt		<10k km²	Mount Gorongosa
Fabaceae	*Tephrosia praecana* Brummitt	VU B1ab(iii)+2ab(iii)	<10k km²	Tsetserra
Fabaceae	*Tephrosia reptans* Baker var. *microfoliata* (Pires da Lima) Brummitt		Endemic	Lower Rovuma Escarpment

Family	Taxon	A(i) species IUCN Red List Assessment	B(ii) species endemism status	IPAs
Fabaceae	*Tephrosia whyteana* Baker f. subsp. *gemina* Brummitt	CR B1ab(iii)+2ab(iii)	Endemic	Mount Namuli
Fabaceae	*Xylia africana* Harms	EN B2ab(iii)		Lower Rovuma Escarpment
Fabaceae	*Xylia mendoncae* Torre	VU B1ab(iii)+2ab(iii)	Endemic	Inhassoro-Vilanculos, Mapinhane
Gentianaceae	*Faroa involucrata* (Klotzsch) Knobl.		Endemic	
Geraniaceae	*Geranium exellii* J.R.Laundon	EN B1ab(iii)+2ab(iii)	<10k km²	Tsetserra
Geraniaceae	*Pelargonium mossambicense* Engl.		<10k km²	Mount Gorongosa, Tsetserra
Gesneriaceae	*Streptocarpus acicularis* I.Darbysh. & Massingue	CR B2ab(iii)	Endemic	Chimanimani Lowlands
Gesneriaceae	*Streptocarpus brachynema* Hilliard & B.L.Burtt	EN B1ab(iii,v) +2ab(iii,v)	Endemic	Mount Gorongosa
Gesneriaceae	*Streptocarpus erubescens* Hilliard & B.L.Burtt	EN B1ab(i,ii,iii,iv,v) +2ab(i,ii,iii,iv,v)	<10k km²	Mount Massangulo
Gesneriaceae	*Streptocarpus grandis* N.E.Br. subsp. *septentrionalis* Hilliard & B.L.Burtt		<10k km²	Chimanimani Mountains
Gesneriaceae	*Streptocarpus hirticapsa* B.L.Burtt	VU D2	<10k km²	Chimanimani Mountains
Gesneriaceae	*Streptocarpus leptopus* Hilliard & B.L.Burtt	EN B1ab(iii)+2ab(iii)	<10k km²	Serra Tumbine
Gesneriaceae	*Streptocarpus michelmorei* B.L.Burtt		<10k km²	Serra Mocuta, Tsetserra
Gesneriaceae	*Streptocarpus milanjianus* Hilliard & B.L.Burtt	VU D2	<10k km²	Mount Mabu
Gesneriaceae	*Streptocarpus montis-bingae* Hilliard & B.L.Burtt		Endemic	Chimanimani Mountains
Gesneriaceae	*Streptocarpus myoporoides* Hilliard & B.L.Burtt	EN B1ab(iii)+2ab(iii)	Endemic	Mount Nállume, Ribáuè-M'paluwe
Gesneriaceae	*Streptocarpus umtaliensis* B.L.Burtt		<10k km²	Tsetserra
Hydrocharitaceae	*Halophila ovalis* (R.Br.) Hook.f. subsp. *linearis* (Hartog) Hartog		Endemic	Inhaca (KaNyaka) Island
Hypericaceae	*Vismia pauciflora* Milne-Redh.	EN B2ab(iii)		Lower Rovuma Escarpment, Quiterajo
Iridaceae	*Dierama inyangense* Hilliard	EN B1ab(iii)+2ab(iii)	<10k km²	Tsetserra
Iridaceae	*Dierama plowesii* Hilliard	VU B1(iii)+2ab(iii)	<10k km²	Chimanimani Mountains
Iridaceae	*Freesia grandiflora* (Baker) Klatt subsp. *occulta* J.C.Manning & Goldblatt		Endemic	Mount Mabu
Iridaceae	*Gladiolus brachyphyllus* F.Bolus		<10k km²	Goba

Family	Taxon	A(i) species IUCN Red List Assessment	B(ii) species endemism status	IPAs
Iridaceae	*Gladiolus zambesiacus* Baker	VU B2ab(i,ii,iii,iv)		Mount Inago and Serra Merripa, Mount Namuli
Iridaceae	*Gladiolus zimbabweensis* Goldblatt	VU B1ab(iii)+2ab(iii)	<10k km²	Chimanimani Mountains, Tsetserra
Iridaceae	*Hesperantha ballii* Wild		<10k km²	Chimanimani Mountains
Iridaceae	*Moraea niassensis* Goldblatt & J.C.Manning	VU D1+2	Endemic	Mount Yao
Iridaceae	*Tritonia moggii* Oberm.		Endemic	Bilene-Calanga, Inhaca (KaNyaka) Island, São Sebastião Peninsula
Lamiaceae	*Acrotome mozambiquensis* G.Taylor		Endemic	
Lamiaceae	*Aeollanthus viscosus* Ryding		<10k km²	Chimanimani Mountains
Lamiaceae	*Clerodendrum abilioi* R.Fern.		Endemic	
Lamiaceae	*Clerodendrum cephalanthum* Oliv. subsp. *cephalanthum* var. *torrei* R.Fern.		Endemic	
Lamiaceae	*Clerodendrum lutambense* Verdc.	VU B1ab(iii)		Lower Rovuma Escarpment
Lamiaceae	*Clerodendrum robustum* Klotzsch var. *macrocalyx* R.Fern.		Endemic	
Lamiaceae	*Coleus caudatus* (S.Moore) E.Downes & I.Darbysh.		<10k km²	Chimanimani Mountains
Lamiaceae	*Coleus cucullatus* (A.J.Paton) A.J.Paton	VU D2	Endemic	Ribáuè-M'paluwe
Lamiaceae	*Coleus namuliensis* E.Downes & I.Darbysh.		Endemic	Mount Namuli
Lamiaceae	*Coleus sessilifolius* (A.J.Paton) A.J.Paton		<10k km²	Chimanimani Mountains, Tsetserra
Lamiaceae	*Leucas nyassae* Gürke var. *velutina* (C.H.Wright ex Baker) Sebald		Endemic	
Lamiaceae	*Ocimum natalense* Ayob. ex A.J. Paton		<10k km²	
Lamiaceae	*Ocimum reclinatum* (S.D.Will. & K.Balkwill) A.J.Paton		<10k km²	
Lamiaceae	*Orthosiphon scedastophyllus* A.J.Paton	CR(PE) B2ab(iii)	<10k km²	Quiterajo
Lamiaceae	*Plectranthus guruensis* A.J.Paton	EN B1ab(iii)+2ab(iii)	Endemic	Mount Namuli
Lamiaceae	*Plectranthus mandalensis* Baker	VU B1ab(iii)+2ab(iii)	<10k km²	Mount Namuli, Ribáuè-M'paluwe
Lamiaceae	*Premna hans-joachimii* Verdc.	VU B1ab(iii)+2ab(iii)	<10k km²	Lower Rovuma Escarpment

Family	Taxon	A(i) species IUCN Red List Assessment	B(ii) species endemism status	IPAs
Lamiaceae	*Premna schliebenii* Werderm.	VU B1+2b		Lupangua Peninsula, Muàgámula River, Quiterajo
Lamiaceae	*Premna tanganyikensis* Moldenke	VU B1ab(iii)+2ab(iii)		Lower Rovuma Escarpment, Matibane Forest
Lamiaceae	*Rotheca luembensis* (De Wild.) R.Fern. subsp. *niassensis* (R.Fern.) R.Fern.		Endemic	Serra Mecula and Mbatamila
Lamiaceae	*Rotheca sansibarensis* (Gürke) Steane & Mabb. subsp. *sansibarensis* var. *eratensis* (R.Fern.) R.Fern.		Endemic	Eráti
Lamiaceae	*Rotheca teaguei* (Hutch.) R.Fern.		<10k km²	
Lamiaceae	*Rotheca verdcourtii* (R.Fern.) R.Fern.		<10k km²	
Lamiaceae	*Stachys didymantha* Brenan		<10k km²	Chimanimani Mountains, Mount Namuli
Lamiaceae	*Syncolostemon flabellifolius* (S.Moore) A.J.Paton		<10k km²	Chimanimani Lowlands, Chimanimani Mountains
Lamiaceae	*Syncolostemon namapaensis* D.F.Otieno		<10k km²	
Lamiaceae	*Syncolostemon oritrephes* (Wild) D.F.Otieno	VU D2	<10k km²	Chimanimani Mountains
Lamiaceae	*Vitex carvalhi* Gürke	VU B2ab(iii)		Lower Rovuma Escarpment, Matibane Forest, Pemba, Quiterajo
Lamiaceae	*Vitex francesiana* I.Darbysh. & Goyder	EN B1ab(iii)+2ab(iii)	Endemic	Lower Rovuma Escarpment
Lamiaceae	*Vitex mossambicensis* Gürke	VU B2ab(iii)		Pemba, Quiterajo
Lauraceae	*Ocotea kenyensis* (Chiov.) Robyns & R.Wilczek	VU A1cd		Mount Gorongosa
Lentibulariaceae	*Utricularia podadena* P.Taylor		<10k km²	
Linaceae	*Hugonia elliptica* N.Robson		Endemic	
Linaceae	*Hugonia grandiflora* N.Robson	EN B2ab(iii)		Mueda Plateau and Escarpments
Linderniaceae	*Crepidorhopalon flavus* (S.Moore) I.Darbysh. & Eb.Fisch.	VU B1ab(iii)+2ab(iii)	<10k km²	Chimanimani Lowlands
Linderniaceae	*Crepidorhopalon namuliensis* I.Darbysh. & Eb.Fisch.		Endemic	Mount Namuli
Linderniaceae	*Hartliella txitongensis* Osborne & Eb.Fisch.		Endemic	Txitonga Mountains
Loganiaceace	*Strychnos xylophylla* Gilg	EN B2ab(iii)		Lower Rovuma Escarpment, Quiterajo
Loranthaceae	*Agelanthus deltae* (Baker & Sprague) Polhill & Wiens		Endemic	

Family	Taxon	A(i) species IUCN Red List Assessment	B(ii) species endemism status	IPAs
Loranthaceae	*Agelanthus igneus* (Danser) Polhill & Wiens	EN B2ab(iii)		
Loranthaceae	*Agelanthus longipes* (Baker & Sprague) Polhill & Wiens	VU B2ab(iii)		Matibane Forest
Loranthaceae	*Agelanthus patelii* Polhill & Timberlake	EN B1ab(iii)+2ab(iii)	<10k km²	Mount Namuli
Loranthaceae	*Englerina oedostemon* (Danser) Polhill & Wiens		<10k km²	Tsetserra
Loranthaceae	*Englerina schlechteri* (Engl.) Polhill & Wiens		Endemic	Mapinhane
Loranthaceae	*Englerina swynnertonii* (Sprague) Polhill & Wiens		<10k km²	Chimanimani Lowlands
Loranthaceae	*Englerina triplinervia* (Baker & Sprague) Polhill & Wiens	VU B2ab(iii)		Quirimbas Inselbergs
Loranthaceae	*Erianthemum lindense* (Sprague) Danser	VU B2ab(iii)		Lower Rovuma Escarpment, Mueda Plateau and Escarpments
Loranthaceae	*Helixanthera schizocalyx* T.Harris, I.Darbysh. & Polhill	EN B1ab(iii)+2ab(iii)	Endemic	Mount Mabu, Mount Namuli
Loranthaceae	*Oncella curviramea* (Engl.) Danser	VU B2ab(iii)		Mount Massangulo, Pemba
Lythraceae	*Ammannia elata* R.Fern.		Endemic	
Lythraceae	*Ammannia fernandesiana* S.A.Graham & Gandhi		Endemic	Inhassoro-Vilanculos
Lythraceae	*Ammannia gazensis* (A.Fern.) S.A.Graham & Gandhi	VU D2	Endemic	
Lythraceae	*Ammannia moggii* (A.Fern.) S.A.Graham & Gandhi	CR B2ab(iii)	Endemic	
Lythraceae	*Ammannia parvula* S.A.Graham & Gandhi	VU D2	Endemic	Mount Inago and Serra Merripa
Lythraceae	*Ammannia pedroi* (A.Fern. & Diniz) S.A.Graham & Gandhi	VU D2	Endemic	
Lythraceae	*Ammannia polycephala* (Peter ex A.Fern.) S.A.Graham & Gandhi		Endemic	
Lythraceae	*Ammannia ramosissima* (A.Fern.& Diniz) S.A.Graham & Gandhi		Endemic	
Lythraceae	*Ammannia spathulata* (A.Fern.) S.A.Graham & Gandhi		Endemic	
Malpighiaceae	*Acridocarpus natalitius* A.Juss. var. *linearifolius* Launert	VU A4cd; C1		Licuáti Forest, Panda-Manjacaze
Malpighiaceae	*Triaspis hypericoides* (DC.) Burch. subsp. *canescens* (Engl.) Immelman		<10k km²	

Family	Taxon	A(i) species IUCN Red List Assessment	B(ii) species endemism status	IPAs
Malpighiaceae	*Triaspis suffulta* Launert	EN B2ab(iii)	Endemic	Inhassoro-Vilanculos, Temane
Malvaceae	*Cola cheringoma* Cheek	EN B1ab(iii)+2ab(iii)	Endemic	Cheringoma Limestone Gorges
Malvaceae	*Cola clavata* Mast.	EN B1ab(iii,v) +2ab(iii,v)	Endemic	Catapú, Inhamitanga Forest
Malvaceae	*Cola dorrii* Cheek	EN B2ab(i,ii,iii,iv,v)		Panda-Manjacaze
Malvaceae	*Eriolaena rulkensii* Dorr	EN B1ab(iii,v) +2ab(iii,v)	Endemic	Pemba
Malvaceae	*Grewia filipes* Burret	EN B2ab(iii)	<10k km²	
Malvaceae	*Grewia limae* Wild	EN B1ab(ii,iii)	Endemic	Lower Rovuma Escarpment, Quiterajo
Malvaceae	*Hibiscus rupicola* Exell		Endemic	
Malvaceae	*Hibiscus torrei* Baker f.	EN B2ab(iii)	Endemic	
Malvaceae	*Hildegardia migeodii* (Exell) Kosterm.	EN B2ab(i,ii,iii,iv,v)		Lupangua Peninsula, Pemba
Malvaceae	*Sterculia schliebenii* Mildbr.	VU D2		Lower Rovuma Escarpment, Mueda Plateau and Escarpments, Quiterajo
Malvaceae	*Thespesia mossambicensis* (Exell & Hillc.) Fryxell		Endemic	Muàgámula River, Pemba, Quiterajo
Malvaceae	*Dombeya lastii* K.Schum.	EN B1ab(iii)+2ab(iii)	Endemic	Mount Namuli
Malvaceae	*Dombeya leachii* Wild	EN B1ab(iii)+2ab(iii)	Endemic	Ribáuè-M'paluwe
Malvaceae	*Grewia occidentalis* L. var. *littoralis* Wild		Endemic	Chidenguele, Inharrime-Závora, Pomene
Malvaceae	*Grewia transzambesica* Wild		Endemic	Urema Valley and Sangarassa Forest
Malvaceae	*Hermannia torrei* Wild	CR B2ab(ii,iii,iv)	Endemic	
Melastomataceae	*Antherotoma angustifolia* (A.Fern. & R.Fern.) Jacq.-Fél.		Endemic	
Melastomataceae	*Dissotis johnstoniana* Baker f. var. *johnstoniana*		<10k km²	Mount Namuli
Melastomataceae	*Dissotis pulchra* A.Fern. & R.Fern.	VU D2	<10k km²	Chimanimani Mountains
Melastomataceae	*Dissotis swynnertonii* (Baker f.) A.Fern. & R.Fern.	VU D2	<10k km²	Chimanimani Mountains
Melastomataceae	*Memecylon aenigmaticum* R.D.Stone	CR B2ab(iii)	Endemic	Quiterajo
Melastomataceae	*Memecylon incisilobum* R.D.Stone & I.G.Mona	CR A3c; B1ab(ii,iii,v) +2ab(ii,iii,v); C2a(ii)	Endemic	Bilene-Calanga
Melastomataceae	*Memecylon insulare* A.Fern. & R.Fern.	CR B1ab(iii)+2ab(iii)	Endemic	Bazaruto Archipelago

Family	Taxon	A(i) species IUCN Red List Assessment	B(ii) species endemism status	IPAs
Melastomataceae	*Memecylon nubigenum* R.D.Stone & I.G.Mona	EN B1ab(iii)+B2ab(iii)	<10k km²	Mount Namuli, Ribáuè-M'paluwe
Melastomataceae	*Memecylon rovumense* R.D.Stone & I.G.Mona	EN B2ab(iii)	<10k km²	Quiterajo
Melastomataceae	*Memecylon torrei* A.Fern. & R.Fern.	EN B2ab(iii)	Endemic	Lower Rovuma Escarpment, Quiterajo
Melastomataceae	*Warneckea albiflora* R.D.Stone & N.P.Tenza	CR B1ab(iii)	Endemic	Quiterajo
Melastomataceae	*Warneckea cordiformis* R.D.Stone	CR B1ab(i,ii,iii,iv,v)	Endemic	Quiterajo
Melastomataceae	*Warneckea parvifolia* R.D.Stone & Ntetha	EN A3c+4c; B1ab(ii,iii,v)+ 2ab(ii,iii,v)	<10k km²	Licuáti Forest
Melastomataceae	*Warneckea sessilicarpa* (A.Fern. & R.Fern.) Jacq.-Fel.	CR B1ab(iii,v)	Endemic	Moebase, Quinga
Meliaceae	*Khaya anthotheca* (Welw.) C.DC.	VU A1cd		Catapú, Cheringoma Limestone Gorges, Inhamitanga Forest, Mount Gorongosa, Mount Inago and Serra Merripa
Melianthaceae	*Bersama swynnertonii* Baker f.		<10k km²	Chimanimani Mountains
Moraceae	*Dorstenia zambesiaca* Hijman	VU D2	Endemic	Catapú, Inhamitanga Forest
Moraceae	*Ficus muelleriana* C.C.Berg	EN B1ab(iii)+2ab(iii)	Endemic	Chimanimani Lowlands
Myricaceae	*Myrica chimanimaniana* (Verdc. & Polhill) Christenh. & Byng	EN B1ab(iii,v) +2ab(iii,v)	<10k km²	Chimanimani Mountains, Tsetserra
Myrtaceae	*Eugenia* sp. A of T.S.M.		Endemic	Bilene-Calanga, Chidenguele, Inharrime-Závora
Myrtaceae	*Syzygium komatiense* Byng & Pahlad.		<10k km²	
Ochnaceae	*Ochna angustata* N.Robson		Endemic	Inhamitanga Forest, Quirimbas Archipelago
Ochnaceae	*Ochna beirensis* N.Robson	EN B2ab(iii)	Endemic	Bazaruto Archipelago
Ochnaceae	*Ochna dolicharthros* F.M.Crawford & I.Darbysh.	VU D2	Endemic	Lower Rovuma Escarpment
Oleaceae	*Olea chimanimani* Kupicha		<10k km²	Chimanimani Mountains
Oleaceae	*Olea woodiana* Knobl. subsp. *disjuncta* P.S.Green	EN B1ab(iii)+2ab(iii)		Vamizi Island
Orchidaceae	*Ansellia africana* Lindl.	VU A2cd+3cd+4cd		
Orchidaceae	*Cynorkis anisoloba* Summerh.		<10k km²	Mount Gorongosa
Orchidaceae	*Cyrtorchis glaucifolia* Summerh.	EN B2ab(iii)	Endemic	
Orchidaceae	*Disa chimanimaniensis* (H.P.Linder) H.P.Linder		<10k km²	Chimanimani Mountains

Family	Taxon	A(i) species IUCN Red List Assessment	B(ii) species endemism status	IPAs
Orchidaceae	*Disa zimbabweensis* H.P.Linder	VU B1ab(iii)+2ab(iii)	<10k km²	Tsetserra
Orchidaceae	*Disperis mozambicensis* Schltr.	CR B2ab(iii)	Endemic	
Orchidaceae	*Eulophia biloba* Schltr.		Endemic	
Orchidaceae	*Eulophia bisaccata* Kraenzl.		Endemic	
Orchidaceae	*Habenaria hirsutissima* Summerh.	VU D2	Endemic	
Orchidaceae	*Habenaria mosambicensis* Schltr.		Endemic	
Orchidaceae	*Habenaria stylites* Rchb.f. & S.Moore	VU B2ab(iii)		Catapú
Orchidaceae	*Habenaria stylites* Rchb.f. & S.Moore subsp. *johnsonii* (Rolfe) Summerh.		<10k km²	
Orchidaceae	*Liparis hemipilioides* Schltr.		Endemic	
Orchidaceae	*Neobolusia ciliata* Summerh.	EN B1ab(ii,iii,v) +2ab(ii,iii,v)	<10k km²	Chimanimani Mountains
Orchidaceae	*Oligophyton drummondii* H.P.Linder & G.Will.		<10k km²	
Orchidaceae	*Polystachya songaniensis* G.Will.		<10k km²	Mount Mabu, Ribáuè-M'paluwe
Orchidaceae	*Polystachya subumbellata* P.J.Cribb & Podz.		<10k km²	Chimanimani Mountains, Mount Gorongosa
Orchidaceae	*Satyrium flavum* la Croix		<10k km²	
Orchidaceae	*Schizochilus lepidus* Summerh.	VU D2	<10k km²	Chimanimani Mountains, Tsetserra
Orobanchaceae	*Buchnera chimanimaniensis* Philcox		<10k km²	Chimanimani Mountains
Orobanchaceae	*Buchnera namuliensis* Skan		Endemic	Mount Namuli
Orobanchaceae	*Buchnera subglabra* Philcox	VU D2	<10k km²	Chimanimani Mountains
Orobanchaceae	*Buchnera wildii* Philcox		<10k km²	
Orobanchaceae	*Striga diversifolia* Pires de Lima		Endemic	
Passifloraceae	*Adenia dolichosiphon* Harms	EN B2ab(iii)		
Passifloraceae	*Adenia mossambicensis* W.J.de Wilde		Endemic	
Passifloraceae	*Adenia schliebenii* Harms	EN B2ab(iii)		
Passifloraceae	*Adenia zambesiensis* R.Fern. & A.Fern.		Endemic	
Passifloraceae	*Paropsia grewioides* Mast. var. *orientalis* Sleumer	EN B2ab(iii)		Mueda Plateau and Escarpments

Family	Taxon	A(i) species IUCN Red List Assessment	B(ii) species endemism status	IPAs
Passifloraceae	*Tricliceras auriculatum* (A.Fern. & R.Fern.) R.Fern.		Endemic	
Passifloraceae	*Tricliceras elatum* (A.Fern. & R.Fern.) R.Fern.	EN B1ab(ii,iii) +2ab(ii,iii)	Endemic	
Passifloraceae	*Tricliceras lanceolatum* (A.Fern. & R.Fern.) R.Fern.	VU D2	Endemic	
Passifloraceae	*Tricliceras longepedunculatum* (Mast.) R.Fern. var. *eratense* R.Fern.		Endemic	
Pedaliaceae	*Dicerocaryum forbesii* (Decne.) A.E. van Wyk		<10k km²	Licuáti Forest
Penaeaceae	*Olinia chimanimani* T.Shah & I.Darbysh.	EN B1ab(iii,v) +2ab(iii,v)	<10k km²	Chimanimani Mountains
Peraceae	*Clutia sessilifolia* Radcl.-Sm.		<10k km²	Chimanimani Mountains
Phyllanthaceae	*Phyllanthus bernierianus* Müll. Arg. var. *glaber* Radcl.-Sm.		<10k km²	Chimanimani Lowlands, Chimanimani Mountains
Phyllanthaceae	*Phyllanthus manicaensis* Brunel ex Radcl.-Sm.	VU D2	Endemic	Tsetserra
Phyllanthaceae	*Phyllanthus reticulatus* Poir. var. *orae-solis* Radcl.-Sm.		Endemic	
Phyllanthaceae	*Phyllanthus tsetserrae* Brunel ex Radcl.-Sm.	CR B2ab(iii)	Endemic	Tsetserra
Poaceae	*Alloeochaete namuliensis* Chippind.	VU D2	Endemic	Mount Namuli
Poaceae	*Baptorhachis foliacea* (Clayton) Clayton		Endemic	Ribáuè-M'paluwe
Poaceae	*Brachychloa fragilis* S.M.Phillips		<10k km²	
Poaceae	*Danthoniopsis chimanimaniensis* (J.B.Phipps) Clayton	EN B1ab(iii)+2ab(iii)	<10k km²	Chimanimani Lowlands, Chimanimani Mountains
Poaceae	*Digitaria appropinquata* Goetgh.		Endemic	Mount Namuli
Poaceae	*Digitaria fuscopilosa* Goetgh.		Endemic	Tsetserra
Poaceae	*Digitaria megasthenes* Goetgh.	EN B1ab(iii)+2ab(iii)	Endemic	Mount Namuli
Poaceae	*Eragrostis desolata* Launert		<10k km²	Chimanimani Mountains
Poaceae	*Eragrostis sericata* Cope		Endemic	
Podostemaceae	*Inversodicraea torrei* (C.Cusset) Cheek	VU D2	Endemic	Mount Namuli
Polygalaceae	*Carpolobia suaveolens* Meikle		Endemic	Catapú, São Sebastião Peninsula, Urema Valley and Sangarassa Forest
Polygalaceae	*Polygala adamsonii* Exell		<10k km²	Mount Namuli, Ribáuè-M'paluwe

Family	Taxon	A(i) species IUCN Red List Assessment	B(ii) species endemism status	IPAs
Polygalaceae	*Polygala franciscii* Exell	VU B1ab(iii)	Endemic	Licuáti Forest
Polygalaceae	*Polygala limae* Exell		Endemic	
Polygalaceae	*Polygala torrei* Exell		Endemic	
Polygalaceae	*Polygala zambesiaca* Paiva	VU B1ab(iii)+2ab(iii)	<10k km²	Chimanimani Mountains, Tsetserra
Primulaceae	*Lysimachia gracilipes* (P.Taylor) U.Manns & Anderb.		<10k km²	Mount Gorongosa
Proteaceae	*Faurea racemosa* Farmar	EN B2ab(iii,v)		Mount Mabu, Mount Namuli
Proteaceae	*Faurea rubriflora* Marner		<10k km²	Chimanimani Mountains, Tsetserra
Proteaceae	*Protea caffra* Meisn. subsp. *gazensis* (Beard) Chisumpa & Brummitt		<10k km²	Chimanimani Mountains, Mount Gorongosa
Proteaceae	*Protea enervis* Wild	VU D2	<10k km²	Chimanimani Mountains
Putranjivaceae	*Drypetes gerrardii* Hutch. var. *angustifolia* Radcl.-Sm.		Endemic	
Putranjivaceae	*Drypetes sclerophylla* Mildbr.	VU B1+B2		Quiterajo
Restionaceae	*Platycaulos quartziticola* (H.P.Linder) H.P.Linder & C.R.Hardy		<10k km²	Chimanimani Mountains
Rosaceae	*Prunus africana* (Hook.f.) Kalkman	VU A1cd		Mount Chiperone, Mount Mabu, Mount Namuli, Tsetserra
Rubiaceae	*Afrocanthium ngonii* (Bridson) Lantz	VU B1ab(iii)+2ab(iii)	<10k km²	Chimanimani Lowlands
Rubiaceae	*Afrocanthium racemulosum* (S.Moore) Lantz var. *nanguanum* (Tennant) Bridson	VU B1+2b		
Rubiaceae	*Afrocanthium vollesenii* (Bridson) Lantz	VU B2ab(iii)		Pemba
Rubiaceae	*Anthospermum zimbabwense* Puff		<10k km²	Tsetserra
Rubiaceae	*Chassalia colorata* J.E.Burrows	EN B1ab(i,iii,v); C2a(i)	<10k km²	Lower Rovuma Escarpment, Quiterajo
Rubiaceae	*Coffea salvatrix* Swynn. & Phillipson	EN B2ab(i,ii,iii)		Chimanimani Lowlands, Mount Chiperone, Mount Zembe
Rubiaceae	*Coffea schliebenii* Bridson	VU B1ab(i,ii,iii) +2ab(i,ii,iii)	<10k km²	Lower Rovuma Escarpment
Rubiaceae	*Coffea zanguebariae* Lour.	VU B2ab(iii)		Eráti, Quirimbas Inselbergs
Rubiaceae	*Conostomium gazense* Verdc.		Endemic	
Rubiaceae	*Cuviera schliebenii* Verdc.	EN B2ab(iii)		Mueda Plateau and Escarpments

Family	Taxon	A(i) species IUCN Red List Assessment	B(ii) species endemism status	IPAs
Rubiaceae	*Cuviera tomentosa* Verdc.	EN B2ab(iii)		Mueda Plateau and Escarpments
Rubiaceae	*Didymosalpinx callianthus* J.E.Burrows & S.M.Burrows	EN B1ab(ii,iii) +2ab(ii,iii)	<10k km²	Lower Rovuma Escarpment
Rubiaceae	*Empogona jenniferae* Cheek	EN B1ab(iii)+2ab(iii)	<10k km²	Chimanimani Mountains
Rubiaceae	*Empogona maputensis* (Bridson & A.E.van Wyk) J.Tosh & Robbr.	EN B1ab(i,ii,iii, v) +2ab(i,ii,iii,v)	<10k km²	Licuáti Forest
Rubiaceae	*Heinsia mozambicensis* (Verdc.) J.E.Burrows & S.M.Burrows	EN C2a(i)	Endemic	Muàgámula River
Rubiaceae	*Hymenodictyon austro-africanum* J.E.Burrows & S.M.Burrows		<10k km²	
Rubiaceae	*Leptactina papyrophloea* Verdc.	EN B2ab(iii)		Lower Rovuma Escarpment, Quiterajo
Rubiaceae	*Oldenlandia cana* Bremek.		<10k km²	Chimanimani Mountains
Rubiaceae	*Oldenlandia verrucitesta* Verdc.		Endemic	
Rubiaceae	*Otiophora inyangana* N.E.Br. subsp. *inyangana*		<10k km²	Chimanimani Mountains, Tsetserra
Rubiaceae	*Otiophora inyangana* N.E.Br. subsp. *parvifolia* (Verdc.) Puff		<10k km²	Chimanimani Mountains
Rubiaceae	*Otiophora lanceolata* Verdc.	VU B1ab(iii)+2ab(iii)	<10k km²	Chimanimani Lowlands
Rubiaceae	*Oxyanthus biflorus* J.E.Burrows & S.M.Burrows	EN B2ab(ii,iii)	<10k km²	Lower Rovuma Escarpment, Mueda Plateau and Escarpments
Rubiaceae	*Oxyanthus strigosus* Bridson & J.E.Burrows	EN C2a(i)	<10k km²	Lower Rovuma Escarpment, Muàgámula River, Quiterajo
Rubiaceae	*Pachystigma* sp. A of F.Z.		Endemic	Bilene-Calanga
Rubiaceae	*Paracephaelis trichantha* (Baker) De Block	VU B2ab(iii)		Matibane Forest
Rubiaceae	*Pavetta chapmanii* Bridson	VU B1ab(iii)	<10k km²	
Rubiaceae	*Pavetta comostyla* S.Moore subsp. *comostyla* var. *inyangensis* (Bremek.) Bridson		<10k km²	Mount Gorongosa, Serra Garuzo, Tsetserra
Rubiaceae	*Pavetta curalicola* J.E.Burrows		Endemic	Matibane Forest
Rubiaceae	*Pavetta dianeae* J.E.Burrows & S.M.Burrows	EN B2ab(iii)	Endemic	Matibane Forest
Rubiaceae	*Pavetta gardeniifolia* A.Rich. var. *appendiculata* (De Wild.) Bridson		Endemic	Mount Massangulo, Mount Morrumbala
Rubiaceae	*Pavetta gurueensis* Bridson	VU D2	Endemic	Mount Mabu, Mount Namuli
Rubiaceae	*Pavetta incana* Klotzsch		Endemic	

Family	Taxon	A(i) species IUCN Red List Assessment	B(ii) species endemism status	IPAs
Rubiaceae	*Pavetta lindina* Bremek.	EN C2a(i)	<10k km²	Quiterajo
Rubiaceae	*Pavetta macrosepala* Hiern var. *macrosepala*	VU B1+2b		Lower Rovuma Escarpment
Rubiaceae	*Pavetta micropunctata* Bridson		<10k km²	Eráti
Rubiaceae	*Pavetta mocambicensis* Bremek.	EN B2ab(i,ii,iii)	Endemic	Matibane Forest, Pemba, Quirimbas Archipelago
Rubiaceae	*Pavetta pumila* N.E.Br.	VU B1ab(ii,iii) +2ab(ii,iii)	Endemic	
Rubiaceae	*Pavetta umtalensis* Bremek.		<10k km²	Chimanimani Mountains, Tsetserra
Rubiaceae	*Pavetta vanwykiana* Bridson		<10k km²	Licuáti Forest
Rubiaceae	*Polysphaeria harrisii* I.Darbysh. & C.Langa	EN B1ab(iii)+2ab(iii)	Endemic	Mount Mabu
Rubiaceae	*Polysphaeria ribauensis* I. Darbysh. & C.Langa	EN B1ab(iii)+2ab(iii)	Endemic	Ribáuè-M'paluwe
Rubiaceae	*Psychotria amboniana* K.Schum. subsp. *mosambicensis* (E.M.A.Petit) Verdc.	VU B2ab(ii,iii,iv,v)	Endemic	Bilene-Calanga, Inhassoro-Vilanculos, Inhaca (KaNyaka) Island, Licuáti Forest
Rubiaceae	*Psydrax micans* (Bullock) Bridson	VU B1+2b		Lower Rovuma Escarpment, Matibane Forest, Quiterajo
Rubiaceae	*Psydrax moggii* Bridson		Endemic	Bazaruto Archipelago, Bilene-Calanga, Chidenguele, Inhaca (KaNyaka) Island, Panda-Manjacaze, Pomene, Urema Valley and Sangarassa Forest
Rubiaceae	*Pyrostria chapmanii* Bridson	EN B1ab(iii)+2ab(iii)	<10k km²	Mount Namuli, Ribáuè-M'paluwe
Rubiaceae	*Pyrostria* sp. D. of F.T.E.A. *"makovui"* ined.	EN B1ab(iii)+2ab(iii)	<10k km²	Lower Rovuma Escarpment
Rubiaceae	*Rothmannia macrosiphon* (K.Schum.) Bridson	VU B1+2b		Lower Rovuma Escarpment, Mueda Plateau and Escarpments
Rubiaceae	*Rytigynia adenodonta* (K.Schum.) Robyns var. *reticulata* (Robyns) Verdc.	VU B1+2b		
Rubiaceae	*Rytigynia celastroides* (Baill.) Verdc. var *australis* Verdc.	VU B1ab(ii,iii,v) +B2ab(ii,iii,v)		Licuáti Forest
Rubiaceae	*Rytigynia* sp. C of F.Z.	CR B2ab(iii)	Endemic	Ribáuè-M'paluwe
Rubiaceae	*Rytigynia torrei* Verdc.	EN B2ab(iii)	Endemic	Mount Inago and Serra Merripa, Quirimbas Inselbergs
Rubiaceae	*Sericanthe chimanimaniensis* Wursten & De Block	VU B1ab(iii)+2ab(iii)	<10k km²	Chimanimani Lowlands, Chimanimani Mountains

Family	Taxon	A(i) species IUCN Red List Assessment	B(ii) species endemism status	IPAs
Rubiaceae	*Spermacoce kirkii* (Hiern.) Verdc.		Endemic	Bazaruto Archipelago, Pomene
Rubiaceae	*Spermacoce schlechteri* K.Schum. ex Verdc.		Endemic	
Rubiaceae	*Tarenna drummondii* Bridson	VU B1+2b		
Rubiaceae	*Tarenna longipedicellata* (J.G.García) Bridson	VU B1ab(iii)+2ab(iii)	Endemic	Catapú, Inhamitanga Forest
Rubiaceae	*Tarenna pembensis* J.E.Burrows	EN B1ab(ii,iii,v); C2a(i)	Endemic	Matibane Forest, Muàgámula River, Pemba
Rubiaceae	*Tarenna* sp. 53 of Degreef (= *Cladoceras rovumense* I.Darbysh., J.E.Burrows & Q.Luke)		<10k km²	Mueda Plateau and Escarpments, Quiterajo
Rubiaceae	*Triainolepis sancta* Verdc.		Endemic	Bazaruto Archipelago, Inhassoro-Vilanculos, Pomene, São Sebastião Peninsula
Rubiaceae	*Tricalysia ignota* Bridson		<10k km²	Tsetserra
Rubiaceae	*Tricalysia jasminiflora* (Klotzsch) Benth. & Hook.f. ex Hiern var. *hypotephros* Brenan		Endemic	
Rubiaceae	*Tricalysia schliebenii* Robbr.	VU B1+2b		Lower Rovuma Escarpment, Quiterajo
Rubiaceae	*Tricalysia semidecidua* Bridson	VU B1ab(iii)+2ab(iii)		Lower Rovuma Escarpment, Mueda Plateau and Escarpments, Quiterajo
Rubiaceae	*Vangueria domatiosa* J.E.Burrows	EN B1ab(iii)+2ab(iii)	Endemic	Lower Rovuma Escarpment
Rutaceae	*Teclea crenulata* (Engl.) Engl.		Endemic	
Rutaceae	*Vepris allenii* I.Verd.	EN B1ab(i,ii,iii,iv,v) +2ab(i,ii,iii,iv,v)	Endemic	Lower Rovuma Escarpment
Rutaceae	*Vepris drummondii* Mendonça	VU B1ab(iii)+2ab(iii)	<10k km²	Chimanimani Lowlands
Rutaceae	*Vepris macedoi* (Exell & Mendonça) Mziray	EN B1ab(iii)+2ab(iii)	Endemic	Mount Nállume, Ribáuè-M'paluwe
Rutaceae	*Vepris myrei* (Exell & Mendonça) Mziray	EN B2ab(iii)		Catapú, Urema Valley and Sangarassa Forest
Rutaceae	*Vepris sansibarensis* (Engl.) Mziray	VU B1+B2		Quiterajo
Rutaceae	*Vepris* sp. nov. (Mount Mabu)		Endemic	Mount Mabu
Rutaceae	*Zanthoxylum delagoense* P.G.Waterman		Endemic	Bazaruto Archipelago, Inhassoro-Vilanculos, Inhaca (KaNyaka) Island, Licuáti Forest, Pomene, São Sebastião Peninsula
Rutaceae	*Zanthoxylum holtzianum* (Engl.)	VU B1+2d		

Family	Taxon	A(i) species IUCN Red List Assessment	B(ii) species endemism status	IPAs
Rutaceae	*Zanthoxylum lindense* (Engl.) Kokwaro	VU B1+B2		Lower Rovuma Escarpment, Quiterajo, Vamizi Island
Rutaceae	*Zanthoxylum tenuipedicellatum* (Kokwaro) Vollesen	EN B2ab(ii)	<10k km²	Matibane Forest
Salicaceae	*Casearia rovumensis* I.Darbysh. & J.E.Burrows	EN B1ab(iii)+2ab(iii)	Endemic	Lower Rovuma Escarpment
Salicaceae	*Dovyalis* sp. A of T.S.M.		Endemic	Catapú
Santalaceae	*Thesium chimanimaniense* Brenan		<10k km²	Chimanimani Mountains
Santalaceae	*Thesium dolichomeres* Brenan		<10k km²	Chimanimani Mountains
Santalaceae	*Thesium inhambanense* Hilliard	CR B1ab(ii,iii,v) +2ab(ii,iii,v)	Endemic	
Santalaceae	*Thesium jeaniae* Brenan		<10k km²	Goba
Santalaceae	*Thesium pygmeum* Hilliard		<10k km²	Chimanimani Mountains
Santalaceae	*Viscum littorum* Polhill & Wiens		Endemic	Pemba, Quirimbas Archipelago
Sapindaceae	*Allophylus chirindensis* Baker f.	VU D2		Mount Gorongosa, Tsetserra
Sapindaceae	*Allophylus mossambicensis* Exell	VU B1ab(iii)+2ab(iii)	Endemic	Inharrime-Závora
Sapindaceae	*Allophylus torrei* Exell & Mend.	EN B1ab(iii)+2ab(iii)	Endemic	Eráti
Sapotaceae	*Pouteria pseudoracemosa* (J.H.Hemsl.) L.Gaut.	VU B1+2b, D2		Quirimbas Inselbergs
Sapotaceae	*Synsepalum chimanimani* Rokni & I.Darbysh.	EN B1ab(iii)+2ab(iii)	<10k km²	Chimanimani Lowlands
Sapotaceae	*Vitellariopsis kirkii* (Baker) Dubard	VU B1+2b		Lower Rovuma Escarpment, Matibane Forest
Scrophulariaceae	*Jamesbrittenia carvalhoi* (Engl.) Hilliard		<10k km²	Mount Gorongosa, Tsetserra
Scrophulariaceae	*Selago anatrichota* Hilliard		<10k km²	Chimanimani Mountains
Scrophulariaceae	*Selago swynnertonii* (S.Moore) Eyles var. *leiophylla* (Brenan) Hilliard		<10k km²	
Solanaceae	*Solanum litoraneum* A.E.Gonç.	EN B2ab(iii)	Endemic	Inhaca (KaNyaka) Island, Inhassoro-Vilanculos, Pomene
Stangeriaceae	*Stangeria eriopus* (Kunze) Baill.	VU A2acd+4acd		
Thymelaeaceae	*Gnidia chapmanii* B.Peterson		<10k km²	Mount Namuli
Thymelaeaceae	*Struthiola montana* B.Peterson		<10k km²	Chimanimani Mountains
Vahliaceae	*Vahlia capensis* (L.f.) Thunb. subsp. *macrantha* (Klotzsch) Bridson		Endemic	

Family	Taxon	A(i) species IUCN Red List Assessment	B(ii) species endemism status	IPAs
Velloziaceae	*Xerophyta argentea* (Wild) L.B.Smith & Ayensu		<10k km²	Chimanimani Mountains
Velloziaceae	*Xerophyta splendens* (Rendle) N.L.Menezes		<10k km²	Mount Namuli
Verbenaceae	*Chascanum angolense* Moldenke subsp. *zambesiacum* (R.Fern.) R.Fern.		<10k km²	
Verbenaceae	*Chascanum schlechteri* (Gürke) Moldenke var. *torrei* Moldenke		Endemic	
Vitaceae	*Cissus aristolochiifolia* Planch.	VU B1ab(iii)+2ab(iii)		Mount Namuli, Ribáuè-M'paluwe
Xyridaceae	*Xyris asterotricha* Lock	VU D2	<10k km²	Chimanimani Mountains
Xyridaceae	*Xyris makuensis* N.E.Br.		<10k km²	Mount Namuli
Zamiaceae	*Encephalartos aplanatus* Vorster	VU A2acd; B1ab(i,ii,iii,iv,v) +2ab(i,ii,iii,iv,v); C1	<10k km²	Goba
Zamiaceae	*Encephalartos chimanimaniensis* R.A.Dyer & I.Verd.	EN B1ab(i,ii,iv,v) +2ab(i,ii,iv,v); C1	<10k km²	Chimanimani Lowlands
Zamiaceae	*Encephalartos ferox* G.Bertol subsp. *emersus* P.Rousseau, Vorster & A.E.van Wyk		Endemic	Inhassoro-Vilanculos
Zamiaceae	*Encephalartos gratus* Prain	EN A4cd		Mount Inago and Serra Merripa, Mount Namuli, Serra Tumbine
Zamiaceae	*Encephalartos lebomboensis* I.Verd.	EN A2acd; B1ab(ii,iii,iv,v) +2ab(ii,iii,iv,v)		Goba
Zamiaceae	*Encephalartos manikensis* (Gilliland) Gilliland	VU A2acd		Serra Garuzo
Zamiaceae	*Encephalartos munchii* R.A.Dyer & I.Verd.	CR B1ab(ii,iv,v) +2ab(ii,iv,v); C2a(ii)	Endemic	Mount Zembe
Zamiaceae	*Encephalartos ngoyanus* I.Verd.	VU A4acd	<10k km²	
Zamiaceae	*Encephalartos pterogonus* R.A.Dyer & I.Verd.	CR B1ab(ii,iv,v) +2ab(ii,iv,v); C1+2a(ii)	Endemic	Mount Muruwere-Bossa
Zamiaceae	*Encephalartos senticosus* Vorster	VU A2acd; C1	<10k km²	Goba
Zamiaceae	*Encephalartos turneri* Lavranos & D.L.Goode		Endemic	Mount Inago and Serra Merripa, Mount Nállume, Ribáuè-M'paluwe
Zamiaceae	*Encephalartos umbeluziensis* R.A.Dyer	EN B1ab(i,ii,iii,iv,v) +2ab(i,ii,iii,iv,v); C1	<10k km²	Goba, Namaacha
Zingiberaceae	*Siphonochilus kilimanensis* (Gagnep.) B.L.Burtt	VU B2ab(ii,iii)	Endemic	
Zosteraceae	*Zostera capensis* Setch.	VU B2ab(ii,iii)		Bazaruto Archipelago, Inhaca (KaNyaka) Island